GPS SATELLITE SURVEYING

GPS SATELLITE SURVEYING

Second Edition

ALFRED LEICK
Department of Surveying Engineering
University of Maine
Orono, Maine

A WILEY-INTERSCIENCE PUBLICATION

JOHN WILEY & SONS, INC.

New York / Chichester / Toronto / Brisbane / Singapore

Library of Congress Cataloging in Publication Data:
Leick, Alfred.
    GPS satellite surveying / Alfred Leick. — 2nd ed.
        p.   cm.
    Includes bibliographical references and index.
    ISBN 0-471-30626-6
    1. Artificial satellites in surveying.   2. Global Positioning
System.   I. Title.
TA595.5.L45   1994
526.9'82—dc20   94-23517

CONTENTS

PREFACE

The Navigation Satellite Timing and Ranging (NAVSTAR) Global Positioning System (GPS) achieved its initial operating capability (IOC) on 8 December 1993 when 24 GPS satellites were successfully operating simultaneously. This satellite system is deployed and operated by the U.S. Department of Defense. During the developmental and experimental phases, the global positioning system gained rapid growth in popularity as it served a wide array of positioning needs. This popularity continues to increase as new uses and users seem to emerge almost continuously. The advances in electronic components and computer technology have resulted in smaller receivers that consume less power but are more capable than their predecessors. Above all, the theory of positioning with GPS has been significantly refined since the first edition of this book was written in 1988/89. Perhaps most important are the successes in dealing with antispoofing (AS), which was activated on 31 January 1994 when the P-codes were encrypted. Most manufacturers have proprietary solutions for dealing with antispoofing. It is important to note that it is still possible to determine accurate dual-frequency pseudoranges and full-wavelength carrier phases despite antispoofing. Another major advancement is the increase in accuracy of C/A-code pseudoranges by using narrow correlator techniques. In terms of algorithmic developments it is observed that ambiguity fixing on-the-fly (OTF) and rapid static techniques have achieved a high degree of maturity.

Major additions were made in Chapters 3, 8, 9, 10, and 11. An overview of the global positioning system is given in Chapter 3. Chapter 3 has been expanded to include a discussion of the GPS signal structure, new hardware developments, and solutions to antispoofing. The pseudorange and GPS carrier phase equations are derived in Chapter 8. A large section has been added that addresses commonly asked questions on GPS positioning using the mathemat-

xiii

ical development of the pseudorange and the carrier phase expressions. The ionosphere, the troposphere, and the multipath are dealt with in Chapter 9. The material on the ionosphere and the troposphere has been significantly expanded and the section on multipath is new. The processing of GPS carrier phases for the purpose of relative positioning and orbital relaxation is detailed in Chapter 10. This chapter has been completely revised. It begins with a discussion on the geometric content of dual-frequency epoch solutions, contains details on popular functions of the observables, such as wide laning, and contains a major section on ambiguity fixing on-the-fly and rapid static techniques. The chapter illustrates how modern GPS positioning techniques tend to be common to surveying and precision navigation. In Chapter 11 the formulation and examples are given for the combined processing of terrestrial observations and GPS vectors. This chapter also includes a section on the precise computation of geoid undulation differences. It has been expanded to include several GPS vector networks and a discussion of network reliability. A new Chapter 14 on Datums, Standards, and Specifications has been added. A discussion of the GLONASS satellite system is found in Appendix G.

Chapters 2, 4, 5, 6, 7, 12, and 13 were modified and updated. Following the updated introductory chapter, Chapter 2 gives an elementary exposition of satellite orbital theory. Least-squares adjustment is treated in detail in Chapter 4. The basic tools of least-squares adjustment are demonstrated in Chapter 5 by means of adjusting plane networks. Chapter 5 was written for the first edition by Steven Lambert, who was at that time a graduate student at the University of Maine, and it has been retained without change. Despite occasional suggestion to eliminate Chapter 5, it has been found useful when the book (Chapter 4) is used as a textbook for an introductory course on adjustments. Chapter 6 addresses the linkage between physical observations and mathematical (model) observations, and Chapter 7 provides the formulation of the three-dimensional geodetic model, assuming that the observations have been reduced to the ellipsoidal normal (if applicable). The expressions for the two-dimensional ellipsoidal and conformal mapping models are summarized in Chapter 12; a complete listing for transverse Mercator and Lambert conformal mapping is given, including specifications of the state plane coordinate systems. Chapter 13 provides two simple but useful transformations between coordinate systems. The appendices summarize elements from linear algebra and statistics in support of the material on least squares in Chapter 4.

The GPS system is revolutionizing the practice of surveying and positioning, making measurement techniques available to professionals from various disciplines that were previously only available to the best trained and best equipped just a few years ago. Fullest use of GPS technology, and of the resulting measurements, requires not only knowledge for operating the receivers, which is becoming easier day by day, but also a firm understanding of the mathematical foundation and of objective quality control. Much thought was given so as not to overload this book with the theoretical detail often found in geodetic texts, but also not to include the oversimplifications that are frequently found in surveying texts. The material presented in the book is an abbreviated synopsis of courses on GPS and satellite geodesy, adjustment computations, geodetic models, and integrated geodesy as taught at the University of Maine. The depth in which the material is presented is intended to be that needed by professionals who deal with spatial data of all levels of accuracy, and who desire to operate within the geodetic reference frame. This book is intended to be of help to anyone who, whether voluntarily or by virtue of economic need, ventures into GPS surveying and positioning. It is also intended to help engineers, scientists, lawyers, and politicians, for whom measurement might merely be a secondary or tertiary tool, to understand the depth of knowledge and skill required for modern surveying and spatial referencing, and to demonstrate the utility of positioning within the geodetic reference frame.

I would like to express my appreciation to P. Segantine, visiting researcher from the University of São Paulo, for helping with figures and tables, the students of the Spring 1994 class on Satellite Geodesy for a first reading of the manuscript, P. Connors, S. Gourevitch, J. Li, R. Lorenz, C. Mundo, X. Quin, and T. Soler for valuable suggestions. I especially appreciate the financial support of DeLorme Mapping of Freeport, Maine, during 1994.

My appreciation goes to C. Counselman of the Massachusetts Institute of Technology, who provided the opportunity for me to participate in the testing of the Macrometer during the summers of 1982 and 1983, D. Eckhardt, who made it possible to spend one year of sabbatical leave at Air Force Geophysics Laboratory at Hanscom AFB in 1984, and E. Grafarend for arranging for me to spend one year at the University of Stuttgart as part of the Alexander von Humboldt grant in 1985. These early exposures to GPS were very helpful.

ALFRED LEICK

Orono, Maine

FREQUENTLY USED ABBREVIATIONS

ACP	Active control point
ACSM	American Congress on Surveying and Mapping
AGU	American Geophysical Union
AS	Antispoofing
AST	Apparent sidereal time
AST1	Apparent sidereal time corrected for polar motion
BIPM	International Bureau of Weights and Measures
BPSK	Binary phase shift keying
C/A-code	Coarse/acquisition code (1.023 MHz)
CBN	Cooperative Base Network
CDMA	Code division multiple access
CEP	Celestial ephemeris pole
CGSIC	Civil GPS Service Interface Committee
CIGNET	Cooperative International GPS Network
CORS	Continuously Operating Reference Station
CTP	Conventional terrestrial pole
CTRS	Conventional Terrestrial Reference System
DGPS	Differential GPS
DLL	Code delay lock loop
DMA	Defense Mapping Agency
DOY	Day of year

Eq.E	Equation of equinox
FAA	Federal Aviation Administration
FBN	Federal Base Network
FBSR	Feedback shift register
FGCS	Federal Geodetic Control Subcommittee
FRP	Federal Radionavigation Plan
GAST	Greenwich apparent sidereal time
GDOP	Geometric dilution of precision
GIS	Geographic information system
GLONASS	Global Navigation Satellite System
GNSS	Global Navigation Satellite System
GMST	Greenwich mean sidereal time
GPS	Global Positioning System
GPSIC	GPS Information Center
GRS 80	Geodetic Reference System of 1980
HARN	High-Accuracy Reference Network
HDOP	Horizontal dilution of precision
HOW	Handover word
IAG	International Association of Geodesy
IAU	International Astronomical Union
IERS	International Earth Rotation Service
ITRF	International terrestrial reference frame
IOC	Initial operational capability
JD	Julian date
L1	L1 carrier (1575.42 MHz)
L2	L2 carrier (1227.6 MHz)
LLR	Lunar laser ranging
MJD	Modified Julian date
MST	Mean sidereal time
MST1	Mean sidereal time corrected for polar motion
NAD 83	North American Datum of 1983
NANU	Notice advisory to NAVSTAR users
NAVSTAR	Navigation Satellite Timing and Ranging
NCO	Numerically controlled oscillator
NEP	North Ecliptic Pole
NGRS	National Geodetic Reference System

NGS	National Geodetic Survey
NIST	National Institute of Standards and Technology
NOAA	National Oceanic and Atmospheric Administration
OCS	GPS operational control system
OTF	On-the-fly ambiguity resolution
P-code	Precision code (10.23 MHz)
PDOP	Positional dilution of precision
PLL	Phase lock loop
PPS	Precise Positioning Service
PRN	Pseudorandom noise
PSK	Phase-shift keying
QPSK	Quadri-phase-shift keying
RDOP	Relative dilution of precision
RTCM	Radio Technical Commission for Maritime Services
SA	Selective availability
SLR	Satellite laser ranging
SPS	Standard Positioning Service
SRP	Solar radiation pressure
SVN	Space vehicle launch number
TAI	International atomic time
TDOP	Time dilution of precision
TEC	Total electron content
TLM	Telemetry word
TOW	Time of week
TRANSIT	Navy Navigation Satellite System
UDN	User Densification Network
USNO	U.S. Naval Observatory
UT1	Universal time corrected for polar motion
UTC	Universal coordinated time
VDOP	Vertical dilution of precision
VLBI	Very long baseline interferometry
WAAS	Wide-area augmentation service
WADGPS	Wide-area differential GPS
WGS 84	World Geodetic System of 1984
WVR	Water vapor radiometer
Y-code	Encrypted P-code

CHAPTER 1

INTRODUCTION

The Global Positioning System (GPS) is being developed and operated to support military navigation and timing needs at an estimated cost of about \$8–10 billion. GPS represents an almost ideal dual-use technology and enjoys increased attention by civilians to explore its suitability for civil applications. Figure 1.1 lists some of the nonmilitary uses of GPS. The complete GPS system consists of 24 operational satellites and provides 24-hour, all-weather navigation and surveying capability worldwide. Figure 1.2 shows an artist's conception of the orbital constellation. A major milestone in the development of GPS was achieved on 8 December 1993, when the Initial Operational Capability (IOC) was declared as 24 satellites (Blocks I, II, IIA) were successfully operating. Full Operational Capability (FOC) will be declared when 24 satellites of the Block II and IIA types become operational. The Block I satellites are experimental and were used only for the first phase of the buildup of the constellation.

The implication of IOC is that commercial, national, and international civil users can henceforth rely on the availability of the Standard Positioning Service (SPS). Current policies quantify SPS as 100-m (95%) position accuracy for a single user. Authorized (military) users will have access to the Precise Positioning Service (PPS), which provides a greater degree of accuracy; however, access is controlled by cryptographic techniques.

The satellites transmit at frequencies L1 = 1575.42 MHz and L2 = 1227.6 MHz modulated with two types of codes and with a navigation message. The two types of codes are the C/A-code and the P-code. SPS is based on the C/A-code, whereas PPS is provided by the P-code portion of the GPS signal. The current authorized level of SPS follows from an intentional degradation of the full C/A-code capability. This measure is called selective availability

1

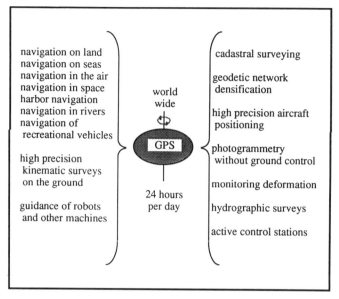

navigation on land
navigation on seas
navigation in the air
navigation in space
harbor navigation
navigation in rivers
navigation of
 recreational vehicles

high precision
 kinematic surveys
 on the ground

guidance of robots
 and other machines

world
wide

GPS

24 hours
per day

cadastral surveying

geodetic network
 densification

high precision aircraft
 positioning

photogrammetry
 without ground control

monitoring deformation

hydrographic surveys

active control stations

Figure 1.1 Some civilian uses of GPS.

Figure 1.2 Artist's conception of GPS satellite constellation.

(SA) and entails falsification of the satellite clock (SA-dither) and the broadcast satellite ephemeris (SA-epsilon), which is part of the navigation message. Despite selective availability, the C/A-code is fully accessible by civilians. On 31 January 1994 the long-anticipated antispoofing (AS) was finally (and apparently permanently) implemented. The purpose of AS is to make the P-codes available only to authorized (military) users. Users must be equipped with a decryption device or the "key" in order to lock on to the P-codes. AS is implemented through a modification of the mathematical formula of the P-code using a classified rule. The encrypted P-code is referred to as the Y-code.

Two types of observables are of interest to users. One is the pseudorange, which equals the distance between the satellite and the receiver plus small corrective terms due to clock errors, the ionosphere, the troposphere, and multipath. Given the geometric positions of the satellites (satellite ephemeris), four pseudoranges are sufficient to compute the position of the receiver and its clock error. Pseudoranges are a measure of the travel time of the codes (C/A, P, or Y). The second observable, the carrier phase, is the difference between the received phase and the phase of the receiver oscillator at the epoch of measurement. Receivers are programmed to make phase observations at the same equally spaced epochs. In addition, receivers keep track of the number of complete cycles received since the beginning of a measurement. Thus, the actual output is the accumulated phase observable at preset epochs.

The Standard Positioning Service and even the Precise Positioning Service mentioned above address "classical" navigation, where just one receiver observes the satellites to determine its geocentric position. Typically, a position is computed for every epoch of observation. Since the beginning of GPS, primarily driven by the high-accuracy requirements of surveying and geodesy, the advantages of relative positioning have been fully explored. In relative positioning, also called differential positioning, the relative location between receivers is determined. In this case, many of the common mode errors cancel or their impact is significantly reduced. This is particularly important in the presence of selective availability. Other error sources can be reduced by observing over a certain period of time while the receivers are in a static position.

During the pioneering years of GPS, there appeared to be a clear distinction between applications in navigation and surveying. This distinction, if ever real, is rapidly disappearing. While navigation solutions used to incorporate primarily pseudorange observations, and surveying solutions were formerly based primarily on carrier phase observations, modern approaches combine both types of observables in an optimal manner. The result is a unified GPS positioning theory for both surveying and navigation. Like GPS surveying, GPS navigation will increasingly be carried out relative to known stations. Powerful processing algorithms reduce the required time period for data collection so as to render even the distinction between static and kinematic techniques unnecessary.

One of the major achievements over the last couple of years is the development of observing and processing techniques to deal with the encrypted

P-code. Various solutions exist. While the best solutions would be certainly obtained in the absence of encryption, it is heartening to know that P-code type of pseudoranges can still be determined for both carriers including full-wavelength carrier phases. It is because of this remarkable achievement that those solutions which require dual-frequency code and full-wavelength carrier phase observations are extensively discussed in Chapter 10.

1.1 THE OUTCOME OF GPS POSITIONING

GPS positioning provides the geocentric position for individual receivers or the relative positions between co-observing receivers. The achievable accuracy depends on many factors, as will be detailed throughout this book. To emphasize the characteristic difference between geocentric and relative positional accuracy, let it be stated that geocentric position accuracy is typically in the meter range, whereas the relative position accuracy is typically at the centimeter level. Note the difference of several orders of magnitude between geocentric and relative position accuracy.

The secrets of GPS measurement capability can be readily explained. At the base is the ability to measure carrier phases to about 1/100 of a cycle, which equals 2–3 mm in linear distance. The high frequencies L1 and L2 penetrate the ionosphere relatively well. Because the time delay caused by the ionosphere is inversely proportional to the square of the frequency, carrier phase observations at both frequencies can be used successfully to model and thus eliminate most ionospheric effects. Dual-frequency observations are particularly important for large station separation and are also useful for the reduction in observation time. There has been significant progress in the design of stable clocks and their miniaturization in recent years, providing a precise reference for satellite clocks. GPS satellite orbits are stable because of their high altitudes where only major gravitational forces affect their motion. There are no atmospheric drag effects acting on GPS satellites. The impact of the sun and the moon on the orbits is significant but can be computed accurately. Thus, the remaining worrisome physical aspects are solar radiation pressure on the satellites, and the tropospheric delay and multipath effects on the propagated signals. Much is gained by forming linear combinations of the basic phase observable; for example, unwanted parameters are eliminated, and certain effects need not be modeled. Assume that the receivers k and m have observed the same satellite p at the same time. The difference between these two phases is called a single difference. It can readily be shown that single-difference observations are largely independent of satellite frequency offset and linear drift. Next assume that two single differences refer to two satellites p and q, respectively. The difference between these two single differences, called the double-difference observation, is largely independent of receiver clock errors. Finally, taking the difference of two double differences referring to measurements at different epochs yields the triple-difference observations. This last type of observation has desirable features for initial processing and screening of the data.

Processing of the single, double, or triple differences yields the relative location between the co-observing receivers, usually referred to as the vector between the stations. However, one accurate vector by itself is generally not of much use, at least in surveying applications. Of course, one can add the vector to the geocentric position of the "known" station and formally compute the geocentric position of the new station. The problem with this procedure is that the uncertainty of the known station is transferred in full to the new station. Also, despite modern technology, the vector observations themselves can be in error. Possibilities of misidentifying stations, centering errors, misreading antenna heights, etc., can never be completely disregarded. Like any other observations, the GPS vector observations are most effectively controlled by a least-squares adjustment in a network of stations whose relative positions are determined by a set of redundant vectors. Such a network adjustment makes it possible to assess the quality of the observations, validate the correctness of the statistical data, and detect (and possibly remove) blunders. Thus, the primary result of a GPS survey is a polyhedron of stations whose accurate relative locations have been controlled by a least-squares adjustment.

1.2   A PERFORMANCE REVIEW

Even though the Initial Operating Capability was only declared in December 1993, the GPS system has already astounded even its staunchest believers. A summary of GPS development and performance to date is detailed in Table 1.1. Because GPS development activities are so broadly based, dynamic, and conducted by many researchers all over the globe, it is impossible to give a comprehensive listing. Table 1.1, therefore, should merely demonstrate the extraordinarily rapid development of GPS positioning and some of the ancillary activities. The author gained his first experience with GPS during the summer of 1982 when he joined Charles C. Counselman's research group at M.I.T. to test the prototype Macrometer receiver. Baselines were measured repeatedly over several hours to study this new surveying technique (static GPS), to evaluate the prototype receiver, and to gain initial experience with the GPS. Even at this early stage, these tests revealed a GPS surveying accuracy of 1–2 parts per million (ppm) for the station separation. This was a shocking revelation. Suddenly, here was a measurement system capable of exceeding traditional first-order geodetic network accuracy. Was there anything else to do?

During 1983 and 1984, GPS surveying was verified beyond any doubt. A 30 (plus) station first-order network densification in the Eifel region of West Germany was observed (Bock et al. 1985). This project was a joint effort by the State Surveying Office of North Rhein–Westfalia, a private U.S. firm, and scientists from M.I.T. In early 1984, the geodetic network densification of Montgomery County (Pennsylvania) was carried out. The sole guidance of this project rested with a private GPS surveying firm (Collins and Leick 1985). For these network densifications, an accuracy of 1–2 ppm was achieved, which is

TABLE 1.1 GPS Development and Performance at a Glance

1978	Launch of first GPS satellite
1982	Prototype Macrometer testing at M.I.T.
1983	Geodetic network densification (Eifel, Germany)
1984	Geodetic network densification (Montgomery County, Pennsylvania)
	Engineering survey at Stanford
1985	Precise geoid undulation differences for Eifel network
	Codeless dual band observations
	Kinematic GPS surveying
	Antenna swap for ambiguity initialization
1986	Challenger accident (January 25)
	10 cm aircraft positioning
1987	JPL repeatability tests to 0.2 - 0.04 ppm
1989	OTF solution
	Wide Area Differential GPS concepts (WADGPS)
	U.S. Coast Guard GPS Information Center (GPSIC)
	Deflection of the vertical from GPS and leveling
1990	GEOID 90
1991	NGS ephemeris service
	GIG 91 experiment (January 22-February 13)
1992	IGS campaign (June 21- September 23)
	Initial solutions to deal with antispoofing (AS)
	Narrow correlator spacing C/A-code receiver
	Attitude determination system
1993	Real-time kinematic GPS
	ACSM ad hoc committee on accuracy standards
	Orange County GIS/cadastral densification
	Initial operating capability (December 8)
	1-2 part per billion repeatability
1994	IGS service (January 1)
	Antispoofing implementation (January 31)
	RTCM recommendations on differential GPS (Version 2.1)
	National Spatial Reference System Committee (NGS)
	Multiple (single-frequency) receiver experiments for OTF
	Proposal for monitoring the earth's atmosphere with GPS

higher than the existing networks in the respective regions. The station separations were about 10 km. In 1984, GPS was used at Stanford University for
a high-precision GPS engineering survey to support construction for extending
the Stanford Linear Accelerator (SLAC). Terrestrial observations (angles and
distances) were combined with GPS vectors. The Stanford project yielded a
truly millimeter accurate GPS network, demonstrating, among other things,
the high quality of the Macrometer antenna. This accuracy was confirmed

through comparison with the alignment laser of the accelerator, which reproduces a straight line within one-tenth of a millimeter (Ruland and Leick 1985).

Thus, by the middle of 1984, 1–2 ppm GPS surveying had been demonstrated beyond any doubt. No visibility was required between the stations. Data processing could be done on a microcomputer. Hands-on experience was sufficient to acquire most of the skills to process the data; that is, first-order geodetic network densification became, suddenly, within the capability of individual surveyors. Engelis et al. (1985) computed accurate geoid undulation differences for the Eifel network, demonstrating how GPS results can be combined with orthometric heights and what it takes to carry out such combinations accurately. New receivers became available, such as the dual-frequency P-code receiver TI-4100 from Texas Instruments, which was developed with the support of several federal agencies. Ladd et al. (1985) reported on a survey using codeless dual-frequency receivers and claimed 1 ppm in all three components of a vector, in as little as 15 minutes of observation time. Thus, the trend toward rapid static surveying had begun. Today static GPS surveying to 1–2 ppm is routinely carried out around the world. Manufacturers typically provide the software for phase processing, datum transformations, conformal mapping computations, and even geoid undulation computations.

Around 1985, kinematic GPS was developed (Remondi 1985b). Kinematic GPS yields centimeter (and better) relative accuracy in seconds for a moving antenna. While one antenna remains stationary at an initial point, the other antenna may continuously move, or move from one station to another. The only constraint on the path of the moving antenna is visibility of the same four (at least) satellites at both receivers. Both receivers record the carrier phase observations continuously. Remondi developed the antenna-swapping technique for rapid initialization of the ambiguities. Antenna swapping is a very useful method for kinematic GPS surveys on land (which was the primary area of application at that time).

The development of GPS satellites came to a sudden halt due to the tragic 25 January 1986 *Challenger* accident. Several years passed until the Delta II launch vehicle was modified to carry GPS satellites. However, the theoretical developments continued at full speed and were certainly facilitated with the publication of Remondi's dissertation (1984) and the very successful First International Symposium on Precise Positioning with the Global Positioning System (Goad 1985).

In the following two years, the first successes were reported in pushing the frontiers of GPS surveying further. Kinematic GPS was used for decimeter positioning of airplanes (Mader 1986; Krabill and Martin 1987). High-precision airplane positions can replace or reduce the need for traditional and expensive ground control in photogrammetry. Although densification for photo control has been a frequent use of GPS surveying, these early successes made it clear that precise airplane positioning would play a major role in photogrammetry and in other remote sensing applications. Lichten and Border (1987) report repeatability of 2–5 parts in 10^8 in all three components for static baselines. Note that 1 part in 10^8 corresponds to 1 mm in 100 kilometers! Such highly accurate solutions require satellite positions of about 1 m and better. Because such accurate orbits were not yet available, researchers were forced to estimate improved GPS orbital parameters simultaneously with their baselines, and the need for a precise orbital service became apparent. Other limitations, such as the uncertainty in the tropospheric delay over long baselines, also became apparent and created an interest in exploring water vapor radiometers to measure the wet part of the troposphere along the path of the satellite transmissions. The geophysical community continues to push the limits of GPS accuracy because, for example, the detection of (slow-moving) crustal motions requires accurate measurements over long distances; the more accurate the measurements are, the shorter is the interval between measurements. However, GPS capability of a few parts in 10^8 was also noticed by surveyors for its potential to change well-established methods of spatial referencing and geodetic network design.

Perhaps the year 1989 could be labeled the year when modern GPS positioning began in earnest. Seeber and Wuebbena (1989) discussed a kinematic technique that uses carrier phases and resolves the ambiguity "on the way." Today this technique is usually called ambiguity resolution "on the fly" (OTF). No static initialization is required to determine the ambiguities. The OTF technique works for postprocessing and real-time applications. OTF is one of the modern techniques that applies equally well to navigation and surveying. In 1989, the navigation community began to take advantage of relative positioning to eliminate errors common to co-observing receivers and to make attempts to extend the range of relative positioning. Brown (1989) referred to it as extended differential GPS, whereas today it is usually called wide-area differential GPS (WADGPS). Many efforts were made to standardize real-time differential GPS procedures. The efforts resulted in the most recent publication of recommended standards for differential GPS by the Radio Technical Commission for Maritime Services (RTCM-104, 1994). Because of the ever increasing interest of the civilian community in GPS, the U.S. Coast Guard established the GPS Information Center (GPSIC) to serve nonmilitary user needs for GPS information. Also, during 1989, one of the more unique and unusual geodetic applications was reported. Soler et al. (1989) used GPS in combination with geodetic leveling to determine the deflection of the vertical. Their technique does not depend on classical astronomical observations. The introduction of the geoid model GEOID 90 represented a major advancement for combining GPS (ellipsoidal) and orthometric height differences.

During 1991 and 1992, the geodetic community embarked on major efforts to explore the limits of GPS on a global scale. The efforts began with the GIG'91 campaign and continued the following year with the IGS campaign. GIG'91 [GPS experiment for IERS (International Earth Rotation Service) and Geodynamics] resulted in very accurate polar motion coordinates and earth rotation parameters. Of particular interest is the fact that geocentric coordinates derived from GPS agreed to a level of 10–15 cm, with estimates determined from satellite laser ranging, and that ambiguities could be fixed on a global scale providing daily repeatability of about 1 part in 10^9. Such results are possible because of the truly global distribution of the tracking stations. The primary purpose of the IGS (International GPS Service for Geodynamics) campaign was to prove that the scientific community is able to produce high-accuracy orbits on an operational basis. The campaign was successful beyond all expectations, thus confirming that the concept of IGS is possible. The IGS service formally began 1 January 1994. Blewitt (1993) gives an excellent review of the development of GPS technology and techniques during the years 1978–1992 with focus on high-accuracy applications for geodynamic investigations.

For many years, users worried about what impact antispoofing would have on the practical uses of GPS. This anxiety was considerably relieved when Hatch et al. (1992) reported on the code-aided squaring technique to be used when AS is active. In the meantime, most manufacturers have developed proprietary solutions for dealing with AS. When antispoofing was actually implemented on 31 January 1994, it presented no insurmountable hindrance to the continued use of GPS and particularly the use of modern techniques such as OTF. GPS users became even less dependent on antispoofing with the introduction of accurate narrow correlator spacing C/A-code receivers (van Dierendonck et al. 1992), since the C/A-code is not subject to antispoofing measures.

Technical developments for GPS applications are continuing at a rapid pace. The determination of attitude/orientation parameters using GPS has been given attention for quite some time. Qin (1992b) reports on a commercial product for attitude determination. Talbot (1993) reports on a real-time kinematic centimeter accuracy surveying system. Lachapelle et al. (1994) experimented with multiple (single-frequency) receiver configurations in order to accelerate the ambiguity resolution on the fly by means of imposing length constraints and conditions between the ambiguities. Whereas the possibilities of monitoring the ionosphere with dual-frequency and single-frequency code or carrier phase observations have been thoroughly discussed, Kursinski (1994) discusses the applicability of radio occultation techniques to use (or include) GPS in a general earth's atmospheric monitoring system. Among other information, this system could possibly provide precise, stable, and high vertical-resolution profiles of atmospheric temperatures across the globe. Will GPS become the meterological satellite system of the future?

The surveying community is responding to the challenges and opportunities that GPS provides. The American Congress on Surveying and Mapping (ACSM) tasked an ad hoc committee to study the appropriate accuracy standards to be used in the era of GPS. The committee addressed questions concerning relative and absolute accuracy standards. The National Geodetic Survey (NGS) enlisted the advice of experts regarding the shape and content of the geodetic reference frame, addressing options such as a Federal Base Network, a Cooperative Base Network, a User Densification Network, and Continuously Operating Reference Stations (CORS). The Orange County (California) densification of 2000 (plus) stations to support Geographic Information Systems (GIS) and cadastral activities is an example of user networks being established.

1.3 SOME GEODETIC CONSIDERATIONS

As a by-product, GPS has the potential of replacing leveling in many applications. As mentioned above, the primary result of a GPS survey is a polyhedron of stations with precisely known relative locations. These geometric positions can be expressed in terms of ellipsoidal (geodetic) coordinates, that is, latitude, longitude, and height. As is well known, the ellipsoidal height differs from the orthometric height (used by surveyors) by the geoid undulation. Thus, it would be desired to compute geoid undulations as accurately as ellipsoidal heights can be derived from GPS measurements. Unfortunately, this is not yet possible. However, it is possible to compute undulation differences accurately to about 2–3 cm, as demonstrated by several small and regional projects over the last couple of years. Such accurate geoid models are available for many regions of the world. For example, GEOID 93 covers the conterminous United States and is available from the National Geodetic Survey. All of these accurate geoid models incorporate gravity data in order to capture the local features, and rely on spherical harmonic expansion of the gravity field to obtain regional and global features. Determining spherical harmonic expansions of the gravity field, or simply measuring the gravity field, has been the subject of geodetic activities for many years. The fruition of these research efforts finally trickles down to surveyors. Spherical harmonic coefficients by themselves, for example, the latest 360×360 solutions, are already suitable for many low-accuracy applications.

GPS surveying has once again highlighted difficulties with the concept of coordinate invariance. One needs only to think about the well-known problem of implementing new observations into an existing network. This topic has been given much attention in the past and is still very relevant. Old positions change in a more or less large neighborhood around new stations as new observations are added to the network (Figure 1.3). The minimum depth of such a neighborhood can be determined from analysis and is a function of the quality of the observations, existing network distortions, and the number of significant digits of the coordinates. In general, the GPS network accuracy of 1–2 ppm exceeds that of existing first-order geodetic networks. Implementing new observations through readjustment of old and new observations is the preferred method from the theoretical point of view. Thanks to modern software engineering, such as new developments in human-computer interfacing and increased computational power at the desktop, readjustment is certainly practical for a local region the size of a county or a state. Such local adjustments could be made on the surveyor's desktop microcomputer whenever new observations, either terrestrial or GPS, become available. This technique would lead

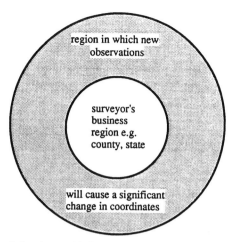

Figure 1.3   Area of influence in network adjustments.

to a gradual upgrading of the network rather than to a degradation of high-quality GPS observations to fit the existing network of lesser quality. If such a procedure is followed, it will entail learning to work with changing coordinates for the same station location, recognizing that observations are the primary result of the measurement process and that coordinates are derived quantities. Software in the classical sense is only part of the solution. Because the correct treatment of observations requires a deep but definable knowledge of adjustment theory and geodetic concepts, there is welcome opportunity here to experiment and eventually use modern tools such as expert systems. There is a need to identify the best database designs and computer-supported telecommunication requirements for the exchange of measurements among surveyors, possibly with a national or a county-wide database. Databases must have optimal capability for making available all observations and related data within neighborhoods defined by the surveyor. The challenge is to create a correct and user-friendly network maintenance capability with objective quality control built into it.

To the creative mind, the GPS capability of 1 part in 10^8 opens the door to new forms of geodetic networks. One often hears, "Put a receiver on the roof of the courthouse and let it record measurements continuously. All boundaries will be referenced right to the courthouse." The fact that data collection at one end would be controlled by authorized personnel could possibly be used advantageously in formulating objective quality control and in gaining legal acceptance or a higher legal weight for measurements. Other people speak of supernetworks covering the whole nation and providing super reference stations (superstations). These could be operated by local authorities, such as state highway departments. These kinds of unconventional approaches have the distinct advantage of allowing the surveyor to invest in only one receiver at a minimum. The superstations would be equipped with dual-frequency receivers, would operate autonomously, and would make the observations available in a database directly accessible to the surveyor.

The superstation idea could conceivably eliminate or at least reduce the need to occupy existing geodetic monuments in order to tie new GPS surveys to the geodetic network. Locating and verifying old control and then including it in new GPS surveys merely to obtain a tie is a major expense. This is particularly true in remote areas that have not been geodetically surveyed very frequently in the past. Since there is still a significant price differential for dual- versus single-frequency receivers, one could rent a dual-frequency receiver to establish the precise connection to the next superstation and then use one's own single-frequency receiver(s) to complete the local survey. Pure economic considerations may indeed soon force surveying in this direction. In this sense, and given the recent advances in kinematic, rapid static, and OTF techniques, GPS surveying is attractive and suitable for parcel surveying. While GPS will not replace classical traversing, it can provide the foundation for a unified system of parcel surveying, description, and registration. This is, in a sense, the foundation of a modern cadastre. The transition from "point of beginning" to "points of beginning" could finally be made economically.

The key to dealing successfully with a database of measurements is automated and objective quality control (Leick 1984). The control mechanism for this is least-squares adjustment and, in particular, the so-called inner or minimal constraint solutions. Data to be shipped (electronically) among surveyors or between a surveyor and an agency (such as a future cadastral-type office) should be subjected to a minimal constraint adjustment before exchange or deposit. Adjustment techniques lend themselves to meaningful and clear specifications of accuracy and precision. Some of the most frequently used elements in this context are the ellipses of standard deviation. Figure 1.4 shows two sets of ellipses generated from the same set of observations, thus containing the same level of information about the network. A thorough understanding of the statistical meaning of these ellipses and of the criteria for judgment of quality is necessary when dealing with measurements. To accomplish this, least-squares adjustment is discussed in great detail in this book. The derivation is based on the so-called mixed adjustment model, which combines observations and parameters by implicit mathematical expressions. The observation equation model and the condition equation model are derived from the solution of the mixed model through appropriate specifications. Because inner and minimal constraint solutions are so important for quality control of observations, a special section is included on identifying those quantities that remain invariant as different minimal constraints are selected.

Each adjustment requires a mathematical model that, in general, relates the observations to a set of parameters (unknowns). After considerable pondering of the issues involved, I decided to focus on the three-dimensional geodetic model as the relevant model. This model includes the classical terrestrial observations, such as slant distance, horizontal angle, azimuth, vertical angle,

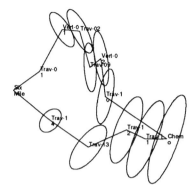

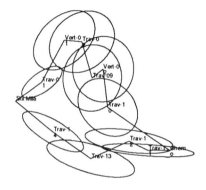

Figure 1.4 Alternative sets of standard ellipses. The size and shape of the ellipses depend on the geometry (design matrix), definition of the coordinate system (minimal constraints), and stochastic component.

and, with some restrictions, leveled height differences. It is definitely the preferred model for processing the three-dimensional GPS vector observations and for combining these vector observations with the classical terrestrial observations in one adjustment. The three-dimensional model is applicable with equal ease to small surveys of the size of a parcel or smaller, large surveys covering whole regions and nations, three-dimensional surveys for measuring and monitoring engineering structures, and the "pseudo three-dimensional" surveys typical of classical geodetic networks or "plane surveying." Application of simple concepts from the theory of adjustments, such as "weighted parameters" and "significance of parameters," make it possible to use the three-dimensional model in all of these applications in a uniform manner. Perhaps the most important point in favor of the three-dimensional model is that the geodesic line on the ellipsoid is not needed at all. Everyone who has studied the mathematics related to geodesics will certainly appreciate this simplification of surveying theory.

To clarify matters even more, I take the approach that, whenever applicable, the observations are reduced for polar motion and the deflection of the vertical.

It is well known that the theodolite senses the local plumb line and thus measures with respect to the local vertical and the local astronomic horizon. It is further known that astronomic observations depend on the position of the pole. The goal is to reduce angular observations measured with the theodolite to the ellipsoidal normal (deflection of vertical reduction) and to reduce the astronomic quantities to the conventional terrestrial pole (CTP), formerly known as conventional international origin (CIO). Having said this, I would like to comfort worried surveyors by reminding them that the most popular observations do not depend critically on polar motion and deflection of the vertical. Horizontal angles depend very little on the deflection of the vertical, because horizontal angles are the difference between two azimuths where the largest term cancels. The GPS vector observations (which refer to the WGS 84 coordinate system, whose third axis coincides with the CTP) and distances measured with the electronic distance meter (EDM) do not depend on either polar motion or deflection of the vertical. Furthermore, modern surveyors are unlikely to make astronomic observations in view of GPS surveying capability.

Another factor to keep in mind is that in most surveying applications there will be no need to improve on the deflection of the vertical already available from, for example, the National Geodetic Survey. Even if astronomic positions and thus deflections of the vertical were included as parameters in the adjustment, they could not accurately be determined with the typical observations, such as horizontal angles, distances, and GPS vectors. In fact, the resulting normal equations might be numerically singular. Besides, surveyors can simply and conveniently introduce their own local ellipsoid tangent to the equipotential surface at the center of the survey area. The deflections of vertical are then zero for all practical purposes within the geographical region of interest. The adjustment and the quality control of the observations can be carried out in this system. The controlled observations can be deposited in the surveyors' database or in the database of the cadastral type of agency, or can be sent to the National Geodetic Survey for inclusion in the national geodetic database. This scheme works fine if the participants recognize that a set of observations (not a set of coordinates) is the primary outcome of a survey. Coordinates can be computed at anytime and used however desired.

The approach followed in this book is shown in Figure 1.5. The scheme starts with observations, which are reduced for polar motion and deflection of the vertical (if applicable), adjusted in the three-dimensional model, and then corrected (with the opposite sign) for deflection of the vertical. The results are quality-controlled observations that refer to the local plumb line and the conventional terrestrial coordinate system. The remaining two loops in Figure 1.5 are actually redundant but are still of much historical interest. In this book, the expressions for the ellipsoidal surface model and the conformal mapping model are only summarized.

The mere conformal mapping of the ellipsoidal surface to the mapping plane does not require the geodesic line. Only if the mapping reductions are introduced to obtain the orientation and distance of the chord on the map is the

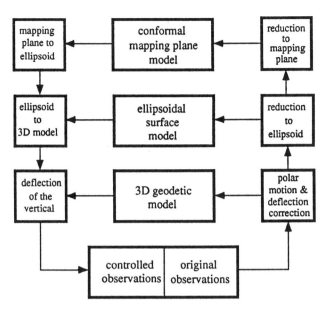

Figure 1.5 Geodetic models. At least three mathematically equivalent models are available to surveyors.

geodesic required. Thus, even the "geodesic-free" theory allows us to make it look to the innocent bystander as if only "plane theory" is at work. Start with approximate plane coordinates (to be interpreted as approximate mapping coordinates) and heights, conformally map the plane coordinate to the ellipsoid, carry out the three-dimensional adjustment, and conformally map the adjusted ellipsoidal latitude and longitude back to the conformal mapping plane. If the conformal mapping is defined by a central meridian or parallel passing through the center of the survey area, then the resulting coordinates are look-alikes of "plane coordinates." Of course, one can adopt the state plane coordinate systems as well for these transformations.

CHAPTER 2

ELEMENTS OF SATELLITE
SURVEYING

Satellite surveying requires an integrated understanding of such disciplines as statistics, astronomy, geodesy, and electronics. Because the satellites move in space, there is a need to relate space-fixed and earth-fixed coordinate systems. Timing enters in at least two ways: First, it relates the coordinate systems, and, second, timing the signals transmitted by the satellites is the basis for measurement. Understanding the capabilities and the limitations of GPS surveying also requires knowledge of the mechanics of orbital motion and those phenomena that affect the path of satellites.

2.1 GEOCENTRIC COORDINATE SYSTEMS AND THEIR MOTIONS

The more advanced and accurate measurement systems become, the more difficult it will be to find suitable definitions for the coordinate systems. There are basically two types of coordinate systems: space-fixed and earth-fixed. In the early days of geodetic astronomy it was sufficient (or at least marginally sufficient) to compute the motion of the earth's rotation axis in space (space-fixed coordinate system) on the basis of a rigid earth model being driven by the gravitational pull of the sun and the moon. It is now necessary to use more realistic earth models, such as elastic models and models with a liquid core, that take the nonrigid behavior of the earth into account. Lunar laser ranging (LLR), satellite laser ranging (SLR), very long baseline interferometry (VLBI), and GPS techniques require precisely defined coordinate systems. Much research has been carried out regarding the definition of coordinate systems. Several international symposia were held on this subject under the direction of the International Astronomical Union (IAU). The current set of nutations, that is, the mathematical expressions describing the position of the earth's rotation axis in space as a function of time, is the "1980 IAU Theory of Nutation." This theory was developed by Wahr (1981). The theory includes all externally forced motions of the earth's rotation axis. As a consequence, there are no longer any forced diurnal motion components with respect to either space-fixed or earth-fixed coordinate systems (Leick and Mueller 1979). This new reference pole is referred to as the celestial ephemeris pole (CEP) and not, as was the practice in earlier times, as the north celestial pole (NCP).

Complicating the definition of the earth-fixed reference system is the division of the lithosphere (solid outer 100 km of the earth) into about 20 nearly rigid plates that seem to move independently. These plates move slowly (1–10 cm per year), responding to driving forces that result from motions in the earth's interior. Although these motions are not of much concern to local and regional surveying, they are of prime interest to geodesy, as the measurement of these motions contributes to the understanding of geophysics and earthquake occurrence. Ultimately, then, those observatories that contribute to the determination of the position of the earth's rotation axis with respect to the earth's crust (polar motion) must take into account their own motion, as they are rigidly attached to their respective plates. An excellent introduction to modern theories of earth orientation and coordinate system definition is given by Moritz and Mueller (1987).

2.1.1 Precession and Nutation

Precession and nutation refer to the motion of the earth's rotation axis in space. To describe and visualize these motions, I use the concept of a sphere of directions (Figure 2.1). The location of the sphere is irrelevant, because the sphere is used only to represent directions. Each direction, such as the direction of the earth's rotation axis, can be shifted parallel such that it passes through the center of the sphere, wherever one envisions the center of the sphere to be located. The intersection of the (thus shifted) direction of the earth's rotation axis with the sphere is denoted by CEP. The actual size of the sphere and the distance to an object is also irrelevant. It is only the intersection of the sphere with the direction that is important. It is therefore logical to call the CEP, for example, the position of the celestial ephemeris pole on the sphere of direction. In this sense, the NEP denotes the position of the north ecliptic pole. The planes of the ecliptic and the equator, which are orthogonal to the directions of the NEP and CEP, respectively, are shifted parallel so as to contain the center of the sphere and intersect the sphere in two distinct great circles. The intersection of the ecliptic and the equator defines, in turn, the direction of the vernal equinox γ. The angle ϵ between the ecliptic and the equator is called the obliquity.

The directions and planes thus described serve as the fundamental reference for the definition of the celestial (space-fixed) coordinate systems. Let S denote the direction of an object, such as a star. The plane defined by arc S-CEP and

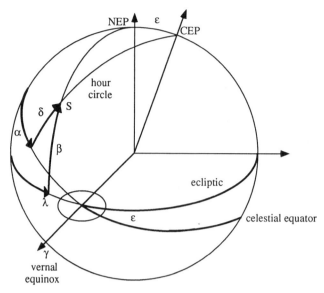

Figure 2.1 Celestial coordinate system. The equator and the ecliptic and their intersection are the basic references for celestial coordinate systems.

the center of the sphere intersects the sphere of directions in another great circle called the hour circle. The position of the star (or any other natural or artificial space object) can now be defined by the right ascension α and the declination δ. The right ascension is the angle in the equatorial plane measured counterclockwise from the vernal equinox to the hour circle. The declination is the angle of the object above or below the equatorial plane, measured in the plane of the hour circle; it is positive for positions in the northern hemisphere and negative for the southern hemisphere. Analogously, the position of the celestial object can be given with respect to the ecliptic system by means of the ecliptic longitude λ and ecliptic latitude β. Both sets of coordinates are related through the obliquity ϵ.

If the center of the sphere of directions is located at the sun, one speaks of heliocentric positions of the celestial object. If the center is located at the center of the earth or at the observing station on the surface of the earth, one speaks of geocentric and topocentric positions, respectively. These positions do not differ very much for stars because of the large distance to them. However, for earth satellites there is a large difference between geocentric and topocentric positions.

Unfortunately, the ecliptic latitude and longitude and the right ascension and declination are functions of time; that is, the respective directions of the NEP and CEP are not constant in space. The motion of the NEP, and thus of the ecliptic, is called planetary precession. This is a very slow motion, indeed.

The CEP and thus the celestial equator exhibit a long periodic (lunisolar precession) and many short periodic motions (nutations). The cause of precession and nutation is the ever-changing gravitational attraction of the sun and the moon (and to a small extent also the planets) on the earth. Newton's law of gravitation states that the gravitational force between two bodies is proportional to their masses and is inversely proportional to the square of their separation. Because of orbital motions of the earth and the moon, the separation between the sun, the moon, and the earth changes continuously. Since these changes in distance are periodic, the resulting precession and nutations are periodic functions of time as well, reflecting the periodicities in the orbital motions of the sun and the moon; the only exception is (the small) planetary precession. Because of Newton's law of gravitation, the distribution of the earth's mass is critical for the precise computation of precession and nutation. The most important elements include the flattening of the earth and the noncoincidence of the equatorial plane with the ecliptic (and the noncoincidence of the orbital plane of the moon with the ecliptic). Nonrigidity effects of the earth on the nutations can be observed with today's high-precision measurement systems. A spherical earth with homogeneous density distribution would neither precess nor nutate.

Figure 2.2, which is an enlargement of a small portion of Figure 2.1 as indicated, shows the location of the ecliptic and equator as ''driven by precession only'' for some starting epoch T_s. Considering precession only, the equator and ecliptic at epoch T_e will differ from the respective positions at T_s. Consider a Cartesian coordinate system whose first and third axes coincide with the vernal equinox and the CEP, respectively, at epoch T_s; the second axis completes a right-handed coordinate system. It is readily seen that the trans-

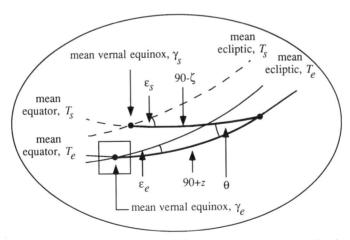

Figure 2.2 Precession. The direction of the mean vernal equinox is given by the intersection of the ecliptic and the mean equator.

formation between both epochs is given by

$$
\begin{bmatrix} x \\ y \\ z \end{bmatrix}_{\alpha_{m_e}, \delta_{m_e}} = P \begin{bmatrix} x \\ y \\ z \end{bmatrix}_{\alpha_{m_s}, \delta_{m_s}}
\tag{2.1}
$$

where

$$
P = R_3(-90 - z) R_1(\theta) R_3(90 - \zeta)
\tag{2.2}
$$

See Appendix D for a definition of the rotation matrices R_i. The subscript m denotes mean position as obtained from precession only. Thus, m_s and m_e denote the mean positions at epochs T_s and T_e, respectively. The right ascension and declination are computed from the general expression

$$
\alpha_m = \tan^{-1} \frac{y}{x}
\tag{2.3}
$$

$$
\delta_m = \tan^{-1} \frac{z}{\sqrt{x^2 + y^2}}
\tag{2.4}
$$

noting that the algebraic signs of the numerator and denominator of (2.3) determines the quadrant of the right ascension, which is $0° \le \alpha < 360°$. The coordinate system driven only by precession is called the mean celestial coordinate system and the respective positions are referred to as mean celestial position; thus, one speaks of mean right ascension and mean declination.

The angles (ζ, z, θ) are called the equatorial precession parameters. They can be computed as follows (Kaplan 1981):

$$
\begin{aligned}
\zeta = (2306''.2181 &+ 1''.39656T - 0''.000139T^2)t \\
&+ (0''.30188 - 0''.000344T)t^2 + 0''.017998t^3
\end{aligned}
\tag{2.5}
$$

$$
\begin{aligned}
z = (2306''.2181 &+ 1''.39656T - 0''.000139T^2)t \\
&+ (1''.09468 + 0''.000066T)t^2 + 0''.018203t^3
\end{aligned}
\tag{2.6}
$$

$$
\begin{aligned}
\theta = (2004''.3109 &- 0''.85330T - 0''.000217T^2)t \\
&- (0''.42665 + 0''.000217T)t^2 - 0''.041833t^3
\end{aligned}
\tag{2.7}
$$

where t is the interval, measured in Julian centuries of barycentric dynamic time (TDB), between the starting epoch JD_s and the ending epoch JD_e (JD is Julian date), and T is the interval measured in Julian centuries of TDB between the reference (standard) epoch J2000.0 and epoch JD_s. Thus,

$$t = \frac{JD_e - JD_s}{36,525} \qquad (2.8)$$

$$T = \frac{JD_s - 2,451,545.0}{36,525} \qquad (2.9)$$

The terms Julian date, Julian century, and barycentric dynamic time are explained in Section 2.2.

The mean obliquity, that is, the angle between the ecliptic and the mean celestial equator as driven by precession, is given by

$$\epsilon = 23°26'21''.448 - 46''.8150T - 0''.00059T^2 + 0''.001813T^3 \quad (2.10)$$

The true celestial equator is orthogonal to the rotation axis of the earth and is driven by precession and nutation. Thus the true celestial equator differs from the mean equator by the nutations computed for the desired epoch JD_e. Figure 2.3 is an enlargement of a small portion of Figure 2.2 showing the nutation in longitude, $\Delta\psi$, and in obliquity, $\Delta\epsilon$. Consider again a Cartesian coordinate system whose first axis coincides with the mean vernal equinox of date, whose third axis is orthogonal to the mean equator of date, and whose second axis completes a right-handed coordinate system. It follows that

$$\begin{bmatrix} x \\ y \\ z \end{bmatrix}_{\alpha, \delta} = N \begin{bmatrix} x \\ y \\ z \end{bmatrix}_{\alpha_{m_e}, \delta_{m_e}} \qquad (2.11)$$

where

$$N = R_1(-\epsilon - \Delta\epsilon) R_3(-\Delta\psi) R_1(\epsilon) \qquad (2.12)$$

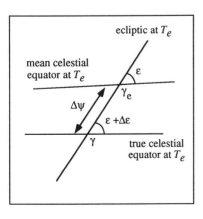

Figure 2.3 Nutation. The direction of the true vernal equinox is defined by the intersection of the ecliptic and the true celestial equator.

The true right ascension α and true declination δ can be computed from equations (2.11) and (2.12). The complete transformation from the mean position at epoch JD_s to the true position at JD_e is

$$\begin{bmatrix} x \\ y \\ z \end{bmatrix}_{\alpha, \delta} = NP \begin{bmatrix} x \\ y \\ z \end{bmatrix}_{\alpha_{m_s}, \delta_{m_s}} \tag{2.13}$$

The elements of nutation are given by the following expression (Kaplan 1981):

$$\Delta\psi = -(17''.1996 + 0''.01742T) \sin (\Omega)$$
$$+ (0''.2062 + 0''.00002T) \sin (2\Omega) \tag{2.14}$$
$$- (1''.3187 + 0''.00016T) \sin (2F - 2D + 2\Omega) + \cdots$$

$$\Delta\epsilon = (9''.2025 + 0''.00089T) \cos (\Omega)$$
$$+ (-0''.0895 + 0''.00005T) \cos (2\Omega) \tag{2.15}$$
$$+ (0''.5736 - 0''.00031T) \cos (2F - 2D + 2\Omega) + \cdots$$

Table 2.1 contains all terms with coefficients greater than $0''.01$. The complete set of nutations contains 106 entries. The periods of the nutations vary from about 18.6 years (6798.4 days) to 5 days. The elements (l, l', F, D, Ω) describe the mean positions of the sun and the moon. Of particular interest is Ω, which appears as an argument in the first term in equations (2.14) and (2.15). This is the largest nutation, with a period of 18.6 years, corresponding to a complete rotation of the lunar orbital plane around the ecliptic pole (see Figure 2.4). The same physical phenomenon is responsible for creating the 18.6-year tidal period. Because tides and nutation are caused by the same gravitational attraction of the sun and the moon, it is actually possible to transform the mathematical series of nutations into the corresponding series of tides. Figure 2.5 summarizes the motion of the CEP due to lunisolar precession and nutation. Completion of the precessional circle takes about 26,000 years. The pattern of nutation repeats every 18.6 years.

2.1.2 The Conventional Terrestrial Reference System

The definition of a Conventional Terrestrial Reference System (CTRS) must account for the motion of the rotation axis of the earth with respect to the crust (lithosphere). This motion is called the polar motion. The path of the rotation axis is shown in Figure 2.6 for the time 1984 to 1988. As can be seen, the motion is somewhat periodic. The complete spectrum of polar motion contains a period of about 434 days, called the Chandler period. The amplitude varies but does not seem to exceed 10 m. The causes for polar motion are not yet satisfactorily explained. Modern techniques from space geodesy should lead to a better understanding of this phenomenon.

TABLE 2.1 Partial Listing of 1980 IAU Nutations

	Argument					Periods	Longitude		Obliquity	
	l	l'	F	D	Ω	[days]	[0".0001]		[0".0001]	
1	0	0	0	0	1	6798.4	-171996	-174.2T	92025	8.9T
2	0	0	0	0	2	3399.2	2062	0.2T	-895	0.5T
3	0	0	2	-2	2	182.6	-13187	-1.6T	5736	-3.1T
4	0	1	0	0	0	365.3	1426	-3.4T	54	-0.1T
5	0	1	2	-2	2	121.7	-517	1.2T	224	-0.6T
6	0	-1	2	-2	2	365.2	217	-0.5T	-95	0.3T
7	0	0	2	-2	1	177.8	129	0.1T	-70	0.0T
8	0	0	2	0	2	13.7	-2274	-0.2T	977	-0.5T
9	1	0	0	0	0	27.6	712	0.1T	-7	0.0T
10	0	0	2	0	1	13.6	-386	-0.4T	200	0.0T
11	1	0	2	0	2	9.1	-301	0.0T	129	-0.1T
12	1	0	0	-2	0	31.8	-158	0.0T	-1	0.0T
13	-1	0	2	0	2	27.1	123	0.0T	-53	0.0T

$$l = 485866''.733 + \left(1325^r + 715922''.633\right)T + 31''.310T^2 + 0''.064T^3 \tag{a}$$

$$l' = 1287099''.804 + \left(99^r + 1292581''.224\right)T - 0''.577T^2 - 0''.012T^3 \tag{b}$$

$$F = 335778''.877 + \left(1342^r + 295263''.137\right)T - 13''.257T^2 + 0''.011T^3 \tag{c}$$

$$D = 1072261''.307 + \left(1236^r + 1105601''.328\right)T - 6''.891T^2 + 0''.019T^3 \tag{d}$$

$$\Omega = 450160''.280 + \left(5^r + 482890''.539\right)T + 7''.455T^2 + 0''.008T^3 \tag{e}$$

$$1^r = 360^0 = 1296000'' \tag{f}$$

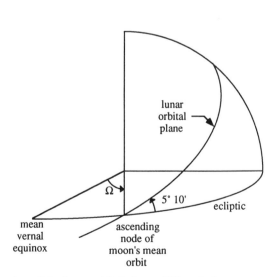

Figure 2.4 Rotation of the lunar orbital plane. Although the moon completes 1 orbital pass in 1 month, it takes 18.6 years for a complete circuit of the orbital plane around the ecliptic pole.

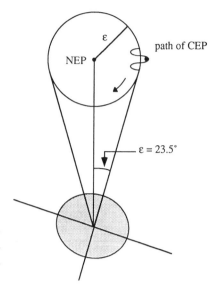

Figure 2.5 Lunisolar precession and nutation. The CEP is the direction of the earth's rotation axis; its spatial motion is a function of precession and nutation.

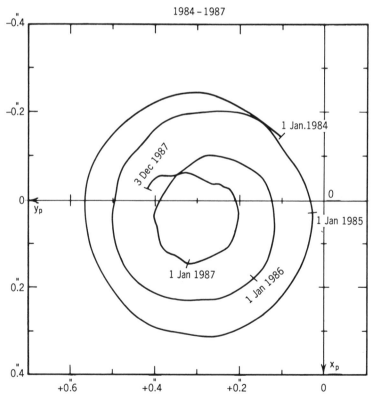

Figure 2.6 Polar motion. The earth's rotation axis moves with respect to the crust. Precise determination of polar motion is one of the objectives of geodesy. (Courtesy of the Central Bureau of the International Earth Rotation Service.)

The center of figure of polar motion for the years 1900–1905, until recently called the conventional international origin (CIO), has been used to position the third axis of the CTRS. With the introduction of the improved set of nutations, this point became known as the conventional terrestrial pole (CTP). The CEP is referenced to the CTP by means of the polar motion coordinate (x_p, y_p). The origin of the polar motion coordinate system is at the CTP, the x axis is along the Greenwich meridian, and the y axis is positive along the 270° meridian. The polar coordinates had been determined by optical star observations, but are now determined through geodetic space techniques, such as lunar laser ranging, satellite laser ranging, and very long baseline interferometry. Note that the center of figure of today's polar motion does not contain the CTP. There appears to be "polar wander" (gradual shifting of the center of figure away from the CTP). This phenomenon is awaiting a satisfactory scientific explanation as well.

The conventional terrestrial equator is orthogonal to the direction of the CTP. The third axis of the conventional terrestrial coordinate system coincides with the CTP; the first axis is defined by the intersection of the terrestrial equator and the Greenwich meridian. This point is denoted by the symbol A in Figure 2.7. The next section provides additional detail on the definition of

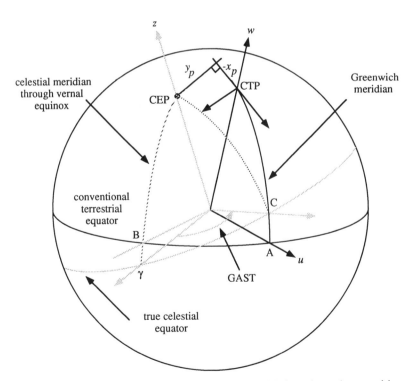

Figure 2.7 Conventional terrestrial system. To avoid time-dependent positions of earth stations, the CTRS is referenced to the CTP and the Greenwich meridian.

the Greenwich meridian. The second axis completes a right-handed coordinate system. Because the polar motion coordinates are small compared to the 90° arc from the pole to the equator, it can readily be verified that the angle from the vernal equinox to point C, measured on the true celestial equator, and the angle from B to A, measured on the terrestrial equator, can be considered the same for all practical purposes. This angle represents Greenwich apparent sidereal time (GAST). It follows that

$$
\begin{bmatrix} x \\ y \\ z \end{bmatrix}_{Gr} = R_3(\text{GAST}) \begin{bmatrix} x \\ y \\ z \end{bmatrix}_{\alpha, \delta} \tag{2.16}
$$

$$
\begin{bmatrix} u \\ v \\ w \end{bmatrix} = R_2(-x_p) R_1(-y_p) \begin{bmatrix} x \\ y \\ z \end{bmatrix}_{Gr} \tag{2.17}
$$

The coordinate system $(X)_{Gr}$ is not completely crust-fixed, because the third axis moves with polar motion; it is sometimes referred to as the instantaneous terrestrial coordinate system.

2.2 TIME SYSTEMS

Any device capable of generating a sequence of constant intervals in time can serve as a clock. The best timing scales available today make use of quartz crystal oscillations or state transitions of atoms. In surveying and navigation these time scales must be related to the earth's rotation. After all, the topocentric distance to a satellite is directly related to the rotational position of the earth, which, in turn, is a function of time. Because atomic clocks generate independent time scales, there is a need to coordinate these times. This is the role of the earth rotation and time service authorities.

2.2.1 Time in Orders of Magnitude

Progress, particularly over the last couple of decades, in producing precise timekeeping devices has profoundly influenced surveying. The critical role of timing in surveying is frequently overlooked. Figure 2.8 relates the progress in measuring the length of a day-long interval to some of the major events in clock technology. Timekeeping has shown about six to seven orders of magnitude improvement during this century. During the era of pendulum clocks, the timing accuracy that could be maintained over one day corresponded more or less to the irregularities in the variation of the daily earth rotation. Thus, it

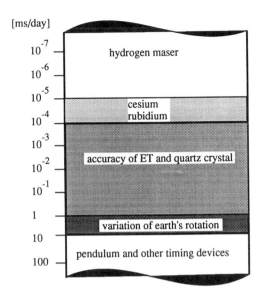

Figure 2.8 Illustration of progress in timing accuracy per day. Atomic clocks make GPS surveying possible.

made sense to use the daily earth rotation as a time standard. This led to the rotational time scales of sidereal time (ST) and solar time (UT). However, observations to celestial objects have shown consistent discrepancies between the observed and predicted positions using these rotational time scales. It was soon correctly determined that the earth rotation was not a constant and therefore was not suitable for defining a constant time scale. Astronomers introduced a new time scale, called the ephemeris time (ET). This time scale is based solely on the orbital motion of the sun and the moon, as implied by the law of gravitation. As vibrating quartz crystals became available for timekeeping, variations at the millisecond level in the daily earth rotation became apparent. A breakthrough of about three orders of magnitude in timing was achieved with atomic frequency standards. The physical principle of atomic time is related to atomic energy levels (transition levels) and electromagnetic radiation. The primary elements used today are rubidium, cesium, and hydrogen. The research and development of even better atomic clocks is still continuing. Blair (1974) gives an extensive treatment of modern timing devices. One often distinguishes the short-term, medium-term, and long-term stability characteristics of atomic clocks by quoting the ratio $\Delta f/f$, where Δf denotes the change in frequency f over a specific interval. Table 2.2 shows the one-day stability and the time it takes for the atomic clock to accumulate a 1-sec error, assuming (for reasons of simplicity) that the frequency stability does not change with time and equals the one-day stability. The time accumulation for a 1-sec error is on the order of thousands of years for atomic clocks. More sophisti-

TABLE 2.2   Accumulation of Timing Errors for Various Oscillators

Type clock	Oscillation frequency [GHz]	Stability per day [$\Delta f/f$]	Time to lose one second
Quartz crystal oscillator	.005 (typical)	10^{-9}	30 years
Rubidium	6,834,682,613	10^{-12}	30,000 years
Cesium	9,192,631,770	10^{-13}	300,000 years
Hydrogen maser	1,420,405,751	10^{-15}	30,000,000 years

cated methods to describe the stability and the performance of atomic clocks are found in the specialized literature.

Table 2.3 shows the distance traveled by light (column 1) during short intervals (column 2). Column 3 shows a rough estimate of the time it takes for a cesium standard to accumulate the respective time errors, assuming again a stability equal to the daily stability given previously. Because we do not wish to use expensive cesium clocks in the receivers, special attention must be given to eliminating clock errors. One could attempt to estimate clock errors for each epoch. However, differencing approaches have been developed so that most of the clock errors cancel. These developments are discussed in Chapter 8 in great detail.

TABLE 2.3   GPS Timing Requirements

Light travel [m]	Time error [nsec]	Time for cesium clock to accumulate error [sec]
300.0	1000	10^7
0.3	1	10^4
0.003	0.01	10^2

2.2.2 Astronomic Time Scales

There are three astronomic time scales: sidereal time (ST), universal (solar) time (UT), and ephemeris time (ET). Sidereal time and universal time are based on the earth's rotation, whereas ephemeris time is based on the orbital motion of the sun (and the moon) following the law of gravitation. Figure 2.9 shows the true celestial meridian of the observer, the true vernal equinox, and the mean vernal equinox of date. The hour angles of the true and mean vernal equinox are called the apparent sidereal time (AST) and mean sidereal time (MST), respectively. Thus, because sidereal time refers to the local celestial meridian attached to the earth, it is a rotational time scale. The small difference between AST and MST is called the equation of equinox (Eq. E). This difference is the result of the nutations of the rotation axis. Figure 2.10 shows that

$$\text{Eq. E} = \text{AST} - \text{MST} \qquad (2.18)$$
$$= \Delta\psi \cos(\epsilon + \Delta\epsilon)$$

The latter part of (2.18) follows from the expressions of spherical trigonometry as applied to small spherical triangles.

Apparent sidereal time is determined by registering the epoch for transits of a star through the celestial meridian. At that instant of transit the apparent sidereal time equals the true right ascension. The latter quantity is available from star catalogs. This method of time determination has one pitfall, however: Because of polar motion, the celestial meridian moves continuously. A time determination carried out on two consecutive days will result in different times for the transit. Figure 2.11 shows this situation for the case of the CEP and the CTP. Choosing two consecutive positions of the CEP does not make practical sense, because the objective is to refer all observations to a point fixed to the crust, which happens to be the CTP. For the sake of completeness,

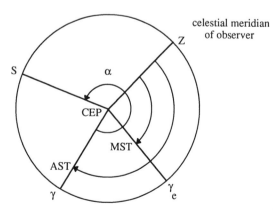

Figure 2.9 The apparent sidereal time (AST).

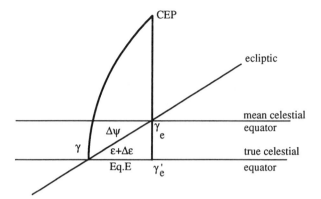

Figure 2.10 The equation of equinox (Eq. E).

Figure 2.11 also shows the other astronomic quantities that are effected by polar motion. The figure may be interpreted as a sphere of direction located at the topocenter. The direction Z is that of the plumb line at the observing station. The symbol T denotes a distant station (target). The observed astronomic longitude, latitude, and azimuth (Λ, Φ, A) refer to the true celestial meridian CEP-Z-E, whereas the reduced astronomic longitude, latitude and azimuth $(\Lambda_{CTP}, \Phi_{CTP}, A_{CTP})$ refer to the "reduced" meridian CTP-Z-B. The latter quantities are no longer a function of time. The apparent sidereal time (AST) corresponds to the angle $\gamma - E$ in Figure 2.11 and the MST equals the angle $\gamma_e - E$. Taking into consideration that the separation of CEP and CTP is much smaller than, say, the distance Z-B or Z-CEP, we can readily verify that $BC = DE$ and $BD = CE$ are valid with negligible error. We can further verify it

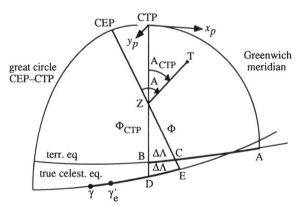

Figure 2.11 Polar motion reductions. Astronomic azimuth, latitude, and longitude must be reduced to the CTP. Reduction in longitude and time are conceptually the same.

by applying either spherical trigonometry or rotation matrices (Leick 1980a; Moritz and Mueller 1987) that

$$\Phi_{CTP} = \Phi + y_p \sin \Lambda - x_p \cos \Lambda \qquad (2.19)$$

$$\Lambda_{CTP} = \Lambda - (y_p \cos \Lambda + x_p \sin \Lambda) \tan \Phi \qquad (2.20)$$

$$A_{CTP} = A - \frac{y_p \cos \Lambda + x_p \sin \Lambda}{\cos \Phi} \qquad (2.21)$$

The subscript CTP refers to the reduced astronomic observables in the CTRS, that is, the reduced astronomic latitude, longitude, and azimuth that would have been observed if the earth's rotation axis had coincided with the direction of the CTP at the instant of observation.

Returning to the problem of time determination, we see that the apparent sidereal time corrected for polar motion, denoted by AST1, which would have been obtained if the earth's rotation axis had coincided with the direction of the CTP at the instant of observation, is

$$
\begin{aligned}
AST1 &= AST - \Delta\Lambda \\
&= AST - (y_p \cos \Lambda + x_p \sin \Lambda) \tan \Phi
\end{aligned} \qquad (2.22)
$$

The correction to sidereal time, $\Delta\Lambda$, is independent of the declination of the star; what matters is the change in right ascension when switching from the CEP system to the CTP. The same time correction can be applied to the mean sidereal time; thus,

$$MST1 = AST1 - Eq. E \qquad (2.23)$$

Next, the longitude of the observing station in the CTRS is subtracted from MST1, yielding an expression for the Greenwich mean sidereal time (GMST):

$$GMST = MST1 - \Lambda_{CTP} \qquad (2.24)$$

Note that the longitude is counted positive eastward of Greenwich. Assume that several observatories with known terrestrial longitudes observe at the same time the meridian transit of a star and that they compute GMST using the procedure explained. They all will obtain the same numerical value for GMST with the endpoint (or beginning) at A. Thus, the location of point A depends on the adopted longitudes for a set of observing stations. The great circle CTP $- A$ is called the Greenwich meridian.

Sidereal time and universal time are mathematically related through the mean time scale. Specifically (Kaplan 1981),

$$GMST = UT1 + \alpha_m - 12^h \qquad (2.25)$$

with

$$\alpha_m = 18^h 41^m 50^s.54841 + 8640184^s.812866 T_U$$
$$+ 0^s.093104 T_U^2 - 6^s.2 \times 10^{-6} T_U^3$$

(2.26)

where T_U is the number of Julian centuries of 36,525 days of universal time elapsed since 2000 Greenwich noon, 1 January, 12^h UT1 (JD 2,451,545.0). UT1 is the universal time of the Greenwich meridian. The geometric interpretation of (2.25) is shown in Figure 2.12. The symbol α_m is the mean right ascension of the mean (fictitious) sun, S_m. The mean sun can be thought of as moving in the ecliptic or in the terrestrial equator. The angular velocity of the mean sun is constant and corresponds to the average angular velocity of the true sun, which is the same as the average angular velocity of the earth in the ecliptic. Because of the ellipticity of the earth's orbit, the actual angular velocity of the earth's orbital motion, as measured from the foci (the sun), varies according to Kepler's second law. The reader is referred to Section 2.3 for a comparison with the satellite motions.

Given two events e_1 and e_2, the rate between sidereal time and mean solar time is (Kaplan 1981)

$$r = \frac{\text{Measure of UT1 between } e_1 \text{ and } e_2}{\text{Measure of GMST between } e_1 \text{ and } e_2}$$
$$= 0.997269566329084 - 5.8684 \times 10^{-11} T_U + 5.9 \times 10^{-15} T_U^2$$

(2.27)

The mean solar day is defined as two consecutive transits of the mean sun over the meridian. The mean solar day begins at lower transit so that civil noon can be labeled 12 o'clock. The length of the mean solar day is a function of the motion of the mean sun and the mean vernal equinox. The Julian date is a

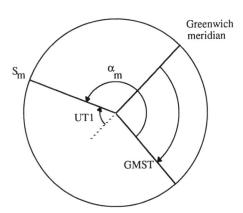

Figure 2.12 Greenwich mean sidereal time versus UT1. Both time scales are related by a mathematical expression.

counter for mean solar days. Conversion of any Gregorian calendar date (Y = year, M = month, D = day) to JD is accomplished by (van Flandern and Pulkkinen 1979)

$$JD = 367 \times Y - 7 \times [Y + (M + 9)/12]/4$$
$$+ 275 \times M/9 + D + 1,721,014 \tag{2.28}$$

for Greenwich noon. This expression is valid for dates since March 1900. The expression is read as a Fortran-type statement; division by integers implies truncation of the quotients of integers (no decimals are carried).

To deal with a smaller number, the modified Julian date (MJD) is used. It is defined as

$$MJD = JD - 2,400,000.5 \tag{2.29}$$

The Julian century consists of 36,525 mean solar days. The tropical (mean solar) year equals the time required for the fictitious sun to make two consecutive passages over the mean vernal equinox. The tropical year consists of $365^d.24219879$ mean solar days. Note that the tropical year is not an integer number of mean solar days. To deal with this fact in the civilian calendar, the leap year system was introduced (Mueller 1969).

In summary, UT1 measures the true angular rotation of the earth as corrected for the rotational component induced by polar motion. UT1 gives the measure of the rotation as if the earth were rotating around the CTP. UT1 is mathematically related to GMST and GAST (=GAST1 because point A is located on the terrestrial equator and Greenwich time always refers to the Greenwich meridian). GAST and polar motion coordinates are the three elements that completely relate the true celestial coordinate system and the conventional terrestrial coordinate system. There are many reasons for the variations in the earth's rotation and therefore the nonlinearities in UT1. For example, change in the angular momentum due to tides results in rotational variations with periods similar to those of the nutations, or seasonal variations resulting from global wind pattern may occur. The ephemeris time, which is based on the orbital motion of the sun and the moon and not on the rotation of the earth, is not without problems either. The primary concern about ET is that the orbital positions of the earth and the moon do not completely follow the law of gravitation; one thinks of energy loss from tidal friction. There are also considerable difficulties in observing the orbital positions of the sun and the moon with the required accuracy, and making the results available without delay. ET has now been replaced with the atomic time scale.

2.2.3 Atomic Time

The length of the atomic second was defined to correspond with the ET second. At the 13th General Conference of Weights and Measures (CGPM) in Paris in

1967, the definition of the atomic second, also called International System (SI) second, was defined as the duration of 9,192,631,770 periods of the radiation corresponding to the transition between two hyperfine levels of the ground state of the cesium-133 atom. The epoch of the new international atomic time (TAI) was set such that $ET - TAI = 32^s.184$ on 1 January 1977. Ever since, ET has been replaced by TAI. For the most accurate ephemeris computations it is necessary to take relativistic effects into account. If the equations of motion refer to the barycenter, the appropriate time scale is barycentric dynamical time (TDB); in the geocentric reference frame the terrestrial dynamical time (TDT) must be used. TDB and TDT differ only by small periodic terms. In the terminology of general relatively, TDT corresponds to the proper time, whereas TDB corresponds to the coordinate time. The terrestrial dynamical time is essentially equal to ET. The complete relations are as follows (Kaplan 1981):

$$TDT = TAI + 32^s.184 \tag{2.30}$$

$$TDB = TDT + 0^s.001658 \sin(g + 0.0167 \sin g) \tag{2.31}$$

with

$$g = (357°.528 + 35,999°.050T)\left(\frac{2\pi}{360°}\right) \tag{2.32}$$

The symbol T is the interval measured in Julian centuries of TDB between J2000.0 and the epoch.

2.2.3.1 Need for Coordination Various laboratories and agencies around the world operate their own atomic clocks, producing independent atomic scales. For example, the atomic time generated by the U.S. Naval Observatory (USNO) is based on a set of cesium beam frequency standards. The combined time of these atomic clocks is called A.1(USNO). The National Institute of Standards and Technology (NIST) also operates several atomic clocks generating the time scale TA(NIST). These agencies and many others around the world cooperate with the International Bureau of Weights and Measures (Bureau International des Poids et Measures—BIPM). The BIPM is solely responsible for combining the various atomic times and for computing the international atomic time. Formerly, the TAI was computed by the Bureau International de l'Heure (BIH). Since 1 January 1988, the services of the BIH have been provided by the BIPM and by the Central Bureau of the new International Earth Rotation Service (IERS/CB). Time comparisons are regularly made by means of LORAN-C links, various satellites, and clock visits. GPS is already the backbone for global time transfer (Miranian and Klepczynski 1991). GPS is especially suitable for time transfer because the satellites can be seen from widely separated stations. The goal for worldwide synchronization is 100 nsec and better. There are currently about 180 cesium beam atomic clocks in some 50 establishments around the world contributing to the BIPM. At this time the USNO clocks contribute about 20% of the weight in the system. Because TAI is a computed time system, it is sometimes referred to as the paper clock. As part of the time computations the BIPM makes the time residuals available to the individual contributors in order to allow them to steer their clocks within prescribed limits of TAI.

2.2.3.2 Link to Earth Rotation   The atomic time is not related to the earth rotation but to the laws of nature governing the transition of energy levels in atoms. For navigation and surveying, however, it is necessary to relate time to the rotation of the earth. After all, one would like to compute the topocentric position of satellites as a function of time. The motion of satellites in their orbits is independent of the earth's rotational velocity, just as the orbital motion of the earth around the sun is not related to the daily rotations of the earth. Atomic time is tied to earth rotation by introducing the universal coordinated time (UTC) scale. This is a hybrid time scale in the sense that the UTC second is the SI second as established and made available through the highly stable atomic time TAI, and the epoch of UTC is such that

$$|UT1 - UTC| < 0^s.9 \tag{2.33}$$

This definition takes advantage of the stable atomic time and yet follows the rotational time UT1. UTC is changed in steps of a full second (leap second) if the difference to UT1 exceeds the specified limit. Adjustments are made on either 30 June or 31 December. It is the responsibility of the International Earth Rotation Service (IERS) to determine the need to introduce a leap second and to announce it. The instability of TAI is about six orders of magnitude smaller than that of the rotational time scale UT1. Accurate differences such as

$$\Delta UT1 = UT1 - UTC \tag{2.34}$$

$$\Delta AT = TAI - UTC \tag{2.35}$$

are available from various time services. In 1972 the UTC system was adjusted by a once-only step such that ΔAT is an exact integer number of seconds. Before 1972 a different procedure was followed in establishing the relationship between TAI and UTC. Figure 2.13 shows the history of the leap second adjustments of UTC until 1988. On 30 June 1994, ΔAT was set to 29 sec. The difference UT1 $-$ TAI was approximately zero on 1 January 1958. An approximate value of $\Delta UT1$ with a precision of ± 0.1 sec is broadcast with the time signals. Changes in this broadcast value are also decided by the IERS.

UTC is also referred to as broadcast time and is the basis for all daily civilian timing requirements. The time announced on the radio or television is UTC. The divergence (or convergence as the case may be) of the UTC and TAI can be expected. After all, the TAI second equals the ET second, which

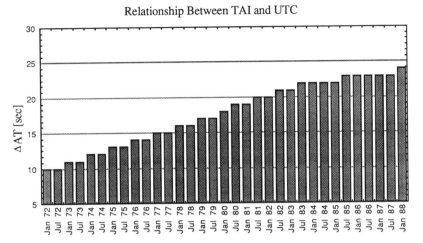

Figure 2.13   Leap second corrections of UTC.

in turn is based on gravitational orbital motion, not on the earth rotation (as is UT1). The deviation of the earth's rotation from a constant velocity is shown in Figure 2.14 for the year 1986 (BIH 1986). The figure shows a strong slope, indicating that soon another leap second adjustment will be required. There was no leap second adjustment in 1986. In Figure 2.15 the linear trend is removed to show the seasonal variations. This figure also indicates the daily variations of the earth's rotation at the 1-msec level. A review of recent prog-

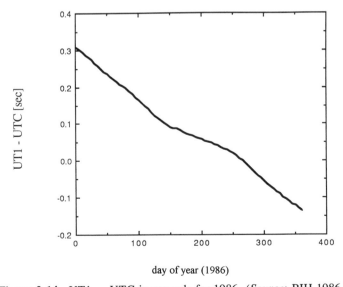

day of year (1986)

Figure 2.14   UT1 − UTC in seconds for 1986. (*Source*: BIH 1986.)

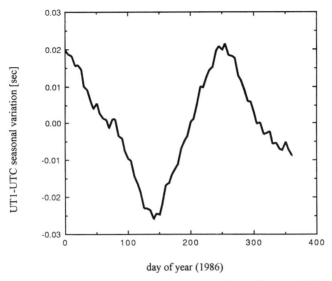

day of year (1986)

Figure 2.15 Variation in the earth rotation for 1986. (*Source*: BIH 1986.)

ress in earth rotation research, assessment of the current state of knowledge in this field, theory of orientation changes, and studies of the excitation of the rotational variation by irregular fluid motion in the atmosphere, oceans, and the liquid outer core of the earth is given in Eubanks (1993).

2.2.4 Time and Earth Orientation Services

The U.S. Naval Observatory Time Service Division provides to the Department of Defense, other Government agencies, and the general public a precise time standard as a reference. Because satellites are an integral part of today's navigation systems, the USNO also provides time and frequency information on satellites. In addition to A.1(UNSO), as derived from the assembly of USNO atomic clocks, the USNO also makes UTC(USNO) available. These two time scales are related by a noninteger value that changes by full seconds as UTC is adjusted for leap seconds. The A.1 time scale is not corrected for leap seconds. Beginning in 1993 the difference was A.1(USNO) − UTC(USNO) = 28.034817 sec. The physical realization of universal coordinated time at the USNO is called UTC(USNO, MC); this is a physical clock that is now steered to be within 100 nsec of UTC. GPS time equals to TAI − 19 sec within one microsecond. The more precise definition of GPS time is that it is steered to be in agreement with UTC(USNO) modula one second. The required steering time accuracy is 1 μsec, but in practice it is typically within 100 nsec. In addition, the GPS broadcasts the corrections necessary to estimate UTC(USNO). The specifications state that this conformity shall be within 100 nsec. In practice, the conformity is typically better than 20 nsec.

The Time and Frequency Division of the National Institute of Standards and Technology, located in Boulder, Colorado, is responsible for NIST's timing function. Time services are available from stations WWV and WWVB in Fort Collins, Colorado, and from WWVH in Kauai, Hawaii. In addition, services using network television and satellites are also available. For example, NIST disseminates time via the GOES (Geostationary Operational Environmental Satellite) satellites. NIST time is based on a number of commercial cesium beam clocks. The primary time scale is TA(NIST). The time UTC(NIST) is generated by adding leap seconds and small corrections to TA(NIST) as needed to keep UTC(NIST) synchronized with the internationally coordinated time UTC computed by the BIPM.

The International Bureau of Weights and Measures and the International Earth Rotation Service compute the TAI, UTC, and polar motion parameters. The IERS has the sole authority to determine the need for and to announce leap seconds. The results produced by these coordinating offices are available through a series of bulletins. Some of these bulletins are disseminated by electronic means or can be accessed electronically from the IERS database. IERS relies on three observation methods: lunar laser ranging, satellite laser ranging, and very long baseline interferometry.

Both the IERS and the UNSO maintain large databases of time-dependent quantities. These include the polar motion coordinates and the UT1 − UTC differences required in surveying. The IERS data are summarized in the IERS annual reports. Winkler (1986) and Withington (1985) explain the USNO database and the telecommunication techniques needed to access it. Much of the UNSO database is directly related to GPS. Both databases can be accessed via the Internet.

2.3   SATELLITE ORBITAL MOTION

The orbital motion of satellites is a result of the earth's gravitational attraction and a number of other forces acting on the satellite. Examples are the attraction of the sun and the moon and the pressure on the satellite caused by impacting solar radiation particles. For high orbiting satellites the atmospheric drag is negligible. Mathematically, the equations of motions for satellites are differential equations that are solved by numerical integration over time. The integration begins with initial conditions, such as the position and velocity of the satellite at some initial epoch. The computed (predicted) satellite positions can be compared with actual observations. Possible discrepancies are useful to improve the force function, the initial conditions, or the station position of the observer.

2.3.1   Kepler Elements

Six Kepler elements are often used to describe the position of satellites in space. To simplify attempts to study satellite motions, we study so-called normal orbits. For normal orbits the satellites move in an orbital plane that is fixed in space; the actual path of the satellite in the orbital plane is an ellipse in the mathematically strict sense. The focal point of the orbital ellipse is at the center of the earth. The conditions leading to such a simple orbital motion are as follows:

1. The earth is treated as a point mass, or, equivalently, as a sphere with constant density distribution. The gravitational field of such a body is radially symmetric; that is, the plumb lines are all straight lines and point toward the center of the sphere.
2. The mass of the satellite is negligible compared to the mass of the earth.
3. The motion of the satellite takes place in a vacuum; that is, there is no atmospheric drag acting on the satellite and no solar radiation pressure.
4. No sun, moon, or other celestial body exerts a gravitational attraction on the satellite.

The orbital plane of a satellite moving under such conditions is shown in Figure 2.16. The ellipse denotes the path of the satellite. The shape of the ellipse is determined by the semimajor axis a and the semiminor axis b. The symbol e denotes the eccentricity of the ellipse. The ellipse is enclosed by an auxiliary circle with radius a. The principle axes of the ellipse form the coordinate system (ξ, η). The current position of the satellite is denoted by S; the line SS' is in the orbital plane and is parallel to the η axis. The coordinate system (q_1, q_2) is in the orbital plane with its origin at the focal point F of the ellipse that coincides with the center of the earth. The third axis q_3, not shown in Figure 2.16, completes the right-handed coordinate system. The geocentric distance from the center of the earth to the satellite is denoted by r. The orbital locations closest to and farthest from the focal point are called the perigee and

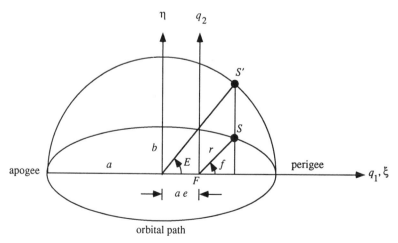

Figure 2.16 Coordinate systems in the orbital plane.

apogee, respectively. The true anomaly f and the eccentric anomaly E are measured counterclockwise, as shown in Figure 2.16.

The orbital plane is shown in Figure 2.17 with respect to the true celestial coordinate system. The center of the sphere of directions is located at the focal point F. The x axis is in the direction of the vernal equinox, the z axis coincides with the celestial ephemeris pole, and y is located in the equator completing the right-handed coordinate system. The intersection of the orbital plane with the equator is called the nodal line. The point at which the satellite ascends the equator is the ascending node. The right ascension of the ascending node is denoted by Ω. The line of apsides connects the focal point F and the perigee. The angle subtended by the nodal line and the line of apsides is called the argument of perigee, ω. The true anomaly f and the argument of perigee ω lie in the orbital plane. Finally, the angle between the orbital plane and the equator is the inclination i. Figure 2.17 shows that (Ω, i) determine the position of the orbital plane in the true celestial system, (Ω, ω, i) position the orbital ellipse in space, and (a, e, f) determine the position of the satellite within the orbital plane.

The six Kepler elements are

$$\{\Omega, \omega, i, a, e, f\}$$

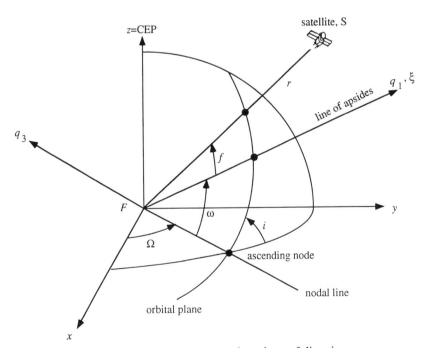

Figure 2.17   Orbital plane on the sphere of direction.

The true anomaly f is the only Kepler element that is a function of time in the case of normal orbits; the remaining five Kepler elements are constant. For actual satellite orbits, which are not subject to the conditions of normal orbits, all Kepler elements are a function of time. They are then called osculating Kepler elements.

2.3.2 Normal Orbital Theory

Normal orbits are particularly useful for the understanding and visualizing of the spatial motions of satellites. The solutions of the respective equations of motions can be given by simple, analytical expressions. Since normal orbits are a function of the central portion of the earth's gravitational field, which is by far the largest force acting on the satellite, normal orbits are indeed usable for orbital predictions over short periods of time when low accuracy is sufficient. Thus, one of the popular uses of normal orbits is for the construction of satellite visibility charts.

The normal motion of satellites is completely determined by Newton's law of gravitation:

$$F = \frac{k^2 mM}{r^2} \tag{2.36}$$

In (2.36) M and m denote the mass of the earth and the satellite, respectively, k is the universal gravitational constant, r is the geocentric distance to the satellite, and F is the gravitational force between the two bodies. This force can also be written as

$$F = ma \tag{2.37}$$

where a denotes the acceleration experienced by the satellite. Combining (2.36) and (2.37) gives

$$a = \frac{k^2 M}{r^2} \tag{2.38}$$

This equation can be written in vector form as

$$\ddot{\mathbf{r}} = -k^2 M \frac{\mathbf{r}}{r^3} = -\mu \frac{\mathbf{r}}{r^3} \tag{2.39}$$

where

$$\mu = k^2 M \tag{2.40}$$

is the earth's gravitational constant. Including the earth's atmosphere it has the value $\mu = 3,986,005 \times 10^8$ m^3 sec^{-2}. The vector $\mathbf{r}$ is directed from the central body (earth) to the satellite. The sign has been chosen such that the acceleration is directed toward the earth. The colinearity of the acceleration and the position vector as in (2.39) is a characteristic of central gravity fields. A particle would fall along a straight line toward the earth (straight plumb line).

Equation (2.39) is valid for the motion with respect to an inertial origin. In general, one is interested in determining the motion of the satellite with respect to the earth. The modified equation of motion for accomplishing this is (Escobal 1965, p. 37)

$$\ddot{\mathbf{r}} = -k^2 (M + m)\, \frac{\mathbf{r}}{r^3} \tag{2.41}$$

Because $m \ll M$ the second term is often neglected and (2.41) becomes (2.39).

Figure 2.16 gives the position of the satellite in the (q) coordinate system as

$$\begin{bmatrix} q_1 \\ q_2 \\ q_3 \end{bmatrix} = r \begin{bmatrix} \cos f \\ \sin f \\ 0 \end{bmatrix} \tag{2.42}$$

Because the geocentric distance and the true anomaly are functions of time, the derivative with respect to time, denoted by a dot, is

$$\dot{\mathbf{q}} = \dot{r} \begin{bmatrix} \cos f \\ \sin f \\ 0 \end{bmatrix} + r\dot{f} \begin{bmatrix} -\sin f \\ \cos f \\ 0 \end{bmatrix} \tag{2.43}$$

The second derivatives with respect to time are

$$\ddot{\mathbf{q}} = \ddot{r} \begin{bmatrix} \cos f \\ \sin f \\ 0 \end{bmatrix} + 2\dot{r}\dot{f} \begin{bmatrix} -\sin f \\ \cos f \\ 0 \end{bmatrix} + r\ddot{f} \begin{bmatrix} -\sin f \\ \cos f \\ 0 \end{bmatrix} - r(\dot{f})^2 \begin{bmatrix} \cos f \\ \sin f \\ 0 \end{bmatrix}$$

$$\tag{2.44}$$

The second derivative is written according to (2.39) and (2.42) as

$$\ddot{\mathbf{r}} = \frac{-\mu}{r^2} \begin{bmatrix} \cos f \\ \sin f \\ 0 \end{bmatrix} \tag{2.45}$$

Evaluating (2.44) and (2.45) at $f = 0$ (perigee) and substituting (2.45) for the left-hand side of (2.44) gives

$$\ddot{r} - r(\dot{f})^2 = \frac{-\mu}{r^2} \tag{2.46}$$

$$r\ddot{f} + 2\dot{r}\dot{f} = 0 \tag{2.47}$$

Equation (2.47) is further developed by multiplying with r and integrating

$$\int (r^2\ddot{f} + 2r\dot{r}\dot{f})\, dt = C \tag{2.48}$$

The result of the integration is

$$r^2\dot{f} + 2r^2\dot{f} = C \tag{2.49}$$

as can be readily verified through differentiation. Combining both terms yields

$$r^2\dot{f} = h \tag{2.50}$$

where h is a new constant. Equation (2.50) is identified as an angular momentum equation, implying that the angular momentum for the orbiting satellite is conserved. To integrate (2.46), we define a new variable:

$$u \equiv \frac{1}{r} \tag{2.51}$$

By using equation (2.50) for dt/df, the differential of (2.51) becomes

$$\frac{du}{df} = \frac{du}{dr}\frac{dr}{dt}\frac{dt}{df} = -\frac{\dot{r}}{h} \tag{2.52}$$

Differentiating again gives

$$\frac{d^2u}{df^2} = \frac{d}{dt}\left(-\frac{\dot{r}}{h}\right)\frac{dt}{df} = -\frac{\ddot{r}}{u^2h^2} \tag{2.53}$$

or

$$\ddot{r} = -h^2u^2\frac{d^2u}{df^2} \tag{2.54}$$

By substituting (2.54) in (2.46), substituting $\dot{f}$ from (2.50) in (2.46), and replacing r by u according to (2.51), equation (2.46) becomes

$$\frac{d^2u}{df^2} + u = \frac{\mu}{h^2} \tag{2.55}$$

which can readily be integrated as

$$\frac{1}{r} \equiv u = C \cos f + \frac{\mu}{h^2} \tag{2.56}$$

where C is a constant. Equation (2.56) is the equation of an ellipse. This is verified by writing the equation for the orbital ellipse in Figure 2.16 in the principle axis form:

$$\frac{\xi^2}{a^2} + \frac{\eta^2}{b^2} = 1 \tag{2.57}$$

where

$$\xi = ae + r \cos f \tag{2.58}$$

$$\eta = r \sin f \tag{2.59}$$

$$b^2 = a^2(1 - e^2) \tag{2.60}$$

Equation (2.60) is valid for any ellipse. Substituting equations (2.58) through (2.60) into (2.57) and solving the resulting second-order equation for r gives

$$\frac{1}{r} = \frac{1}{a(1 - e^2)} + \frac{e}{a(1 - e^2)} \cos f \tag{2.61}$$

With

$$C = \frac{e}{a(1 - e^2)} \tag{2.62}$$

and

$$h = \sqrt{\mu a(1 - e^2)} \tag{2.63}$$

the identity between the expression for the ellipse (2.61) and equation (2.56) is established. Thus the motion of a satellite under the condition of a normal orbit is an ellipse. This is the content of Kepler's first law. The focus of the ellipse is at the center of mass. Kepler's second law states that the geocentric vector **r** sweeps equal areas during equal times. Because the area swept for the differential angle df is

$$dA = \tfrac{1}{2}r^2 \, df \tag{2.64}$$

it follows from (2.50) and (2.63) that

$$\frac{dA}{dt} = \frac{1}{2} \sqrt{\mu a (1 - e^2)} \qquad (2.65)$$

which is a constant. The derivation of Kepler's third law requires the introduction of the eccentric anomaly E. From Figure 2.16 we see that

$$q_1 = \xi - ae = a(\cos E - e) \qquad (2.66)$$

where

$$\xi = a \cos E \qquad (2.67)$$

From equation (2.57)

$$q_2 \equiv \eta = \sqrt{\left(1 - \frac{\xi^2}{a^2}\right) b^2} \qquad (2.68)$$

Substitute (2.67) in (2.68); then

$$q_2 \equiv \eta = b \sin E \qquad (2.69)$$

With (2.66), (2.69), and (2.60) the geocentric satellite distance becomes

$$r = \sqrt{q_1^2 + q_2^2} = a(1 - e \cos E) \qquad (2.70)$$

Differentiating equations (2.70) and (2.61) gives

$$dr = ae \sin E \, dE \qquad (2.71)$$

$$dr = \frac{r^2 e}{a(1 - e^2)} \sin f \, df \qquad (2.72)$$

Equating (2.72) and (2.71) and using (2.59), (2.60), (2.69), and (2.42) and multiplying the resulting equation by r gives

$$rb \, dE = r^2 \, df \qquad (2.73)$$

Substituting (2.60) for b and (2.70) for r, replacing df by dt using (2.50), using (2.63) for h, and then integrating, we obtain

$$\int_{E=0}^{E} (1 - e \cos E) \, dE = \int_{t_0}^{t} \sqrt{\frac{\mu}{a^3}} \, dt \qquad (2.74)$$

Integrating both sides gives

$$E - e \sin E = M \tag{2.75}$$

$$M = n(t - t_0) \tag{2.76}$$

$$n = \sqrt{\frac{\mu}{a^3}} \tag{2.77}$$

Equation (2.77) is Kepler's third law. Equation (2.75) is called the Kepler equation. The symbol n denotes the mean motion, M is the mean anomaly, and t_0 denotes the time of perigee passage of the satellite. The mean anomaly M should not be confused with the same symbol used for the mass of the central body in (2.36). Let P denote the orbital period, that is, the time required for one complete revolution; then

$$P = \frac{2\pi}{n} \tag{2.78}$$

The mean motion n equals the average angular velocity of the satellite. Equation (2.77) shows that the semimajor axis completely determines the mean motion and thus the period of the orbit.

With the Kepler laws in place, one can identify alternative sets of Kepler elements, such as

$$\{\Omega, \omega, i, a, e, M\}$$

or

$$\{\Omega, \omega, i, a, e, E\}$$

Often the orbit is not specified by the Kepler elements but by the vector $\mathbf{r} = (x, y, z) = \mathbf{X}$ and the velocity $\dot{\mathbf{r}} = (\dot{x}, \dot{y}, \dot{z}) = \dot{\mathbf{X}}$, expressed in the true celestial coordinate system (X). Figure 2.17 shows that

$$\mathbf{q} = R_3(\omega) R_1(i) R_3(\Omega) \mathbf{X}$$

$$= R_{qx}(\Omega, i, \omega) \mathbf{X} \tag{2.79}$$

where R_i denotes a rotation around axis i. The inverse transformation is

$$\mathbf{X} = R_{xq}(\Omega, i, \omega) \mathbf{q} \tag{2.80}$$

Differentiating (2.80) once gives

$$\dot{\mathbf{X}} = R_{xq}(\Omega, i, \omega) \dot{\mathbf{q}} \tag{2.81}$$

Note that the elements of $R_{xq}(\Omega, i, \omega)$ are constants, because the orbital ellipse does not change its position in space. Using relations (2.60), (2.66), and (2.69), it follows that

$$
\mathbf{q} = \begin{bmatrix} a(\cos E - e) \\ a\sqrt{1 - e^2} \sin E \\ 0 \end{bmatrix} = \begin{bmatrix} r \cos f \\ r \sin f \\ 0 \end{bmatrix}
\tag{2.82}
$$

The velocity becomes

$$
\dot{\mathbf{q}} = \frac{na}{1 - e \cos E} \begin{bmatrix} -\sin E \\ \sqrt{1 - e^2} \cos E \\ 0 \end{bmatrix} = \frac{na}{\sqrt{1 - e^2}} \begin{bmatrix} -\sin f \\ e + \cos f \\ 0 \end{bmatrix}
\tag{2.83}
$$

The first part of (2.83) follows from (2.74), and the second part can be verified using known relations between the anomalies E and f. Equations (2.81) to (2.83) transform the Kepler elements into Cartesian coordinates and their velocities $(X, \dot{X})$.

The transformation from $(X, \dot{X})$ to Kepler elements starts with the computation of both the magnitude and direction of the angular momentum vector

$$
\mathbf{h} = \mathbf{X} \times \dot{\mathbf{X}} = \begin{bmatrix} h_x \\ h_y \\ h_z \end{bmatrix}
\tag{2.84}
$$

which is the vector form of equation (2.50). The various components of $\mathbf{h}$ are shown in Figure 2.18. The right ascension of the ascending node and the inclination of the orbital plane are, according to Figure 2.18,

$$
\Omega = \tan^{-1}\left(\frac{h_x}{-h_y}\right)
\tag{2.85}
$$

$$
i = \tan^{-1}\left(\frac{\sqrt{h_x^2 + h_y^2}}{h_z}\right)
\tag{2.86}
$$

By defining the auxiliary coordinate system (p) such that the p_1 axis is along the nodal line, p_3 is along the angular momentum vector, and p_2 completes a right-handed coordinate system, we obtain

$$
\mathbf{p} = R_1(i)R_3(\Omega)\mathbf{X}
\tag{2.87}
$$

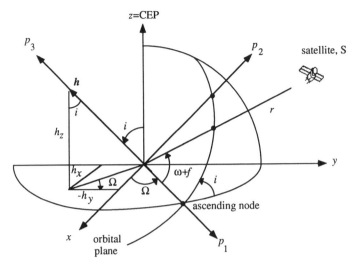

Figure 2.18 Angular momentum vector and Kepler elements. The angular momentum vector is orthogonal to the orbital plane.

The sum of the argument of perigee and the true anomaly becomes

$$\omega + f = \tan^{-1}\left(\frac{p_2}{p_1}\right) \tag{2.88}$$

Thus far, the orbital plane and the orientation of the orbital ellipse have been determined. The shape and size of the ellipse depend on the velocity of the satellite. The velocity, geocentric distance, and magnitude of the angular momentum are

$$v = \sqrt{\dot{\mathbf{X}} \cdot \dot{\mathbf{X}}} = \|\dot{\mathbf{X}}\| \tag{2.89}$$

$$r = \sqrt{\mathbf{X} \cdot \mathbf{X}} = \|\mathbf{X}\| \tag{2.90}$$

$$h = \sqrt{\mathbf{h} \cdot \mathbf{h}} = \|\mathbf{h}\| \tag{2.91}$$

The velocity expressed in the (q) coordinate system can be written as follows, using (2.61), (2.77), and (2.83):

$$
\begin{aligned}
v^2 &= \dot{q}_1^2 + \dot{q}_2^2 \\
&= \frac{n^2 a^2}{1 - e^2}\, (\sin^2 f + e^2 + 2e \cos f + \cos^2 f) \\
&= \frac{\mu}{a(1 - e^2)}\, [2 + 2e \cos f - (1 - e^2)] \\
&= \mu \left(\frac{2}{r} - \frac{1}{a}\right)
\end{aligned}
\tag{2.92}
$$

Equation (2.92) yields the expression for the semimajor axis

$$a = \frac{r}{2 - rv^2/\mu} \tag{2.93}$$

From equation (2.63) it follows that

$$e = \left(1 - \frac{h^2}{\mu a}\right)^{1/2} \tag{2.94}$$

and equations (2.70), (2.82), and (2.83) give an expression for the eccentric anomaly:

$$\cos E = \frac{a - r}{ae} \tag{2.95}$$

$$\sin E = \frac{\mathbf{q} \cdot \dot{\mathbf{q}}}{e\sqrt{\mu a}} \tag{2.96}$$

Equations (2.95) and (2.96) together determine the quadrant of the eccentric anomaly. Having E, the true anomaly follows from (2.82):

$$f = \tan^{-1} \frac{\sqrt{1 - e^2} \sin E}{(\cos E - e)} \tag{2.97}$$

Finally, Kepler's equation yields the mean anomaly:

$$M = E - e \sin E \tag{2.98}$$

Equations (2.85) to (2.98) comprise the transformation from $(X, \dot{X})$ to the Kepler elements.

2.3.3 Satellite Visibility and Topocentric Motion

The topocentric motion of a satellite is the motion of the satellite as it is seen by an observer on the surface of the earth. Because the satellite is relatively close to the earth, as compared to the stars, the topocentric and geocentric motions of satellites must be carefully distinguished. Furthermore, time enters not only for determining the position of the satellite in the celestial reference system, but also to fix the rotational position of the earth. Figure 2.19 shows the right ascension and declination of the satellite. Because the center of the sphere of direction is located at the center of the earth, the positions in Figure 2.19 denote geocentric positions as would be observed by an observer located at the center of the earth. The transformation to the coordinate system whose

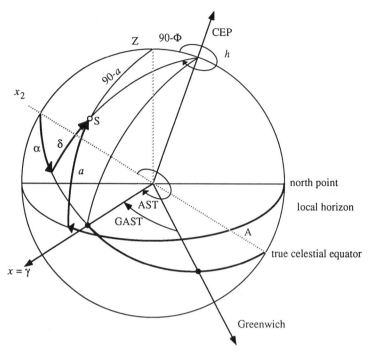

Figure 2.19   Geocentric azimuth and altitude.

first axis is fixed to the Greenwich meridian and whose third axis still coincides with the CEP is

$$
\begin{bmatrix} x \\ y \\ z \end{bmatrix}_{Gr} = R_3(\text{GAST}) \begin{bmatrix} x \\ y \\ z \end{bmatrix}_{\alpha, \delta}
\qquad (2.99)
$$

[see also equation (2.16)]. Figure 2.19 also shows the location of the local horizon. The horizon is the plane perpendicular to the plumb line at the observing station. This plane is shifted parallel until it passes through the geocenter. The intersection of this shifted plane with the sphere of direction is the great circle labeled as the local horizon in Figure 2.19. The intersection of the horizon with the lower meridian is called the north point. The azimuth is measured in the horizon in the clockwise direction, starting at the north point to the vertical circle of satellite S. The altitude a is the angle of the satellite above the horizon. For visible satellites the altitude is greater than zero.

The expressions for azimuth and altitude are obtained by applying several consecutive rotations to the celestial coordinate system (X). The first rotation is

$$
\begin{bmatrix} x \\ y \\ z \end{bmatrix}_{h,\delta} = \begin{bmatrix} 1 & 0 & 0 \\ 0 & -1 & 0 \\ 0 & 0 & 1 \end{bmatrix} R_3(\text{AST}) \begin{bmatrix} x \\ y \\ z \end{bmatrix}_{\alpha,\delta} \tag{2.100}
$$

This rotation brings the first axis of the new coordinate system in the direction of the intersection of meridian and equator; this axis is labeled x_2 in Figure 2.19. The coefficient -1 causes the direction of the new y axis to be reversed, creating, in effect, a left-handed coordinate system. The reason for this change is that the hour angle h is computed in the correct quadrant when equations such as (2.3) and (2.4) are used, and the sign of the numerator and denominator of the inverse tangent function is used to determine the quadrant. Note that the hour angle is counted positive from the upper meridian to the hour circle in a clockwise sense. Next, the coordinate system is rotated into the horizon, and is then rotated by $180°$, putting the new first axis into the direction of the north point. Thus,

$$
\begin{bmatrix} x \\ y \\ z \end{bmatrix}_{A,a} = R_3(-180) R_2(90 - \Phi) \begin{bmatrix} x \\ y \\ z \end{bmatrix}_{h,\delta} \tag{2.101}
$$

determines the geocentric azimuth and altitude of the satellite. The symbol Φ denotes the spherical latitude.

In Figure 2.20 the topocentric position of the satellite is denoted by an asterisk. The sphere of direction in Figure 2.20 is identical with the one in Figure 2.19 except that the center has been shifted to the topocentric position P. The topocentric coordinate system (X^*) is parallel to the celestial coordinate system (X). The topocentric position of the satellite is

$$
\mathbf{X}^* = \mathbf{X} - \mathbf{X}_P \tag{2.102}
$$

Usually the positions of stations are given in the earth-fixed conventional terrestrial coordinate system. Applying (2.16) and (2.17) gives the position of the station in the true celestial coordinate system (X):

$$
\mathbf{X}_P = R_3(-\text{GAST}) R_2(x_p) R_1(y_p) \begin{bmatrix} u \\ v \\ w \end{bmatrix} \tag{2.103}
$$

Substituting (2.103) into (2.102) and using (2.100) and (2.101) gives for the

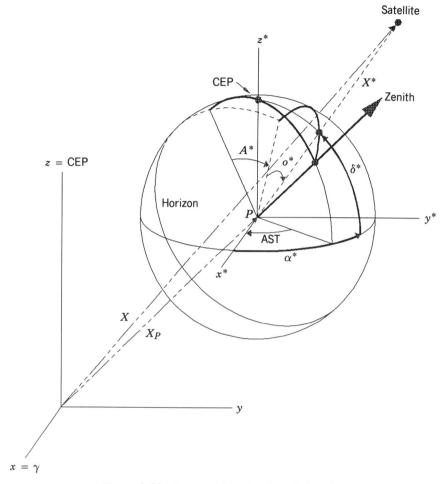

Figure 2.20   Topocentric azimuth and altitude.

topocentric azimuth and altitude

$$
\begin{bmatrix} x \\ y \\ z \end{bmatrix}_{A^*,a^*} = R_3(-180)\,R_2(90-\Phi)\begin{bmatrix} 1 & 0 & 0 \\ 0 & -1 & 0 \\ 0 & 0 & 1 \end{bmatrix}R_3(\mathrm{AST})\begin{bmatrix} x \\ y \\ z \end{bmatrix}_{\alpha^*,\delta^*}
$$

$$(2.104)$$

The topocentric distance to the satellite follows readily from (2.102).

Figure 2.21 shows a polar plot of topocentric azimuths and altitudes of some GPS satellites as a function of time. For satellite positions at the center of the

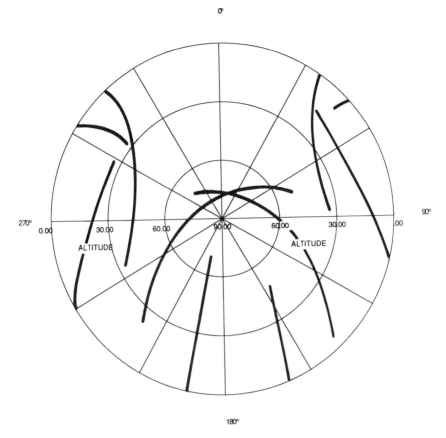

Figure 2.21 Satellite visibility chart.

chart, the zenith angle is zero; the outer circle represents a zenith angle of 90°. Visibility charts like this can be prepared well in advance of a planned GPS survey and taken to the field to help scan the sky for obstructions. The pattern shown in the figure is, of course, a function of the station location and time. Normal orbits are adequate to construct high-quality visibility charts.

2.3.4 Disturbed Satellite Motion

The accurate computation of satellite positions must take the disturbing forces into consideration. Disturbing forces are all those forces causing the satellite to deviate from the simple normal orbit. The disturbance of the orbit is caused primarily by the nonsphericity of the gravitational potential, the attraction of the sun and the moon, and the solar radiation pressure. These are the main disturbances for GPS satellites. Satellites closer to the earth are affected by additional forces, such as residual atmospheric drag.

The equations of motion are expressed in an inertial coordinate system, corresponding to the epoch at which the initial conditions are given. The initial conditions are either $(X, \dot{X})$ or the Kepler elements at a specified epoch. Because of the disturbing forces, all Kepler elements are functions of time. If necessary, the transformation in Section 2.3.2 can be used to transform the initial conditions from $(X, \dot{X})$ to Kepler elements and vice versa. The equations of motion, as expressed in Cartesian coordinates, are

$$\frac{d\mathbf{X}}{dt} = \dot{\mathbf{X}} \qquad (2.105)$$

$$\frac{d\mathbf{X}}{dt} = -\frac{\mu \mathbf{X}}{\|\mathbf{X}\|^3} + \ddot{\mathbf{X}}_g + \ddot{\mathbf{X}}_s + \ddot{\mathbf{X}}_m + \ddot{\mathbf{X}}_{\text{SRP}} \qquad (2.106)$$

These are six first-order differential equations. The symbol μ denotes the geocentric gravitational constant (2.40). The first term in (2.106) is the acceleration caused by the central gravity field generating the normal orbits discussed in the previous section. Compare (2.106) with (2.39). The remaining accelerations are discussed in what follows. The simplest way to solve (2.105) and (2.106) is to carry out a simultaneous numerical integration. Most of the high-quality engineering or mathematical software packages have such integration routines available. Kaula (1966) expresses the equations of motion in terms of Kepler elements and provides analytical expressions for the disturbing potential in terms of Kepler elements. Kaula (1962) gives similar expressions for the disturbing functions of the sun and the moon.

2.3.4.1 Gravitational Field of the Earth

The acceleration of the noncentral portion of the gravity field of the earth is given by

$$\ddot{\mathbf{X}}_g = \begin{bmatrix} \partial R/\partial x \\ \partial R/\partial y \\ \partial R/\partial z \end{bmatrix} \qquad (2.107)$$

with the disturbing potential

$$R = \sum_{n=2}^{\infty} \sum_{m=0}^{n} \frac{\mu a_e^n}{r^{n+1}} \overline{P}_{nm} (\cos \theta)(\overline{C}_{nm} \cos m\lambda + \overline{S}_{nm} \sin m\lambda) \qquad (2.108)$$

$$P_{nm}(\cos \theta) = \frac{(1 - \cos^2 \theta)^{m/2}}{2^n n!} \frac{d^{(n+m)}}{d(\cos \theta)^{(n+m)}} (\cos^2 \theta - 1)^n \qquad (2.109)$$

$$\overline{P}_n = \sqrt{2n + 1} P_n \qquad (2.110)$$

$$\overline{P}_{nm} = \cfrac{1}{\sqrt{\cfrac{(n+m)!}{2(2n+1)(n-m)!}}} \, P_{nm} \qquad (2.111)$$

$$\begin{bmatrix} x \\ y \\ z \end{bmatrix} = R_3(-\text{GAST}) \begin{bmatrix} r \sin\theta \cos\lambda \\ r \sin\theta \sin\lambda \\ r \cos\theta \end{bmatrix} \qquad (2.112)$$

Equation (2.108) expresses the disturbing potential (as used in satellite orbital computations) in terms of a spherical harmonic expansion. The symbol a_e denotes the mean earth radius, r is the geocentric distance to the satellite, and θ and λ are the spherical colatitude and longitude of the satellite position in the earth-fixed coordinate system. $\overline{P}_{nm}$ denotes the associated Legendre functions, which are known mathematical functions of the latitude. $\overline{C}_{nm}$ and $\overline{S}_{nm}$ are the spherical harmonic coefficients of degree n and order m. The bar indicates fully normalized potential coefficients. Note that the summation over n in (2.108) starts at 2. The term $n = 0$ equals the central component of the gravitational field. It can be shown that the coefficients for $n = 1$ are zero for coordinate systems whose origin is at the center of mass. Equation (2.108) shows that the disturbing potential decreases exponentially with the power of n. The largest coefficient in (2.108) is $\overline{C}_{20}$. This coefficient represents the effect of the flattening of the earth on the gravitational field. Its magnitude is about 1000 times larger than any of the other spherical harmonic coefficients. The most complete solutions available today for the spherical harmonic coefficients give coefficients up to 360 for n and m. Only the coefficients of lower degree and order, say, up to degree and order 36, are significant for satellite orbital computations. Table 2.4 shows those spherical harmonic coefficients that are important for the orbital motion of GPS satellites.

2.3.4.2 Acceleration Due to the Sun and the Moon
The lunar and solar accelerations on the satellites are (Escobal 1965, p. 37)

$$\ddot{\mathbf{X}}_m = \frac{\mu m_m}{m_e} \left(\frac{\mathbf{X}_m - \mathbf{X}}{\|\mathbf{X}_m - \mathbf{X}\|^3} - \frac{\mathbf{X}_m}{\|\mathbf{X}_m\|^3} \right) \qquad (2.113)$$

$$\ddot{\mathbf{X}}_s = \frac{\mu m_s}{m_e} \left(\frac{\mathbf{X}_s - \mathbf{X}}{\|\mathbf{X}_s - \mathbf{X}\|^3} - \frac{\mathbf{X}_s}{\|\mathbf{X}_s\|^3} \right) \qquad (2.114)$$

The mass ratios are $m_m/m_e = 0.0123002$ and $m_s/m_e = 332{,}946.0$ (Kaplan 1981). Mathematical expressions for the geocentric positions of the moon $\mathbf{X}_m$ and the sun $\mathbf{X}_s$ (in the true celestial coordinate system) are given in van Flandern and Pulkkinen (1979).

TABLE 2.4 Spherical Harmonics of Low Degree and Order

Degree and order	Normalized gravitational coefficients		Degree and order	Normalized gravitational coefficients	
n m	$\overline{C}_{n,m}$	$\overline{S}_{n,m}$	n m	$\overline{C}_{n,m}$	$\overline{S}_{n,m}$
2 0	-0.48416685E-03		6 3	0.53370577E-07	0.61334720E-08
2 1	- - - - - - - - - -	- - - - - - - - - -	6 4	0.88694856E-07	-0.47260945E-06
2 2	0.24395796E-05	-0.13979548E-05	6 5	-0.26818820E-06	-0.53491073E-06
3 0	0.95706390E-06		6 6	0.10237832E-07	-0.23741002E-06
3 1	0.20318729E-05	0.25085759E-06	7 0	0.85819217E-07	
3 2	0.90666113E-06	-0.62102428E-06	7 1	0.27905196E-06	0.94231346E-07
3 3	0.71770352E-06	0.14152388E-05	7 2	0.32873832E-06	0.88835092E-07
4 0	0.53699587E-06		7 3	0.24940240E-06	-0.21223369E-06
4 1	-0.53548044E-06	-0.47420394E-06	7 4	-0.27123034E-06	-0.12696607E-06
4 2	0.34797519E-06	0.65579158E-06	7 5	0.10246290E-08	0.17321672E-07
4 3	0.99172321E-06	-0.19912491E-06	7 6	-0.35843745E-06	0.15202633E-06
4 4	-0.18686124E-06	0.30953114E-06	7 7	-0.20991457E-08	0.22805664E-07
5 0	0.71092048E-07		8 0	0.42979835E-07	
5 1	-0.64185265E-07	-0.92492959E-07	8 1	0.18889342E-07	0.47856967E-07
5 2	0.65184984E-06	-0.32007416E-06	8 2	0.73553952E-07	0.47867693E-07
5 3	-0.44903639E-06	-0.21328272E-06	8 3	-0.12132459E-07	-0.83461853E-07
5 4	-0.29719055E-06	0.53213480E-07	8 4	-0.24208264E-06	0.71603924E-07
5 5	0.17523221E-06	-0.67059456E-06	8 5	-0.24966587E-07	0.87751047E-07
6 0	-0.15064821E-06		8 6	-0.65093424E-07	0.30904202E-06
6 1	-0.74180259E-07	0.32780040E-07	8 7	0.66323292E-07	0.74661766E-07
6 2	0.51824409E-07	-0.35866634E-06	8 8	-0.12372281E-06	0.12210258E-06

2.3.4.3 *Solar Radiation Pressure*

Solar radiation pressure (SRP) is a result of the impact of light photons emitted from the sun on the satellite's surface. The basic parameters of the SRP are the effective area (surface normal to the incident radiation), the surface reflectivity, the luminosity of the sun, and the distance to the sun. Computing SRP requires the evaluation of surface integrals over the illuminated regions. Even if these regions are known, the evaluation of these surface integrals can still be difficult because of the complex shape of the satellite. The ROCK4 and ROCK42 models (Fliegel et al. 1985; Fliegel and Gallini 1989) attempt to take most of these complex relations and properties into consideration for GPS Block I and Block II satellites, respectively. For satellites in the earth shadow region (eclipse), the SRP is zero. Thus, in precise computations the shadow region must be carefully delineated by considering the relative positions of sun, earth, and satellite.

In Figure 2.22 the z axis is along the antenna and is therefore in the direction of the earth. The y axis is along the axis of the solar panels, and x completes the right-handed coordinate system. The satellites are always aligned such that the y axis is perpendicular to the plane of the earth, sun, and satellite. The solar panels are rotated around the y axis such that their surface is perpendicular to the direction of the sun. This direction is denoted by **e** in Figure 2.22. One of the simplest models for estimating the solar radiation pressure is

$$\mathbf{X}_{SRP} = -p\,\frac{\mathbf{X}_s - \mathbf{X}}{\|\mathbf{X}_s - \mathbf{X}\|} + Y\,\frac{\mathbf{X}_s \times \mathbf{X}}{\|\mathbf{X}_s \times \mathbf{X}\|} \qquad (2.115)$$

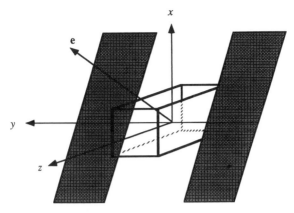

Figure 2.22 Defining the coordinate system for solar radiation pressure.

The symbol p denotes the solar radiation pressure in the direction of the sun. With the sign convention of equation (2.115), p should be a positive constant. The other parameter is called the Y bias. The reasons for the existence of such a bias parameter could be structural misalignments and thermal phenomena. The fact that a Y bias seems to exist demonstrates the complexity of accurate solar radiation pressure modeling. Dealing with solar radiation pressure is always necessary if high-precision orbits in the submeter range or long arcs are considered. The options are either to compute the respective accelerations of the vehicle using models such as ROCK4 and ROCK42, or to estimate the respective SRP parameters. If the solar radiation pressure remains unmodeled, GPS orbital errors can grow over 1 km after 1–2 weeks of integration. Additional details on the solar radiation pressure are found in Fliegel et al. (1992).

CHAPTER 3

THE GLOBAL POSITIONING SYSTEM

Space-based positioning has been pursued by the U.S. military and NASA since the early 1960s. One of the early and very successful satellite positioning systems was the Navy Navigation Satellite System (TRANSIT). Since its release for commercial use in 1967, the TRANSIT positioning system found many applications in surveying and geodesy for the establishment of widely spaced network stations over large regions or even over the globe. The TRANSIT satellite system was instrumental in establishing modern geocentric datums and in connecting various national datums to a geocentric reference frame. Unfortunately, this satellite system is unable to provide the accuracy for surveying at the "parcel and traverse" level. The TRANSIT satellites are in a polar orbit and are only about 1100 km above the earth. Their positions are affected more by local gravity field variations than are the much higher-orbiting GPS satellites. TRANSIT satellite transmissions at 150 and 400 MHz are more susceptible to ionospheric delays and disturbances than the higher GPS frequencies. Satellites are visible for only about 20 min. A position fix can be obtained every 1–3 hours when a satellite comes into view. It is expected that the TRANSIT system will be terminated by the end of 1996.

The NAVSTAR (Navigation Satellite Timing and Ranging) GPS is the answer to the need for a comprehensive and cohesive approach satisfying weather-independent future military navigation and timing needs worldwide. The policies governing access to GPS are stated in the Federal Radionavigation Plan (FRP), which is jointly prepared by the Department of Defense and the Department of Transportation. It is published every two years and is available from the National Technical Information Service, Springfield, VA 22161. As is well known, the respective policies, or GPS services, include the Standard Positioning Service (SPS) and the Precise Positioning Service (PPS). SPS is a result of implementing selective availability (SA), which simply means intentional degradation of the C/A-code positioning capability. In keeping with the policy, SA was implemented on 25 March 1990 on all Block II satellites. The level of degradation was reduced in September 1990 during the Gulf conflict, and was reactivated to its full level on 1 July 1991. The PPS is based on the P-code, which was cryptographically changed to the Y-code beginning 31 January 1994. Thus, PPS is available only to authorized (military) users who have decryption capability. Encryption is a measure of antispoofing (AS) that prevents a deception jammer from broadcasting a beacon that mimics the actual GPS signals. Because the jammer cannot generate the (classified) Y-code needed to mimic the satellite signals, spoofing is thus prevented.

For many civil applications the SA degradation is effectively eliminated or reduced by relative positioning of co-observing receivers and by making ample use of carrier phase observations (in addition to code observations). Several solutions are even available for dealing with the antispoofing measure to allow ranging on both carriers. Thus, even with SA/AS, the most important features of GPS include its high positional accuracy in three dimensions, global coverage, all-weather capability, continuous availability to an unlimited number of users, accurate timing capability, and its ability to meet the needs of a broad spectrum of users. It is the stated policy of the Department of Defense to phase out other electronic navigation systems, such as TACAN, VOR/DME, OMEGA, LORAN-C, and TRANSIT, on military vehicles. The Federal Avionics Administration (FAA) is studying the use of GPS for precise civilian navigation in the national airspace. On 17 February 1994, the FAA announced approval of the first GPS receivers capable of stand-alone use for en route and nonprecision instrument approach flight anywhere in U.S. airspace. Many efforts are under way in the study of GPS, and new applications are reported almost daily.

Figure 3.1 and Table 3.1 give an overview of the major components of

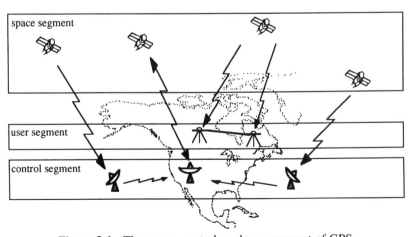

Figure 3.1   The space, control, and user segment of GPS.

TABLE 3.1 Functions and Products of Space, Control, and User Segment

Segment	Input	Function	Product
Space	navigation message	generate and transmit code and carrier phases, navigation message	P(Y)-codes, C/A-codes, L1, L2 carrier, navigation message
Control	P(Y)-code, observations, time (UTC)	produce GPS time, ephemerides, manage space vehicles	navigation message
User	code and carrier phase observations, navigation message	navigation solution, relative positioning, OTF, etc	position, velocity, time

GPS. The space segment consists of the GPS satellites, which transmit signals on two phase modulated frequencies. These transmissions are carefully controlled by highly stable atomic clocks inside the satellites. The satellites also transmit a navigation message that contains, among other things, orbital data for computing the positions of all satellites. The control segment consists of a Master Control Station located near Colorado Springs and several monitoring stations located around the world. The purpose of the control segment is to monitor the satellite transmissions continuously, to predict the satellite ephemerides, to calibrate the satellite clocks, and to update the navigation message periodically. The user segment simply stands for the total user community. The user will typically observe and record the transmissions of several satellites and will apply solution algorithms to obtain position, velocity, and time.

3.1 SPACE SEGMENT

This section contains a brief description of the GPS satellites, the orbital satellite constellation, the signal structure, and the contents of the navigation message.

3.1.1 GPS Satellites

The buildup of the satellite constellation began with a series of space vehicles called the Block I satellites. These concept validation satellites do not have SA/AS capability. At the time of writing, only three Block I satellites are still in service. These satellites were launched into 63° declined orbital planes and were positioned within the planes such that optimal observing geometry was available over certain military proving grounds in the continental United States. They were designed to provide 3–4 days of positioning service without contact with the control center. This initial phase has long since been completed.

The first operational GPS satellite of the Block II series was launched in February 1989. Figure 3.2 shows a Block II satellite on the assembly line at Rockwell International, Satellite Systems Division, Seal Beach, California. The size of the satellite can readily be appreciated when compared to the assembly worker. The design life goal of the satellites is 7.5 years. The weight of the GPS satellite is 1860 lb at insertion into the final orbit. The single-degree solar arrays each cover a surface area of 7.2 m^2 when deployed. The surface of the solar panels is kept perpendicular to the direction of the sun. There are batteries on board to provide energy during eclipse periods. The time generated within the satellite is based on several rubidium and cesium atomic clocks. The Block II satellites have radiation-hardened electronics, full SA/AS capability, and at least 14 days worth of navigation message storage. The first satellite of the new Block IIA series was launched in November 1990. These satellites are slightly modified Block II satellites having 180 days memory capability. Block IIA satellites can provide 180 days of positioning service with-

Figure 3.2 GPS satellite. (Courtesy of Rockwell International, Satellite Systems Division, Seal Beach, California)

out contact from the control segment. Under normal operation, however, the control segment will provide daily uploads to each satellite. If the uploads are not possible daily, the satellites will individually change to a short-term extended operational mode and eventually to the long-term extended operational mode based on the time since the last upload. In either extended mode the positional accuracy will decrease. All satellites of the Block II series are in 55° inclined orbital planes.

Currently there are 20 replenishment satellites in development, called the Block IIR series. The first satellite of this series is expected to be launched in 1996. These satellites will have the capability to autonomously determine their orbits through crosslink ranging and to generate their own navigation message by onboard processing capability. They will be able to measure ranges between themselves and to transmit observations to other satellites and to ground control. Once fully deployed these satellites can operate for half a year without ground control support and without losing accuracy in the broadcast ephemeris. Production of a follow-on series, the Block IIF series, is expected to start around the year 2000.

3.1.2 Orbital Configuration

The orbital constellation is shown in Figure 3.3. Additional information is provided in Table 3.2, including three identifiers for the satellites: the PRN

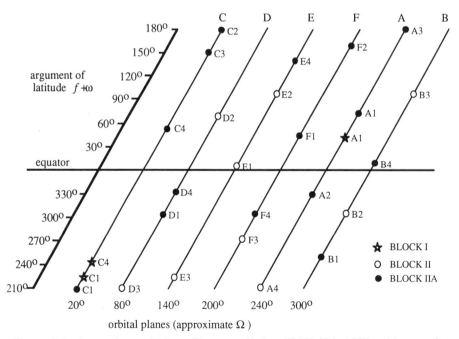

Figure 3.3 Approximate GPS satellite constellation (DOY 084, 1994, arbitrary reference epoch).

TABLE 3.2 Launch Dates and Satellite Identification

PRN	SVN	Launch	Plane	Block	NASA	Remark
4	1	22 Feb 78		I		off
7	2	13 May 78		I		off
6	3	6 Oct 78		I		off
8	4	10 Dec 78		I		off
5	5	9 Feb 80		I		off
9	6	26 Apr 80		I		off
	7	18 Dec 81		I		launch failed
11	8	4 May 83		I		off
13	9	13 Jun 84	C1	I	15039	
12	10	8 Sep 84	A1	I	15271	
3	11	9 Oct 85	C4	I	16129	
14	14	14 Feb 89	E1	II	19802	
2	13	10 Jun 89	B3	II	20061	
16	16	18 Aug 89	E3	II	20185	
19	19	21 Oct 89	A4	II	20302	
17	17	11 Dec 89	D3	II	20361	
18	18	24 Jan 90	F3	II	20452	
20	20	26 Mar 90	B2	II	20533	
21	21	2 Aug 90	E2	II	20724	
15	15	1 Oct 90	D2	II	20830	
23	23	26 Nov 90	E4	IIA	20959	
24	24	4 Jul 91	D1	IIA	21552	
25	25	23 Feb 92	A2	IIA	21890	
28	28	10 Apr 92	C2	IIA	21930	
26	26	7 Jul 92	F2	IIA	22014	
27	27	9 Sep 92	A3	IIA	22108	
1	32	22 Nov 92	F1	IIA	22231	
29	29	18 Dec 92	F4	IIA	22275	
22	22	3 Feb 93	B1	IIA	22446	
31	31	30 Mar 93	C3	IIA	22581	
7	37	13 May 93	C4	IIA	22657	
9	39	26 Jun 93	A1	IIA	22700	
5	35	30 Aug 93	B4	IIA	22779	
4	34	26 Oct 93	D4	IIA	22877	
6	36	03 Mar 94	C1	IIA	23027	

number, the launch number, and NASA's catalog number. The six orbital planes, labeled A–F, are evenly spaced in right ascension and are inclined by 55° with respect to the equator. Because of the flattening of the earth the nodal regression is about −0.04187° per day; an annual orbital adjustment carried out by the control center keeps the orbit in place. Each orbital plane contains four satellites; however, to optimize global satellite visibility, the satellites are not evenly spaced within the orbital plane. The orbital path is close to circular, with a semimajor axis of about 26,000 km. By applying Kepler's third law [equation (2.77)], one obtains an orbital period of slightly less than 12 hours. The satellites will complete two orbital revolutions while the earth rotates 360° (one sidereal day). This means the satellites will rise about 4 min earlier each day. Because the orbital period is an exact multiple of the period of the earth's rotation, the satellite trajectory on the earth (i.e., the trace of the geocentric satellite vector on the earth's surface) repeats itself daily.

Because of their high altitude, the GPS satellites can be viewed simultaneously from a large portion of the earth. Usually the satellites are observed only once they are above a certain vertical angle, referred to as the mask angle, usually $10-15°$. The reason for using a mask angle is that tropospheric effects on the signal propagation are especially unpredictable for altitudes within the mask region. Figure 3.4 shows how the mask angle ϵ limits the region of visibility. The earth center angle α is directly related to the mask angle. Simple computations give the following relations between both angles: ($\epsilon = 0°$, $\alpha = 152°$), ($\epsilon = 5°$, $\alpha = 142°$), ($\epsilon = 10°$, $\alpha = 132°$). The viewing angle from the satellite is about $27°$ in all cases.

Finally, as the earth orbits around the sun, there will be two periods per year when the satellite travels through the earth's shadow. This occurs whenever the sun is in or near the orbital plane. See Figures 3.5 and 3.6 for a graphical presentation. The umbra is that portion of the shadow cone that no light from the sun can reach. The penumbra is the region of partial shadowing; it surrounds the umbra cone. Although the transit through the shadow regions is not of immediate interest to the user, it is of importance to those computing precise orbital ephemerides. While the satellite transits through the shadow regions, the solar radiation force acting on the satellite either is zero (umbra)

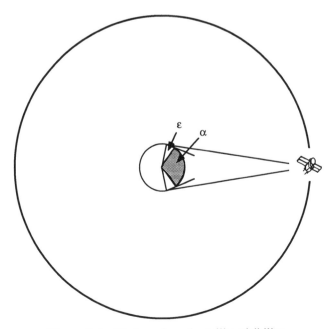

Figure 3.4 Mask angle and satellite visibility.

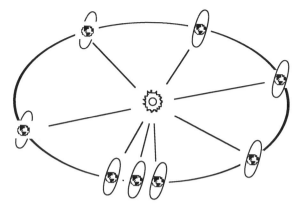

Figure 3.5 Biannual eclipse periods.

or changing (penumbra). These changes in force must be taken into consideration in precise orbital modeling. The satellites move through the shadow regions in less than 60 min.

3.1.3 Satellite Transmissions

This section contains a discussion on the GPS signal structure, signal generation, and GPS time. Because an in-depth exposition of the subject requires knowledge of many concepts from the field of electronic communication, a detailed treatment is not within the scope of this book. However, the possibilities and constraints of GPS for positioning depend very much on the types of signals. Details on the signal structure are found in the interface control document ICD-GPS-200 (1991) and in Spilker (1980). In-depth knowledge on

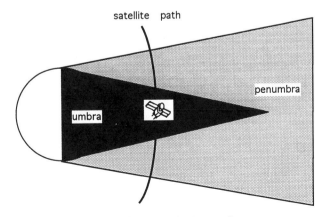

Figure 3.6 Earth shadow regions.

digital signal transmission is given in Bissell and Chapman (1992), Winch (1993), and other specialized texts.

3.1.3.1 Overview All satellite transmissions are coherently derived from the fundamental frequency of 10.23 MHz, and are made available by the set of onboard atomic clocks. Multiplying the fundamental frequency by 154 gives L1 = 1575.42 MHz, which is the frequency for the L1 carrier; and multiplying by 120 gives L2 = 1227.60 MHz, the frequency for the L2 carrier. The chipping rate of the P(Y)-code is that of the fundamental frequency, that is, 10.23 MHz, whereas the chipping rate of the C/A-code is 1.023 MHz, which is one-tenth of the fundamental frequency. The navigation message, also called the telemetry, is modulated on both the L1 and the L2 carriers at a chipping rate of 50 bps. It contains information on the ephemerides of the satellites, GPS time, clock behavior, and system status messages. Figure 3.7 and Table 3.3 summarize the transmissions of the satellites.

The precision P(Y)-code is the principal code used for military navigation. It is a pseudorandom noise (PRN) code generated mathematically by mixing two other pseudorandom codes. The P(Y)-code does not repeat itself for 37 weeks. Thus, it is possible to assign weekly portions of this code to the various satellites. As a result, all satellites can transmit on the same carrier frequency, and yet can be distinguished because of the mutually exclusive code sequences

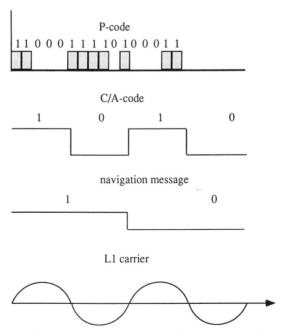

Figure 3.7 Schematic presentation of GPS codes and carrier.

TABLE 3.3 Summary of Satellite Transmission

	C/A	P(Y)	Navigation data
Chipping rate	1.023 Mbps	10.23 Mbps	50 bps
Length per chip	293 m	29.3 m	5950 km
Repetition	1 ms	1 week	N/A
Code type	Gold	pseudo random	N/A
Carried on	L1	L1, L2	L1, L2
Feature	easy to acquire	precise positioning, jam resistant	time, ephemeris, HOW

being transmitted. All codes are initialized once per GPS week at midnight from Saturday to Sunday, thus creating, in effect, the GPS week as a major unit of time. Because there are less than 37 GPS satellites in orbit, some of the P(Y)-code weekly sequences remain unused; however, they are available for transmission ground stations. The L1 and L2 carriers are both modulated with the same P(Y)-code.

The period of the coarse/acquisition (C/A) code is merely 1 msec and consists of 1023 bits. Each satellite transmits a different set of C/A-codes. The code is available on L1 and, as a ground-controlled option, can also be put on L2. The C/A-codes belong to the family of Gold codes, which characteristically have low cross-correlation between all members of the family. This property makes it possible to rapidly distinguish the signals received simultaneously from different satellites. The C/A-code is the basis for the Standard Positioning Service (SPS). Because the epochs of the C/A-code and the P(Y)-code are synchronized, the C/A-code makes the rapid acquisition of the P(Y)-code possible.

One of the satellite identification systems makes use of the PRN weekly number. For example, if one refers to PRN 13, one refers to the satellite that transmits the 13th weekly portion of the PRN-code. The short version of PRN 13 is SV 13 (SV = space vehicle). Another identification system uses the launch sequence number. For example, the identification of PRN 13 in terms of launch sequence number is NAVSTAR 9 or SVN 9.

3.1.3.2 *Space Vehicle and GPS Time Systems* The satellite or space vehicle time is defined by the onboard atomic clocks of each satellite. The satellite operates on its own time system, that is, all satellite transmissions such as the C/A-code, the P(Y)-codes, and the navigation message are initiated on satellite time. The data in the navigation message, however, are relative to GPS time. The GPS time is established by the Control Segment and is related to UTC(USNO) within 1 μsec. However, GPS time is continuous and is not adjusted for leap seconds. The largest unit of GPS time is 1 week, defined as 604,800 sec. The last common epoch between GPS time and UTC(USNO) was midnight 5/6 January 1980. The navigation message contains corrections to convert space vehicle time to GPS time. In case of PPS the corrected time will be within 176 nsec (95%) to UTC(USNO). For the SPS the accuracy is 363 nsec (95%). Additional details on the satellite clock correction are given in Section 8.3.1.3.

Relativistic effects are important in GPS surveying, but fortunately can be accurately computed. The atomic frequency standards in the satellites are affected by both special relativity (the satellite's velocity) and general relativity (the difference in the gravitational potential at the satellite's position relative to the potential at the earth's surface). Jorgensen (1986) gives a discussion in lay terms of these effects and identifies two distinct parts in the relativity correction. The predominant portion is common to all satellites and is independent of the orbital eccentricity. The respective relative frequency offset is $\Delta f / f = -4.4647 \times 10^{-10}$. This offset corresponds to an increase in time of 38.3 μsec per day; the clocks in orbit appear to run faster. The apparent change in frequency is $\Delta f = 0.0045674$ Hz at the fundamental frequency of 10.23 MHz. The frequency is corrected by adjusting the frequency of the satellite clocks in the factory before launch to 10.22999999543 MHz. The second portion of the relativistic effect is proportional to the eccentricity of the satellite's orbit. For exact circular orbits this correction is zero. For GPS orbits with an eccentricity of 0.02 this effect can be as large as 45 nsec, corresponding to a ranging error of about 14 m. This relativistic effect fortunately can be computed from a simple mathematical expression that is a function of the semimajor axis, the eccentricity, and the eccentric anomaly (see Section 8.3.1.3). In relative positioning as typically carried out in surveying, the relativistic effects cancel for all practical purposes.

3.1.3.3 *GPS Signal Structure*

A simple sine wave, called the carrier, can be modulated in order to carry information such as voice communication and digital images. In the case of GPS, the modulation makes ranging possible by measuring the travel time of the modulation. There are at least three commonly used digital modulation methods: amplitude-shift keying (ASK), frequency-shift keying (FSK) and phase-shift keying (PSK).

Figure 3.8 shows an arbitrary digital data stream consisting of binary digits 0 and 1. These binary digits are also called chips, bits, codes, or pulses. The sequence of the binary digits in the data stream is usually a pseudorandom sequence which might "locally" look like a random signal but which in reality follows a mathematical formula. Amplitude-shift keying corresponds to an on/off operation. The digit 1 might represent turning the carrier on and 0 might mean turning it off. Frequency-shift keying implies transmission on one or the other frequency. The transmitting oscillator is required to switch back and forth between two distinct frequencies. In the case of phase-shift keying, the same carrier frequency is used, but the phase changes abruptly. With binary phase-shift keying (BPSK) the phase shifts 0 and 180°. The BPSK method is used with GPS signals.

A popular technique is to combine two binary data streams into one by

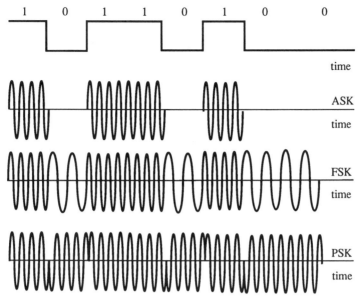

Figure 3.8 Digital modulation methods.

means of a modulo-2 addition. In Figure 3.9 there are two such data streams. To put this example in the context of GPS, let us assume that the sequence (a) represents the navigation data at a chipping rate of 50 bits per second (bps) and that the (b) stream is the C/A-code or the P(Y)-code at the 1.023- or 10.23-MHz chipping rate, respectively. Although it is not relevant in the present context, one should notice that telemetry and the code streams have significantly different chipping rates and that the times of bit transition are aligned. A chipping rate of 50 bps implies 50 opportunities per second for the digital

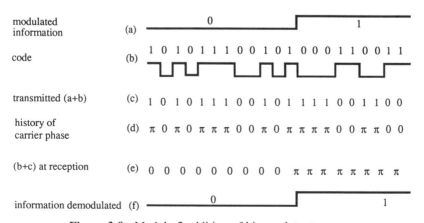

Figure 3.9 Modulo-2 addition of binary data streams.

stream to change from 1 to 0 and vice versa. There are 31,508,400 L1 carrier cycles during the time of one telemetry chip, 20,460 C/A-code chips fill one telemetry chip, and 204,600 P(Y)-code chips are needed to do the same. Modulo-2 addition follows the well-known rules: $0 + 0 = 0$, $1 + 0 = 1$, $0 + 1 = 1$, and $1 + 1 = 0$. The result of this modulo-2 superposition is shown. The figure also shows the phase history of the transmitted carrier. Thus, whenever a binary 1 occurs in the 50-bps navigation data stream, the modulo-2 addition inverts 20,460 adjacent digits of the C/A-code. A binary 1 becomes 0 and vice versa. A binary 0 leaves the next 20,460 C/A-codes unchanged.

Figure 3.9 shows another interesting phenomenon that is quite important to the way GPS works. Assume that the receiver reproduces the original C/A-code sequence. We can then modulo-2 add the receiver-generated C/A-code with the received, phase-modulated carrier. The sum is the demodulated 50-bps telemetry data stream.

The modulo-2 addition method must be generalized one additional step because the L1 carrier is modulated by three data streams: the navigation data, the C/A-codes and P(Y)-codes. Thus, the problem of superimposing both code streams on the navigation data stream arises. Two sequential superimpositions are not unique, because the C/A-code and the P(Y)-code have identical bit transition epochs (although their length is different). The solution is called quadri-phase-shift keying (QPSK). The carrier is split into two components by a phase splitter. One of the components is electronically advanced by 90°. Each component is binary phase-modulated. Both streams are electronically combined before transmission. This operation is represented by the general mathematical expression

$$y(t) = y_1(t) + y_2(t) = x_1(t) \cos \omega t + x_2(t) \sin \omega t \qquad (3.1)$$

The two signals $y_1(t)$ and $y_2(t)$ are said to be in phase (I) and quandrature (Q) respectively. All spectral components of $y_1(t)$ are 90° out of phase with those of $y_2(t)$, which makes it possible to separate them in the receiver.

The codes on L1 are transmitted in phase and in quadrature. The general expression is

$$S_1^p(t) = A_P P^p(t) D^p(t) \cos(2\pi f_1 t) + A_C G^p(t) D^p(t) \sin(2\pi f_1 t) \qquad (3.2)$$

where

A_p, A_C = amplitudes (power) of P(Y)-code and C/A-code
$P^p(t)$ = pseudorandom P(Y)-code
$G^p(t)$ = C/A-code (Gold code)
$D^p(t)$ = navigation data stream

The integer superscript p denotes the PRN number of the satellite. The P(Y)-code on L2 is

$$S_2^p(t) = B_P P^p(t) D^p(t) \cos(2\pi f_2 t) \qquad (3.3)$$

Here the products $P^p(t)D^p(t)$ and $G^p(t)D^p(t)$ imply modulo-2 addition. The P(Y)-code by itself is a modulo-2 sum of two pseudorandom data streams as follows:

$$P^p(t) = X_1(t)X_2(t - pT) \qquad (3.4)$$

$$0 \le p \le 36 \qquad (3.5)$$

$$\frac{1}{T} = 10.23 \text{ MHz} \qquad (3.6)$$

Expression (3.4) defines 37 different codes that are used by individual satellites in the constellation. The implication is that all satellites are using the same code; however, there is a phase difference in the code that is transmitted by each satellite at a given time. That difference is determined by the PRN number p, which identifies the satellite and which determines the phase of X_2. At the beginning of the GPS week, the P(Y)-codes are reset. Similarly, the C/A-codes are the modulo-2 sum of two 1023 pseudorandom bit codes as follows:

$$G^p(t) = G_1(t)G_2[t - N^p(10T)] \qquad (3.7)$$

$G^p(t)$ is 1023 bits long or has a 1-msec duration at a 1.023-Mbps bit rate. The $G^p(t)$ chip is ten times as long as the X_1 chip. The G_2-code is selectively delayed by an integer number of chips, expressed by the integer N^p, to produce 36 unique Gold codes, $G^p(t)$, one to correspond to each of the 36 different P(Y)-codes. Because the cross-correlation between two Gold codes is low, they are good codes for achieving rapid lock.

The actual generation of the codes X_1, X_2, G_1, and G_2 is accomplished by a device called a feedback shift register (FBSR). Such devices can generate a large variety of pseudorandom codes. These codes look random over a certain interval, but the feedback mechanism causes the codes to repeat after some time. Figure 3.10 shows a very simple register. A block represents a stage register whose content is in either a one or a zero state. When the clock pulse is input to the register, each block has its state shifted one block to the right.

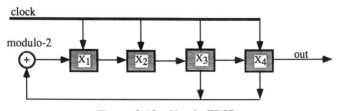

Figure 3.10 Simple FBSR.

TABLE 3.4   Output of FBSR

x_1	0	1	0	0	....	1	0	0	0
x_2	0	0	1	0	....	1	1	0	0
x_3	0	0	0	1	....	1	1	1	0
x_4	1	0	0	0	....	1	1	1	1
Output	1	0	0	0	....	1	1	1	1

In this particular example, the output of the last two stages is modulo-2 added and the result is fed back into the first stage and modulo-2 added to the old state to create the new state. The successive states of the individual blocks as the FBSR is stepped through a complete cycle are shown in Table 3.4. The elements of the column represent the state of each block and the successive columns represent the behavior of the shift register as the succession of timing pulses cause it to shift from state to state. In this example the initial state is (0001). For n blocks, $2^n - 1$ states are possible before repetition occurs. The output corresponds to the state of the last block, and would represent the PRN code generated by such a 4-stage FBSR.

The shift registers that are used in GPS are much more complex. They have many more feedback loops and they have many more blocks in the sequence. The P(Y)-code is derived from two 12-stage shift registers, $X_1(t)$ and $X_2(t)$, having 15,345,000 and 15,345,037 stages (chips), respectively. Both registers continuously recycle. The modulo-2 sum of both registers has the length of 15,345,000 times 15,345,037 chips. At the chipping rate of 10.23 MHz it takes 266.4 days to complete the whole P(Y)-code cycle. It takes 1.5 sec for the X_1 register to go though 1 cycle. The X_1 cycles (epochs) are known as the Z count.

3.1.3.4   The Navigation Message

Each satellite transmits a data stream, called the navigation message, on L1 and L2 at a rate of 50 bps. An overview of the message structure is seen in Figure 3.11. A complete message consists of 25 frames, each containing 1500 bits. Each frame is subdivided into five 300 bit subframes, and each subframe consists of 10 words of 30 bits each. At the 50-bps rate it takes 6 sec to transmit a subframe, 30 sec to complete a frame, and 12.5 min for one complete transmission of the navigation message. The subframes 1, 2, and 3 are transmitted with each frame; these three frames repeat every 30 sec. Subframes 4 and 5 are each subcommutated 25 times. The 25 versions of subframes 4 and 5 are referred to as pages 1 through 25. Thus, each of these pages repeats every 750 sec or every 12.5 min.

Each subframe begins with the telemetry word (TLM) and the handover word (HOW). The telemetry word begins with a preamble, and otherwise contains only information that is needed by the authorized user. The HOW represents a truncation of the GPS time of week (TOW). HOW, when multiplied

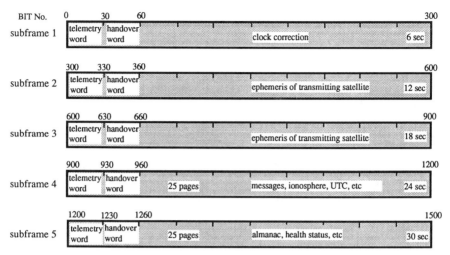

Figure 3.11 Structure of navigation message frame.

with 4, gives the X_1 count at the start of the following subframe. As soon as the receivers have locked to the C/A-code, the HOW word is extracted and is used to identify the X_1 count at the start of the following subframe. In this way the receiver knows exactly which part of the long P(Y)-code is being transmitted. P(Y)-code tracking can then readily begin, thus the name, "handover word." To rapidly lock to the P(Y)-code, the HOW is included on each subframe (see Figure 3.12).

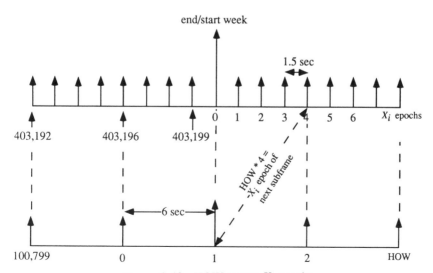

Figure 3.12 HOW versus X_1 epochs.

GPS time is directly related to the X_1 counts of the P(Y)-code. The Z count is a 29-bit number that contains several pieces of timing information. One part of the Z count is the HOW, which relates to the X_1 count as discussed above. Other parts of the Z count can be converted to time of week (TOW), which represents the number of seconds since the beginning of the GPS week. A full week equals $403,199$ X_1 counts. Yet another part of the Z count gives the current GPS week number (modulo 1024). The first GPS week began 5/6 January 1980. The beginning of the current GPS week is offset from midnight UTC by the current number of leap seconds.

The week number portion of the Z count is part of word 3 of subframe 1. The remainder of the first subframe contains, among others, the satellite clock correction terms and the clock reference time t_{oc} (Section 8.3.1.3), the differential group delay, T_{GD} (Chapter 9), and the IODC (issue of date, clock) term. The latter term indicates the issue number of the clock data set to alert users of changes in the clock parameters. The messages are updated usually every 4 hours.

Subframes 2 and 3 contain the ephemeris parameters for the transmitting satellite. The various elements are listed in Table 3.5. Note that terminology typical of Kepler parameterization is used, and that the parameters seem to have a Keplerian appearance. However, these elements are a result of least-squares curve fitting of the predicted ephemeris over a well-specified interval of time; therefore, they are not valid for the total orbit. The IODE (issue of data, ephemeris) term allows users to detect changes in the ephemeris param-

TABLE 3.5   Ephemeris Parameterization of the Navigation Message

M_0	Mean anomaly at reference time
Δn	Mean motion difference from computed value
e	Eccentricity
$\sqrt{a}$	Square root of the semimajor axis
Ω_0	Longitude of ascending node of orbit plane at weekly epoch
i_0	Inclination angle at reference time
ω	Argument of perigee
$\dot{\Omega}$	Rate of right ascension
IDOT	Rate of inclination angle
C_{uc}	Amplitude of the cosine harmonic correction term to the argument of latitude
C_{us}	Amplitude of the sine harmonic correction term to the argument of latitude
C_{rc}	Amplitude of the cosine harmonic correction term to the orbit radius
C_{rs}	Amplitude of the sine harmonic correction term to the orbit radius
C_{ic}	Amplitude of the cosine harmonic correction term to the angle of inclination
C_{is}	Amplitude of the sine harmonic correction term to the angle of inclination
t_{oe}	Ephemeris reference time
IODE	Issue of Data (Ephemeris)

eters. For each upload the control center assigns a new number. The IODE is given in both subframes. During the time of an upload, both IODEs will have different values. Users should download ephemeris data only when both IODEs have the same value. The broadcast elements should be used with the algorithm of Table 3.6. The results are coordinates in the WGS 84 coordinate system.

TABLE 3.6   Ephemeris Algorithm

$\mu = 3.986005 \times 10^{14} \text{ meters}^3/\text{sec}^2$	WGS 84 value of the earth's universal gravitational parameter
$\dot{\Omega}_e = 7.292115167 \times 10^5 \text{ rad/sec}$	WGS 84 value of earth's rotation rate
$a = \left(\sqrt{a}\right)^2$	Semimajor axis
$n_0 = \sqrt{\dfrac{\mu}{a^3}}$	Computed mean motion - rad/sec
$t_k = t - t_{oe}^{\bullet}$	Time from ephemeris reference epoch
$n = n_0 + \Delta_n$	Corrected mean motion
$M_k = M_0 + n t_k$	Mean anomaly
$M_k = E_k - e \sin E_k$	Kepler's equation for eccentric anomaly
$\left.\begin{array}{l}\cos f_k = \left(\cos E_k - e\right)/\left(1 - e \cos E_k\right) \\ \sin f_k = \sqrt{1 - e^2} \sin E_k /\left(1 - e \cos E_k\right)\end{array}\right\}$	True anomaly
$E_k = \cos^{-1}\left[\dfrac{e + \cos f_k}{1 + e \cos f_k}\right]$	Eccentricity anomaly
$\phi_k = f_k + \omega$	Argument of latitude
$\begin{array}{l}\delta u_k = C_{us} \sin 2\phi_k + C_{uc} \cos 2\phi_k \\ \delta r_k = C_{rc} \cos 2\phi_k + C_{rs} \sin 2\phi_k \\ \delta i_k = C_{ic} \cos 2\phi_k + C_{is} \sin 2\phi_k\end{array}$	Argument of latitude correction Radius corrections Correction to inclination $\left.\vphantom{\begin{array}{c}a\\b\\c\end{array}}\right\}$ 2nd harmonic pertubations
$u_k = \phi_k + \delta u_k$	Corrected argument of latitude
$r_k = a\left(1 - e \cos E_k\right) + \delta r_k$	Corrected radius
$i_k = i_0 + \delta i_k + (\text{IDOT}) t_k$	Corrected inclination
$\left.\begin{array}{l}x_k' = r_k \cos u_k \\ y_k' = r_k \sin u_k\end{array}\right\}$	Positions in orbital plane
$\Omega_k = \Omega_0 + \left(\dot{\Omega} - \dot{\Omega}_e\right) t_k - \dot{\Omega}_e t_{oe}$	Corrected longitude of ascending node
$\left.\begin{array}{l}x_k = x_k' \cos \Omega_k - y_k' \cos i_k \sin \Omega_k \\ y_k = x_k' \sin \Omega_k + y_k' \cos i_k \cos \Omega_k \\ z_k = y_k' \sin i_k\end{array}\right\}$	Earth fixed coordinates

* t is GPS system time at time of transmission, i.e., GPS time corrected for transit time (range/speed of light). Furthemore, t_k shall be the actual total time difference between the time t and the epoch time t_{oe}, and must account for beginning or end of week crossovers. That is, if t_k is greater than 302,400, subtract 604,800 from t_k. If t_k is less than -302,400 sec, add 604,800 sec to t_k.

This is a crust-fixed and earth-centered coordinate system whose third axis coincides with the conventional terrestrial pole (CTP). There is no need for polar motion rotations, since the respective rotations are incorporated in the representation parameters. However, when computing the topocentric distance, the user must account for the rotation of the earth during the signal travel time from satellite to receiver. Two methods are discussed in Section 8.1 to account for this rotation.

The 25 pages of subframe 4 contain special messages, the ionospheric correction terms, coefficients to convert GPS time to UTC, and the almanac for satellites with SVN number 25 and higher. The ionospheric terms are the eight coefficients $\{\alpha_n, \beta_n\}$ referenced in Table 9.2 of Chapter 9. For deriving UTC as accurately as possible from GPS time, the message provides a constant offset term, a linear polynomial term, the reference time t_{ot}, and the current value of the leap second. With these data a correction can be computed and added to the GPS time (as computed in Section 8.3.1.3). Note, however, that the thus computed UTC is still biased by atmospheric delay and possibly SA-dither effects.

Subframe 5 contains the almanac of satellites 1 through 24. The almanac provides the data necessary to compute the positions of satellites other than the transmitting satellite. The almanac essentially is a truncated form of the ephemeris data listed in subframes 1 to 3. For each satellite the almanac contains the following elements: t_{oa}, δ_i, a_{f0}, a_{f1}, e, $\dot{\Omega}$, $a^{1/2}$, Ω_0, ω, and M_0. The almanac reference time is t_{oa}. The correction to the inclination δ_i is given with respect to the fixed value $i_0 = 0.30$ semicircles ($=54°$). The clock polynomial coefficients a_{f0} and a_{f1} are used to convert SV time to GPS time following equation 8.52. The remaining elements of the almanac are identical to those listed in Table 3.5. The algorithm of Table 3.6 applies, using zero for all elements that are not included in the almanac; the reference time t_{oe} is replaced with t_{oa}.

In summary, the mean anomaly, the longitude of the ascending node, the inclination, and UTC (if desired) are formulated as a polynomial in time; the time argument is GPS time. The polynomial coefficients are, of course, a function of the epoch of expansion. The respective epochs are t_{oc}, t_{oe}, t_{oa}, and t_{ot}. The clock reference epoch t_{oc} is chosen 2 h after the first valid transmission for a 4-h fit interval data set, and 3 h for a 6-h fit interval data set. The same rule applies to the ephemeris reference epoch t_{oe}. The almanac and UTC reference epochs t_{oa} and t_{ot} are 3.5 days after the first valid transmission. Thus, the time arguments $(t - t_{oc})$ etc. can be negative. Block II and IIA satellite memory can store uploaded data for 14 and 180 days of operations, respectively. For Block II satellites the reference epochs and the fit intervals are a function of time when the satellite is in extended operational mode.

The navigation message contains other relevant information such as the User Range Error (URE). This is equal to the projection of the ephemeris curve fit errors onto the user range, and includes effects of satellite timing errors and SA. Also within the navigation message are flags that indicate the health status of the satellites.

3.2 CONTROL SEGMENT AND OTHER SUPPORT

The control segment is responsible for operating the GPS system. The master control station (MCS) is located at Falcon Air Force base near Colorado Springs, Colorado. From the user's point of view, the primary mission of the control segment is to update the navigation message transmitted by the satellites. To accomplish this, the Operational Control System (OCS) consists of monitoring stations distributed around the world (Colorado Springs, Hawaii, Florida, Ascension Islands, Diego Garcia, Kwajalein) that continuously track all satellites in view. These data are transmitted to the master control station at Colorado Springs, where extensive computations are performed, with the ultimate goal of creating up-to-date navigation messages for upload to the satellites.

It is important to note that the control segment is required only to monitor the satellites and to provide navigation messages of such an accuracy that a specified navigation accuracy is obtained [Standard Position Service (SPS) and Precise Positioning Service (PPS)]. To aid civil applications of GPS, many additional services have become available. These services include such diverse items as precise ephemerides, mission planning software, general positioning software, ionospheric and tropospheric data, differential corrections, receiver rentals, and campaign schedules. This expansion of GPS services is continuing. Most of these services are readily accessible by electronic means. For example, several key databases can be accessed via the Internet.

To support their mapping, charting, and geodetic activities, the Defense Mapping Agency (DMA) operates its own worldwide GPS tracking network. Tracking stations are located in Argentina, Australia, Bahrain, Ecuador, and England. DMA combines these data with those of the OCS network to compute precise ephemerides. The National Geodetic Survey (NGS) has assumed responsibility for providing these precise ephemerides to civil users.

The number of GPS tracking stations has increased significantly in recent years. Various national mapping agencies, universities, and other research groups around the world have been operating such stations in their "backyards" to support high-accuracy applications. These efforts began with the loosely organized CIGNET (Cooperative International GPS Network) and have resulted in the International GPS service for Geodynamics (IGS), which formally began operation 1 January 1994. Whereas the IGS is the brainchild of individual scientists, its materialization grew out of discussions, resolutions, etc., of the International Union of Geodesy and Geophysics (IUGG) and the International Association of Geodesy (IAG). The proof of the concept and the operability of the IGS were verified during several test campaigns. IGS is a service for supporting, through GPS data products, geodetic and geophysical research activities and high-accuracy applications of GPS. There is close cooperation and coordination between the IGS and the IERS (International Earth Rotation Service). The IGS organizational structure consists of the Central Bureau, currently located at the Jet Propulsion Laboratory (JPL) and of three data processing centers. Data exchange is primarily carried out on the Internet.

A summary on IGS is found in *Bulletin Géodésique*, Vol. 68, No. 1, 1994, pp. 43–46.

The U.S. Coast Guard is the leading agency within the U.S. government for making GPS information available to nonmilitary users. It runs the Civil GPS Service (CGS), which currently consists of the GPS Information Center (GPSIC), a Civil GPS Service Interface Committee (CGSIC), a Differential GPS (DGPS) service for U.S. harbors and harbor approach areas, and a PPS Program Office (PPSPO) for administering a program to allow qualified civil users access to the PPS. Of special interest to users is the GPSIC, whose database also can be accessed electronically. The GPSIC makes NANUs (notice advisory to NAVSTAR users) available, which contain critical information such as health status, planned outages, and system adjustments. The NANUs contain both predicted interruptions and details on unscheduled satellite events. The GPSIC obtains its information directly from the Master Control Station and from other supporting agencies such as the National Geodetic Survey. Barndt (1993) provides a detailed discussion on the U.S. Coast Guard's role and the various services it provides to the civil community.

3.3 RECEIVERS

Because new receivers are continuously being developed, it seems inappropriate to expend much effort in describing individual receivers in detail. However, the technology inside the receiver is intellectually challenging, stimulating, and truly in line with today's "high tech." Since receivers draw their technical roots from the field of digital communication systems, learning about the working of receivers means gaining contemporary knowledge at the same time. The material in this section is intended for non-electrical engineering students as an introduction, and to entice them to fill their science electives with courses on digital communication and related subjects. Because this book focuses on the utility of the geometric observables, which are the outcome of signal processing inside the receiver, it is not appropriate to present the mathematics of signal processing step by step. Consequently, this section contains only a few simplified explanations for the most important terms and concepts used in connection with GPS receivers.

Major recent advances in receiver technology include narrow correlation for C/A-code receivers and several solutions for coping with antispoofing. Receivers are now available that yield full-wavelength carrier phases on L1 and L2, the C/A-code pseudorange, and two P(Y)-code pseudoranges. Receivers are being developed to simultaneously observe signals from both the GPS and GLONASS satellites. Also, more and more receivers have real-time capability. Most receivers today are of the multichannel type; that is, each channel tracks one satellite. They all use C/A-code correlation to lock to the L1 carrier. In a few cases, signal squaring is used to recover the L2 carrier in the presence of antispoofing. All receivers read the navigation message to make the broadcast ephemeris available for postprocessing.

The tracking is done principally by the code delay lock loop (DLL) and the phase lock loop (PLL). These devices, generally embodied as hardware and software, assure that the incoming codes and carrier phases are matched (locked) to the receiver-generated codes and phases, and remain locked throughout the tracking of continuously received signals. Code matching directly generates the pseudorange, while phase matching generates the ambiguous carrier phase observable. The carrier phases are measured to an accuracy of a couple of millimeters, whereas the measurement accuracy of the pseudoranges depends on whether C/A-codes or P(Y)-codes are used. The interested reader should consult the electrical engineering literature for details on DLL and PLL. Examples include Blanchard (1976) for studying PLL in coherent receiver designs, Lindsey and Chie (1986) serving as an introduction to digital PLL, and Ascheid and Meyr (1986) obtaining a tutorial survey on cycle slips in PLL.

3.3.1 Some Fundamentals

3.3.1.1 Bandwidth Bandwidth terminology is used in connection with range (or spread) in frequency. If T denotes the duration of the chip (pulse), then an approximate relationship to the bandwidth is

$$B \approx \frac{1}{T} \tag{3.8}$$

As T becomes larger the signal spreads further in the time domain while the bandwidth (in the frequency domain) becomes smaller. Shorter chips (pulses) require greater bandwidth. An example for demonstrating bandwidth is given in Figure 3.13 for the P-code and the C/A-code. The plot shows the Fourier transform of these binary pseudorandom signals (power spectral density) generating the so-called "sinc" function:

$$S(f) = S(f_c) \times \frac{sin^2 \ (\pi \Delta f / 10.23 \ \text{MHz})}{(\Delta f / 10.23 \ \text{MHz})^2} \tag{3.9}$$

The symbol Δf denotes the difference in frequency with respect to the carrier frequency f_c (either L1 or L2). The chipping frequency 10.23 MHz applies to the P-code, and is replaced with 1.023 MHz to get the respective function of the C/A-code. The first lobe stretches over the bandwidth, covering the range of ± 10.23 MHz with respect to the center frequency (P-code). This symmetric function is zero at multiples of the chipping rate. For GPS, the spectral portion of the P-code signal beyond this bandwidth region is filtered out at the satellite and is not transmitted.

3.3.1.2 Power Due to the large variations in power requirements encountered in electronic systems, the decibel unit was introduced (based on the logarithmic scale). The instantaneous power of a real-valued periodic signal $y(t)$

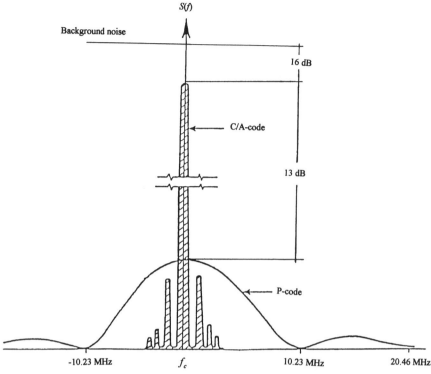

Figure 3.13 GPS signal power density. (Spilker, 1980.)

is given by

$$p(t) = y^2(t) \qquad (3.10)$$

and the ratio of two powers in units of decibel, dB, is

$$G_{dB} = 10 \log \frac{p_2(t)}{p_1(t)} \qquad (3.11)$$

If $p_2 > p_1$ the decibel gain is positive. The decibel is not an absolute, but is rather a relative unit. Indicating a certain signal level is meaningful only if a reference level has been established. Several decibel reference levels have become standard and have been incorporated into the abbreviation. For example, if power is given in milliwatts (mW) and the reference power $p_1(t)$ equals 1 mW, then the abbreviation dbm is used. If the units are watts (W) and the reference power is 1 W, one uses the abbreviation dBW.

The power of the signal can be represented in the time domain or the frequency domain. In the frequency domain the average power is obtained in the simplest case by summing the power levels associated with each frequency.

Such graphs are called the power density spectrum, the power spectral density, or simply the power spectrum. It can readily be visualized that periodic signals generate a line power spectrum in the frequency domain and that the height of individual lines represents power. For nonperiodic signals the power spectrum is continuous.

Figure 3.13 shows the power density of the P(Y)-code and the C/A-code; see also equation (3.9). The strength of the C/A-code is clearly seen. Of particular interest is that the received C/A-code density is lower than the background noise (thermal noise). To track the signal, the receiver multiplies the incoming signal by a locally generated replica of the code. This results in the GPS signal collapsing to a narrow bandwidth signal (bandwidth equal to that of the navigation message modulation). The result of the multiplication is filtered and used to track the GPS signal. Multiplying (correlating) with the locally generated code and filtering results in a signal that is well above the noise level. The subsequent tracking must occur within the narrow bandwidth.

3.3.1.3 *Filtering*
Assume that a signal consists of a desired and an undesired part. Filtering is a technique that allows the separation of these two signal components. The goal is always to preserve the desired signal in its original form as much as possible. For example, if one wishes to keep the low-frequency content of the signal by letting the low frequency pass, one speaks of a low-pass filter. The opposite is the high-pass filter, through which only high-frequency signals pass. If frequencies within a certain bandwidth should pass, one speaks of a band-pass filter. For example, one might design a filter through which only frequencies in the range $f \pm 10.23$ MHz will pass. The frequency region that lets signals pass is called the passband; the region that does not let signals through is called the stopband. In real filters, both of these bands are connected by a transition band in which signals lose increasingly more power as the stopband is approached. Figure 3.13 shows no power at the multiples of the chipping rate.

3.3.1.4 *Frequency Translation and Multiplication*
Frequency translation is a technique for shifting the frequency spectrum for a given signal up or down in the frequency domain while the shape of the spectrum remains invariant. If the signal is shifted to a higher frequency range, one speaks of upconversion, while shifting to a lower frequency range is called down-conversion.

Frequency translation might be advantageous when signals are easier to generate at the transmitter and easier to process at the receiver, at a different frequency than the transmission frequency. Down-conversion is generally carried out in GPS receivers. Multiplication of a signal by a sinusoid results in a frequency translation. The electronic device that performs this is called a mixer. A simple trigonometric computation demonstrates this. Consider the function

$$y_1(t) = A(t) \cos [\omega_1 t + \Phi(t)] \tag{3.12}$$

which should be mixed (multiplied) with a simple cosine wave

$$y_0(t) = \cos \omega_0 t \qquad (3.13)$$

giving

$$v(t) = A(t) \cos [\omega_1 t + \Phi(t)] \cos (\omega_0 t)$$

$$= \frac{A(t)}{2} \cos [(\omega_1 - \omega_0)t + \Phi(t)]$$

$$+ \frac{A(t)}{2} \cos [(\omega_1 + \omega_0)t + \Phi(t)] \qquad (3.14)$$

This multiplication preserves the amplitude modulation function $A(t)$ and the phase modulation function $\Phi(t)$. The resulting signal consists of two frequencies $(\omega_1 - \omega_0)$ and $(\omega_1 + \omega_0)$. If the original signal is mixed with a cosine wave of frequency ω_0, the resulting spectrum consists of two sidebands whose centers are separated by $2\omega_0$. Depending on whether up-conversion or down-conversion is desired, an appropriate filter is used to block the undesired sideband.

Frequency multiplication implies that the frequency of a given signal is multiplied by a constant, generally an integer. This technique is straightforward when applied to an unmodulated signal. Of special interest in GPS applications is the case in which the signal frequency increases by a factor of two as a result of squaring the original signal, that is, multiplying the signal by itself. Consider

$$y^2(t) = \{A(t) \cos [\omega_1 t + \Phi(t)]\}^2$$

$$= \frac{[A(t)]^2}{2} + \frac{[A(t)]^2}{2} \cos [2\omega_1 t + 2\Phi(t)] \qquad (3.15)$$

The first term can be filtered out. The remaining signal experiences a change in the amplitude modulation (squaring of the amplitude). However, in many applications squaring is applied to derived signals which have constant amplitudes. In addition to the frequency, the phase modulation is also multiplied by the factor 2. For binary phase modulation, where $\Phi(t)$ is 0 or π, then

$$y^2(t) = \frac{[A(t)]^2}{2} \cos (2\omega_1 t) \qquad (3.16)$$

The binary phase modulation has been eliminated in (3.16). Signal squaring is one of several methods in recovering the pure carrier from the biphase-modulated ranging signal.

3.3.1.5 *Spread Spectrum* The spread spectrum can be broadly defined as a mechanism by which the bandwidth of the transmitted code is much greater than the bandwidth required by the baseband information signal (navigation message). Several methods are available for employing the spread spectrum. For example, the FDMA (frequency division multiple access) method requires different carriers. Each carrier has its own modulation. The GLONASS satellite system is an example of this approach. Details on GLONASS are given in Appendix G. In the case of TDMA (time division multiple access) several channels share the same transmission link. LORAN is an example of TDMA.

GPS uses the CDMA (code division multiple access) method, which requires that pseudorandom codes be transmitted by the satellites and that these pseudorandom codes also be generated inside the receiver. The information stream can contain spoken words, digital images, or, as in case of GPS, the navigation message. The information stream is combined with a pseudorandom code, which causes the transmissions to occupy a large bandwidth. The pseudorandom signal must be separable at the receiver; the receiver duplicates the pseudorandom code and extracts the information stream from the bandwidth. The technique is advantageous for military applications because the spread spectrum makes "hardening the receiver" against jamming possible. Because receiver signals are combined with pseudorandom codes, narrow-band interfering signals are spread over a wide bandwidth. This results in improved interference rejection or jamming immunity.

Thus, in the case of GPS, the information bandwidth comes from the 50-bps navigation data stream. The P(Y)-codes and C/A-codes spread the navigation message. In communication systems terminology, the P(Y)-codes and C/A-codes are the spreading functions. Regarding positioning, the codes have the additional purpose of serving as time makes which can be clocked to measure the range. The bandwidth of the pseudorandom codes is much larger than that of the information message.

3.3.1.6 *Cross-correlation* Correlating received signals with signals generated in the receiver is a fundamental task for every receiver. For real-valued signals, the cross-correlation function is

$$C_{ij}(\Delta t) = \frac{1}{\tau} \int_{t_0}^{t_0 + \tau} y_i(t)\, y_j(t + \Delta t)\, dt = \begin{cases} 1 & \text{if} \quad \Delta t = 0 \\ 1 - \dfrac{|\Delta t|}{T} & \text{if} \quad |\Delta t| \le T \\ \approx 0 & \text{if} \quad |\Delta t| > T \end{cases} \quad (3.17)$$

This expression refers to continuous functions $y_i(t)$ and $y_j(t)$. The symbol τ denotes the integration time. In digital receivers, the signal processing is often considered a sampled data version of an analog system. The cross-correlation function of the locally generated code with the incoming signal is a triangle function (Bracewell triangle), the height being a function of the signal to noise ratio (SNR), the width being two chips. Band-limiting tends to round the peak of the function and to make the zeros more broad. It is important to note that the correlation function is continuous; this allows the receiver to track the code phase.

Figure 3.14 shows a simplified situation in which only the C/A-code is considered. An emitted binary C/A-code stream arrives at the receiver after a travel time of about 70 msec. The receiver generates the same code (in time synchronization with the code generation at the satellite). The received code and the receiver-generated code streams are cross-correlated. One can visualize the receiver-generated code shifted in time until there is a complete match with the arriving code. At that instant the autocorrelation is 1. The amount of shift equals the sum of the signal travel time and the receiver clock offset. The example shows the receiver running two chips too early. The satellite clock error is neglected in this simple example. Since the C/A-code is only 1023 chips long and repeats every millisecond, correlation peak is achieved each time the receiver code is shifted 1/1000 sec. It follows that the travel time as measured by the C/A-code is inherently ambiguous by 1 msec or 300 km. There are several ways to eliminate this ambiguity. The simplest way is by observing the transitions in the telemetry data and detecting the preamble in the telemetry data.

Since the C/A-code and the navigation stream are binary phase modulated, the code ranging is more complicated in reality. Ignoring the P(Y)-code for simplicity, the correlation coefficient for the pth satellite could be formulated

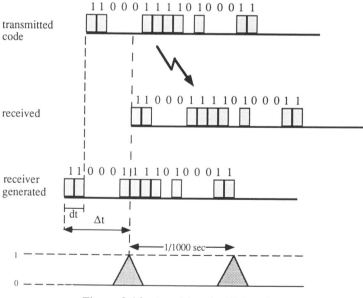

Figure 3.14 Acquiring the C/A-code.

as follows:

$$C_p(\Delta t) = \frac{1}{\tau} \int_{t_0}^{t_0 + \tau} \left[\sum_{i=1}^{n} y_i(t) S_i(t) + \text{noise} \right] y_p(t + \Delta t) \, dt \qquad (3.18)$$

where n is the number of satellites whose signals are being received. The product (modulo-2 sum) of $y_i(t) S_i(t)$ is received, where y_i denotes the C/A-code, and S_i is the binary navigation message. This cross-correlation must be carried out for each satellite p, with i running from 1 to n. The shift $(t + \Delta t)$ equals the travel time of the signal plus the clock correction. Since each satellite uses a unique C/A-code, the satellite can be identified by correlating the received signals with multiple locally generated C/A-codes, y_p. The cross-correlation coefficient for the Gold codes, $i \neq p$ in (3.18), are small enough to be ignored only if the integration extends over the complete code sequence.

In multichannel receivers one satellite and one computation are assigned to each channel. In each case the respective navigation stream is retrieved. See also Figure 3.9, which shows that by modulo-2 adding of the received and receiver-generated codes, the navigation stream is retrieved. The wide-bandwidth C/A-code signal collapses into the narrow bandwidth of the telemetry signal, giving a signal with SNR of about 40–50 dB in a 1-Hz bandwidth.

The spread spectrum as shown in Figure 3.13 indicates that the signals of the satellites are hidden in the noise by 16 dB. The spread spectrum technique allows the receiver to find the signal from the satellite. One can view the satellite signal as one that has the noise spectrum added to it. The noise plus satellite signal is positively biased by the satellite signal during the interval of 1 (ones). If this composite signal is cross-correlated with the signal from the code in the receiver, there is a small positive correlation during that chip. Adding statistically many samples effectively increases that correlation to a level sufficient for detection.

3.3.1.7 Locking and Tracking
The receiver is said to be code-locked if the receiver-generated codes have been aligned with the received codes. Tracking refers to the situation in which the receiver is continuously locked, i.e., the receiver code is continuously shifted to match the incoming signal. The shifting of the locally generated receiver code is controlled by the delay lock loop.

DLL implementation might use two correlators to measure the time shift of the received code. Sometimes the terminology of measuring the "phase of the received code" is used. A numerically controlled oscillator (NCO) generates the C/A-code or P-code signal. This signal is put in a digital delay line which generates three phases of the code: early, punctual, and late. The incoming signal with be correlated with all three. The output of the late correlator is subtracted from that of the early correlator, and the difference is used to drive the code generator, which advances or retards the code in a manner to zero the difference output signal. The code correlator also generates a punctual code signal that corresponds to the estimate of the phase of the satellite signal, i.e., the time delay that one is trying to measure.

The correlation between the punctual and incoming codes is used to control the carrier tracking loop and to recover the navigation message. The carrier phase lock loop is similar in many respects to the DLL. The NCO produces an output signal that drives the receiver frequency so that the locally generated code is matched in phase with the incoming code. A phase comparator generates an output value that is proportional to the difference in phase between the incoming carrier and the receiver carrier phase.

3.3.2 Modern Geodetic Receivers

All receivers filter the received GPS signals sensed by the antenna and subject them to a low-noise preamplifier. Most noise in the system comes from this preamplification. The amplifier adds enough gain so that the noise added by subsequent stages is not significant. Therefore, the first preamplification must generate as little noise as possible.

Figure 3.15 depicts an analog system until the signal reaches the digital sampler. One step in modern receivers is to convert the analog signal to a

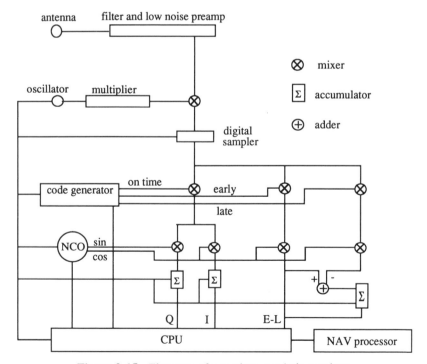

Figure 3.15 Elements of a modern geodetic receiver.

digital data stream. However, not all receivers are the same, and designers have choices. For example, one may chose to down-convert the received signals to an intermediary frequency f_{IF}. If one prefers to use the baseband signals, $f_{IF} = 0$, the first mixer should preferably be complex, i.e., generate in-phase and quadrature signals. Otherwise the f_{IF} is usually several megahertz. The signal is usually sampled at more than the Nyquist rate (twice the intermediary frequency) to make up for losses caused by the digital sampling.

The remainder of the receiver essentially consists of a digital implementation of the DLL and the PLL, both loops being controlled by a CPU. The three correlation coefficients to the CPU are on-time in-phase (I), the quadrature phase (Q), and the early minus late in-phase signal (E − L). These signals are used in the PLL and the DLL. When the loops are tracking, on time (I) is the cosine of the difference between the numerically controlled oscillator and the received signal carrier phase, while (Q) is the sine of that difference. (E − L) is the difference between the correlation of the signal with an early code replica and a late code replica (see description of the code delay lock loop given above). The CPU loop filters keep the NCO phase and the code generator output synchronous with the incoming signals. Both loops are usually second-order loops, that is, they have two internal states. For example, the states for the PLL are phase and frequency. The error signal for the carrier phase loop is derived from (I) and (Q). The phase loop filter controls the NCO so as to reduce the error signal (Q). For precise geodetic receivers, the loop bandwidth is kept constant (to guarantee that the receivers at both ends of the baseline have the same bandwidth and, thus, the same measurement resolution). The purpose of the (E − L) is to control the code generator so that the error signal is minimized.

The output of the accumulators, which typically sum from 1 to 10 msec, is used to steer the loops. For example, the digital data streams that are mixed with the digital version of the code and carrier might be sampled at 20 MHz by the digital sampler. Thus, there is a bit stream of ones and zeros at 1 bit per 50 nsec. These samples are multiplied at 20 MHz by the sine and cosine of the NCO-driven synthesizer. The result of these multiplication goes in the accumulator, which sums them, most likely, over a millisecond. If the incoming signal and the (digital) sine or cosine correlate well, the accumulated sum will be large. Thus, this sum can be fed to the CPU to steer the loops. The loops controlled by the CPU have a bandwidth generally in the range of 1–10 Hz; thus, the bandwidth of the tracking loops is independent of the hardware integration time. The settling time of the loops, which is the time it takes for error signals to decay to $1/e$ of the original value, is proportional to the inverse of the loop bandwidths. Figure 3.16 shows an example of a modern geodetic receiver.

3.3.3 Pseudorange Measurement Accuracy

No single formula describes the range measurement accuracy of a receiver, because such a formula would depend on the various electronic components

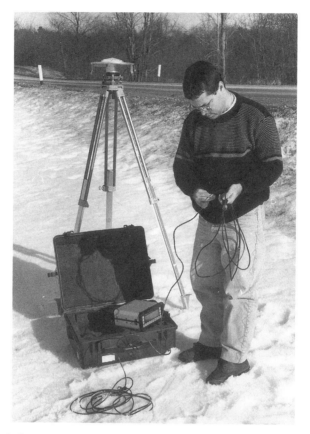

Figure 3.16 Laboratory assistant with the Ashtech P-XII receiver at the University of Maine. The follow-up receiver, called the Z receiver, gives dual-frequency pseudoranges and full wavelength carrier phases even with AS turned on.

and strategies used in the receiver design. Often an expression is given for the standard deviation of the DLL time-update error. Van Dierendonck et al. (1992) and RTCM-104 (1994) list several slightly differing expressions depending on the choices made for the DLL, such as coherent and noncoherent DLL. At present, most receivers use coherent DLL, in which case the carrier is incorporated into the demodulation process. Several schemes for noncoherent DLL are available that eliminate the need for generating a phase-synchronous reference in the receiver. For the present discussion, the expression

$$\sigma_C = \frac{c}{R_C} \sqrt{\frac{B_L d}{2S/N_0}} \qquad (3.19)$$

suffices. This expression, of course, does not include any errors due to the propagation media or multipath. It represents the errors in range due to the presence of the received noise, called the thermal (background) noise. The code is band-limited at the satellite, so (3.19) well describes the standard deviation of the code phase error for $d > 1/20$ of a chip. The chip spacing d cannot be made arbitrarily small; however, expression (3.19) is good for identifying the characteristics. It is seen that measurement resolution is proportional to the square root of the product of the loop bandwidth B_L and the chip spacing d, and is inversely proportional to the square root of the SNR, S/N_0. The symbol c denotes the velocity of light, and R_C denotes the chipping rate of the C/A-code or P-code, respectively.

In more precise terms, B_L is the equivalent single-sided noise bandwidth of the pseudorange smoothing process and is given in units of chips. In practical applications the bandwidth must be wide enough to track the received code as it changes with respect to the receiver-generated code due to the Doppler and possibly ionospheric effects. To reduce B_L, some receivers implement carrier-aided tracking to shift the center of bandwidth as predicted by the Doppler. The bandwidth still must be wide enough to handle the ionospheric divergence between the code and carrier phases. While smoothing reduces B_L, it may cause the receiver to lose lock due to ionospheric disturbances.

Because the P(Y)-code chipping rate is 10 times that of the C/A-code chip, the superiority in measurement accuracy of P-codes is readily seen. Equation (3.19) also shows how a smaller chip spacing increases the measurement accuracy even if the other parameters remain unchanged. For narrow correlator receivers, the chip spacing is $d < 1$. The use of 1.0 chip correlator spacing in DLLs has been the usual practice in GPS receivers until NovAtel Communications Ltd. of Canada began a new trend in C/A-code receiver technology by introducing a narrow correlator spacing of as little as 0.05 of the C/A-code chip width (Fenton et al. 1991). Their correlator spacing is variable, depending on the dynamic situation of the receiver. Referring to Figure 3.13, we can say that narrow correlator receivers make use of more than one C/A-code lobe. The range accuracy of these receivers is about a factor $\sqrt{5}$ worse than current P(Y)-code receivers. Narrow correlation technology does not reduce the multipath effect on the carrier-phase measurement. The introduction of narrow correlator spacing can surely be considered the first major advancement of C/A-code technology since its introduction in the early years of GPS. Current literature seems to suggest that the narrow correlator approach will not improve P(Y)-code ranging appreciably.

3.3.4 Coping With Antispoofing (AS)

A major accomplishment in the continued development of GPS technology, and at the same time a clear sign of an innovation-happy GPS industry, are the various solutions that have become available over the last couple of years for dealing with AS. The widespread fear of a decade ago that AS might severely limit the use of advanced positioning technique for civilians fortunately did not materialize. Even such techniques as kinematic GPS positioning and ambiguity fixing on the fly are possible in the presence of AS. The following is a very brief overview of the techniques currently used to neutralize or reduce the effects of AS. Because of their complexity, several patents are referenced for the interested reader.

3.3.4.1 *Squaring*

Equation (3.16) shows that squaring of the signal removes the binary phase modulation but produces a half-wavelength signal of the carrier, also called the second harmonic. The major drawbacks of this technique are a significant loss of SNR ratio, roughly -30 dB compared to direct correlation with the P-code, and half-wavelength carrier phase observations on L2. Signal squaring was implemented in the Macrometer, which was one of the first successful GPS receivers that did not use P(Y)-codes at all (Ladd et al. 1985; Counselman 1987).

3.3.4.2 *Code-Aided Squaring*

Hatch et al. (1992) present a technique that uses the receiver-generated P-code to aid in the tracking of the Y-code. This technique is possible because of existing similarities between the P-codes and the Y-codes, even though the precise form of the Y-codes is not publicly known. The received L1 signal is down-converted to an intermediary frequency and then correlated (multiplied) with a locally generated replica of the underlying P-code present in the Y-code. The result of this correlation is a reduced-bandwidth signal with a bandwidth of about ± 500 kHz, instead of ± 10.23 MHz for the P-code prior to squaring. The reduced-bandwidth portion of the signal is squared. The narrowing of the bandwidth prior to squaring results in an increased signal-to-noise ratio after the squaring. Thus, the code-aiding technique provides improved signal-to-noise ratio as compared to regular squaring. It yields L2 P(Y)-code ranges, but, because of the squaring, still gives only half-wavelength L2 carrier phases. The method was patented by Keegan (1990).

3.3.4.3 *Cross-correlation*

This technique takes advantage of the fact that both L1 and L2 carriers are modulated with the same P(Y)-code, although the Y-code is not known. Because each carrier has the same P(Y)-code phase modulations, both signal streams can be cross-correlated (Meehan et al. 1992). The result of this process is the difference of the pseudoranges, $P_1(Y) - P_2(Y)$, and the phase difference between the L1 and L2 carrier phases. These difference observables are combined with the L1 C/A-code pseudorange and the L1 carrier phase. Cross-correlation has a 3-dB SNR improvement over squaring, but the available SNR limits the bandwidth of the L2 carrier loop and subsequently the dynamics of the ionosphere the loop can track, especially at low elevation angles.

3.3.4.4 *Z Tracking*

This technique takes advantage of the fact that the Y-code is the modulo-2 sum of the P-code and is a substantially lower rate encryption code of approximately 500 kHz. The L1 and L2 signals are correlated with locally generated P-codes, and the bandwidth is reduced to that of the encryption code (Ashjaee and Lorenz 1992). The encryption code is estimated and removed from the received signal to allow locally generated code replicas to be locked with the P-code signals of L1 and L2. The method does not require knowledge of the encryption code, but requires only some aspects of the timing. Z tracking gives L1 and L2 Y-code pseudoranges, and full-wavelength carrier phases of L1 and L2. Thus, the same observables are available as obtained by correlating with the P-code directly. Z tracking gives best signal-to-noise ratios compared to the other techniques mentioned above. Multipath performance of Z tracking and code-aided squaring is better than cross-correlation and squaring, because the incoming signals are correlated with the code. Technical details are found in Lorenz et al. (1992).

Table 3.7 summarizes the characteristic features of the various methods developed to cope with antispoofing. Full-wavelength carrier-phase observables are necessary to perform wide laning, the case in which a simple function of the two phases is used to form the mathematical equivalent of an 86-cm lane (wavelength). Wide laning is very popular in rapid static and kinematic GPS. It is particularly easy to determine the wide lane if both P(Y)-code pseudoranges are available. See Chapter 10 for additional details on the respective positioning techniques.

A high SNR is a desirable quality. The higher this ratio, the better the receiver will perform in rapid static and kinematic applications. Nolan et al. (1992) discuss various practical tests suitable for characterizing the performance of receivers. They point out that the standard deviations of the pseudorange must always be viewed in the context of averaging time. The signal-to-noise ratio translates directly into how long one has to integrate or average, i.e., the time it takes for the tracking loops to settle (to match the incoming signal). If one receiver has 10 times the signal-to-noise ratio of another, it only has to average for one-tenth the time to get the same noise in the observables. The averaging time is important for two reasons;

1. Receivers with long averaging time might have difficulty tracking during a rapidly changing ionosphere. The code loop usually is aided by the

TABLE 3.7 System Comparison of AS Technologies (Ashjaee and Lorenz 1992)

Parameters	Squaring	Code-aided squaring	Cross-correlation	Z-tracking
C/A-code	no	yes	yes	yes
Pseudorange (Y)	no	Y1	Y2-Y1	Y1, Y2
L2 wave	half	half	full	full
Signal-to-noise	-16dB	-3dB	-13dB	0dB

carrier loop. The guiding will be in error if the ionosphere changes rapidly. The L2 carrier loop is the most sensitive.

2. The time it takes for a transient (such as on acquisition or loss of lock) to settle is directly proportional to the averaging time. This time can range from a few seconds to 100 sec or more, depending on the receiver.

The SNR in Table 3.7 are 6–14 dB worse than when correlating with the P-code directly. Thus, having access to the P-code in the clear yields the best solutions.

CHAPTER 4

ADJUSTMENT COMPUTATIONS

Least-squares adjustment is a device for carrying out objective quality control of surveying measurements by processing sets of redundant observations according to mathematically well-defined rules. The objectivity of least-squares quality control will become increasingly important as sets of observations are deposited into public databases in support of a multipurpose cadastre or used to verify the consistency of the observations in an interdisciplinary user community. The term "redundant observations" implies that more observations are available than necessary to determine a set of unknowns, for example, the station coordinates of a network. Details will be given as to what constitutes optimal redundancy. Because of the overriding importance of least-squares estimation, this chapter contains compact but complete derivations of least-squares algorithms. First, the statistical nature of measurements in surveying is analyzed. This is followed by a general discussion of stochastic and mathematical models and the law of variance–covariance propagation of random errors. The so-called mixed adjustment model is derived in detail, and the observation equation model and the condition equation model are deduced from the mixed model through appropriate specification. The case of observed, or weighted, parameters is presented as well. A special section is devoted to minimal and inner constraint solutions and to the identification of those quantities that remain invariant with respect to a change in minimal constraints. Whenever the goal is to perform quality control on the observations, minimal or inner constraint solutions are used. Statistical testing is yet another important element for judging the observations or for making qualitative statements about the unknowns (estimated parameters). A complete section is devoted to statistics in least-squares adjustments. Finally, additional quality measures in terms of internal and external reliability are given. The chapter concludes with some considerations on blunder detection.

4.1 ELEMENTS

Objective quality control is a requirement for any survey independent of the actual accuracy of the survey. Unfortunately, least-squares adjustment has most often been associated with high-precision surveying. In a low-precision survey of 1 m, it might be just as important to ensure that there is no 10-m blunder, just, as in a millimeter survey one does not want a centimeter error. One reason for the frequent use of approximate solutions rather than rigorous least-squares adjustment in earlier times was the lack of adequate computing capabilities. Given today's desktop microcomputer power, inexpensive mass storage devices, and broad programming skills, new opportunities exist for least-squares adjustment to provide quality control in all types of surveys.

Least-squares adjustment allows the combination of different types of observations, such as angles, distances, and height differences, into one solution and permits simultaneous statistical analysis of that solution. Surveying theory can be simplified because there is no need to distinguish between computational rules and expressions for traverses, intersections, and resections. Since all of these specialized geometric figures basically consist of angle and distance measurements, they can be treated alike, and the same least-squares rules apply to all of them.

Least-squares adjustment techniques are also useful in planning surveys to ensure that accuracy specifications are met once the observations are made. This general area of survey simulation can be extended to include consideration of the optimal choice of observations where alternatives exist. For example, should a surveyor primarily measure angles or rely on distances? Taking the firm's pool of available equipment into consideration, what is the optimal use of the equipment under the constraints of a particular project? Many of these questions are answered intuitively by the experienced surveyor. Even in these cases, an objective verification using least-squares simulation and the concept of internal and external reliability of networks is a welcome assurance to those who carry the responsibility. In deformation monitoring the detection of motion can be verified only with the help of the least-squares formalism and inherent statistical analysis capability. This is particularly true if the magnitude of the suspected motions is nearly that of the measurement precision.

4.1.1 Statistical Nature of Surveying Measurement

Assume that a tape with centimeter divisions is used to measure a distance of exactly 100 m repeatedly. Initial measurements could be 99.99, 100.02, 100.01, etc. Because of the centimeter subdivision of the tape, the surveyor is likely to record the observed distances to two decimal places. The result of the measurements is a series of numbers ending with two decimal places. One could wrongly conclude that measuring a distance is a case of discrete statistics yielding discrete values (distances) to two decimal places. In reality, however, the two decimal places are only a consequence of the existing centimeter division of the tape and the fact that the surveyor chose not to estimate the millimeters. If we imagine a reading device that would allow the surveyor to read the tape to as many decimal places desired, we see that the process of measuring a distance belongs to the realm of continuous statistics. The measurement process can result in any value, since there would be no limit on the number of decimal digits in an observation. A similar thought experiment can be carried out with other measurement types in surveying, such as angles and heights. An important conclusion, therefore, is that the measurement processes in surveying should be analyzed using continuous statistics rather than discrete statistics.

To amplify the difference between continuous and discrete statistics, consider the familiar textbook example of throwing a die. As is well known, the probability of throwing any of the six numbers is the same. Only one of the numbers from 1 to 6 can be the outcome of the experiment, that is, throwing the die. The outcome is a discrete integer number and not a real number. This is a classical example taken from discrete statistics. In this case all possible outcomes of the experiment are limited to the numbers 1 to 6. This set of numbers constitutes the population (in the statistical sense).

Continuing the example of measuring the distance, one recognizes that, in principle, any value could be obtained, although experience dictates that values close to 100.00 are most likely. Values such as 99.90 or 100.25 are very unlikely. Assume that n measurements are made. Plot along the x axis in Figure 4.1 the numerical value of each measurement. Subdivide the x axis into intervals of equal length Δx, and count the number n_i of measurements within each interval. The relative frequency n_i/n is plotted along the ordinate. Next, a rectangle whose height equals the relative frequency is constructed, as shown in Figure 4.1. This plot is called a histogram. The smoothed step function of the rectangles of the histogram has a bell-like shape. The maximum occurs around the sample mean. The larger the deviation from the mean, the smaller the relative frequency, that is, the probability that such a measurement will actually be obtained. A goodness-of-fit test, as explained in Appendix C, can be carried out. The hypothesis of a normal distribution is confirmed. Thus, the typical measurement process in surveying follows the statistical law of normal distribution.

4.1.2 Elementary Statistical Concepts

Several concepts from statistics are required in least-squares adjustment. The following is a summary:

- An observation or a statistical event is the outcome of a statistical experiment, for example, throwing a die or measuring an angle or a distance.

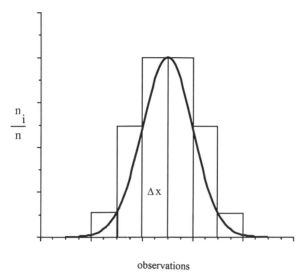

observations

Figure 4.1   Relative frequency of measurements.

- A random variable is the name for the outcome of an event. The random variable is denoted by a tilde. Thus, $\tilde{x}$ is a random variable and $\tilde{X}$ is a vector of random variables.
- The population is the totality of all events. It includes all possible values that the random variable can have. The population is described by a finite set of parameters, called the population parameters. The normal distribution, for example, is such a population, and it is completely described by the mean and the variance.
- A sample is a subset of the population. For example, if the same distance is measured 10 times, then these 10 measurements are a sample of all the possible measurements.
- A statistic represents an estimate of the population parameters or functions of these parameters. It is computed from a sample. Thus, the 10 measurements of the same distance can be used to estimate the mean and the variance of the (normal) distribution.
- Probability is related to the frequency of occurrence of a specific event. Each value of a random variable has an associated probability.
- The probability density function describes the probabilities as related to the possible values of the random variable. If $f(x)$ denotes the probability density function, then

$$P(a \leq \tilde{x} \leq b) = \int_a^b f(x)\, dx \qquad (4.1)$$

gives the probability that the random variable $\tilde{x}$ assumes a value in the interval $[a, b]$.

4.1.3 Observational Errors

Field observations are not perfect, nor is the recording and management of observations. The measurement process suffers from several error sources. As explained, repeated measurement of the same distance does not yield identical numerical values each time. The reason is the presence of random errors in the measurements. These errors are usually small: That is, the probability of a positive or negative error of a given magnitude is the same (equal frequency of occurrence). Random errors are inherent in the nature of measurements and can never be completely overcome. Random errors are dealt with in least-squares adjustment.

Systematic errors are errors that vary systematically in sign and/or magnitude. The variation is usually over a longer period of time. Examples are a tape that is 10 cm too short or the failure to correct for vertical or lateral refraction in angular measurement. Systematic errors are particularly dangerous because they tend to accumulate. Attempts should be made to avoid systematic errors through adequate instrument calibration, compensation through careful observation technique, such as double centering in surveying, and observation under various conditions. If errors are known, the observations can be corrected before making the adjustment or the systematic errors can be modeled in the adjustment by adding additional unknowns. Discovering and dealing with systematic errors during the least-squares computations requires a great deal of experience and advanced knowledge. Success is not guaranteed!

Blunders are usually large errors resulting from carelessness by the observer. Examples of blunders are counting errors in a whole tape length, transposing digits when recording field observations, continuing measurements after upsetting the tripod, and so on. Blunders largely can be avoided through careful observation; however, there can never be absolute certainty that all blunders are removed. Therefore, an important part of least-squares adjustment is to discover and remove remaining blunders in the observations.

4.1.4 Accuracy and Precision

Accuracy refers to the closeness of the observations (or the quantities derived from the observations) to the true value. Precision refers to the closeness of repeated observations (or quantities derived from repeated sets of observations) to the sample mean. Figure 4.2 shows four density functions that represent four distinctly different measurement processes of the same quantity. Curves 1 and 2 are symmetric with respect to the true value x_T. These measurements have a high accuracy, because the sample mean, x_S coincides or is very close to the true value. However, the shapes of both curves are quite different. Curve 1 is tall and narrow, whereas curve 2 is short and broad. The observations of process 1 are clustered closely around the mean (true value), whereas the spread of observations around the mean is larger for process 2. The relative frequency of larger deviations from the true value are larger for process 2 than for process 1. Thus, process 1 is more precise than process 2; however, both processes are equally accurate. Curves 3 and 4 are symmetric with respect to the sample

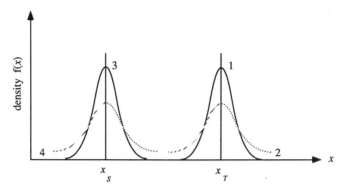

Figure 4.2   Accuracy and precision.

mean x_S, which differs from the true value x_T. Both sequences have equally low accuracy, but the precision of process 3 is higher than that of process 4. The difference $x_T - x_S$ is caused by a systematic error. An increase in the number of observations does not reduce this difference.

4.2   STOCHASTIC AND MATHEMATICAL MODELS

Any least-squares adjustment deals with two equally important components: the stochastic model and the mathematical model. Both components are indispensable parts of the adjustment and contribute to the adjustment algorithm (see Figure 4.3). The observations are the primary input to the adjustment. It is customary to denote the vector of observation with L_b, and the number of observations by the letter n. The observations are random variables; thus the complete notation for the $(n \times 1)$ vector of observations is $\tilde{L}_b$. To simplify the notation, one does not usually use the tilde in connection with L_b. The true value of the observations, or the means of the respective populations, are estimated during the adjustment on the basis of the sample, that is, the particular set of observations made. Each observation belongs to a different population; that is, the sample size is usually 1! The variances of these distributions comprise the stochastic model. The stochastic model introduces information about the (relative) precision of the observations. This is accomplished through the variance–covariance matrix of the observations Σ_{L_b} (note that the arrangement of elements in the variance–covariance matrix must correspond to the sequence of elements in the observation vector L_b). In most cases the variance–covariance matrix of the observations is diagonal; that is, the observations are not correlated. Occasionally, when so-called derived observations are used (which are the outcome from a previous adjustment) or when linear combinations of original observations are adjusted, the variance–covariance matrix contains off-diagonal elements. Because in surveying the observations belong to a normal distribution, the vector of observations is a sample from

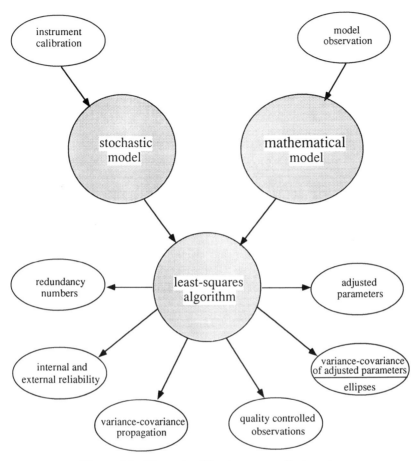

Figure 4.3 Elements of least-squares adjustment.

the multivariate normal distribution, which is expressed by the following notation:

$$L_b \sim N(L_T, \Sigma_{L_b}) \qquad (4.2)$$

where L_T denotes the vector mean of the population. Strictly, the adjustment requires as input not the variance–covariance matrix but the cofactor matrix Q_{L_b} of the observations, which corresponds to the scaled variance–covariance matrix

$$Q_{L_b} = \frac{1}{\sigma_0^2} \Sigma_{L_b} \qquad (4.3)$$

or the weight matrix, defined as

$$P = P_{L_b} = Q_{L_b}^{-1} = \sigma_0^2 \Sigma_{L_b}^{-1} \tag{4.4}$$

The scale factor σ_0^2 is called the a priori variance of unit weight. It is an arbitrary scale factor by which the inverted covariance matrix is multiplied to get the weight matrix. One of the fundamental capabilities of least-squares adjustment is the estimation of σ_0^2 based on the observations and the weight matrix P. This estimate is denoted by $\hat{\sigma}_0^2$ and is called the a posteriori variance of unit weight. If the a priori and a posteriori variance of unit weight are statistically equal, the adjustment is said to be correct. More on this fundamental statistical test and its implications will follow in later sections. In general, the a priori variance of unit weight σ_0^2 is set to one; that is, the weight matrix is equated with the inverse of the variance–covariance matrix of the observations. The name ''variance of unit weight'' is derived from the fact that, if the variance of an observation equals σ_0^2, then the weight for this observation equals unity. The special case that $P = I$, that is, the weight matrix equals the identity matrix, can be considered the classical least-squares case. These cases frequently allow a simple and geometrically intuitive interpretation of what is minimized during the adjustment.

The mathematical model expresses a simplification of existing physical reality. It attempts to express mathematically the relations between observations, and between observations and parameters or unknowns of the adjustment. Examples of parameters are coordinates, heights, and refraction coefficients. Naturally, the objective is to find a mathematical model to represent the relationships in nature as well as possible. Finding the right mathematical model is always a challenge. Even though the mathematical model is known for many routine applications, there are always new cases in which a new formulation of the mathematical model is required. Least-squares adjustment is a very general tool and can be used whenever a relationship between observations or between observations and parameters can be established.

Much research has gone into identifying or constructing a mathematical model that is applicable to all cases in surveying. The model of integrated geodesy is becoming increasingly popular because it combines all surveying and geodetic measurement in a unified model. The possible observations include distances, angles, heights, gravity anomalies, gravity gradients, geopotential differences, and astronomical observations. The model of integrated geodesy is indeed universal, for it includes an accurate formulation of the dependence of observations on gravity and is inherently three dimensional. The accuracy that can be achieved with today's instrumentation and observing techniques, particularly in view of the capabilities of the Global Position System, requires full incorporation of the gravity field into the mathematical model.

Frequently the original observations are modified (reduced) to yield so-called model observations. For example, it is well known that refraction significantly affects vertical angle measurement. If the original observations, that is, the observed values, are used, the mathematical model must be general enough to estimate the refraction parameters. One now has the option of estimating refraction parameters at each station, for subsets of stations, or for the whole network. On the other hand, the original observations can be corrected for refraction on the basis of some atmospheric model. The thus reduced observation can be modeled. There are many other examples for which options in modeling exist. Generally, the fewer reductions applied to the original observation, the more general the respective mathematical model must be. The final form of the model also depends on the purpose of the adjustment. For example, if refraction is to be studied, one would preferably have refraction parameters explicitly included in the mathematical model.

In most surveying applications the mathematical model in nonlinear. In the most general case, the observations and the parameters are related by an implicit nonlinear function, as follows:

$$F(X_a, L_a) = 0 \qquad (4.5)$$

This is the *mixed adjustment model*. The subscript a denotes the adjusted quantity. Usually the number of observations is denoted by n and the number of parameters by u. Thus

$$L_a = \text{vector of } n \text{ adjusted observations}$$

$$X_a = \text{vector of } u \text{ adjusted parameters (unknowns)}$$

$$F = r \text{ nonlinear mathematical functions}$$

In many cases a modification of the model (4.5) is used if the observations are explicitly related to the parameters, such as in

$$L_a = F(X_a) \qquad (4.6)$$

This is the *observation equation model*. The number of equations in (4.6) is denoted by n. The three-dimensional model developed in Chapter 7 is of this type. A further modification would be the total elimination of the parameters. Thus,

$$F(L_a) = 0 \qquad (4.7)$$

is the *condition equation model*. The number of equations in (4.7) is denoted by r.

The decision as to which of the three models should be used depends on the application. Sometimes it is the complexity of the mathematics required for formulating the mathematical model that determines which model is used. In other cases the ease of computer implementation is important. The observation equation model has the particular advantage that "each observation generates one equation." This allows the observation equation model to be implemented relatively easily on a computer.

Figure 4.3 indicates some potential outcomes of an adjustment. Several statistical tests are available to verify the quality of the observations, and various approaches are possible to discover and remove the effect of blunders in the observations. The adjustment provides probability regions for the estimated parameters and allows variance–covariance propagation to determine functions of the estimated parameters and their standard deviations. Of particular interest in surveying applications is the ability of least-squares adjustment to perform internal and external reliability analysis to identify marginally detectable blunders and to determine their potential influence to estimated parameters.

4.3 PROPAGATION OF MEAN AND VARIANCE–COVARIANCE

Statistical concepts enter the least-squares formulation in two distinct ways. The least-squares solution in the narrow sense merely requires a variance–covariance matrix to exist; there is no need to specify a particular distribution for the observations. The covariance matrix, or equivalently the cofactor matrix or the weight matrix, is sufficient to estimate the parameters and to compute adjusted observations and the variances of functions of the parameters through variance–covariance propagation. If statistical tests are required, then the distribution of the observations must be known. In most cases, one indeed desires to carry out some fundamental testing.

The purpose of variance–covariance propagation is to compute the variances and covariances of functions of the original random variables or of functions of parameters estimated from the original observations. These functional relationships must be linear. Nonlinear functions are first linearized. Variance–covariance propagation is directly related to multivariate distributions, that is, distributions of several random variables. To deal with such a general case, some of the pertinent concepts from statistics are reviewed.

Probability Density and Cumulative Probability   For $f(x)$ to be a probability function of the random variable $\tilde{x}$, it has to fulfill certain conditions. First, $f(x)$ must be a nonnegative function, because there is *always* an outcome of an experiment; that is, the observations can be positive, negative, or even zero. Second, the probability that a sample (observation) is one of all possible outcomes should be 1. Thus the density function $f(x)$ must fulfill the following conditions:

$$f(x) \geq 0 \tag{4.8}$$

$$\int_{-\infty}^{\infty} f(x) \, dx = 1 \tag{4.9}$$

The integration is taken over the whole range (population) of the random variable. Conditions (4.8) and (4.9) imply that the density function is zero at minus infinity and plus infinity. The probability

$$P(\tilde{x} \leq x) = F(x) = \int_{-\infty}^{x} f(t) \, dt \qquad (4.10)$$

is called the cumulative distribution function. It is a nondecreasing function because of condition (4.8).

The Mean   The mean, also called the expected value of a continuously distributed random variable, is defined as

$$\mu_x = E(\tilde{x}) = \int_{-\infty}^{\infty} x f(x) \, dx \qquad (4.11)$$

The mean is a function of the density function of the random variable. The integration is extended over the whole population. Equation (4.11) is the analogy to the weighted mean in the case of discrete distributions.

The Variance   The variance is defined by

$$\sigma_x^2 = E(\tilde{x} - \mu_x)^2 = \int_{-\infty}^{\infty} (x - \mu_x)^2 f(x) \, dx \qquad (4.12)$$

The variance measures the spread of the probability density in the sense that it gives the expected value of the squared deviations from the mean. A small variance therefore indicates that most of the probability density is located around the mean.

Multivariate Distribution   Any function $f(x_1, x_2, \ldots, x_n)$ of n continuous variables $\tilde{x}_i$ can be a joint density function provided that

$$f(x_1, x_2, \ldots, x_n) \geq 0 \qquad (4.13)$$

$$\int_{-\infty}^{\infty} \int_{-\infty}^{\infty} \int_{-\infty}^{\infty} \cdots \int_{-\infty}^{\infty} f(x_1, x_2, \ldots, x_n) \, dx_1 \, dx_2 \cdots dx_n = 1 \qquad (4.14)$$

It follows as a natural extension from (4.10) that

$$P(\tilde{x}_1 < a_1, \tilde{x}_2 < a_2, \ldots, \tilde{x}_n < a_n)$$

$$= \int_{-\infty}^{a_1} \cdots \int_{-\infty}^{a_n} f(x_1, x_2, \ldots, x_n) \, dx_1 \, dx_2 \cdots dx_n \qquad (4.15)$$

The marginal density of a subset of random variables $(x_1, x_2, \ldots, x_p)$ is

$$g(x_1, x_2, \ldots, x_p)$$

$$= \int_{-\infty}^{\infty} \cdots \int_{-\infty}^{\infty} f(x_1, x_2, \ldots, x_p) \, dx_{p+1} \, dx_{p+2} \cdots dx_n \quad (4.16)$$

Stochastic Independence   The concept of stochastic independence is required when dealing with multivariate distributions. Two sets of random variables, $(\tilde{x}_1, \ldots, \tilde{x}_p)$ and $(\tilde{x}_{p+1}, \ldots, \tilde{x}_n)$, are stochastically independent if the joint density function can be written as a product of the two respective marginal density functions, for example,

$$f(x_1, x_2, \ldots, x_n) = g_1(x_1, x_2, \ldots, x_p) g_2(x_{p+1}, x_{p+2}, \ldots, x_n) \quad (4.17)$$

Vector of Means   The expected value for the individual parameter x_i is

$$\mu_{x_i} = E(\tilde{x}_i) = \int_{-\infty}^{\infty} \cdots \int_{-\infty}^{\infty} x_i f(x_1, x_2, \ldots, x_n) \, dx_1 \, dx_2 \cdots dx_n \quad (4.18)$$

In vector notation the expected values of all parameters are

$$E(\tilde{X}) = \begin{bmatrix} E(\tilde{x}_1) \\ E(\tilde{x}_2) \\ E(\tilde{x}_3) \\ \vdots \\ E(\tilde{x}_n) \end{bmatrix} \quad (4.19)$$

Variance   The variance of an individual parameter is given by

$$\sigma_{x_i}^2 = E(\tilde{x}_i - \mu_{x_i})^2$$

$$= \int_{-\infty}^{\infty} \cdots \int_{-\infty}^{\infty} (x_i - \mu_{x_i})^2 \, f(x_1, x_2, \ldots, x_n) \, dx_1 \, dx_2 \cdots dx_n \quad (4.20)$$

Covariance   For multivariate distributions another quantity, called the covariance, becomes important. The covariance describes the statistical relationship between two random variables. The covariance is

$$\sigma_{x_i, x_j} = E[(\tilde{x}_i - \mu_{x_i})(\tilde{x}_j - \mu_{x_j})]$$

$$= \int_{-\infty}^{\infty} \cdots \int_{-\infty}^{\infty} (x_i - \mu_{x_i})(x_j - \mu_{x_j}) f(x_1, x_2, \ldots, x_n) \, dx_1 \, dx_2 \cdots dx_n$$

$$(4.21)$$

Whereas the variance is always larger than or equal to zero, the covariance can be negative, positive, or even zero.

Correlation Coefficients   The correlation coefficient of two random variables is defined as

$$\rho_{x_i, x_j} = \frac{E[(\tilde{x}_i - \mu_{x_i})(\tilde{x}_j - \mu_{x_j})]}{\sigma_{x_i}\sigma_{x_j}} = \frac{\sigma_{x_i, x_j}}{\sigma_{x_i}\sigma_{x_j}} \qquad (4.22)$$

Therefore, the correlation coefficient equals the covariance divided by the respective standard deviations. An important property of the correlation coefficient is that

$$-1 \le \rho_{x_i, x_j} \le 1 \qquad (4.23)$$

If two random variables are stochastically independent, then the covariance (and thus the correlation coefficient) is zero. By making use of (4.17) for the density function of stochastically independent random variables, we can write (4.21) as

$$
\begin{aligned}
\sigma_{x_i, x_j} &= \int_{-\infty}^{\infty} \cdots \int_{-\infty}^{\infty} (x_i - \mu_{x_i})(x_j - \mu_{x_j}) f(x_1, x_2, \ldots, x_n) \, dx_1 \, dx_2 \cdots dx_n \\
&= \int_{-\infty}^{\infty} \int_{-\infty}^{\infty} (x_i - \mu_{x_i})(x_j - \mu_{x_j}) f_i(x_i) f_j(x_j) \, dx_i \, dx_j \\
&= \int_{-\infty}^{\infty} (x_i - \mu_{x_i}) f_i(x_i) \, dx_i \int_{-\infty}^{\infty} (x_j - \mu_{x_j}) f_j(x_j) \, dx_j \qquad (4.24)
\end{aligned}
$$

These integrals are zero because of the definition of the mean. The converse—zero correlation implies stochastic independence—is valid only for the multivariate normal distribution.

Variance–Covariance Matrix   Equations (4.20) and (4.21) can be used to compute the variances and covariances for all components in the random vector $\tilde{X}$. Arranging the result in the form of a matrix yields the variance–covariance matrix. Thus, for the random vector

$$(\tilde{X} - M_X) = \begin{bmatrix} (\tilde{x}_1 - \mu_{x_1}) \\ (\tilde{x}_2 - \mu_{x_2}) \\ \vdots \\ (\tilde{x}_n - \mu_{x_n}) \end{bmatrix} \qquad (4.25)$$

the $(n \times n)$ variance–covariance matrix becomes

$$\Sigma_x = E[(\tilde{X} - M_X)(\tilde{X} - M_X)^T]$$

$$= \begin{bmatrix} \sigma^2_{x_1} & \sigma_{x_1,x_2} & \cdots & \sigma_{x_1,x_n} \\ & \sigma^2_{x_2} & \cdots & \sigma_{x_2,x_n} \\ & & \ddots & \vdots \\ \text{symmetric} & & & \sigma^2_{x_n} \end{bmatrix} \tag{4.26}$$

The variance–covariance matrix is symmetric because of (4.21). The expectation operator E is applied to each matrix element. In what follows, the variance–covariance matrix is simply called the covariance matrix for the sake of brevity. The correlations are computed according to equation (4.22) and can be arranged in the same order. Thus, the correlation matrix is

$$C = \begin{bmatrix} 1 & \rho_{x_1,x_2} & \cdots & \rho_{x_1,x_n} \\ & 1 & \cdots & \rho_{x_2,x_n} \\ & & \ddots & \vdots \\ \text{symmetric} & & & 1 \end{bmatrix} \tag{4.27}$$

The correlation matrix is symmetric, the diagonal elements equal 1, and the off-diagonal elements are between -1 and $+1$.

Propagation   Usually we are more interested in a linear function of the random variables than in the random variables themselves. Typical examples are the adjusted coordinates used to compute distances and angles. From the definition of the mean (4.11), it follows that for a constant c

$$E(c) = c \int_{-\infty}^{\infty} f(x) \, dx = c \tag{4.28}$$

and

$$E(c\tilde{x}) = cE(\tilde{x}) \tag{4.29}$$

The expected value (mean) of a constant equals the constant. Because the mean is a constant, it follows that

$$E[E(\tilde{x})] = \mu_x \tag{4.30}$$

Relations (4.28) and (4.29) also hold for multivariate density functions, as can be seen from (4.18). Let $\tilde{y} = \tilde{x}_1 + \tilde{x}_2$ be a linear function of random variables, then

$$
\begin{aligned}
E(\tilde{x}_1 + \tilde{x}_2) &= \int_{-\infty}^{\infty} \int_{-\infty}^{\infty} (x_1 + x_2)\, f(x_1, x_1)\, dx_1\, dx_2 \\
&= \int_{-\infty}^{\infty} \int_{-\infty}^{\infty} x_1 f(x_1, x_2)\, dx_1\, dx_2 + \int_{-\infty}^{\infty} \int_{-\infty}^{\infty} x_2 f(x_1, x_2)\, dx_1\, dx_2 \\
&= E(\tilde{x}_1) + E(\tilde{x}_2)
\end{aligned}
\tag{4.31}
$$

Thus, the expected value of the sum of two random variables equals the sum of the individual expected values. By combining (4.28) and (4.31), we can compute the expected value of a general linear function of random variables. Thus, if the elements of the $(n \times u)$ matrix A and the $(n \times 1)$ vector A_0 are constants and

$$
\tilde{Y} = A_0 + A\tilde{X}
\tag{4.32}
$$

then the expected value is

$$
E(\tilde{Y}) = A_0 + AE(\tilde{X})
\tag{4.33}
$$

This is the law for propagating the mean. The law of variance–covariance propagation is as follows:

$$
\begin{aligned}
\Sigma_Y &\equiv E[(\tilde{Y} - M_Y)(\tilde{Y} - M_Y)^{\mathrm{T}}] \\
&= E\{[\tilde{Y} - E(\tilde{Y})][\tilde{Y} - E(\tilde{Y})]^{T}\} \\
&= E\{[\tilde{Y} - A_0 - AE(\tilde{X})][\tilde{Y} - A_0 - AE(\tilde{X})]^{\mathrm{T}}\} \\
&= E\{[A\tilde{X} - AE(\tilde{X})][A\tilde{X} - AE(\tilde{X})]^{\mathrm{T}}\} \\
&= AE\{[\tilde{X} - E(\tilde{X})][\tilde{X} - E(\tilde{X})]^{\mathrm{T}}\}\, A^{\mathrm{T}} \\
&= A\Sigma_X A^{\mathrm{T}}
\end{aligned}
\tag{4.34}
$$

The first line in expression (4.34) is the general expression for the variance–covariance matrix of the random variable $\tilde{Y}$ according to definition (4.26); M_Y is the expected value of $\tilde{Y}$. The third line follows by substituting (4.33) for the expected value of $\tilde{Y}$. Equation (4.32) has been substituted in the third line for $\tilde{Y}$, and, finally, the expected value operator and the matrix operator A have been interchanged. Thus, the variance–covariance matrix of the random variable $\tilde{Y}$ is obtained by pre- and postmultiplying the variance–covariance matrix of the original random variable $\tilde{X}$ by the coefficient matrix A. The constant term A_0 cancels. This is the law of variance–covariance propagation for linear functions of random variables. The covariance matrix Σ_Y is a full matrix in general. Only for very special cases is this matrix of diagonal form. The same is true for the covariance matrix Σ_X.

Finally, the expression for the covariance matrix (4.26) is rewritten as

$$\begin{aligned} \Sigma_X &= E[(\tilde{X} - M_X)(\tilde{X} - M_X)^T] \\ &= E[\tilde{X}\tilde{X}^T - \tilde{X}M_X^T - M_X\tilde{X}^T + M_XM_X^T] \qquad (4.35) \\ &= E[\tilde{X}\tilde{X}^T - M_XM_X^T] \end{aligned}$$

4.4 MIXED ADJUSTMENT MODEL

To simplify the notation, the tilde will not be used in this section to identify random variables. Observations or functions of observations are always random variables. A key role in least-squares adjustment is played by the product V^TPV, where V denotes the vector of residuals and P is the weight matrix; thus the product is a 1×1 matrix. A caret is used to identify quantities estimated by least squares, that is, those quantities that are a solution of a specific minimization. The caret quantities are random variables because all of them are functions of observations. As a final step in simplifying the notation, the caret is used only in connection with the parameters X. Throughout this chapter and whenever adjustments are involved, the so-called matrix notation is used; that is, even one-dimensional matrices such as X and V are symbolized by capital letters.

In the mixed adjustment model the observations and the parameters are implicitly related. If L_a denotes the vector of n adjusted observations and X_a denotes u adjusted parameters (unknowns), the mathematical model is given by

$$F(L_a, X_a) = 0 \qquad (4.36)$$

The total number of equations in (4.36) is denoted by r. The stochastic model is

$$P = \sigma_0^2 \Sigma_{L_b}^{-1} \qquad (4.37)$$

where P denotes the $(n \times n)$ weight matrix, and Σ_{L_b} denotes the covariance matrix of the observations. The objective is to estimate the parameters. It should be noted that the observations are the stochastic (random) variables, whereas the parameters are deterministic quantities. The parameters exist, but their values are unknown. The estimated parameters, however, are functions of the observations and thus are random variables.

4.4.1 Linearization

Let X_0 denote a vector of known approximate values of the parameters, then the parameter corrections X are

$$X = X_a - X_0 \tag{4.38}$$

If L_b denotes the vector of observations, then the residuals are defined by

$$V = L_a - L_b \tag{4.39}$$

With (4.38) and (4.39) the mathematical model can be written as

$$F(L_b + V, X_0 + X) = 0 \tag{4.40}$$

The nonlinear mathematical model is linearized around the known point of expansion (L_b, X_0), giving

$$_rB_n \, _nV_1 + _rA_u \, _uX_1 + _rW_1 = 0 \tag{4.41}$$

where

$$_rB_n = \left. \frac{\partial F}{\partial L} \right|_{X_0, L_b} \tag{4.42}$$

$$_rA_u = \left. \frac{\partial F}{\partial X} \right|_{X_0, L_b} \tag{4.43}$$

$$_rW_1 = F(L_b, X_0) \tag{4.44}$$

See Appendix B for linearization of multivariable functions. The coefficient matrices must be evaluated for the observations and the approximate parameters. The symbol W denotes the value of the mathematical function (4.36) evaluated at the point of expansion. The better the approximate values X_0, the smaller are the parameter corrections X.

4.4.2 Minimization and Solution

The least-squares estimate $\tilde{X}$ is based on the minimization of the function V^TPV. A solution is obtained by introducing a vector of Lagrange multipliers, K, and minimizing the function

$$\phi(V, K, X) = V^TPV - 2K^T(BV + AX + W) \tag{4.45}$$

Equation (4.45) is a function of three variables, namely V, K, and X. A necessary condition for the minimum is that the partial derivatives must be zero.

It can readily be shown that this condition is also sufficient. Differentiating (4.45) and using the rules of Appendix A gives

$$\frac{1}{2}\left(\frac{\partial \phi}{\partial V}\right)^{\mathrm{T}} = PV - B^{\mathrm{T}}K = 0 \tag{4.46}$$

$$\frac{1}{2}\left(\frac{\partial \phi}{\partial K}\right)^{\mathrm{T}} = BV + A\hat{X} + W = 0 \tag{4.47}$$

$$\frac{1}{2}\left(\frac{\partial \phi}{\partial X}\right)^{\mathrm{T}} = -A^{\mathrm{T}}K = 0 \tag{4.48}$$

To clarify the notation once again, the caret could be used for V and K in (4.46) to (4.48), because their specific values (estimates) result from the minimization of $V^{\mathrm{T}}PV$. However, as stated, the caret is used only for the estimated parameters X. The solution of (4.46) to (4.48) starts with the recognition that P is a square matrix and can be inverted. Thus, the expression for the residuals follows from (4.46):

$$V = P^{-1}B^{\mathrm{T}}K \tag{4.49}$$

Substituting (4.49) into (4.47), we obtain the solution for the Lagrange multiplier:

$$K = -M^{-1}(A\hat{X} + W) \tag{4.50}$$

with

$$_rM_r = {}_rB_u{}_uP_u^{-1}{}_uB_r^{\mathrm{T}} \tag{4.51}$$

Finally, the estimate $\hat{X}$ follows from (4.48) and (4.50)

$$\hat{X} = -(A^{\mathrm{T}}M^{-1}A)^{-1} A^{\mathrm{T}}M^{-1}W \tag{4.52}$$

The estimates $\hat{X}$ and V are independent of the a priori variance of unit weight. In actual computations we compute the parameters $\hat{X}$ first [equation (4.52)], then the Lagrange multipliers K [equation (4.50)], and then the residuals V [equation (4.49)]. The adjusted parameters and adjusted observations follow from (4.38) and (4.39).

4.4.3 Cofactor Matrices

Equation (4.44) shows that W is a random variable because it is a function of the observation L_b. With (4.37) and the law of variance–covariance propagation (4.34), the cofactor matrix Q_W becomes

$$Q_W = BP^{-1}B^{\mathrm{T}} = M \tag{4.53}$$

From (4.53) and (4.52) it follows that

$$Q_X = (A^{\mathrm{T}}M^{-1}A)^{-1} \tag{4.54}$$

Because

$$\begin{aligned}
V &= -P^{-1}B^{\mathrm{T}}M^{-1}(A\hat{X} + W) \\
&= -P^{-1}B^{\mathrm{T}}M^{-1}[-A(A^{\mathrm{T}}M^{-1}A)^{-1}A^{\mathrm{T}}M^{-1}W + W] \\
&= [P^{-1}B^{\mathrm{T}}M^{-1}A(A^{\mathrm{T}}M^{-1}A)^{-1}A^{\mathrm{T}}M^{-1} - P^{-1}B^{\mathrm{T}}M^{-1}]W
\end{aligned} \tag{4.55}$$

it follows from the law of variance propagation and (4.53) that

$$Q_V = P^{-1}B^{\mathrm{T}}M^{-1}[M - A(A^{\mathrm{T}}M^{-1}A)^{-1}A^{\mathrm{T}}]M^{-1}BP^{-1} \tag{4.56}$$

The adjusted observations are

$$\begin{aligned}
L_a &= L_b + V \\
&= L_b + [P^{-1}B^{\mathrm{T}}M^{-1}A(A^{\mathrm{T}}M^{-1}A)^{-1}A^{\mathrm{T}}M^{-1} - P^{-1}B^{\mathrm{T}}M^{-1}]W
\end{aligned} \tag{4.57}$$

Because

$$\frac{\partial L_a}{\partial L_b} = I + P^{-1}B^{\mathrm{T}}M^{-1}A(A^{\mathrm{T}}M^{-1}A)^{-1}A^{\mathrm{T}}M^{-1}B - P^{-1}B^{\mathrm{T}}M^{-1}B \tag{4.58}$$

it follows that

$$Q_{L_a} = P^{-1} - Q_V \tag{4.59}$$

4.4.4 A Posteriori Variance of Unit Weight

The minimum of the quadratic form $V^{\mathrm{T}}PV$ follows from (4.48) through (4.50):

$$\begin{aligned}
V^{\mathrm{T}}PV &= -K^{\mathrm{T}}W \\
&= (A\hat{X} + W)^{\mathrm{T}} M^{-1}W \\
&= \hat{X}^{\mathrm{T}}A^{\mathrm{T}}M^{-1}W + W^{\mathrm{T}}M^{-1}W \\
&= -W^{\mathrm{T}}M^{-1}A(A^{\mathrm{T}}M^{-1}A)^{-1}A^{\mathrm{T}}M^{-1}W + W^{\mathrm{T}}M^{-1}W \\
&= W^{\mathrm{T}}[M^{-1} - M^{-1}A(A^{\mathrm{T}}M^{-1}A)^{-1}A^{\mathrm{T}}M^{-1}]W
\end{aligned} \tag{4.60}$$

This is a random variable with an expected value

$$
\begin{aligned}
E(V^{T}PV) &= E(\text{Tr } V^{T}PV) \\
&= E\{\text{Tr } [W^{T}(M^{-1} - M^{-1}A(A^{T}M^{-1}A)^{-1}A^{T}M^{-1})W]\} \quad (4.61) \\
&= E\{\text{Tr } [(M^{-1} - M^{-1}A(A^{T}M^{-1}A)^{-1}A^{T}M^{-1})WW^{T}]\} \\
&= \text{Tr } \{[M^{-1} - M^{-1}A(A^{T}M^{-1}A)^{-1}A^{T}M^{-1}]E(WW^{T})\}
\end{aligned}
$$

The trace (Tr) of a matrix equals the sum of its diagonal elements. In the first part of (4.61) the property that the trace of a (1×1) matrix equals the matrix element itself is used. Next, the matrix products are switched, leaving the trace invariant. In the last part of the equation, the expected operator and the trace are switched. The expected value $E(WW^{T})$ can readily be computed. The expected value of the residuals is zero by definition, because the residuals represent random errors for which a positive and negative error of the same magnitude occurs with the same probability; that is,

$$
E(V) = 0 \tag{4.62}
$$

It follows from (4.41) that

$$
E(W) = -AX \tag{4.63}
$$

Note that X in (4.63) or (4.41) is not a random variable. In these expressions X simply denotes the vector of unknown parameters that have fixed values even though they are not known. $\hat{X}$ is a random variable, because it is a function of the observations. By using (4.35) for the definition of the covariance matrix and using (4.53) and (4.63), it follows that

$$
\begin{aligned}
E(WW^{T}) &= \Sigma_{w} + E(W)E(W)^{T} \\
&= \sigma_{0}^{2} M + AXX^{T}A^{T}
\end{aligned} \tag{4.64}
$$

Substituting (4.64) into (4.61) yields the expected value for $V^{T}PV$:

$$
\begin{aligned}
E(V^{T}PV) &= \text{Tr } [_{r}I_{r} - M^{-1}A(A^{T}M^{-1}A)^{-1}A^{T}]\sigma_{0}^{2} \\
&= (r - u)\sigma_{0}^{2}
\end{aligned} \tag{4.65}
$$

The difference $r - u$ is called the degree of freedom and equals the number of redundant observations in the model (4.36). Strictly, the degree of freedom is $r - R(A)$, which can be readily verified by taking into account that the matrix expression in (4.65) is idempotent. The symbol $R(A)$ denotes the rank of the matrix A. The a posteriori variance of unit weight is computed from

$$
\hat{\sigma}_{0}^{2} = \frac{V^{T}PV}{r - u} \tag{4.66}
$$

Using (4.65), we see that

$$E(\hat{\sigma}_0^2) = \sigma_0^2 \qquad (4.67)$$

The expected value of the a posteriori variance of unit weight equals the a priori variance of unit weight. Finally, the estimated covariance matrices are

$$\Sigma_X = \hat{\sigma}_0^2 \, Q_X \qquad (4.68)$$

$$\Sigma_V = \hat{\sigma}_0^2 \, Q_V \qquad (4.69)$$

$$\Sigma_{L_a} = \hat{\sigma}_0^2 \, Q_{L_a} \qquad (4.70)$$

With equation (4.59) it follows that

$$\Sigma_{L_a} = \Sigma_{L_b} - \Sigma_V \qquad (4.71)$$

Because the diagonal elements of all three covariance matrices in (4.71) are positive, it follows that the variances of the adjusted observations are smaller than those of the original observations. The difference is a function of the "geometry" of the adjustment, as implied by the covariance matrix Σ_V.

4.4.5 Iterations

Because the mathematical model is generally nonlinear, the solution must be iterated. Recall that (4.36) is true only for (L_a, X_a). Since neither of these quantities is known before the adjustment, the initial point of expansion is chosen as (L_b, X_0). For the ith iteration, the linearized model can be written (Pope 1972; Leick 1980b)

$$B_{X_{0i}, L_{0i}} \overline{V}_i + A_{X_{0i}, L_{0i}} X_i + W_{X_{0i}, L_{0i}} = 0 \qquad (4.72)$$

where the point of expansion (L_{0i}, X_{0i}) is taken from the previous solution such that

$$\overline{V}_i = L_{ai} - L_{0i} \qquad (4.73)$$

$$X_i = X_{ai} - X_{0i} \qquad (4.74)$$

The symbols L_{ai} and X_{ai} denote the adjusted observations and the adjusted parameters for the current (ith) solution. They are computed from (4.73) and (4.74) once the least-squares solution of (4.72) has been obtained. The iteration starts with $L_{01} = L_b$ and $X_{01} = X_0$. If the adjustment converges properly, then both $\overline{V}_i$ and X_i converge to zero, or, stated differently, L_{ai} and X_{ai} converge toward L_a and X_a, respectively. The quantity $\overline{V}_i$ does not equal the residuals. The residuals express the random difference between the adjusted observations and the original observations according to equation (4.39). Defining

$$V_i \equiv L_{ai} - L_b \qquad (4.75)$$

it follows from (4.73) that

$$\overline{V}_i = V_i + (L_b - L_{0i}) \qquad (4.76)$$

Substituting this expression into (4.72) gives

$$B_{X_{0i}, L_{0i}} V_i + A_{X_{0i}, L_{0i}} X_i + W_{X_{0i}, L_{0i}} + B_{X_{0i}, L_{0i}} (L_b - L_{0i}) = 0 \qquad (4.77)$$

Formulation (4.77) assures that the vector V_i converges toward the vector of residuals V. The additional term in (4.77) will be zero for the first iteration when $L_{0i} = L_b$. The iteration has converged if

$$|V^T P V_i - V^T P V_{i-1}| < \epsilon \qquad (4.78)$$

where ϵ is a small positive number.

4.5 OBSERVATION EQUATION AND CONDITION EQUATION MODELS

Often there is an explicit relationship between the observations and the parameters, such as

$$L_a = F(X_a) \qquad (4.79)$$

This is called the *observation equation model*. By comparing both nonlinear expressions (4.36) and (4.79), we see that the observation equation model follows from the mixed model upon the specification

$$B = \frac{\partial [F(X_a) - L_a]}{\partial L_a} = -I \qquad (4.80)$$

It is customary to use the symbol L instead of W when dealing with the observation equation model. Thus,

$$L = W = F(X_0) - L_b = L_0 - L_b \qquad (4.81)$$

The symbol L_0 equals the value of the observations as computed from the approximate parameters X_0. The point of expansion for the linearization is X_0; the observation vector is not involved in the iteration because of the explicit form of (4.79). The difference L is often called the misclosure. The linearized equations

$$_nV_1 = {_n}A_{uu}X_1 + {_n}L_1 \qquad (4.82)$$

are called the *observation equations*. There is one equation for each observation in (4.82).

If the observations are related by a nonlinear function without use of the additional parameters, we speak of the *condition equation model*. It is written as

$$F(L_a) = 0 \qquad (4.83)$$

By comparing this with the mixed model (4.36), we see that the condition equation model follows upon the specification

$$A = \frac{\partial[F(L_a)]}{\partial X} = 0 \qquad (4.84)$$

The linear equations

$$_rB_n\, {_n}V_1 + {_r}W_1 = 0 \qquad (4.85)$$

are called the *condition equations*. The iteration for the model (4.85) is analogous to a mixed model with the added simplification that there is no A matrix and no parameter vector X.

The significance of these three models (observation, condition, and mixed) is that a specific adjustment problem can usually be formulated more easily in one of the models. Clearly, that model should be chosen. There are situations in which it is equally easy to use any of the models. A typical example is the adjustment of a level network. Most of the time, however, the observation equation model is preferred, because the simple rule "one observation, one equation" is suitable for setting up general software. Table 4.1 lists the important expressions for all three models.

4.6 SEQUENTIAL SOLUTION

Assume that observations are made in two groups, with the second group consisting of one or several observations. Both groups have a common set of parameters. The two mixed adjustment models can be written as

$$F_1(L_{1a}, X_a) = 0 \qquad (4.86)$$

$$F_2(L_{2a}, X_a) = 0 \qquad (4.87)$$

TABLE 4.1 Three Standard Adjustment Models

	Mixed model	Observation model	Condition model
Non-linear model	$F(L_a, X_a) = 0$	$L_a = F(X_a)$	$F(L_a) = 0$
Specifications		$B = -I, \quad L = W,$ $r = n$	$A = 0$
Linear model	$BV + AX + W = 0$	$V = AX + L$	$BV + W = 0$
Normal equation elements	$M = BP^{-1}B^T$ $N = A^T M^{-1} A$ $U = A^T M^{-1} W$	$M = P^{-1}$ $N = A^T PA$ $U = A^T PL$	$M = BP^{-1}B^T$
Normal equations	$N\hat{X} = -U$	$N\hat{X} = -U$	
Minimum $V^T PV$	$V^T PV = -K^T W$ $= U^T \hat{X} + W^T M^{-1} W$ $= -U^T N^{-1} U + W^T M^{-1} W$	$V^T PV = U^T \hat{X} + L^T PL$ $= -U^T N^{-1} U + L^T PL$	$V^T PV = -K^T W$ $= W^T M^{-1} W$
Estimated parameters	$\hat{X} = -N^{-1}U$	$\hat{X} = -N^{-1}U$	
Estimated residuals	$V = P^{-1}B^T K$	$V = A\hat{X} + L$	$V = P^{-1}B^T K$
Estimated variance of unit weight	$\hat{\sigma}_0^2 = \dfrac{V^T PV}{r - u}$	$\hat{\sigma}_0^2 = \dfrac{V^T PV}{n - u}$	$\hat{\sigma}_0^2 = \dfrac{V^T PV}{r}$
Estimated parameter cofactor matrix	$Q_X = N^{-1}$	$Q_X = N^{-1}$	
Estimated residual cofactor matrix	$Q_V = P^{-1}B^T M^{-1}$ $(M - AN^{-1}A^T)M^{-1}BP^{-1}$	$Q_V = (P^{-1} - AN^{-1}A^T)$	$Q_V = P^{-1}B^T M^{-1} BP^{-1}$
Adjusted observation cofactor matrix	$Q_{L_a} = Q_{L_b} - Q_V$	$Q_{L_a} = Q_{L_b} - Q_V$	$Q_{L_a} = Q_{L_b} - Q_V$

Both sets of observations should be uncorrelated, and the a priori variance of unit weight should be the same for both groups; that is,

$$P = \begin{bmatrix} P_1 & 0 \\ 0 & P_2 \end{bmatrix} = \sigma_0^2 \begin{bmatrix} \Sigma_1^{-1} & 0 \\ 0 & \Sigma_2^{-1} \end{bmatrix} \tag{4.88}$$

The number of observations in L_{1a} and L_{2a} are n_1 and n_2 respectively; and r_1 and r_2 are the number of equations in the models F_1 and F_2, respectively. The linearization of (4.86) and (4.87) yields

$$B_1 V_1 + A_1 X + W_1 = 0 \tag{4.89}$$

$$B_2 V_2 + A_2 X + W_2 = 0 \tag{4.90}$$

where

$$B_1 = \left.\frac{\partial F_1}{\partial L_1}\right|_{L_{1b}, X_0} \quad \text{and} \quad B_2 = \left.\frac{\partial F_2}{\partial L_2}\right|_{L_{2b}, X_0} \tag{4.91a}$$

$$A_1 = \left.\frac{\partial F_1}{\partial X}\right|_{L_{1b}, X_0} \quad \text{and} \quad A_2 = \left.\frac{\partial F_2}{\partial X}\right|_{L_{2b}, X_0} \tag{4.91b}$$

$$W_1 = F_1(L_{1b}, X_0) \quad \text{and} \quad W_2 = F_2(L_{2b}, X_0) \tag{4.91c}$$

The function to be minimized is

$$\phi(V_1, V_2, K_1, K_2, X) = V_1^T P_1 V_1 + V_2^T P_2 V_2 - 2K_1^T(B_1 V_1 + A_1 X + W_1)$$
$$- 2K_2^T(B_2 V_2 + A_2 X + W_2) \tag{4.92}$$

The partial derivatives of (4.92) are

$$\frac{1}{2}\left(\frac{\partial \phi}{\partial V_1}\right)^T = P_1 V_1 - B_1^T K_1 = 0 \tag{4.93}$$

$$\frac{1}{2}\left(\frac{\partial \phi}{\partial V_2}\right)^T = P_2 V_2 - B_2^T K_2 = 0 \tag{4.94}$$

$$\frac{1}{2}\left(\frac{\partial \phi}{\partial X}\right)^T = -A_1^T K_1 - A_2^T K_2 = 0 \tag{4.95}$$

$$\frac{1}{2}\left(\frac{\partial \phi}{\partial K_1}\right)^T = B_1 V_1 + A_1 \hat{X} + W_1 = 0 \tag{4.96}$$

$$\frac{1}{2}\left(\frac{\partial \phi}{\partial K_2}\right)^T = B_2 V_2 + A_2 \hat{X} + W_2 = 0 \tag{4.97}$$

Equations (4.93) and (4.94) give the residuals

$$V_1 = P_1^{-1} B_1^T K_1 \tag{4.98}$$

$$V_2 = P_2^{-1} B_2^T K_2 \tag{4.99}$$

Combining (4.98) and (4.96) yields

$$M_1 K_1 + A_1 \hat{X} + W_1 = 0 \tag{4.100}$$

where

$$M_1 = B_1 P_1^{-1} B_1^T \tag{4.101}$$

is an $(r_1 \times r_1)$ symmetric matrix. The Lagrange multiplier becomes

$$K_1 = -M_1^{-1}A_1\hat{X} - M_1^{-1}W_1 \tag{4.102}$$

Equations (4.95) and (4.97) become, after combination with (4.102) and (4.99),

$$A_1^TM_1^{-1}A_1\hat{X} + A_1^TM_1^{-1}W_1 - A_2^TK_2 = 0 \tag{4.103}$$

$$B_2P_2^{-1}B_2^TK_2 + A_2\hat{X} + W_2 = 0 \tag{4.104}$$

By using

$$M_2 = B_2P_2^{-1}B_2^T \tag{4.105}$$

we can write both equations (4.103) and (4.104) in matrix form:

$$\begin{bmatrix} A_1^TM_1^{-1}A_1 & A_2^T \\ A_2 & -M_2 \end{bmatrix} \begin{bmatrix} \hat{X} \\ -K_2 \end{bmatrix} = \begin{bmatrix} -A_1^TM_1^{-1}W_1 \\ -W_2 \end{bmatrix} \tag{4.106}$$

Equation (4.106) shows how the normal matrix of the first group must be augmented in order to find the solution of both groups. It is, of course, possible to invert the whole matrix in one step to give the solution for $\hat{X}$ and K_2. Alternatively, one can apply the matrix partitioning techniques given in (A.41). Thus,

$$\hat{X} = -Q_{11}A_1^TM_1^{-1}W_1 - Q_{12}W_2 \tag{4.107}$$

$$K_2 = Q_{21}A_1^TM_1^{-1}W_1 - Q_{22}W_2 \tag{4.108}$$

with

$$Q_{11} = (A_1^TM_1^{-1}A_1)^{-1}$$
$$- (A_1^TM_1^{-1}A_1)^{-1} A_2^T [M_2 + A_2(A_1^TM_1^{-1}A_1)^{-1} A_2^T]^{-1} A_2(A_1^TM_1^{-1}A_1)^{-1} \tag{4.109}$$

$$Q_{12} = (A_1^TM_1^{-1}A_1)^{-1} A_2^T[M_2 + A_2(A_1^TM_1^{-1}A_1)^{-1} A_2^T]^{-1} \tag{4.110}$$

$$Q_{21} = Q_{12}^T \tag{4.111}$$

$$Q_{22} = -[M_2 + A_2(A_1^TM_1^{-1}A_1)^{-1}A_2^T]^{-1} \tag{4.112}$$

Substituting the Q_{11} and Q_{12} into (4.107) give the sequential solution for the parameters. By denoting the solution of the first group by an asterisk and the contribution of the second group by Δ, that is

$$\hat{X} = X^* + \Delta X \tag{4.113}$$

we see from (4.107) that

$$X^* = -(A_1^T M_1^{-1} A_1)^{-1} A_1^T M_1^{-1} W_1 \qquad (4.114)$$

and

$$\Delta X = -(A_1^T M_1^{-1} A_1)^{-1} A_2^T [M_2 + A_2 (A_1^T M_1^{-1} A_1)^{-1} A_2^T]^{-1} (A_2 X^* + W_2) \quad (4.115)$$

Similarly,

$$K_2 = -[M_2 + A_2 (A_1^T M_1^{-1} A_1)^{-1} A_2^T]^{-1} (A_2 X^* + W_2) \qquad (4.116)$$

An alternative form for the solution of equation (4.106) is obtained by making use of the second expression in (A.41). It follows readily that

$$\begin{aligned}
\hat{X} &= -(A_1^T M_1^{-1} A_1 + A_2^T M_2^{-1} A_2)^{-1} (A_1^T M_1^{-1} W_1 + A_2^T M_2^{-1} W_2) \\
&= -(A_1^T M_1^{-1} A_1 + A_2^T M_2^{-1} A_2)^{-1} (-A_1^T M_1^{-1} A_1 X^* + A_2^T M_2^{-1} W_2) \qquad (4.117) \\
&= X^* - (A_1^T M_1^{-1} A_1 + A_2^T M_2^{-1} A_2)^{-1} (A_2^T M_2^{-1} A_2 X^* + A_2^T M_2^{-1} W_2)
\end{aligned}$$

The procedure implied in the first line of (4.117) is called the *method of adding normal equations*. The contributions of the new observations are simply added appropriately.

Equating both Q_{11} expressions in (A.41) gives two expressions for the cofactor matrix of the parameters:

$$\begin{aligned}
Q_X &= Q_{11} \\
&= (A_1^T M_1^{-1} A_1 + A_2^T M_2^{-1} A_2)^{-1} \\
&= (A_1^T M_1^{-1} A_1)^{-1} \\
&\quad - (A_1^T M_1^{-1} A_1)^{-1} A_2^T [M_2 + A_2 (A_1^T M_1^{-1} A_1)^{-1} A_2^T]^{-1} A_2 (A_1^T M_1^{-1} A_1)^{-1} \\
&= Q_{X^*} - Q_{X^*} A_2^T [M_2 + A_2 Q_{X^*} A_2^T]^{-1} A_2 Q_{X^*} \\
&= Q_{X^*} + \Delta Q_X \qquad (4.118)
\end{aligned}$$

where Q_{X^*} is the cofactor matrix of the first group of observations only, and

$$\Delta Q_X = -Q_{X^*} A_2^T (M_2 + A_2 Q_{X^*} A_2^T)^{-1} A_2 Q_{X^*} \qquad (4.119)$$

is the contribution of the second group of observations to the cofactor matrix of the first group. The change ΔQ_X can be computed without having the actual observations of the second group. This is of great use in simulation studies.

The computation of V^TPV proceeds as usual

$$V^TPV = V_1^TP_1V_1 + V_2^TP_2V_2 \tag{4.120}$$

By using (4.95) through (4.99), we can write this expression as

$$V^TPV = -K_1^TW_1 - K_2^TW_2 \tag{4.121}$$

By substituting the solution (4.102) for K_1, and using (4.113), we obtain

$$V^TPV = (X^*)^TA_1^TM_1^{-1}W_1 + \Delta X^TA_1^TM_1^{-1}W_1 + W_1^TM_1^{-1}W_1 - K_2^TW_2 \tag{4.122}$$
$$= V^TPV^* + \Delta X^TA_1^TM_1^{-1}W_1 - K_2^TW_2$$

The final form for V^TPV is obtained by substituting (4.115) and (4.116) for ΔX and K_2:

$$V^TPV = V^TPV^* + \Delta V^TPV$$
$$= V^TPV^*$$
$$+ (A_2X^* + W_2)^T[M_2 + A_2(A_1^TM_1^{-1}A_1)^{-1}A_2^T]^{-1} (A_2X^* + W_2) \tag{4.123}$$

The a posteriori variance of unit weight is computed in the usual way:

$$\hat{\sigma}_0^2 = \frac{V^TPV}{r_1 + r_2 - u} \tag{4.124}$$

where r_1 and r_2 are the number of equations in (4.86) and (4.87), respectively. The letter u denotes, again, the number of parameters.

The second set of observations contributes to all residuals. From (4.98), (4.102), and (4.113) we obtain

$$V_1 = V_1^* + \Delta V_1$$
$$= -P_1^{-1}B_1^TM_1^{-1}[A_1(X^* + \Delta X) + W_1] \tag{4.125}$$
$$= -P_1^{-1}B_1^TM_1^{-1}(A_1X^* + W_1) - P_1^{-1}B_1^TM_1^{-1}A_1\Delta X$$

The expression for V_2 follows from equations (4.99) and (4.116):

$$V_2 = -P_2^{-1}B_2^TT(A_2X^* + W_2) \tag{4.126}$$

where

$$T = (M_2 + A_2N_1^{-1}A_2^T)^{-1} \tag{4.127}$$

and

$$N_1 = A_1^T M_1^{-1} A_1 \tag{4.128}$$

The cofactor matrices for the residuals follow, again, from the law of variance–covariance propagation. The residuals V_1 are a function of W_1 and W_2, according to (4.125). Substituting the expressions for X^* and ΔX, we obtain, from (4.125)

$$\frac{\partial V_1}{\partial W_1} = -P_1^{-1} B_1^T M_1^{-1} (I - A_1 N_1^{-1} A_1^T M_1^{-1} + A_1 N_1^{-1} A_2^T T A_2 N_1^{-1} A_1^T M_1^{-1}) \tag{4.129}$$

$$\frac{\partial V_1}{\partial W_2} = -P_1^{-1} B_1^T M_1^{-1} A_1 N_1^{-1} A_2^T T \tag{4.130}$$

Applying the law of variance–covariance propagation to W_1 and W_2 of (4.91) and knowing that the observations are uncorrelated gives

$$Q_{W_1, W_2} = \begin{bmatrix} M_1 & 0 \\ 0 & M_2 \end{bmatrix} \tag{4.131}$$

By using the partial derivatives (4.129) and (4.130), expression (4.131), and the law of variance–covariance propagation, we obtain, after some algebraic computations, the cofactor matrices:

$$Q_{V_1} = Q_{V_1^*} + \Delta Q_{V_1} \tag{4.132}$$

where

$$Q_{V_1^*} = P_1^{-1} B_1^T M_1^{-1} (P_1^{-1} B_1^T)^T - (P_1^{-1} B_1^T M_1^{-1} A_1) N_1^{-1} (P_1^{-1} B_1^T M_1^{-1} A_1)^T \tag{4.133}$$

$$\Delta Q_{V_1} = (P_1^{-1} B_1^T M_1^{-1} A_1 N_1^{-1} A_2^T) T (P_1^{-1} B_1^T M_1^{-1} A_1 N_1^{-1} A_2^T)^T \tag{4.134}$$

The partial derivatives of V_2 with respect to W_1 and W_2 follow from (4.126):

$$\frac{\partial V_2}{\partial W_1} = P_2^{-1} B_2^T T A_2 N_1^{-1} A_1^T M_1^{-1} \tag{4.135}$$

$$\frac{\partial V_2}{\partial W_2} = -P_2^{-1} B_2^T T \tag{4.136}$$

By using, again, the law of variance–covariance propagation and (4.131), we obtain the cofactor for V_2:

$$Q_{V_2} = P_2^{-1} B_2^T T B_2 P_2^{-1} \tag{4.137}$$

The estimated variance–covariance matrix is

$$\hat{\Sigma}_{V_2} = \hat{\sigma}_0^2 Q_{V_2} \tag{4.138}$$

The variance–covariance matrix of the adjusted observations is, as usual,

$$\Sigma_{L_a} = \Sigma_{L_b} - \Sigma_V \tag{4.139}$$

As for iterations, one has to make sure that all groups are evaluated for the same approximate parameters. If the first system is iterated, the approximate coordinates for the last iteration must be used as expansion points for the second group. Because there are no observations common to both groups, the iteration with respect to the observations can be done individually for each group.

Occasionally, it is desirable to remove a set of observations from an existing adjustment. Consider again the uncorrelated case in which the set of observations to be removed is uncorrelated with the other observations. The solution is readily seen from (4.117), which shows how normal equations are added when a new set of observations is incorporated. When observations are removed, the corresponding part of the normal matrix and the right-hand term have to be subtracted. Equation (4.117) becomes

$$\begin{aligned}\hat{X} &= -(A_1^T M_1^{-1} A_1 - A_2^T M_2^{-1} A_2)^{-1}(A_1^T M_1^{-1} W_1 - A_2^T M_2^{-1} W_2) \\ &= -(A_1^T M_1^{-1} A_1 + A_2^T(-M_2^{-1})A_2)^{-1}[A_1^T M_1^{-1} W_1 + A_2^T(-M_2^{-1})W_2]\end{aligned} \tag{4.140}$$

One only has to assign a negative sign to the weight matrix of the group of observations that is being removed, because

$$-M_2 = B_2(-P_2^{-1})B_2^T \tag{4.141}$$

For the sequential solution (4.115) we note that the matrix

$$[-M_2 + A_2(A_1^T M_1^{-1} A_1)^{-1}A_2^T]$$

is still symmetric but is no longer positive-definite when observations are subtracted. The inversion cannot be done using Cholesky's algorithm. A more general inversion procedure must be used.

The sequential solution can be used in quite a general manner. One can add or remove any number of groups of observations. The groups can consist of a single observation. The most important formulas are summarized in what follows. The first group produces the solution according to (4.114), (4.118), and (4.122):

$$\hat{X}_1 = -Q_1 A_1^T M_1^{-1} W_1 \tag{4.142}$$

$$Q_1 = Q_{X_1} = (A_1^T M_1^{-1} A_1)^{-1} \tag{4.143}$$

$$V^T P V_1 = W_1^T M_1^{-1} W_1 + W_1^T M_1^{-1} A_1 \hat{X}_1 \tag{4.144}$$

Given the solution for the $(i - 1)$ groups, the parameters estimated from all i groups of observations are, according to (4.115),

$$\hat{X}_i = \hat{X}_{i-1} + \Delta \hat{X}_i$$

$$\Delta \hat{X}_i = -Q_{i-1} A_i^T (M_i + A_i Q_{i-1} A_i^T)^{-1} (A_i \hat{X}_{i-1} + W_i) \tag{4.145}$$

The quadratic form is obtained from equation (4.123):

$$V^T P V_i = V^T P V_{i-1} + \Delta V^T P V_i$$

$$\Delta V^T P V_i = (A_i \hat{X}_{i-1} + W_i)^T (M_i + A_i Q_{i-1} A_i^T)^{-1} (A_i \hat{X}_{i-1} + W_i) \tag{4.146}$$

The new cofactor matrix for the parameters is, according to (4.119),

$$Q_i = Q_{i-1} - Q_{i-1} A_i^T (M_i + A_i Q_{i-1} A_i^T)^{-1} A_i Q_{i-1} \tag{4.147}$$

The residuals of the first group follow from (4.125):

$$\hat{V}_1 = -P_1^{-1} B_1^T M_1^{-1} (A_1 \hat{X}_i + W_1) \tag{4.148}$$

General expressions can be derived for the residuals of the other groups and their respective cofactor matrices. Each new group of observations changes the residuals and the cofactor matrices of the previously included groups. If additional observations are to be added, continue the loop with (4.145), noting that the subscript i denotes the new observations.

Every sequential solution is equivalent to a one-step adjustment that contains the same observations (as the sequential solution). Preference of one or the other technique depends on the application and the quality of the available software. The sequential solution, for example, requires that the inverse of the normal matrix be available (Table 4.2). As explained in Appendix A, computing the inverse of the normal matrix requires much more computation than merely solving the system of normal equations, as is required in the one-step solution.

4.7 WEIGHTED PARAMETERS AND CONDITIONS BETWEEN PARAMETERS

The algorithms developed in the previous section can be used to incorporate exterior information about parameters. This includes weighted functions of parameters, weighted individual parameters, and conditions on parameters. The

TABLE 4.2   Sequential Adjustment Models

	Mixed model	Observation model
Non-linear model	$F_1(L_{1a}, X_a) = 0$ $F_2(L_{2a}, X_a) = 0$ $P = \begin{bmatrix} P_1 & 0 \\ 0 & P_2 \end{bmatrix}$	$L_{1a} = F_1(X_a)$ $L_{2a} = F_2(X_a)$ $P = \begin{bmatrix} P_1 & 0 \\ 0 & P_2 \end{bmatrix}$
Linear model	$B_1 V_1 + A_1 X + W_1 = 0$ $B_2 V_2 + A_2 X + W_2 = 0$	$V_1 = A_1 X + L_1$ $V_2 = A_2 X + L_2$
Normal equation elements	$M_1 = B_1 P_1^{-1} B_1^T \quad M_2 = B_2 P_2^{-1} B_2^T$ $N_1 = A_1^T M_1^{-1} A_1 \quad N_2 = A_2^T M_2^{-1} A_2$ $U_1 = A_1^T M_1^{-1} W_1 \quad U_2 = A_2^T M_2^{-1} W_2$	$M_1 = P_1^{-1} \quad M_2 = P_2^{-1}$ $N_1 = A_1^T P_1 A_1 \quad N_2 = A_2^T P_2 A_2$ $U_1 = A_1^T P_1 L_1 \quad U_2 = A_2^T P_2 L_2$
Normal equations	$\begin{bmatrix} N_1 & A_2^T \\ A_2 & -M_2 \end{bmatrix} \begin{bmatrix} \hat{X} \\ -K_2 \end{bmatrix} = \begin{bmatrix} -U_1 \\ -W_2 \end{bmatrix}$	$\begin{bmatrix} N_1 & A_2^T \\ A_2 & -P_2 \end{bmatrix} \begin{bmatrix} \hat{X} \\ -K_2 \end{bmatrix} = \begin{bmatrix} -U_1 \\ -L_2 \end{bmatrix}$
Minimum $V^T P V$	$V^T P V = V^T P V^* + \Delta V^T P V$ $V^T P V^* = -U_1^T N_1^{-1} U_1 + W_1^T M_1^{-1} W_1$ $\Delta V^T P V = (A_2 X^* + W_2)^T T (A_2 X^* + W_2)$	$V^T P V = V^T P V^* + \Delta V^T P V$ $V^T P V^* = -U_1^T N_1^{-1} U_1 + L_1^T P_1^{-1} L_1$ $\Delta V^T P V = (A_2 X^* + L_2)^T T (A_2 X^* + L_2)$
Estimated parameters	$\hat{X} = X^* + \Delta X$ $X^* = -N_1^{-1} U_1$ $T = \left(M_2 + A_2 N_1^{-1} A_2^T \right)^{-1}$ $\Delta X = -N_1^{-1} A_2^T T \left(A_2 X^* + W_2 \right)$	$\hat{X} = X^* + \Delta X$ $X^* = -N_1^{-1} U_1$ $T = \left(P_2^{-1} + A_2 N_1^{-1} A_2^T \right)^{-1}$ $\Delta X = -N_1^{-1} A_2^T T \left(A_2 X^* + L_2 \right)$
Estimated residuals	$V_1 = V_1^* + \Delta V_1$ $V_1^* = -P_1^{-1} B_1^T M_1^{-1} \left(A_1 X^* + W_1 \right)$ $\Delta V_1 = -P_1^{-1} B_1^T M_1^{-1} A_1 \Delta X$	$V_1 = V_1^* + \Delta V_1$ $V_1^* = A_1 X^* + L_1$ $\Delta V_1 = A_1 \Delta X$
Estimated variance of unit weight	$\hat{\sigma}_0^2 = \dfrac{V^T P V}{r_1 + r_2 - u}$	$\hat{\sigma}_0^2 = \dfrac{V^T P V}{n_1 + n_2 - u}$
Estimated parameter cofactor matrix	$Q_X = Q_{X^*} + \Delta Q$ $Q_{X^*} = N_1^{-1}$ $\Delta Q = -N_1^{-1} A_2^T T A_2 N_1^{-1}$	$Q_X = Q_{X^*} + \Delta Q$ $Q_{X^*} = N_1^{-1}$ $\Delta Q = -N_1^{-1} A_2^T T A_2 N_1^{-1}$
Estimated residual cofactor matrix	$Q_{V_1} = Q_{V_1^*} + \Delta Q_{V_1}$ $Q_{V_1^*} = \left(P_1^{-1} B_1^T \right) M_1^{-1} \left(P_1^{-1} B_1^T \right)^T -$ $\left(P_1^{-1} B_1^T M_1^{-1} A_1 \right) N_1^{-1} \left(P_1^{-1} B_1^T M_1^{-1} A_1 \right)^T$ $\Delta Q_{V_1} = \left(P_1^{-1} B_1^T M_1^{-1} A_1 N_1^{-1} A_2^T \right) T$ $\left(P_1^{-1} B_1^T M_1^{-1} A_1 N_1^{-1} A_2^T \right)^T$ $Q_{V_2} = P_2^{-1} B_2^T T B_2 P_2^{-1}$	$Q_{V_1} = Q_{V_1^*} + \Delta Q_{V_1}$ $Q_{V_1^*} = P_1^{-1} - A_1 N_1^{-1} A_1^T$ $\Delta Q_{V_1} = A_1 N_1^{-1} A_2^T T A_2^T N_1^{-1} A_1^T$ $Q_{V_2} = P_2^{-1} T P_2^{-1}$
Adjusted obs. cofactor matrix	$Q_{L_a} = Q_{L_b} - Q_V$	$Q_{L_a} = Q_{L_b} - Q_V$

objective is to incorporate new types of observations that directly refer to the parameters, to specify parameters in order to avoid singularity of the normal equations, or to incorporate the results of prior adjustments. Evaluating conditions between the parameters is the basis for hypothesis testing. These cases are obtained by specifying the coefficient matrices A and B of the mixed model. For example, the mixed model (4.86) and (4.87) can be specified as

$$F_1(L_a, X_a) = 0 \qquad (4.149)$$

$$L_{2a} = F_2(X_a) \qquad (4.150)$$

The linearized form is

$$B_1 V_1 + A_1 X + W_1 = 0 \qquad (4.151)$$

$$V_2 = A_2 X + L_2 \qquad (4.152)$$

The specifications are $B_2 = -I$ and $L_2 = W_2$. For the observation equation model we obtain

$$L_{1a} = F_1(X_a) \qquad (4.153)$$

$$L_{2a} = F_2(X_a) \qquad (4.154)$$

with the linearized form being

$$V_1 = A_1 X + L_1 \qquad (4.155)$$

$$V_2 = A_2 X + L_2 \qquad (4.156)$$

The stochastic model is given by the matrices P_1 and P_2. With proper choice of the elements of A_2 and P_2, it is possible to introduce a variety of relations about the parameters.

As a first case, consider nonlinear relations between parameters. The design matrix A_2 contains the partial derivatives, and L_{2b} contains the observed value of the function. This is the case of *weighted functions of parameters*. Examples are the area or volume of geometric figures as computed from coordinates, angles in geodetic networks, and differences between parameters (coordinates). Each function contributes one equation to (4.152) or (4.156). The respective expressions are identical with those given in Table 4.2 and require no further discussion.

As a second case, consider information about individual parameters. This is a special case of the general method discussed above. Each row of A_2 contains zeros with the exception of one position, which contains a one. The number of rows in the A_2 matrix corresponds to the number of weighted parameters. The expressions of Table 4.2 are still valid for this case. If information enters into the adjustment in this manner, one speaks of the *method of weighted pa-*

rameters. In the most general case all parameters are observed and weighted, giving

$$L_{2a} = X_a \tag{4.157}$$

$$L_{2b} = X_b \tag{4.158}$$

$$A_2 = I \tag{4.159}$$

$$L_2 = F_2(X_0) - L_{2b} = X_0 - X_b \tag{4.160}$$

The symbols X_b and X_0 denote the observed parameters and approximate parameters. During the iterations X_0 converges toward the solution, whereas X_b remains unchanged just as does the vector L_b. As a special case the vector L_2 can be zero, which implies that the current values for the approximate parameters also serve as observations of the parameters. This can generally be done if the intent is to define the coordinate system by assigning weights to the current approximate parameters. Table 4.3 summarizes the solution for weighted parameters for the observation equation model. The expressions for the sequential solution and those for the mixed adjustment model can be taken from Table 4.2. The expression for the adjusted parameter in Table 4.3 shows that the parameters are weighted simply by adding the respective weights to the diagonal elements of the normal matrix. In this way a subset of the parameters can be conveniently weighted. The parameters not weighted have zeros in the respective diagonal elements of P_2. Parameters can be fixed by assigning a large weight.

It is not necessary that the second group of observations pertain to the observed parameters. Table 4.4 shows the case in which the first group consists of the observed parameters. This approach has the unique feature that all ob-

TABLE 4.3   Observed Parameters in the Observation Equation Model

$L_{1a} = F_1(X_a)$ $L_{2a} = X_a$ $P = \begin{bmatrix} P_1 & 0 \\ 0 & P_2 \end{bmatrix}$
$V_1 = A_1 X + L_1$ $V_2 = X + L_2$ $L_2 = X_0 - X_b$
$N_1 = A_1^T P_1 A_1$ $N_2 = P_2$ $U_1 = A_1^T P_1 L_1$ $U_2 = P_2 L_2$
$\hat{X} = -(N_1 + P_2)^{-1}(U_1 + P_2 L_2)$ $Q_X = (N_1 + P_2)^{-1}$

TABLE 4.4   Sequential Solution without Inversion of the Normal Matrix (Observation Equation Model)

$$L_{1a} = X_a \qquad\qquad P = \begin{bmatrix} P_1 & 0 \\ 0 & P_2 \end{bmatrix}$$
$$L_{2a} = F_{2a}(X_a)$$

$$V_1 = X + L_1$$
$$L_1 = X_0 - X_b$$
$$V_2 = A_2 X + L_2$$

$$N_1 = P_1 \qquad\qquad N_2 = A_2^T P_2 A_2$$
$$U_1 = P_1 L_1 \qquad\qquad U_2 = A_2^T P_2 L_2$$

$$\hat{X}_1 = -(X_0 - X_b)$$
$$Q_1 = P_1^{-1}$$
$$V^T P V_1 = 0$$

$$\hat{X}_i = \hat{X}_{i-1} + \Delta \hat{X}_{i-1}$$
$$V^T P V_i = V^T P V_{i-1} + \Delta V^T P V_{i-1}$$
$$Q_i = Q_{i-1} + \Delta Q_{i-1}$$

$$T = \left(P_i^{-1} + A_i Q_{i-1} A_i^T \right)^{-1}$$
$$\Delta X_{i-1} = -Q_{i-1} A_i^T T \left(A_i \hat{X}_{i-1} + L_i \right)$$
$$\Delta V^T P V_{i-1} = \left(A_i \hat{X}_{i-1} + L_i \right)^T T \left(A_i \hat{X}_{i-1} + L_i \right)$$
$$\Delta Q_{i-1} = -Q_{i-1} A_i^T T A_i Q_{i-1}$$

servations can be added to the adjustment in a sequential manner; the first solution is a nonredundant solution based solely on the values of the observed parameters. It is important, once again, to distinguish the roles of the observed parameters X_b and the approximations X_0. Because, in most applications the P_1 matrix is diagonal, no computations for matrix inversion are required. The size of the matrix T equals the number of observations in the second group. Thus, if one observation is added at a time, only a 1×1 matrix must be inverted. The residuals can be computed directly from the mathematical model whenever desired.

A third case pertains to the role of the weight matrix. The weight matrix expresses the quality of the information known about the observed parameters. For the adjustment to be meaningful, one must make every attempt to obtain a weight matrix that truly reflects the quality of the additional information. Low weights, or, equivalently, large variances, imply low precision. Even low weighted parameters can have, occasionally, a positive effect on the quality of the least-squares solution. If the parameters or functions of the parameters are introduced with an infinitely large weight, one speaks of *conditions between parameters*. The only specifications for implementing conditions are

$$P_2^{-1} = 0 \tag{4.161}$$

and

$$P_2 = \infty \tag{4.162}$$

The respective mathematical models are

$$F(L_{1a}, X_a) = 0 \tag{4.163}$$

$$G(X_a) = 0 \tag{4.164}$$

with

$$B_1 V_1 + A_1 X + W_1 = 0 \tag{4.165}$$

$$A_2 X + L_2 = 0 \tag{4.166}$$

and

$$L_{1a} = F(X_a) \tag{4.167}$$

$$G(X_a) = 0 \tag{4.168}$$

with

$$V_1 = A_1 X + L_1 \tag{4.169}$$

$$A_2 X + L_2 = 0 \tag{4.170}$$

Table 4.5 contains the expression of the sequential solution. If expression (4.162) is used to impose the conditions, the largest numbers that can still be represented in the computer should be used. In most solutions it will readily be clear what constitutes a large weight; the weight must simply be large enough so that the respective observations or parameters do not change during the adjustment. For sequential solution the solution of the first group must exist. Conditions cannot be imposed sequentially to eliminate a singularity in the first group; for example, conditions should not be used sequentially to define the coordinate system. A one-step solution is given by (4.117).

The a posteriori variance of unit weight is always computed from the final set of residuals. The degree of freedom increases by one for every observed parameter function, weighted parameter, or condition. In nonlinear adjust-

TABLE 4.5 Conditions on Parameters

	Mixed model with conditions	Observation model with conditions
Non-linear model	$F_1(L_{1a}, X_a) = 0$ $\quad P_1$ $G(X_a) = 0$	$L_{1a} = F_1(X_a)$ $\quad P_1$ $G(X_a) = 0$
Linear model	$B_1 V_1 + A_1 X + W_1 = 0$ $A_2 X + L_2 = 0$	$V_1 = A_1 X + L_1$ $A_2 X + L_2 = 0$
Normal equation elements	$M_1 = B_1 P_1^{-1} B_1^T$ $N_1 = A_1^T M_1^{-1} A_1$ $U_1 = A_1^T M_1^{-1} W_1$	$M_1 = P_1^{-1}$ $N_1 = A_1^T P_1 A_1$ $U_1 = A_1^T P_1 L_1$
Normal equation	$\begin{bmatrix} N_1 & A_2^T \\ A_2 & 0 \end{bmatrix} \begin{bmatrix} \hat{X} \\ -K_2 \end{bmatrix} = \begin{bmatrix} -U_1 \\ -L_2 \end{bmatrix}$	$\begin{bmatrix} N_1 & A_2^T \\ A_2 & 0 \end{bmatrix} \begin{bmatrix} \hat{X} \\ -K_2 \end{bmatrix} = \begin{bmatrix} -U_1 \\ -L_2 \end{bmatrix}$
Minimum $V^T P V$	$V^T P V = V^T P V^* + \Delta V^T P V$ $V^T P V^* = -U_1^T N_1^{-1} U_1 + W_1^T M_1^{-1} W_1$ $\Delta V^T P V = \left(A_2 X^* + L_2\right)^T T \left(A_2 X^* + L_2\right)$	$V^T P V = V^T P V^* + \Delta V^T P V$ $V^T P V^* = -U_1^T N_1^{-1} U_1 + L_1^T P_1^{-1} L_1$ $\Delta V^T P V = \left(A_2 X^* + L_2\right)^T T \left(A_2 X^* + L_2\right)$
Estimated parameters	$\hat{X} = X^* + \Delta X$ $X^* = -N_1^{-1} U_1$ $T = \left(A_2 N_1^{-1} A_2^T\right)^{-1}$ $\Delta X = -N_1^{-1} A_2^T T \left(A_2 X^* + W_2\right)$	$\hat{X} = X^* + \Delta X$ $X^* = -N_1^{-1} U_1$ $T = \left(A_2 N_1^{-1} A_2^T\right)^{-1}$ $\Delta X = -N_1^{-1} A_2^T T \left(A_2 X^* + L_2\right)$
Estimated residuals	$V_1 = V_1^* + \Delta V_1$ $V_1^* = -P_1^{-1} B_1^T M_1^{-1} \left(A_1 X^* + W_1\right)$ $\Delta V_1 = -P_1^{-1} B_1^T M_1^{-1} A_1 \Delta X$	$V_1 = V_1^* + \Delta V_1$ $V_1^* = A_1 X^* + L_1$ $\Delta V_1 = A_1 \Delta X$
Estimated variance of unit weight	$\hat{\sigma}_0^2 = \dfrac{V^T P V}{r_1 + r_2 - u}$	$\hat{\sigma}_0^2 = \dfrac{V^T P V}{n_1 + n_2 - u}$
Estimated parameter cofactor matrix	$Q_X = Q_{X^*} + \Delta Q$ $Q_{X^*} = N_1^{-1}$ $\Delta Q = -N_1^{-1} A_2^T T A_2 N_1^{-1}$	$Q_X = Q_{X^*} + \Delta Q$ $Q_{X^*} = N_1^{-1}$ $\Delta Q = -N_1^{-1} A_2^T T A_2 N_1^{-1}$
Estimated residual cofactor matrix	$Q_{V_1} = Q_{V_1^*} + \Delta Q_{V_1}$ $Q_{V_1^*} = \left(P_1^{-1} B_1^T\right) M_1^{-1} \left(P_1^{-1} B_1^T\right)^T -$ $\left(P_1^{-1} B_1^T M_1^{-1} A_1\right) N_1^{-1} \left(P_1^{-1} B_1^T M_1^{-1} A_1\right)^T$ $\Delta Q_{V_1} = \left(P_1^{-1} B_1^T M_1^{-1} A_1 N_1^{-1} A_2^T\right) T$ $\left(P_1^{-1} B_1^T M_1^{-1} A_1 N_1^{-1} A_2^T\right)^T$	$Q_{V_1} = Q_{V_1^*} + \Delta Q_{V_1}$ $Q_{V_1^*} = P_1^{-1} - A_1 N_1^{-1} A_1^T$ $\Delta Q_{V_1} = A_1 N_1^{-1} A_2 T A_2^T N_1^{-1} A_1^T$
Adjusted obs. cofactor matrix	$Q_{L_a} = Q_{L_b} - Q_V$	$Q_{L_a} = Q_{L_b} - Q_V$

ments the linearized condition must always be evaluated for the current point of expansion, that is, the point of expansion of the last iteration (current solution).

The expressions in Tables 4.2 and 4.5 are almost identical. The only difference is that the matrix T contains the matrix M_2 in Table 4.2.

4.8 MINIMAL AND INNER CONSTRAINT ADJUSTMENTS

This section deals with the implementation of minimal and inner constraints to the observation equation model. The symbol r denotes the rank of the design matrix, $R({}_nA_u) = R(A^TPA) = r \le u$. Note that the use of the symbol r in this context is entirely different from its use in the mixed model, where r denotes the number of equations. The rank deficiency of $(u - r)$ is generally caused by a lack of coordinate system definition. For example, a network of distances is invariant with respect to translation and rotations, a network of angles is invariant with respect to translation, rotation, and scaling, and a level network (consisting of measured height differences) is invariant with respect to a translation in the vertical. The rank deficiency is dealt with by specifying $(u - r)$ conditions of the parameters. Practical examples are given in Chapter 5. Much of the theory of inner and minimal constraint solution is discussed by Pope (1971). The main reason for dealing with minimal and inner constraint solutions is that this type of adjustment is important for the quality control of observations. Inner constraint solutions have the additional advantage that the standard ellipses (ellipsoids) represent the geometry of the design matrix.

The formulation of the least-squares adjustment for the observation equation model in the presence of a rank deficiency is

$$_nV_1 = {}_nA_uX_B + {}_nL_1 \tag{4.171}$$

$$P = \sigma_0^2\Sigma_{L_b}^{-1} \tag{4.172}$$

$$_{u-r}B_uX_B = 0 \tag{4.173}$$

The subscript B indicates that the solution of the parameters X depends on the special condition implied by the B matrix in (4.173). This is the observation equation model with conditions between the parameters that was treated in Section 4.7. The one-step solution is given by (4.106):

$$\begin{bmatrix} A^TPA & B^T \\ B & 0 \end{bmatrix} \begin{bmatrix} \hat{X}_B \\ -K_2 \end{bmatrix} = \begin{bmatrix} -A^TPL \\ 0 \end{bmatrix} \tag{4.174}$$

The matrix on the left side of (4.174) is a nonsingular matrix if the conditions (4.173) are linearly independent; that is, the $(u - r) \times u$ matrix B has full row rank, and the rows are linearly independent of the rows of the design matrix A. A general expression for the inverse is obtained from

$$
\begin{bmatrix} A^{\mathrm{T}}PA & B^{\mathrm{T}} \\ B & 0 \end{bmatrix} \begin{bmatrix} Q_B & S^{\mathrm{T}} \\ S & R \end{bmatrix} = \begin{bmatrix} I & 0 \\ 0 & I \end{bmatrix} \tag{4.175}
$$

This matrix equation gives the following four equations of submatrices:

$$
A^{\mathrm{T}}PAQ_B + B^{\mathrm{T}}S = I \tag{4.176}
$$

$$
A^{\mathrm{T}}PAS^{\mathrm{T}} + B^{\mathrm{T}}R = 0 \tag{4.177}
$$

$$
BQ_B = 0 \tag{4.178}
$$

$$
BS^{\mathrm{T}} = I \tag{4.179}
$$

The solution of these equations requires the introduction of the $(u - r) \times u$ matrix E, whose rows span the null space of the design matrix A or the null space of the normal matrix. According to (A.21), there is a matrix E such that

$$
(A^{\mathrm{T}}PA)E^{\mathrm{T}} = 0 \tag{4.180}
$$

or

$$
AE^{\mathrm{T}} = 0 \tag{4.181}
$$

Because the rows of B are linearly independent of the rows of A, the $(u - r) \times (u - r)$ matrix BE^{T} has full rank and thus can be inverted. Multiplying (4.176) by E from the left and using (4.181), we get

$$
S = (EB^{\mathrm{T}})^{-1}E \tag{4.182}
$$

This expression also satisfies (4.179). Substituting S into (4.177) gives

$$
A^{\mathrm{T}}PAE^{\mathrm{T}}(BE^{\mathrm{T}})^{-1} + B^{\mathrm{T}}R = 0 \tag{4.183}
$$

Because of (4.181), this expression becomes

$$
B^{\mathrm{T}}R = 0 \tag{4.184}
$$

Because B has full rank, it follows that the matrix $R = 0$. Thus,

$$
\begin{bmatrix} A^{\mathrm{T}}PA & B^{\mathrm{T}} \\ B & 0 \end{bmatrix}^{-1} = \begin{bmatrix} Q_B & E^{\mathrm{T}}(BE^{\mathrm{T}})^{-1} \\ (EB^{\mathrm{T}})^{-1}E & 0 \end{bmatrix} \tag{4.185}
$$

Substituting expression (4.182) for S into (4.176) gives the nonsymmetric matrix

$$T_B \equiv A^\mathrm{T} P A Q_B = I - B^\mathrm{T}(EB^\mathrm{T})^{-1}E \tag{4.186}$$

This expression is modified with the help of (4.178), (4.180), and (4.186):

$$(A^\mathrm{T} P A + B^\mathrm{T} B)[Q_B + E^\mathrm{T}(BE^\mathrm{T})^{-1}(EB^\mathrm{T})^{-1}E] = I \tag{4.187}$$

It can be solved for Q_B:

$$Q_B = (A^\mathrm{T} P A + B^\mathrm{T} B)^{-1} - E^\mathrm{T}(EB^\mathrm{T}BE^\mathrm{T})^{-1}E \tag{4.188}$$

The least-squares solution of $\hat{X}_B$ subject to condition (4.713) is, according to (4.174) and (4.175), and (4.185),

$$\hat{X}_B = -Q_B A^\mathrm{T} P L \tag{4.189}$$

The cofactor matrix of the parameters follows from the law of variance–covariance propagation

$$Q_{X_B} = Q_B A^\mathrm{T} P A Q_B = Q_B \tag{4.190}$$

The latter part of (4.190) follows from (4.186) upon multiplying from the left by Q_B and using (4.178). Multiplying (4.186) from the right by $A^\mathrm{T}PA$ and using (4.181) gives

$$A^\mathrm{T} P A = A^\mathrm{T} P A Q_B A^\mathrm{T} P A \tag{4.191}$$

The relation implied in (4.190) is

$$Q_B A^\mathrm{T} P A Q_B = Q_B \tag{4.192}$$

$u - r$ conditions are necessary to solve the least-squares problem; that is, the minimal number of conditions is equal to the rank defect of the design (or normal) matrix. Any solution derived in this manner is called a minimal constraint solution. There are obviously many different sets of minimal constraints possible for the same adjustment. The only prerequisite on the B matrix is that it have full row rank and that its rows be linearly independent of A. Assume that

$$CX_C = 0 \tag{4.193}$$

is an alternative set of conditions. The solution $\hat{X}_C$ follows from the expressions given by simply replacing the matrix B by C. The pertinent expressions are

$$\hat{X}_C = -Q_C A^\mathrm{T} P L \tag{4.194}$$

$$Q_C = (A^TPA + C^TC)^{-1} - E^T(EC^TCE^T)^{-1}E \qquad (4.195)$$

$$T_C \equiv A^TPAQ_C = I - C^T(EC^T)^{-1}E \qquad (4.196)$$

$$A^TPAQ_CA^TPA = A^TPA \qquad (4.197)$$

$$Q_CA^TPAQ_C = Q_C \qquad (4.198)$$

The solutions pertaining to the various alternative sets of conditions are all related. In particular,

$$\hat{X}_B = T_B^T\hat{X}_C \qquad (4.199)$$

$$Q_B = T_B^TQ_CT_B \qquad (4.200)$$

$$\hat{X}_C = T_C^T\hat{X}_B \qquad (4.201)$$

$$Q_C = T_C^TQ_BT_C \qquad (4.202)$$

Equations (4.199) to (4.202) constitute the transformation of minimal control; that is, they relate the adjusted parameters and the covariance matrix for different minimal constraints. These transformation expressions are readily proven. For example, by using (4.194), (4.186), (4.196), and (4.181), we obtain

$$
\begin{aligned}
T_B^T\hat{X}_C &= -T_B^TQ_CA^TPL \\
&= -Q_BA^TPAQ_CA^TPL \\
&= -Q_B[I - C^T(EC^T)^{-1}E]A^TPL \qquad (4.203) \\
&= -Q_BA^TPL \\
&= \hat{X}_B
\end{aligned}
$$

With (4.196), (4.191), and (4.198), it follows that

$$
\begin{aligned}
T_C^TQ_BT_C &= Q_CA^TPAQ_BA^TPAQ_C \\
&= Q_CA^TPAQ_C \qquad (4.204) \\
&= Q_C
\end{aligned}
$$

Instead of using the general condition (4.193), we can use the condition

$$EX_P = 0 \qquad (4.205)$$

The rows of E are linearly independent of A because of (4.181). Thus, replacing the matrix C by E in equations (4.194) through (4.202) gives this special solution:

$$\hat{X}_P = -Q_P A^T PL \tag{4.206}$$

$$Q_P = (A^T PA + E^T E)^{-1} - E^T (EE^T EE^T)^{-1} E \tag{4.207}$$

$$T_P \equiv A^T PAQ_P = I - E^T (EE^T)^{-1} E \tag{4.208}$$

$$A^T PAQ_P A^T PA = A^T PA \tag{4.209}$$

$$Q_P A^T PAQ_P = Q_P \tag{4.210}$$

$$\hat{X}_B = T_P^T \hat{X}_P \tag{4.211}$$

$$Q_B = T_B^T Q_P T_B \tag{4.212}$$

$$\hat{X}_P = T_B^T \hat{X}_B \tag{4.213}$$

$$Q_P = T_P^T Q_B T_P \tag{4.214}$$

The solution (4.206) is called the inner constraint solution. The matrix T_P in (4.208) is symmetric. The matrix Q_P is a generalized inverse, called the pseudoinverse of the normal matrix; the following notation is used:

$$Q_P = N^+ = (A^T PA)^+ \tag{4.215}$$

The pseudoinverse of the normal matrix is computed from available algorithms of generalized matrix inverses or, equivalently, by finding the E matrix and using equation (4.207). For typical applications in surveying the matrix E can readily be identified. Examples are given in Chapter 5. Because of (4.181), the solution (4.206) can also be written as

$$\hat{X}_P = -(A^T PA + E^T E)^{-1} A^T PL \tag{4.216}$$

Note that the covariance matrix of the adjusted parameters is

$$\Sigma_X = \hat{\sigma}_0^2 Q_{B,C,P} \tag{4.217}$$

depending on whether constraints (4.173), (4.193), or (4.205) are used.

The inner constraint solution is yet another minimal constraint solution, although it has some special features. It can be shown that among all possible minimal constraint solutions, the inner constraint solution also minimizes the sum of the squares of the parameters, that is,

$$X^T X = \text{minimum} \tag{4.218}$$

This property can be used to obtain a geometric interpretation of the inner constraints. For example, it can be shown that the approximate parameters X_0

and the adjusted parameters $\hat{X}_P$ can be related by a similarity transformation whose least-squares estimates of translation and rotation are zero. For inner constraint solutions the standard ellipses show the geometry of the network and are not affected by the definition of the coordinate system. It can also be shown that the trace of Q_P is the smallest compared to the trace of the other cofactor matrices. All minimal constraint solutions yield the same adjusted observations, a posteriori variance of unit weight, covariance matrices for residuals, and adjusted observations and the same values for estimable functions of the parameters and their variances. The next section contains further explanation of quantities invariant with respect to changes in minimal constraints.

4.9 STATISTICS IN LEAST-SQUARES ADJUSTMENTS

Statistics completes the theory of adjustments, because it makes it possible to make objective statements about the data. The basic requirements, however, are that the mathematical model and the stochastic model be correct and that the observations have a multivariate normal distribution. Statistics cannot guarantee the right decision, but it can be helpful in gaining deeper insight into often unconscious motives that lead to certain decisions.

4.9.1 Multivariate Normal Distribution

This section contains a brief introduction to multivariate normal distribution. A few theorems are given that will be helpful in subsequent derivations. The multivariate normal distribution is especially pleasing, because the marginal distributions derived from multivariate normal distributions are also normally distributed. An extensive treatment of this distribution is found in the standard statistical literature. To simplify notation, the tilde is not used to identify random variables. The random nature of variables can readily be deduced from the context.

Let X be a vector with n random components with a mean of

$$E(X) = M \qquad (4.219)$$

and a covariance matrix of

$$E[(X - M)(X - M)^T] = {}_n\Sigma_n \qquad (4.220)$$

If X has a multivariate normal distribution, then the multivariate density function is

$$f(x_1, x_2, \ldots, x_n) = \frac{1}{(2\pi)^{n/2}|\Sigma|^{1/2}} e^{-(X-M)^T\Sigma^{-1}(X-M)/2} \qquad (4.221)$$

The mean and the covariance matrix completely describe the multivariate normal distribution. The notation

$$_nX_1 \sim N_n(_nM_1, \, _n\Sigma_n) \tag{4.222}$$

is used. The dimension of the distribution is n.

In the following, some theorems on multivariate normal distributions are given without proofs. These theorems are useful in deriving the distribution of V^TPV and some of the basic statistical tests in least-squares adjustments.

Theorem 1   If X is multivariate normal

$$X \sim N(M, \Sigma) \tag{4.223}$$

and

$$Z = \, _mD_nX \tag{4.224}$$

is a linear function of the random variable, where D is a $(m \times n)$ matrix of rank $m \leq n$, then

$$Z \sim N_m(DM, D\Sigma D^T) \tag{4.225}$$

is a multivariate normal distribution of dimension m. The mean and variance of the random variable Z follow from the laws for propagating the mean (4.33) and variance–covariances (4.34).

Theorem 2   If X is multivariate normal $X \sim N(M, \Sigma)$, the marginal distribution of any set of components of X is multivariate normal with means, variances, and covariances obtained by taking the proper component of M and Σ. For example, if

$$X = \begin{bmatrix} X_1 \\ X_2 \end{bmatrix} \sim N\left[\begin{pmatrix} M_1 \\ M_2 \end{pmatrix}, \begin{pmatrix} \Sigma_{11} & \Sigma_{12} \\ \Sigma_{21} & \Sigma_{22} \end{pmatrix} \right] \tag{4.226}$$

then the marginal distribution of X_2 is

$$X_2 \sim N(M_2, \Sigma_{22}) \tag{4.227}$$

The same law holds, of course, if the set contains only one component, say x_i. The marginal distribution of x_i is then

$$x_i \sim n(\mu_i, \sigma_i^2) \tag{4.228}$$

Theorem 3 If X is multivariate normal, a necessary and sufficient condition that two subsets of the random variables are stochastically independent is that the covariances be zero. For example, if

$$\begin{bmatrix} X_1 \\ X_2 \end{bmatrix} \sim N\left[\begin{pmatrix} M_1 \\ M_2 \end{pmatrix}, \begin{pmatrix} \Sigma_{11} & 0 \\ 0 & \Sigma_{22} \end{pmatrix}\right] \tag{4.229}$$

then X_1 and X_2 are stochastically independent. If one set of normally distributed random variables is uncorrelated with the remaining variables, the two sets are independent. The proof of the above theorem follows from the fact that the density function can be written as a product of $f_1(X_1)$ and $f_2(X_2)$ because of the special form of the density function (4.221).

4.9.2 Distribution of $V^T PV$

The derivation of the distribution is based on the assumption that the observations have a multivariate normal distribution. The dimension of the distribution equals the number of observations. In the subsequent derivations the observation equation model is used. However, these statistical derivations could just as well have been carried out with the mixed model.

The observation equations are

$$\begin{aligned} V &= AX + L_0 - L_b \\ &= AX + L \end{aligned} \tag{4.230}$$

A first basic assumption is that the residuals are a result of random errors; that is, the probability for a positive or negative residual of equal magnitude is the same. From this definition it follows that

$$E(V) = 0 \tag{4.231}$$

From (4.230) and (4.231) it follows that

$$E(L_b) = L_0 + AX \tag{4.232}$$

with the given covariance matrix

$$E\{[L_b - E(L_b)][L_b - E(L_b)]^T\} = \Sigma_{L_b} = \sigma_0^2 P^{-1} \tag{4.233}$$

The parameters X in the observation equation (4.230) are constants before their estimation. It also follows from (4.230) that

$$E(L) = -AX \tag{4.234}$$

The variance of the residuals before estimation is

$$
\begin{aligned}
E(VV^T) &= E[(AX + L)(AX + L)^T] \\
&= E\{[L - E(L)][L - E(L)]^T\} \\
&= E\{[L_b - E(L_b)][L_b - E(L_b)]^T\} \\
&= \Sigma_{L_b}
\end{aligned}
\tag{4.235}
$$

The second basic assumption refers to the type of distribution of the observations. It is assumed that typical observations in surveying and geodesy have a multivariate normal distribution. The mean is given by (4.232), and the covariance matrix is given by (4.233). Thus,

$$
L_b \sim N_n(L_0 + AX, \Sigma_{L_b})
\tag{4.236}
$$

is multivariate normal with dimension n. Alternative expressions are

$$
L \sim N_n(-AX, \Sigma_{L_b})
\tag{4.237}
$$

$$
\begin{aligned}
V &\sim N_n(0, \Sigma_{L_b}) \\
&\sim N_n(0, \sigma_0^2 P^{-1})
\end{aligned}
\tag{4.238}
$$

Distribution of $V^T P V$ can readily be derived after two orthogonal transformations have been applied and the parameters have been estimated. If Σ_{L_b} is nondiagonal, one can always transform the observations such that the transformed observations are stochastically independent and have a unit variate normal distribution. In Appendix A it is shown that for a positive definite matrix, say P, there exists a nonsingular matrix D such that

$$
D^T P^{-1} D = I
\tag{4.239}
$$

where

$$
D = E\Lambda^{-1/2}
\tag{4.240}
$$

The columns of the orthogonal matrix E consist of the normalized eigenvectors of P^{-1}, and Λ is a diagonal matrix with eigenvalues of P^{-1} at the diagonal. The transformed observations

$$
D^T V = D^T AX + D^T L
\tag{4.241}
$$

or

$$
{}_n\overline{V}_1 = {}_n\overline{A}_u X + {}_n\overline{L}_1
\tag{4.242}
$$

with

$$\overline{L} = D^T L_0 - D^T L_b = \overline{L}_0 - \overline{L}_b \qquad (4.243)$$

are distributed as multivariate normal with zero mean and identity covariance matrix because

$$E(\overline{V}) = D^T E(V) = 0 \qquad (4.244)$$

$$\Sigma_{\overline{V}} = D^T P^{-1} D \sigma_0^2 = \sigma_0^2 I \qquad (4.245)$$

The quadratic form $V^T P V$ remains invariant under this transformation because

$$R \equiv V^T P V = \overline{V}^T \Lambda^{1/2} E^T P E \Lambda^{1/2} \overline{V} = \overline{V}^T \Lambda^{1/2} \Lambda^{-1} \Lambda^{1/2} \overline{V} = \overline{V}^T \overline{V} \qquad (4.246)$$

If the covariance matrix Σ_{L_b} has a rank defect, then one could use matrix F of (A.21) for the transformation. The dimension of the transformed observations $\overline{L}_b$ equals the rank of the covariance matrix.

There is yet another transformation necessary to find the distribution of R. To keep the generality, let the matrix $\overline{A}$ in (4.242) have less than full column rank, that is, $R(\overline{A}) = r < u$. Let the matrix F be an $(n \times r)$ matrix whose columns constitute an orthonormal basis for the column space of $\overline{A}$. One such choice for the columns of F may be to take the normalized eigenvectors of $\overline{A}\,\overline{A}^T$. Let G be an $n \times (n - r)$ matrix such that $(F \vdots G)$ is orthogonal and such that the columns of G constitute an orthonormal basis to the $(n - r)$-dimensional null space of $\overline{A}\,\overline{A}^T$. Such a matrix always exists. There is no need to compute this matrix explicitly. With these specifications we obtain

$$\begin{bmatrix} F^T \\ \cdots \\ G^T \end{bmatrix} (F \vdots G) = \begin{bmatrix} F^T F & F^T G \\ G^T F & G^T G \end{bmatrix} = \begin{bmatrix} {}_r I_r & 0 \\ 0 & {}_{n-r} I_{n-r} \end{bmatrix} \qquad (4.247)$$

$$(F \vdots G) \begin{bmatrix} F^T \\ \cdots \\ G^T \end{bmatrix} = FF^T + GG^T = I \qquad (4.248)$$

$$\overline{A}^T G = 0 \qquad (4.249)$$

$$G^T \overline{A} = 0 \qquad (4.250)$$

The required transformation is

$$\begin{bmatrix} F^T \\ G^T \end{bmatrix} \overline{V} = \begin{bmatrix} F^T \\ G^T \end{bmatrix} \overline{A} X + \begin{bmatrix} F^T \\ G^T \end{bmatrix} \overline{L} \qquad (4.251)$$

or, equivalently,

$$\begin{bmatrix} F^{\mathrm{T}}\overline{V} \\ G^{\mathrm{T}}\overline{V} \end{bmatrix} = \begin{bmatrix} F^{\mathrm{T}}\overline{A}X \\ 0 \end{bmatrix} + \begin{bmatrix} F^{\mathrm{T}}\overline{L} \\ G^{\mathrm{T}}\overline{L} \end{bmatrix} \tag{4.252}$$

Labeling the newly transformed observations by Z, that is,

$$Z = \begin{bmatrix} Z_1 \\ Z_2 \end{bmatrix} = \begin{bmatrix} F^{\mathrm{T}}\overline{L} \\ G^{\mathrm{T}}\overline{L} \end{bmatrix} \tag{4.253}$$

We can write (4.252) as

$$\overline{V}_Z = \begin{bmatrix} \overline{V}_{Z_1} \\ \overline{V}_{Z_2} \end{bmatrix} = \begin{bmatrix} F^{\mathrm{T}}\overline{A}X \\ 0 \end{bmatrix} + \begin{bmatrix} Z_1 \\ Z_2 \end{bmatrix} \tag{4.254}$$

There are r random variables in Z_1 and $n - r$ random variables in Z_2. The quadratic form again remains invariant under the orthogonal transformation, since

$$\overline{V}_Z^{\mathrm{T}}\overline{V}_Z = \overline{V}^{\mathrm{T}}(FF^{\mathrm{T}} + GG^{\mathrm{T}})\overline{V}$$

$$= \overline{V}^{\mathrm{T}}\overline{V} = R \tag{4.255}$$

according to (4.248). The actual quadratic form is obtained from (4.254):

$$R = \overline{V}_Z^{\mathrm{T}}\overline{V}_Z = (F^{\mathrm{T}}\overline{A}X + Z_1)^{\mathrm{T}}(F^{\mathrm{T}}\overline{A}X + Z_1) + Z_2^{\mathrm{T}}Z_2 \tag{4.256}$$

The least-squares solution requires that R be minimized by variation of the parameters. Generally, the minimum is found by equating to zero partial derivatives with respect to X and solving the resulting equations. The special form of (4.256) permits a much simpler approach. The expressions on the right side of equation (4.256) consist of the sum of two positive terms (sum of squares). Because only the first term is a function of the parameters X, the minimum is achieved if the first term is zero, that is,

$$-_r F_n^{\mathrm{T}}\,_n\overline{A}_u\,_u\hat{X}_1 = Z_1 \tag{4.257}$$

Note that the caret identifies the estimated parameters. Consequently, the estimate of the quadratic form is

$$\hat{R} = Z_2^{\mathrm{T}}Z_2 \tag{4.258}$$

Because there are $r < u$ equations for the u parameters in (4.257), there always exists a solution for $\hat{X}$. The simplest approach is to equate $(u - r)$ parameters to zero. This would be identical with having these $(u - r)$ parameters treated as constants in the adjustment. They could be left out when setting up the design matrix and thus the singularity problem would be avoided altogether. Equation (4.257) can be solved subject to $(u - r)$ general conditions between the parameters. The resulting solution is a minimal constraint solution. If the particular condition (4.205) is applied, one obtains the inner constraint solution. If $\overline{A}$ has no rank defect, then the system (4.257) consists of u equations for u unknowns.

The estimate for the quadratic form (4.258) does not depend on the parameters X and thus is invariant with respect to the selection of the minimal constraints for finding the least-square estimate of X. Moreover, the residuals themselves are independent of the minimal constraints. Substituting the solution (4.257) into (4.252) gives

$$\begin{bmatrix} F^T \\ G^T \end{bmatrix} \hat{V} = \begin{bmatrix} 0 \\ G^T \overline{L} \end{bmatrix} \tag{4.259}$$

Since the matrix $(F \ \vdots \ G)$ is orthonormal, the expression for the residuals becomes

$$\hat{V} = [FG] \begin{bmatrix} 0 \\ G^T \overline{L} \end{bmatrix} = GG^T \overline{L} \tag{4.260}$$

Thus, the residuals are independent of the specific solution for $\hat{X}$. The matrix G depends only on the structure of the design matrix $\overline{A}$. By applying the law of variance–covariance propagation to (4.260), we clearly see that the covariance matrix of the adjusted residuals, and thus the covariance matrix of the adjusted observations, does not depend on the specific set of minimal constraints. Note that the transformation (4.241) does not invalidate these statements, since the D matrix is not related to the parameters.

Returning to the derivation of the distribution of $V^T PV$, we find from (4.253) that

$$E(Z) = \begin{bmatrix} -F^T \overline{A} X \\ 0 \end{bmatrix} \tag{4.261}$$

where (4.250) and the fact the $E(\overline{L}) = -\overline{A}X$ according to (4.242) were used. The covariance matrix is

$$\Sigma_Z = \sigma_0^2 \begin{bmatrix} F^T \\ G^T \end{bmatrix} I[FG] = \sigma_0^2 \begin{bmatrix} F^T F & F^T G \\ G^T F & G^T G \end{bmatrix} = \sigma_0^2 \begin{bmatrix} I & 0 \\ 0 & I \end{bmatrix} \tag{4.262}$$

because the covariance matrix of $\overline{L}$ equals the identity matrix, as is readily verified from (4.243), (4.239), and (4.237). Since a linear transformation of a random variable with multivariate normal distribution results in another multivariate normal distribution according to Theorem 1, it follows that Z is distributed as

$$
Z \sim N_n \left[\begin{pmatrix} -F^{\mathrm{T}}\overline{A}X \\ 0 \end{pmatrix}, \sigma_0^2 \begin{pmatrix} {}_rI_r & 0 \\ 0 & {}_{n-r}I_{n-r} \end{pmatrix} \right]
\tag{4.263}
$$

The random variables Z_1 and Z_2 are stochastically independent, as are the individual components according to Theorem 3. From Theorem 2 it follows that

$$
Z_2 \sim N_{n-r}(0, \sigma_0^2 I)
\tag{4.264}
$$

Thus

$$
z_{2i} \sim n(0, \sigma_0^2)
\tag{4.265}
$$

or

$$
\frac{z_{2i}}{\sigma_0} \sim n(0, 1)
\tag{4.266}
$$

Appendix C lists the distribution of a few simple functions of random variables. For example, the square of a standardized normal distributed variable has a chi-square distribution with one degree of freedom, and the sum of chi-square distributed variables is also a chi-square distribution with a degree of freedom equal to the sum of the individual degrees of freedom. Using these relations, it follows that $V^{\mathrm{T}}PV$

$$
\frac{\hat{R}}{\sigma_0^2} = \frac{Z_2^{\mathrm{T}}Z_2}{\sigma_0^2} = \sum_{i=1}^{n-r} \frac{z_{2i}^2}{\sigma_0^2} \sim \chi_{n-r}^2
\tag{4.267}
$$

has a chi-square distribution with $n - r$ degrees of freedom.

4.9.3 Testing $V^{\mathrm{T}}PV$ and $\Delta V^{\mathrm{T}}PV$

Combining the result of (4.267) with the expression for the a posteriori variance of unit weight of Table 4.1, we obtain the formulation for the most fundamental statistical test in least-squares estimation:

$$
\frac{V^{\mathrm{T}}PV}{\sigma_0^2} = \frac{\hat{\sigma}_0^2}{\sigma_0^2} (n - r) \sim \chi_{n-r}^2
\tag{4.268}
$$

Note that $(n - r)$ is the degree of freedom of the adjustment. If there is no rank deficiency in the design matrix, the degree of freedom is $(n - u)$. Based on the statistics (4.268), the test can be performed to find out whether the adjustment is distorted. The formulation of the hypothesis is as follows

$$H_0: \sigma_0^2 = \hat{\sigma}_0^2 \tag{4.269}$$

$$H_1: \sigma_0^2 \neq \hat{\sigma}_0^2 \tag{4.270}$$

The zero hypothesis states that the a priori variance of unit weight statistically equals the a posteriori variance of unit weight. Recall that the a posteriori variance of unit weight is a random variable; the adjustment makes a sample value available for this quantity on the basis of the observations (the samples). Both variances of unit weight do not have to be numerically equal, but they should be statistically equal in the sense of (4.67). If the zero hypothesis is accepted, the adjustment is judged to be correct. If the numerical value

$$\chi^2 = \frac{\hat{\sigma}_0^2}{\sigma_0^2} (n - r) = \frac{V^T P V}{\sigma_0^2} \tag{4.271}$$

is such that

$$\chi^2 < \chi^2_{n-r, 1-\alpha/2} \tag{4.272}$$

$$\chi^2 > \chi^2_{n-r, \alpha/2} \tag{4.273}$$

then the zero hypothesis is rejected. The significance level α, that is, the probability of a type I error, or the probability of rejecting the zero hypothesis even though it is true, is generally fixed to 0.05. Here the significance level is the sum of the probabilities in both tails. Table 4.6 lists selected values from the chi-square distribution $\chi^2_{n-r, \alpha}$. The probability β of the type II error, that is, the probability of rejecting the alternative hypothesis and accepting the zero

TABLE 4.6 Selected Values from the Chi-Square Distribution

Degree of freedom (DF)	Probability α			
	.975	.950	.050	.025
1	0.00	0.00	3.84	5.02
5	0.83	1.15	11.07	12.83
10	3.25	3.94	18.31	20.48
20	9.59	10.85	31.41	34.17
50	32.36	34.76	67.50	71.42
100	74.22	77.93	124.34	129.56

hypothesis although the alternative hypothesis is true, is generally not computed. Rejection of the zero hypothesis is taken as an indication that something is wrong with the adjustment. The cause of the rejection still remains to be clarified. Note that rejection can be caused by an inadequate stochastic model or mathematical model or both. Type II errors are considered in Section 4.10.2 in regards to reliability and in Section 10.2.3.7 in regards to discernibility of estimated ambiguity sets.

Figure 4.4 shows the limits on $V^T PV$ corresponding to a significant level of $\alpha = 0.05$. At higher degrees of freedom there is essentially linear increase. The difference of these limits is shown in Figure 4.5. In Figure 4.6 the corresponding limits are given for the a posteriori variance of unit weight.

The test statistics for testing adjustment of groups of observations is based on $V^T PV^*$, as computed from the first group only, and $\Delta V^T PV$, which is the contribution from the new group, according to Table 4.2:

$$\Delta V^T PV = (A_2 X^* + L_2)^T T (A_2 X^* + L_2) \tag{4.274}$$
$$= Z_3^T T Z_3$$

The new random variable Z_3 is a function of observations L_1 and L_2. Applying the laws of propagation of mean and variance, one finds

$$E(Z_3) = A_2 E(X^*) + E(L_2) \tag{4.275}$$
$$= A_2 X - A_2 X = 0$$

and

$$\Sigma_{Z_3} = T^{-1} \tag{4.276}$$

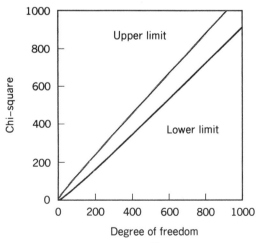

Figure 4.4   Limits on $V^T PV$ ($\alpha = 0.05$).

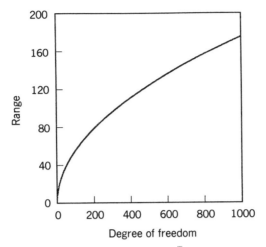

Figure 4.5 Allowable range of $V^T P V$ ($\alpha = 0.05$).

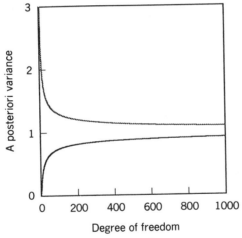

Figure 4.6 Limits on $\hat{\sigma}_0^2$ ($\alpha = 0.05$).

Thus, the new random variable is distributed as

$$Z_3 \sim N(0, \sigma_0^2 T^{-1}) \tag{4.277}$$

Carrying out the orthonormal transformation yields a random vector whose components are stochastically independent and normally distributed. By standardizing these distributions and summing the squares of these random variables, it follows that

$$\frac{\Delta V^T P V}{\sigma_0^2} = \frac{Z_3^T T Z_3}{\sigma_0^2} = \sim \chi_{n_2}^2 \tag{4.278}$$

has a chi-square distribution with n_2 degrees of freedom, where n_2 equals the number of observations in the second group. The random variables (4.278) and (4.268) are stochastically independent. To prove this, consider the new random variable $Z = (Z_1, Z_2, Z_3)^T$, which is a linear functions of the random variables L (first group) and L_2 according to equations (4.241), (4.253), and (4.274). By using the covariance matrix (4.88) and applying the law of variance–covariance propagation, we find that the covariances between the Z_i are zero. Because the distribution of the Z is multivariate normal, it follows that the random variables Z_i are stochastically independent. Since $\Delta V^T P V$ is a function of Z_3 only, it follows that $V^T P V$ in (4.268), which is only a function of Z_2, and $\Delta V^T P V$ in (4.278) are stochastically independent. Thus, it is permissible to form the following ratio of random variables:

$$\frac{\Delta V^T P V(n_1 - r)}{V^T P V^* (n_2)} \sim F_{n_2, n_1 - r} \tag{4.279}$$

which has an F distribution. The symbol (*) denotes the least-squares solution of the first group only.

Thus the fundamental test in sequential adjustment is based on the F distribution. The zero hypothesis states that the second group of observations does not distort the adjustment, or that there is no indication that something is wrong with the second group of observations. The alternative hypothesis states that there is indication that the second group of observations contains errors. The zero hypothesis is rejected, and the alternative hypothesis is accepted if

$$F < F_{n_2, n_1 - r, 1 - \alpha/2} \tag{4.280}$$

$$F > F_{n_2, n_1 - r, \alpha/2} \tag{4.281}$$

Table 4.7 lists selected values from the F distribution as a function of the degrees of freedom. The tabulation refers to the parameters as specified in $F_{n_1, n_2, 0.05}$.

**TABLE 4.7   Selected Values from the F
Distribution**

n_2	n_1			
	1	2	3	4
5	6.61	5.79	5.41	5.19
10	4.96	4.10	3.71	3.48
20	4.35	3.49	3.10	2.87
60	4.00	3.15	2.76	2.53
120	3.92	3.07	2.68	2.45
∞	3.84	3.00	2.60	2.37

4.9.4 General Linear Hypothesis

The general linear hypothesis deals with linear conditions between parameters. Nonlinear conditions are first linearized. The basic idea is to test the change $\Delta V^T P V$ for its statistical significance. Any of the three adjustment models can be used to carry out the general linear hypothesis test. For the observation equation model with additional conditions between the parameters, one has

$$V_1 = A_1 X + L_1 \tag{4.282}$$

$$A_2 X + L_2 = 0 \tag{4.283}$$

Equation (4.283) expresses the zero hypothesis H_0. The solution of the combined adjustment is found in Table 4.5. Adjusting (4.282) alone results in $V^T P V^*$, which has a chi-square distribution with $(n - r)$ degrees of freedom according to (4.268). The change $\Delta V^T P V$ resulting from the condition (4.283) is

$$\Delta V^T P V = (A_2 X^* + L_2)^T T (A_2 X^* + L_2) \tag{4.284}$$

The expression in (4.284) differs from (4.274) in two respects. First, the matrix T differs; that is, the matrix T in (4.284) does not contain the P_2 matrix. Second, the quantity L_2 is not a random variable. These differences, however, do not matter in the proof of stochastic independence of $V^T P V^*$ and $\Delta V^T P V$. Analogously to (4.274), we can express the change $\Delta V^T P V$ in (4.284) as a function of a new random variable Z_3. The proof for stochastic independence follows the same lines of thought as given before (for the case of additional observations). Thus, just as (4.279) is the basis for testing two groups of observations, the basic test for the general linear hypothesis (4.283) is

$$\frac{\Delta V^T P V (n_1 - r)}{V^T P V^* \, n_2} \sim F_{n_2, n_1 - r} \tag{4.285}$$

A small $\Delta V^T P V$ implies that the null hypothesis (4.283) is acceptable; that is, the conditions are in agreement with the observations. The conditions do not impose any distortions on the adjustment. The rejection criterion is based on the one-tail test at the upper end of the distribution. Thus, reject H_0 at a 100 $\alpha\%$ significance level if

$$F > F_{n_2, n_1 - r, \alpha} \tag{4.286}$$

The general formulation of the null hypothesis in (4.283) makes it possible to test any hypothesis on the parameters so long as the hypothesis can be expressed in a mathematical equation. Nonlinear hypotheses must first be linearized. Simple hypotheses could be used to test whether an individual parameter has a certain numerical value, whether two parameters are equal, whether the distance between two stations has a certain length, whether an angle has a certain size, etc. For example, consider the hypothesis

$$H_0: X - X_T = 0 \qquad (4.287)$$

$$H_1: X - X_T \neq 0 \qquad (4.288)$$

The zero hypothesis states that the parameters equal a certain (true) value X_T. From (4.283) it follows that $A_2 = I$ and $L_2 = -X_T$. Using these specifications in (4.284), the statistic (4.285) becomes

$$\frac{(\hat{X}* - X_T)^T N(\hat{X}* - X_T)}{\hat{\sigma}_0^2 r} \sim F_{r, n_1 - r, \alpha} \qquad (4.289)$$

where the a posteriori variance of unit weight (first group only) has been substituted for $V^T P V*$. Once the adjustment of the first group (4.282) is completed, the values for the adjusted parameters and the a posteriori variance of unit weight are entered in (4.289), and the fraction is computed and compared with the F value taking the proper degrees of freedom and the desired significance level into account. Rejection or acceptance of the zero hypothesis follows rule (4.286).

Note that one of the degrees of freedom in (4.289) is $r = R(N) \leq u$, instead of u, which equals the number of parameters, even though equation (4.287) expresses u conditions. Because of the possible rank defect of the normal matrix N, the distribution of $\Delta V^T P V$ in (4.284) is a chi-square distribution with r degrees of freedom. Consider the derivation leading to (4.278). The u components of Z_3 are transformed to r stochastically independent unit variate normal distributions that are then squared and summed to yield the distribution of $\Delta V^T P V$. The interpretation is that (4.287) represents one hypothesis on all parameters X, and not u hypotheses on the u components on X.

Expression (4.289) can be used to define the r-dimensional confidence region. Replace the particular X_T by the unknown parameter X, and drop the asterisk; then

$$P\left[\frac{(\hat{X} - X)^T N(\hat{X} - X)}{\hat{\sigma}_0^2 r} \leq F_{r, n_1 - r, \alpha}\right] = \int_0^{F_{r, n_1 - r, \alpha}} F_{r, n_1 - r} dF = 1 - \alpha$$

$$(4.290)$$

The probability region described by the expression on the left side of equation (4.290) is an $R(N)$-dimensional ellipsoid. The probability region is an ellipsoid, because the normal matrix N is positive definite or, at least, semi-positive definite. If one identifies the center of the ellipsoid with $\hat{X}$, then there is $(1 - \alpha)$ probability that the unknown point X lies within the ellipsoid. The orientation and the size of this ellipsoid are a function of the eigenvectors and eigenvalues of the normal matrix, the rank of the normal matrix, and the degree of freedom. Consider the orthonormal transformation

$$Z = F^T(X - \hat{X}) \tag{4.291}$$

with F as specified in (A.19); then

$$F^T N F = \Lambda \tag{4.292}$$

and

$$(\hat{X} - X)^T N (\hat{X} - X) = Z^T \Lambda Z = \sum_{i=1}^{r} z_i^2 \lambda_i = \sum_{i=1}^{r} \frac{z_i^2}{(1/\sqrt{\lambda_i})^2} \tag{4.293}$$

Combining equations (4.290) and (4.293), we can write the r-dimensional ellipsoid, or the r-dimensional confidence region, in the principal axes form:

$$P\left[\frac{z_1^2}{(\hat{\sigma}_0\sqrt{1/\lambda_1}\, rF_{r,n-r,\alpha})^2} + \cdots + \frac{z_r^2}{(\hat{\sigma}_0\sqrt{1/\lambda_r}\, rF_{r,n-r,\alpha})^2} \leq 1 \right] = 1 - \alpha$$

$$\tag{4.294}$$

The confidence region is centered at $\hat{X}$. Whenever the zero hypothesis H_0 of (4.287) is accepted, the point X_T falls within the confidence region. The probability that the ellipsoid contains the true parameters X_T, is $1 - \alpha$. For these reasons, one naturally would like the ellipsoid to be small. Equation (4.294) shows that the semimajor axes are proportional to the inverse of the eigenvalues of the normal matrix. It is exactly this relationship that makes us choose the eigenvalues of N as large as possible, provided that we have a choice through appropriate network design variation. As an eigenvalue approaches zero, the respective axis of the confidence ellipsoid approaches infinity; this is an undesirable situation, both from a statistical point of view and because of the numerical difficulties encountered during the inversion of the normal matrix.

4.9.5 Ellipses as Confidence Regions

Confidence ellipses are statements of precision. They are frequently used in connection with two-dimensional networks in order to make the directional precision of station location visible. Ellipses of confidence follow from Section 4.9.4 simply by limiting the hypothesis (4.287) to two parameters, that is, the Cartesian coordinates of a station. Of course, in a three-dimensional network one can compute three-dimensional ellipsoids or several ellipses, for example, one for the horizontal and others for the vertical. Confidence ellipses or ellipsoids are not limited to the specific application of networks. However, in networks the confidence regions can be referenced with respect to the coordinate system of the network and thus can provide an integrated view of the geometry of the confidence regions and the network.

Consider the following hypothesis:

$$H_0: X_1 - X_{1,T} = 0 \tag{4.295}$$

where the notation

$$X_1 = \begin{bmatrix} x_1 \\ x_2 \end{bmatrix} \tag{4.296}$$

is used. The symbols x_1 and x_2 denote the Cartesian coordinates of a two-dimensional network station. The test of this hypothesis follows the outline given in the previous section. The A_2 matrix is of size $(2 \times u)$ because there are two separate equations in the hypothesis and u components in X. The elements of A_2 are zero except those elements of rows 1 and 2, which correspond to the respective positions of x_1 and x_2 in X. With these specifications it follows that

$$Q_1 = A_2 N^{-1} A_2^{\mathrm{T}} = \begin{bmatrix} q_{x_1} & q_{x_1,x_2} \\ q_{x_2,x_1} & q_{x_2} \end{bmatrix} \tag{4.297}$$

and expression (4.289) becomes

$$\frac{1}{2\hat{\sigma}_0^2} (\hat{X}_1 - X_{1,\mathrm{T}})^{\mathrm{T}} Q_1^{-1} (\hat{X}_1 - X_{1,\mathrm{T}}) \sim F_{2,n-r} \tag{4.298}$$

The elements q in (4.297) consist of the respective elements of the cofactor matrix Q_X. Given the significance level α, the hypothesis test can be carried out. The two-dimensional confidence region is

$$P\left[\frac{(\hat{X}_1 - X_1)^{\mathrm{T}} Q_1^{-1} (\hat{X}_1 - X_1)}{2\hat{\sigma}_0^2} \le F_{2,n-r,\alpha} \right]$$

$$= \int_0^{F_{2,n-r,\alpha}} F_{2,n-r} \, dF = 1 - \alpha \tag{4.299}$$

The size of the confidence ellipses defined by (4.299) depends on the degree of freedom of the adjustment and the significance level. The ellipses are centered at the adjusted position and delimit the $(1 - \alpha)$ probability area for the true position. The principal axis form of (4.299) is obtained through orthogonal transformation. Let R_i denote the matrix whose rows are the orthonormal eigenvectors of Q_i, then

$$R_1^T Q_1^{-1} R_1 = \Lambda_1^{-1} \tag{4.300}$$

according to (A.15). The matrix Λ_i is diagonal and contains the eigenvalues of Q_i. With

$$Z_1 = R_1^T (X_1 - \hat{X}_1) \tag{4.301}$$

expression (4.299) becomes

$$P\left\{\left[\frac{z_1^2}{(\hat{\sigma}_0 \sqrt{\lambda_1^Q 2F_{2,n-r,\alpha}})^2} + \frac{z_2^2}{(\hat{\sigma}_0 \sqrt{\lambda_2^Q 2F_{2,n-r,\alpha}})^2}\right] \le 1\right\}$$

$$= \int_0^{F_{2,n-r,\alpha}} F_{2,n-r}\, dF = 1 - \alpha \tag{4.302}$$

If $F_{2,n-r,\alpha} = \frac{1}{2}$, the ellipse is called the *standard ellipse* or the *error ellipse*. Thus, the probability enclosed by the standard ellipse is a function of the degree of freedom $(n - r)$ and is computed as follows:

$$P(\text{standard ellipse}) = \int_0^{1/2} F_{2,n-r}\, dF \tag{4.303}$$

The magnification factor, $\sqrt{2F_{2,n-r,\alpha}}$, as a function of the probability and the degree of freedom, is shown in Table 4.8. The table shows immediately that a small degree of freedom requires a large magnification factor to obtain, for example, 95% probability. It is seen that in the range of small degrees of freedom, an increase in the degree of freedom rapidly decreases the magnification factor, whereas with a large degree of freedom, any additional observations cause only a minor reduction of the magnification factor. After a degree of freedom of about 8 or 10, the decrease in the magnification factor slows down noticeably. Thus, based on the speed of decreasing magnification factor, a degree of 10 appears optimal, considering the expense of additional observations and the little gain derived from them in the statistical sense. For a degree of freedom of 10, the magnification factor is about 3 to cover 95% probability.

The hypothesis (4.295) can readily be generalized to three dimensions encompassing the Cartesian coordinates of a three-dimensional network station. The magnification factor of the respective *standard ellipsoid* is $\sqrt{3F_{3,n-r,\alpha}}$ for it to contain $(1 - \alpha)$ probability. Similarly, the standard deviation of an individual coordinate is converted to a $(1 - \alpha)$ probability confidence interval by multiplication with $\sqrt{F_{1,n-r,\alpha}}$. These magnification factors are shown in

TABLE 4.8 Magnification Factor $\sqrt{2F_{2,n-r,\alpha}}$ for Standard Ellipses

	Probability $1-\alpha$		
n-r	95%	98%	99%
1	20.00	50.00	100.00
2	6.16	9.90	14.10
3	4.37	6.14	7.85
4	3.73	4.93	6.00
5	3.40	4.35	5.15
6	3.21	4.01	4.67
8	2.99	3.64	4.16
10	2.86	3.44	3.89
12	2.79	3.32	3.72
15	2.71	3.20	3.57
20	2.64	3.09	3.42
30	2.58	2.99	3.28
50	2.52	2.91	3.18
100	2.49	2.85	3.11
∞	2.45	2.80	3.03

Figure 4.7 for $\alpha = 0.05$. For higher degrees of freedom the magnification factors converge toward the respective chi-square values because of the relationship $rF_{r,\infty} = \chi_r^2$.

In drawing the confidence ellipse for station P_i, the rotation between the (X_i) and (Z_i) coordinate system must be computed. If (Y_i) denotes the translated (X_i) coordinate system through the adjusted point $\hat{X}_i$, then equation (4.301) be-

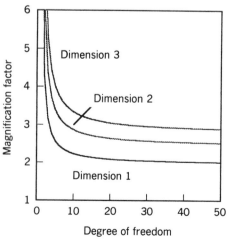

Figure 4.7 Magnification factors for confidence regions.

comes

$$Z_i = R_i^T Y_i \tag{4.304}$$

Rather than computing the eigenvectors of Q_i to get the columns of the R matrix, we can readily obtain the rotational relation from Figure 4.8, as follows:

$$z_1 = y_2 \sin \varphi + y_1 \cos \varphi \tag{4.305}$$

$$z_2 = y_2 \cos \varphi - y_1 \sin \varphi \tag{4.306}$$

From (4.300) and (4.304) it follows that

$$Y_i^T Q_i Y_i = Z_i^T \Lambda_i Z_i \tag{4.307}$$

Substituting equation (4.305) and (4.306) into the right-hand side of (4.307) and the elements of Q_i into the left-hand side gives

$$
\begin{aligned}
y_2^2 q_{x2} &+ y_1^2 q_{x1} + y_1 y_2 2 q_{x1,x2} \\
&= y_2^2 (\lambda_1^Q \sin^2 \varphi + \lambda_2^Q \cos^2 \varphi) + y_1^2 (\lambda_1^Q \cos^2 \varphi + \lambda_2^Q \sin^2 \varphi) \quad (4.308) \\
&+ y_1 y_2 (\lambda_1^Q - \lambda_2^Q) \, 2 \sin \varphi \cos \varphi
\end{aligned}
$$

The superscript Q is used to emphasize that the eigenvalues are those of the submatrix Q_i. Comparing the coefficients in (4.308) yields three equations that

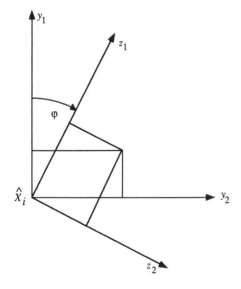

Figure 4.8 Rotation of the principal axis coordinate system.

can be used to compute $(\lambda_1, \lambda_2, \varphi)$ as a function of the elements of Q_i. To reduce the algebraic computations, we compare the coefficients only for the mixed term, giving immediately the expression for the rotation angle:

$$2(\lambda_1^\varrho - \lambda_2^\varrho) \sin \varphi \cos \varphi = 2q_{x1,x2} \tag{4.309}$$

or

$$\sin 2\varphi = \frac{2q_{x1,x2}}{\lambda_1^\varrho - \lambda_2^\varrho} \tag{4.310}$$

$$\cos 2\varphi = \sqrt{1 - \sin^2 2\varphi} = \frac{\sqrt{(\lambda_1^\varrho - \lambda_2^\varrho)^2 - 4q_{x1,x2}^2}}{\lambda_1^\varrho - \lambda_2^\varrho} \tag{4.311}$$

The eigenvalues are computed directly from the characteristic equation

$$|Q_i - \lambda^\varrho I| = \begin{bmatrix} q_{x1} - \lambda^\varrho & q_{x1,x2} \\ q_{x1,x2} & q_{x2} - \lambda^\varrho \end{bmatrix}$$

$$= (q_{x1} - \lambda^\varrho)(q_{x2} - \lambda^\varrho) - q_{x1,x2}^2 = 0 \tag{4.312}$$

The solution of the quadratic equation is

$$\lambda_1^\varrho = \frac{q_{x1} + q_{x2}}{2} + \frac{1}{2} W \tag{4.313}$$

$$\lambda_2^\varrho = \frac{q_{x1} + q_{x2}}{2} - \frac{1}{2} W \tag{4.314}$$

where

$$W = \sqrt{(q_{x1} - q_{x2})^2 + 4q_{x1,x2}^2} \tag{4.315}$$

denoting the difference $\lambda_1^\varrho - \lambda_2^\varrho$. By using (4.313) through (4.315), we can write (4.310) and (4.311) for the angle φ as

$$\sin 2\varphi = \frac{2q_{x1,x2}}{W} \tag{4.316}$$

$$\cos 2\varphi = \frac{q_{x1} - q_{x2}}{W} \tag{4.317}$$

Figure 4.9 shows the defining elements of the standard ellipse. Recall equation (4.302) regarding the interpretation of the standard ellipses as a confidence region. In any adjustment, any two parameters can comprise X_i regardless of

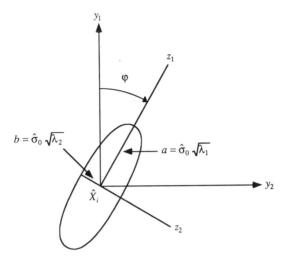

Figure 4.9 Defining elements of the standard ellipse.

the geometric meaning of the parameters. Examples of "nonnetwork" parameters are the intercept and slope in the fitting of a straight line. The components X_i can always be interpreted as Cartesian coordinates for the purpose of drawing the standard ellipse and thus can give a graphical display of the covariance. In the language of linear algebra, X_i simply represents two components of a u-dimensional vector X that lies in the u-dimensional space spanned by the column space of the design matrix. For surveying networks, the vectors X_i contain coordinates of stations in a well-defined coordinate system. If X_i contains latitude and longitude or northing and easting parameters, the horizontal standard ellipse is computed. If X_i contains the vertical coordinate and easting, then the standard ellipse in the prime vertical is obtained. Instead of hypothesis (4.295) one can test coordinate differences. This leads to the definition of relative standard ellipses or relative standard ellipsoids. Because the shape of the standard ellipses and ellipsoids depends on the geometry of the network through the design matrix, the interpretation is enhanced if network geometry and the standard ellipses are displayed in one figure.

4.9.6 Properties of Standard Ellipses

The positional error p of a station is directly related to the standard ellipse, as seen in Figure 4.10. The positional error is the standard deviation of a station in a certain direction, say ψ. It is identical with the standard deviation of the distance to a known (fixed) station along the same direction ψ as computed from the linearized distance equation and variance–covariance propagation. The linear function is

$$r = z_1 \cos \psi + z_2 \sin \psi \tag{4.318}$$

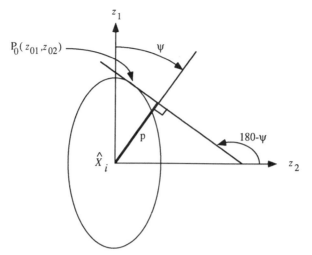

Figure 4.10   Positional error.

Because of equations (4.300) and (4.301), the distribution of the random variable Z_i is multivariate normal with

$$\begin{bmatrix} z_1 \\ z_2 \end{bmatrix} \sim N\left(\begin{bmatrix} 0 \\ 0 \end{bmatrix}, \hat{\sigma}_0^2 \begin{bmatrix} \lambda_1^Q & 0 \\ 0 & \lambda_2^Q \end{bmatrix} \right) = N\left(\begin{bmatrix} 0 \\ 0 \end{bmatrix}, \begin{bmatrix} a^2 & 0 \\ 0 & b^2 \end{bmatrix} \right) \tag{4.319}$$

The variance of the random variable r follows from the law of variance–covariance propagation:

$$\sigma_r^2 = a^2 \cos^2 \psi + b^2 \sin^2 \psi \tag{4.320}$$

The variance (4.320) is geometrically related to the standard ellipse. Let the ellipse be projected onto the direction ψ. The point of tangency is denoted by P_0. Because the equation of the ellipse is

$$\frac{z_1^2}{a^2} + \frac{z_2^2}{b^2} = 1 \tag{4.321}$$

the slope of the tangent is

$$\frac{dz_1}{dz_2} = -\frac{z_2 a^2}{z_1 b^2} = -\tan \psi \tag{4.322}$$

See Figure 4.10 regarding the relation of the slope of the tangent and the angle

ψ. The second part of (4.322) yields

$$\frac{z_{01}}{a^2} \sin \psi - \frac{z_{02}}{b^2} \cos \psi = 0 \qquad (4.323)$$

This equation relates the coordinates of the point of tangency P_0 to the slope of the tangent. The length p of the projection of the ellipse is, according to Figure 4.10,

$$p = z_{01} \cos \psi + z_{02} \sin \psi \qquad (4.324)$$

Next, (4.323) is squared and then multiplied with $a^2 b^2$, and the result is added to the square of (4.324), giving

$$p^2 = a^2 \cos^2 \psi + b^2 \sin^2 \psi \qquad (4.325)$$

By comparing this expression with (4.320), it follows that $\hat{\sigma}_r = p$; that is, the standard deviation in a certain direction is equal to the projection of the standard ellipse onto that direction. Therefore, the standard ellipse is not a standard deviation curve. Figure 4.11 shows the continuous standard deviation curve. We see that for narrow ellipses there is only a small segment for which the standard deviations are close to the length of the semiminor axis. The standard deviation increases rapidly as the direction ψ moves away from the minor axis. Therefore, an extremely narrow ellipse is not desirable if the overall accuracy for the station position is important.

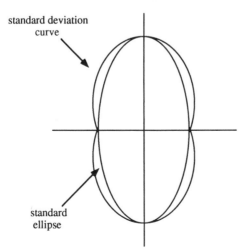

Figure 4.11 Standard deviation curve.

As a by-product of the property discussed, we see that the standard deviations of the parameters x_1 and x_2

$$\hat{\sigma}_{x_1} = \hat{\sigma}_0 \sqrt{q_{x_1}} \tag{4.326}$$

$$\hat{\sigma}_{x_2} = \hat{\sigma}_0 \sqrt{q_{x_2}} \tag{4.327}$$

are the projections of the ellipse onto the directions of the x_1 and x_2 axes. This is shown in Figure 4.12. Equations (4.326) and (4.327) follow from the fact that the diagonal elements of the covariance matrix are the variances of the respective parameters. Equation (4.325) confirms for $\psi = 0$ and $\psi = 90$ that the axes a and b equal the maximum and minimum standard deviations, respectively.

The rectangle formed by the semisides $\hat{\sigma}_{x_1}$ and $\hat{\sigma}_{x_2}$ encloses the ellipse. This rectangle can be used as an approximation for the ellipses. The diagonal itself is sometimes referred to as the mean position error. From equations (4.326), (4.327), (4.313), and (4.314) the mean position error is

$$\hat{\sigma} = \sqrt{\hat{\sigma}_{x_1}^2 + \hat{\sigma}_{x_2}^2} = \hat{\sigma}_0 \sqrt{\text{Tr } Q_i} = \hat{\sigma}_0 \sqrt{q_{x_1} + q_{x_2}} \tag{4.328}$$

The points of contact between the ellipse and the rectangle in Figure 4.12 are functions of the correlation coefficients. For these points the tangent on the ellipse is either horizontal or vertical in the (Y_i) coordinate system. The equation of the ellipse in the (Y) system is, according to (4.299), (4.301), and

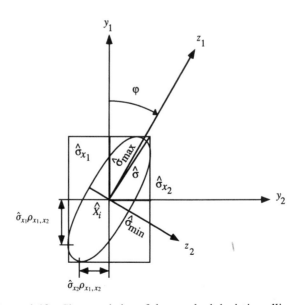

Figure 4.12 Characteristics of the standard deviation ellipse.

(4.304),

$$[y_1 \quad y_2] \begin{bmatrix} q_{x1} & q_{x1,x2} \\ q_{x1,x2} & q_{x2} \end{bmatrix}^{-1} \begin{bmatrix} y_1 \\ y_2 \end{bmatrix} = \hat{\sigma}_0^2 \tag{4.329}$$

By replacing the matrix by its inverse, the expression becomes

$$[y_1 \quad y_2] \begin{bmatrix} q_{x2} & -q_{x1,x2} \\ -q_{x1,x2} & q_{x1} \end{bmatrix} \begin{bmatrix} y_1 \\ y_2 \end{bmatrix} = (q_{x1} q_{x2} - q_{x1,x2}^2) \hat{\sigma}_0^2 \tag{4.330}$$

Evaluating the left-hand side and dividing both sides by $q_{x1} q_{x2}$ gives

$$\frac{y_1^2}{q_{x1}} + \frac{y_2^2}{q_{x2}} - \frac{2y_1 y_2 q_{x1,x2}}{q_{x1} q_{x2}} = \text{constant} \tag{4.331}$$

from which it follows that

$$\frac{dy_1}{dy_2} = \frac{2y_2/q_{x2} - 2y_1 \rho_{x1,x2}/\sqrt{q_{x1} q_{x2}}}{2y_2 \rho_{x1,x2}/\sqrt{q_{x1} q_{x2}} - 2y_1/q_{x1}} \tag{4.332}$$

Consider the tangent for which the slope is infinity. The equation of this tangent line is

$$y_2 = \hat{\sigma}_0 \sqrt{q_{x2}} \tag{4.333}$$

Substituting this expression into the denominator of (4.332) and equating it to zero gives

$$\frac{\hat{\sigma}_0 \sqrt{q_{x2}} \rho_{x1,x2}}{\sqrt{q_{x1} q_{x2}}} = \frac{y_1}{q_{x1}} \tag{4.334}$$

which yields the y_1 coordinate for the point of tangency:

$$y_1 = \hat{\sigma}_0 \sqrt{q_{x1}} \rho_{x1,x2} = \hat{\sigma}_{x1} \rho_{x1,x2} \tag{4.335}$$

The equation for the horizontal tangent is

$$y_1 = \hat{\sigma}_0 \sqrt{q_{x1}} \tag{4.336}$$

It follows from the numerator of (4.332) that

$$y_2 = \hat{\sigma}_0 \sqrt{q_{x2}} \rho_{x1,x2} = \hat{\sigma}_{x2} \rho_{x1,x2} \tag{4.337}$$

Figure 4.12 shows that the standard ellipse becomes narrower the larger the correlation coefficient. If the correlation is one or minus one (linear dependence of the two parameters), the ellipse degenerates into the diagonal of the rectangle. The ellipse becomes a circle if $a = b$, or $\sigma_{x_1} = \sigma_{x_2}$ and $\rho_{x_1,x_2} = 0$.

4.9.7 Other Measures of Precision

In surveying and geodesy the most popular measure of precision is the standard deviation. The confidence regions are usually expressed in terms of ellipses and ellipsoids of standard deviation. These figures are often scaled to contain 95% probability or higher. Because GPS is a popular tool for both surveying and navigation, several of the measures of precision used in navigation are becoming increasingly popular in surveying. Examples include the dilution of precision (DOP) numbers. The DOPs are discussed in detail in Chapter 8. Other single-number measures refer to circular or spherical confidence regions for which the eigenvalues of the cofactor matrix have the same magnitude. In these cases the standard deviations of the coordinates and the semiaxes are of the same size. See equation (4.302). When the standard deviations are not equal, these measures become a function of the ratio of the semiaxes of the standard ellipses. The derivation of the following measures and additional interpretation are given in Greenwalt and Shultz (1962).

The radius of a circle that contains 50% probability is called the circular error probable (CEP). This function is usually approximated by segments of straight lines. The expression

$$\text{CEP} = 0.5887(\hat{\sigma}_{x_1} + \hat{\sigma}_{x_2}) \qquad (4.338)$$

is, strictly speaking, valid in the region $\sigma_{\min}/\sigma_{\max} \geq 0.2$ but it is the function used most often. The 90% probability region

$$\text{CMAS} = 1.8227 \times \text{CEP} \qquad (4.339)$$

is called the circular map accuracy standard. The mean position error (4.328) is also called the mean square positional error (MSPE) or the distance root mean square (DRMS), i.e.,

$$\text{DRMS} = \sqrt{\hat{\sigma}_{x_1}^2 + \hat{\sigma}_{x_2}^2} \qquad (4.340)$$

This measure contains 64 to 77% probability. The related measure

$$2\,\text{DRMS} = 2 \times \text{DRMS} \qquad (4.341)$$

contains about 95 to 98% probability.

The three-dimensional equivalent of CEP is the spherical error probable (SEP) defined as

$$SEP = 0.5127(\hat{\sigma}_{x_1} + \hat{\sigma}_{x_2} + \hat{\sigma}_{x_3}) \tag{4.342}$$

Expression (4.342) is, strictly speaking, valid in the region $\sigma_{min}/\sigma_{max} \geq 0.35$. The corresponding 90% probability region,

$$SAS = 1.626 \times SEP \tag{4.343}$$

is called the spherical accuracy standard. The mean radial spherical error (MRSE) is defined as

$$MRSE = \sqrt{\hat{\sigma}_{x_1}^2 + \hat{\sigma}_{x_2}^2 + \hat{\sigma}_{x_3}^2} \tag{4.344}$$

and contains about 61% probability.

These measures of precision are sometimes used to capture the achieved or anticipated precision conveniently using single numbers. However, the geometry of the adjustment seldom produces covariance matrices that yield circular distribution. Consequently, the probability levels contained in these measures of precision inevitably are a function of the correlations between the parameters.

4.10 RELIABILITY

Small residuals are not necessarily an indication of a quality adjustment. Equally important is the knowledge that all blunders in the data have been identified and removed and that remaining small blunders in the observations do not adversely impact the adjusted parameters. Reliability refers to the controlability of observations, that is, the ability to detect blunders and to estimate the effects that undetected blunders may have on a solution. The theory outlined here follows that of Baarda (1967, 1968) and Kok (1984).

4.10.1 Redundancy Numbers

Table 4.1 contains the following expression for the observation equation model:

$$\hat{V} = Q_V PL \tag{4.345}$$

with a cofactor matrix for the residuals

$$Q_V = P^{-1} - AN^{-1}A^T \tag{4.346}$$

Compute the trace

$$\text{Tr}\,(Q_V P) = \text{Tr}\,(I - AN^{-1}A^{\mathrm{T}}P)$$

$$= \text{Tr}\,(I) - \text{Tr}\,(AN^{-1}A^{\mathrm{T}}P) \qquad (4.347)$$

$$= n - \text{Tr}\,(N^{-1}A^{\mathrm{T}}PA)$$

$$= n - u$$

A more general expression is obtained by noting that the matrices $AN^{-1}A^{\mathrm{T}}P$ and $I - AN^{-1}A^{\mathrm{T}}P$ are idempotent. The trace of an idempotent matrix equals the rank of that matrix. Thus,

$$\text{Tr}\,(AN^{-1}A^{\mathrm{T}}P) = R(A^{\mathrm{T}}PA) = R(A) = r \le u \qquad (4.348)$$

Thus, from equations (4.347) and (4.348)

$$\text{Tr}\,(Q_V P) = \text{Tr}\,(PQ_V) = n - R(A) \qquad (4.349)$$

By denoting the diagonal element of the matrix $Q_V P$ by r_i, we can write

$$\sum_{i=1}^{n} r_i = n - R(A) \qquad (4.350)$$

The sum of the diagonal elements of $Q_V P$ equals the degree of freedom. The element r_i is called the *redundancy number* for the observation i. It is the contribution of the ith observation to the degree of freedom. If the weight matrix P is diagonal, which is usually the case when original observations are adjusted, then

$$r_i = q_i p_i \qquad (4.351)$$

where q_i is the diagonal element of the cofactor matrix Q_V, and p_i denotes the weight of the ith observation. Equation (4.346) implies the inequality

$$0 \le q_i \le \frac{1}{p_i} \qquad (4.352)$$

Multiplying by p_i gives

$$0 \le r_i \le 1 \qquad (3.353)$$

Considering the general relation

$$Q_{L_a} = Q_{L_b} - Q_V \qquad (4.354)$$

given in Table 4.1 and the specification (4.351) for the redundancy number r_i as the diagonal element of $Q_V P$, it follows that if the redundancy number is close to one, then the variance of the residuals is close to the variance of the observations, and the variance of the adjusted observations is close to zero. If the redundancy number is close to zero, then the variance of the residuals is close to zero, and the variance of the adjusted observations is close to the variance of the observations.

Intuitively it is expected that the variance of the residuals and the variance of the observations are close; for this case, the noise in the residuals equals that of the observations, and the adjusted observations are determined with high precision. Thus the case of r_i close to one is preferred, and it is said that the gain of the adjustment is high. If r_i is close to zero, one expects the noise in the residuals to be small. Thus, small residuals as compared to the expected noise of the observations is not necessarily desirable. Because the inequality (4.353) is a result of the geometry as represented by the design matrix A, small residuals can be an indication of a weak part of the network.

Because the weight matrix P is considered diagonal, that is,

$$p_i = \frac{\sigma_0^2}{\sigma_i^2} \tag{4.355}$$

it follows that

$$\hat{\sigma}_{v_i} = \hat{\sigma}_0 \sqrt{q_i} = \hat{\sigma}_0 \sqrt{\frac{r_i}{p_i}} = \hat{\sigma}_0 \sqrt{\frac{r_i \sigma_i^2}{\sigma_0^2}} = \frac{\hat{\sigma}_0}{\sigma_0} \sigma_i \sqrt{r_i} \tag{4.356}$$

From (4.350) it follows that the average redundancy number is

$$r_{av} = \frac{n - R(A)}{n} \tag{4.357}$$

The higher the degree of freedom, the closer the average redundancy number is to one. However as seen from Table 4.6, the gain, in terms of probability enclosed by the standard ellipses, reduces noticeably after a certain degree of freedom.

4.10.2 Controlling Type II Error for a Single Blunder

Baarda's (1967) development of the concept of reliability of networks is based on un-Studentized hypothesis tests, which means that the a priori variance of unit weight is assumed to be known. Consequently, the a priori variance of unit weight (and not the a posteriori variance of unit weight) is used in this section. The alternative hypothesis H_a specifies that the observations contain one blunder, that the blunder be located at observation i, and that its magnitude be ∇_i. Thus, the adjusted residuals for the case of the alternative hypothesis are

$$\hat{V}|H_a = \hat{V} - Q_V P e_i \nabla_i \tag{4.358}$$

where

$$e_i = [0, 0, \ldots, 0, 1, 0, \ldots, 0]^T \tag{4.359}$$

denotes an $(n \times 1)$ vector containing 1 in position i and 0's elsewhere. The expected value and the covariance matrix are

$$E(\hat{V}|H_a) = -Q_V P e_i \nabla_i \tag{4.360}$$

$$\Sigma_{V|H_a} = \hat{\Sigma}_V = \sigma_0^2 Q_V \tag{4.361}$$

It follows from Theorem 1 of Section 4.9.1 that

$$\hat{V}|H_a \sim N(-Q_V P e_i \nabla_i, \sigma_0^2 Q_V) \tag{4.362}$$

Since P is a diagonal matrix, the individual residuals are distributed as

$$\hat{v}_i|H_a \sim n(-q_i p_i \nabla_i, \sigma_0^2 q_i) \tag{4.363}$$

according to Theorem 2. Standardizing gives

$$w_a|H_a = \frac{\hat{v}_i|H_a}{\sigma_0 \sqrt{q_i}} \sim n\left(\frac{-q_i p_i \nabla_i}{\sigma_0 \sqrt{q_i}}, 1\right)$$

$$= n\left(\frac{-\sqrt{q_i} p_i \nabla_i}{\sigma_0}, 1\right) \tag{4.364}$$

or

$$H_a: w_a = \frac{\hat{v}_i|H_a}{\sigma_{v_i}} \sim n\left(\frac{-\nabla_i p_i \sqrt{q_i}}{\sigma_0}, 1\right) \tag{4.365}$$

The zero hypothesis, which states that there is no blunder, is

$$H_0: w_0 = \frac{\hat{v}_i|H_0}{\sigma_{v_i}} \sim n(0, 1) \tag{4.366}$$

The noncentrality parameter in (4.365), that is, the mean of the noncentral normal distribution, is denoted by δ_i and is

$$\delta_i = \frac{-\nabla_i p_i \sqrt{q_i}}{\sigma_0} = \frac{-\nabla_i \sqrt{r_i}}{\sigma_i} \qquad (4.367)$$

The parameter δ_i is a translation parameter of the normal distribution. The situation is shown in Figure 4.13. The probability of committing an error of the first kind, that is, of accepting the alternative hypothesis, equals the significance level α of the test

$$P(|w| \le t_{\alpha/2}) = \int_{-t_{\alpha/2}}^{t_{\alpha/2}} n(0, 1) \, dx = 1 - \alpha \qquad (4.368)$$

or

$$P(|w| \ge t_{\alpha/2}) = \int_{-\infty}^{t_{1-\alpha/2}} n(0, 1) \, dx + \int_{t_{\alpha/2}}^{\infty} n(0, 1) \, dx = \alpha \qquad (4.369)$$

In $100\alpha\%$ of the cases the observations are rejected, and remeasurement or investigations for error sources are performed even though the observations are correct (they do not contain a blunder). From Figure 4.13 it is seen that the probability β_i of a type II error, that is, the probability of rejecting the alternative hypothesis (and accepting the zero hypothesis) even though the alternative hypothesis is correct, depends on the noncentrality factor δ_i. Because the blunder ∇_i is not known, the noncentrality factor is not known either. As a practical matter one can proceed in the reverse: One can assume an acceptable probability β_0 for the error of the second kind and compute the respective noncentrality parameter δ_0. This parameter in turn is used to compute the lower

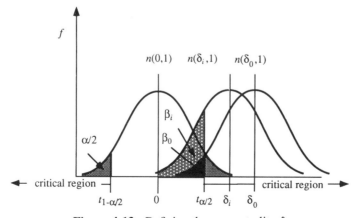

Figure 4.13   Defining the noncentrality δ_0.

limit for the blunder, which can still be detected. Figure 4.13 shows that

$$P(|w_a| \le t_{\alpha/2}) = \int_{-t_{\alpha/2}}^{t_{\alpha/2}} n(\delta_i, 1) \, dx \le \beta_0 \qquad (4.370)$$

if

$$\delta_i \ge \delta_0 \qquad (4.371)$$

Substituting equation (4.367) into (4.371) gives the limit for the marginally detectable blunder, given the probability levels α and β_0:

$$|\nabla_{0i}| \ge \frac{\delta_0}{\sqrt{r_i}} \sigma_i \qquad (4.372)$$

Equations (4.370) and (4.372) state that in $100(1 - \beta_0)\%$ of the cases, blunders greater than those given in (4.372) are detected. In $100\beta_0\%$ of the cases, blunders greater than those given in (4.372) remain undetected. The larger the redundancy number of the observation, the smaller is the marginally detectable blunder (for the same δ_0 and σ_i). It is important to recognize that the marginal detectable blunders (4.372) are based on adopted probabilities of type I and type II errors for the normal distribution. The probability levels α and β_0 refer to the one-dimensional test (4.366) of the individual residual v_i, with the non-centrality being δ_0. The assumption is that only one blunder at a time is present. The geometry is shown in Figure 4.13. It is readily clear that there is a simple functional relationship $\delta_0 = \delta_n(\alpha, \beta_0)$ between two normal distributions. Table 4.9 contains selected probability levels and the respective δ_0 values.

The chi-square test (4.268) of the a posteriori variance of unit weight $\hat{\sigma}_0^2$ is also sensitive to the blunder ∇_i. It fact, the blunder will cause a noncentrality of δ_i^2 for the chi-square distribution of the alternative hypothesis. One can

TABLE 4.9 Selected Probability Levels in Reliability

α	β_0	δ_0
0.05	0.20	2.80
0.025	0.20	3.1
0.001	0.20	4.12
0.05	0.10	3.24
0.025	0.10	3.52
0.001	0.10	4.57

choose the probabilities α_{chi} and β_{chi} for this multidimensional chi-square test such that $\delta_0^2 = \delta_{chi}(\alpha_{chi}, \beta_{chi}, n - u)$. The factor δ_0 depends on the degree of freedom because the chi-square distribution depends on it. Baarda's B method suggests equal traceability of errors through one-dimensional tests of individual residuals, v_i, and the multidimensional test of the a posteriori variance of unit weight $\hat{\sigma}_0^2$. This is achieved by requiring that the one-dimensional and the multidimensional test have the same type II error, i.e., $\beta_0 = \beta_{chi}$. Under this condition there exists a relationship between the type II error, the significance levels, and the degree of freedom expressed symbolically by $\delta_0 = \delta_n(\alpha, \beta_0)$ $= \delta_{chi}(\alpha_{chi}, \beta_0, n - r)$. The B method assures equal traceability but implies different significance levels for the one-dimensional and multidimensional tests, i.e., the probability of rejecting the zero hypothesis when it is true. Baarda (1968, p. 25) shows that for a constant α_{chi} the respective α rapidly decreases as the degree of freedom increases. For example, $\alpha_{chi} = 0.05$, $\beta = \beta_0 = 0.20$, and $n - r = 100$ gives $\alpha = 0.000003$ and $\delta_0 = 6.37$. Such a small significance level makes the one-dimensional test, according to Baarda, "very dubious." However, useful modifications are possible. For example, one can choose the factor δ_0 on the basis of a reasonable value for α and β_0 (and not be concerned with α_{chi}). See remarks regarding internal reliability and data snooping below.

4.10.3 Internal Reliability

Even though the one-dimensional test is based on the assumptions that only one blunder exists in a set of observations, the limit (4.372) is usually computed for all observations. Thus, the marginally detectable errors, computed for all observations, are a measure of the capability of the network to detect blunders with probability $(1 - \beta_0)$. They constitute the *internal reliability of the network*. Because the marginally detectable errors (4.372) do not depend on the observations or on the residuals, they can be computed as soon as the configuration of the network and the stochastic model are known. If the limits (4.372) are of about the same size, the observations are equally well checked, and the internal reliability is said to be consistent. The factor δ_0 is chosen to reflect the type and size of blunders expected from the measurement system. A typical value is $\delta_0 = 4$. The emphasis is then on the variability of the marginal detectable blunders rather on their magnitude.

4.10.4 Absorption

According to (4.345) the residuals in the presence of one blunder are

$$V = Q_V P(L - e_i \nabla_i) \qquad (4.373)$$

The impact on the residual of observation i is

$$\nabla v_i = -r_i \nabla_i \qquad (4.374)$$

Equation (4.374) is used to estimate the blunders that might cause large residuals. Solving for ∇_i gives

$$\nabla_i = -\frac{\nabla v_i}{r_i} \approx -\frac{v_i^* + \nabla v_i}{r_i} \approx -\frac{v_i}{r_i} \tag{4.375}$$

because $v_i^* \ll \nabla v_i$, where v_i^* denotes the residual without the effect of the blunder. The computation (4.375) provides only estimates of possible blunders. Because the matrix $Q_V P$ is not a diagonal matrix, a specific blunder has an impact on all residuals. If several blunders are present, their effects overlap and one blunder can mask others; a blunder may cause rejection of a good observation.

Equation (4.374) demonstrates that the residuals in least-squares adjustments are not robust with respect to blunders in the sense that the effect of a blunder on the residuals is smaller than the blunder itself because r varies between 0 and 1. The absorption, i.e., the portion of the blunder that propagates into the estimated parameters and falsifies the solution, is

$$A_i = \nabla v_i + \nabla_i = (1 - r_i)\nabla_i \tag{4.376}$$

The factor $(1 - r_i)$ is called the absorption number. The larger the redundancy number is, the less absorption of the blunder, i.e., the less falsification. If $r_i = 1$, the observation is called fully controlled, because the residual completely reflects the blunder. A zero redundancy implies uncontrolled observations in that a blunder enters into the solution with its full size. Observations with small redundancy numbers might have small residuals and instill false security in the analyst. Substituting ∇_i from (4.375) expresses the absorption number as a function of the residuals:

$$A_i = -\frac{1 - r_i}{r_i} v_i \tag{4.377}$$

The residuals can be looked on as the visible parts of errors. The factor in (4.377) is required to compute the invisible part from the residuals.

4.10.5 External Reliability

A good and homogeneous internal reliability does not automatically guarantee reliable coordinates. What are the effects of undetectable blunders on the parameters? In deformation analysis, where changes in parameters between adjustments of different epochs indicate existing deformations, it is particularly important that the impact of blunders on the parameters be minimal. The influence of each of the marginally detectable errors on the parameters of the adjustment or on functions of the parameters is called *external reliability*. The estimated parameters in the presence of a blunder are, for the observation equation model,

$$\hat{X} = -N^{-1}A^\mathrm{T}P(L - e_i \nabla_i) \tag{4.378}$$

The effect of the blunder in observation i is

$$\nabla X = N^{-1}A^\mathrm{T}P e_i \nabla_i \tag{4.379}$$

The shifts ∇X are sometimes called local external reliability. The blunder affects all parameters. The impact of the marginally detectable blunder ∇_{0i} is

$$\nabla X_{0i} = N^{-1}A^\mathrm{T}P e_i \nabla_{0i} \tag{4.380}$$

Because there are n observations, one can compute n vectors (4.380), showing the impact of each marginal detectable blunder on the parameters. Graphical representations of these effects can be very helpful in the analysis. The problem with (4.380) is that the effect on the coordinates depends on the definition (minimal constraints) of the coordinate system. Baarda suggested the following alternative expression:

$$\lambda_{0i}^2 = \frac{\nabla X_{0i}^\mathrm{T} N \nabla X_{0i}}{\sigma_0^2} \tag{4.381}$$

By substituting (4.380) and (4.372), we can write this equation as

$$\begin{aligned}
\lambda_{0i}^2 &= \frac{\nabla_{0i} e_i^\mathrm{T} PAN^{-1}A^\mathrm{T}P e_i \nabla_{0i}}{\sigma_0^2} = \frac{\nabla_{0i}^2 e_i^\mathrm{T} P(I - Q_V P)e_i}{\sigma_2^0} \\
&= \frac{\nabla_{0i}^2 p_i(1 - r_i)}{\sigma_0^2} = \frac{1 - r_i}{r_i} \delta_0^2
\end{aligned} \tag{4.382}$$

The values λ_{0i} are a measure of global external reliability. There is one such value for each observation. If the λ_{0i} are the same order of magnitude, the network is homogeneous with respect to external reliability. If r_i is small, the external reliability factor becomes large and the global falsification caused by a blunder can be significant. It follows that very small redundancy numbers are not desirable. The global external reliability number (4.382) and the absorption number (4.377) have the same dependency on the redundancy numbers.

4.10.6 Correlated Cases

The derivation for detectable blunders, internal reliability, absorption, and external reliability assume uncorrelated observations for which the covariance matrix Σ_{L_b} is diagonal. Correlated observations are decorrelated by the transformation (4.241). It can readily be verified that the redundancy numbers for the decorrelated observations $\bar{L}$ are

$$\bar{r}_i = (\bar{Q}_V \bar{P})_{ii} = (I - D^T A N^{-1} A^T D)_{ii} \qquad (4.383)$$

In many applications the covariance matrix Σ_{L_b} is of block-diagonal form. For example, for GPS vector observations, this matrix consists of (3×3) full block-diagonal matrices if the correlations between the vectors are neglected. In this case the matrix D is also block diagonal and the redundancy numbers can be computed vector by vector from (4.383). The sum of the redundancy numbers for the three vector components varies between 0 and 3. Since, in general, the matrix D has a full rank, the degree of freedom $(n - r)$ of the adjustment does not change. Once the redundancy numbers $\bar{r}_i$ are available, the marginal detectable blunders $\bar{\nabla}_{0i}$, the absorption numbers $\bar{A}_i$, and other reliability values can be computed for the decorrelated observations. These quantities, in turn, can be transformed back into the physical observation space by premultiplication with the matrix $(D^T)^{-1}$. Appendix A contains a discussion on the use of the Cholesky factors to decorrelate observations.

4.11 BLUNDER DETECTION

Errors (blunders) made during the recording of field observations, data transfer, computations, etc., can be costly and time-consuming to find and eliminate. Blunder detection can be carried out before the adjustment or as part of the adjustment. Before the adjustment, the misclosures (angle and/or distance of simple figures such as triangles and traverses) are analyzed. A priori blunder detection is helpful in detecting extra-large blunders caused by, for example, erroneous station numbering. Blunder detection in conjunction with the adjustment is based on the analysis of the residuals. The problem with using least-squares adjustments when blunders are present, is that the adjustment tends to hide (reduce) their impact and distribute their effects more or less throughout the entire network [confer (4.373) and (4.374), noting that the redundancy number varies between zero and one]. The prerequisite for any blunder-detection procedure is the availability of a set of redundant observations. Only observations with redundancy numbers greater than zero can be controlled.

It is important to understand that if a residual does not pass a statistical test, it does not mean that there is a blunder in that observation. The observation is merely flagged so that it can be examined and a decision about its retention or rejection can be made. Blind rejection is never recommended. A blunder in one observation usually affects the residuals in other observations. Therefore, the tests will often flag other observations in addition to the ones containing blunders. If one or more observations are flagged, the search begins to determine if there is a blunder. The first step is to check the field notes to confirm that no error occurred during the transfer of the observations to the computer file, and that all observations are reasonable "at face value." If a blunder is not located, the network should be broken down into smaller networks, and each one should be adjusted separately. At the extreme, the entire network may be broken down into triangles or other simple geometric entities, such as traverses, and adjusted separately. Alternatively, the observations can be added sequentially, one at a time, until the blunder is found. This procedure starts with weights assigned to all parameters. The observations are then added sequentially. The sum of the normalized residuals squared is then inspected for unusually large deviations. When searching for blunders, the coordinate system should be defined by minimal constraints.

Blunder detection in conjunction with the adjustment, takes advantage of the total redundancy and the strength provided by the overall geometry of the network, and thus is particularly sensitive to smaller blunders. Only if the existence of a blunder is indicated does action need to be taken to locate the blunder. The flagged observations are the best hint to where to look for troublesome observations and thus avoid unnecessary and disorganized searching of the whole observation data set.

4.11.1 The τ Test

The τ test was introduced by Pope (1976). The type II error is not taken into consideration. The test belongs to the group of Studentized tests, which make use of the a posteriori variance of unit weight as estimated from the observations. The test statistic is

$$\tau_i = \frac{v_i}{\hat{\sigma}_{v_i}} = \frac{v_i}{\hat{\sigma}_0 \sqrt{q_i}} = \frac{v_i}{\hat{\sigma}_0 \sqrt{r_i/p_i}} = \frac{\sigma_0}{\hat{\sigma}_0} \frac{v_i}{\sigma_i \sqrt{r_i}} \sim \tau_{n-r} \qquad (4.384)$$

The symbol τ_{n-r} denotes the τ distribution with $n - r$ degrees of freedom. It is related to Student's t by

$$\tau_{n-r} = \frac{\sqrt{n - r} \, t_{n-r-1}}{\sqrt{n - r - 1 + t_{n-r-1}^2}} \qquad (4.385)$$

For an infinite degree of freedom the τ distribution converges toward the Student distribution or the standardized normal distribution, i.e., $\tau_\infty = t_\infty = n(0, 1)$. Curve 1 of Figure 4.14 shows the critical value such that $P(\tau \geq \tau_{n-r,\alpha/2})$ with $\alpha = 0.05$. The value of 5% is a commonly used level of significance (type I error). The critical value begins at 1.41 and converges rapidly toward the normal distribution value of 1.96 for higher degrees of freedom.

The test statistic (4.384) represents a data-adoptive criterion. For large sys-

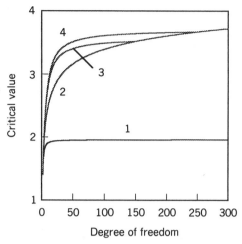

Figure 4.14 Examples of critical values for Pope's test ($\alpha = 0.05$). Curve 1: $\tau(n - r)$; curve 2: $c_{n-r=1}$; curve 3: $c_{n=150}$; curve 4: $c_{n=250}$.

tems, the redundancy numbers are often replaced by the average value according to equation (4.357) in order to reduce computation time; thus,

$$\tau_i = \frac{\sigma_0}{\hat{\sigma}_0} \frac{v_i}{\sigma_i \sqrt{(n - r)/n}} \tag{4.386}$$

Pope's blunder rejection procedure tests the hypothesis $v_i \sim n(0, \hat{\sigma}_{v_i}/\hat{\sigma}_0)$ for all i. The hypothesis is rejected, i.e., the observation is flagged for further investigation and possibly rejection, if

$$|\tau_i| \geq c \tag{4.387}$$

The critical value c is selected such that the significance level $\alpha = P(\tau_{\max} > c)$ using a fixed, preselected value, say $\alpha = 0.05$. The critical value c is related to the number of observations in the adjustment n such that $P(\tau_i > c) = \alpha/n$ (Pope 1976, p. 16). Thus, the critical value c is determined by

$$\int_c^\infty \tau_{n-r} \, dx = \frac{\alpha}{2n} \tag{4.388}$$

The critical value c becomes a function of the degree of freedom and the number of observations. Figure 4.14 shows several examples for c. Curve 2 represents the limiting case of $n - r = 1$. All other curves that refer to more than one degree of freedom (as indicated by the horizontal axis) must lie above this base curve. For example, curve 3 refers to $n = 150$ and approaches curve 2

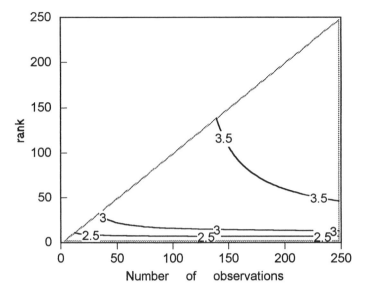

Figure 4.15   Contouring Pope's c as a function of n and r ($\alpha = 0.05$).

at $r = 150$ where it stops. Curve 4 shows the critical values c for $n = 250$. Typically, the critical values c increase rapidly in the range of low degrees of freedom and gradually approach curve 2. The latter curve reaches 4.70 at $r = 20{,}000$. Figure 4.15 shows contour lines of c. The rapid increase of the critical value with the degree of freedom is clearly visible.

4.11.2   Data Snooping

Baarda's data snooping applies to the testing of individual residuals as well. The theory assumes that only one blunder be present in the set of observations. Applying a series of one-dimensional tests, that is, testing consecutively all residuals, is called a data snooping strategy. Baarda's tests belongs to the group of un-Studentized tests that assume that the a priori variance of unit weight is known. The zero hypothesis (4.366) is written as

$$n_i = \frac{v_i}{\sigma_0 \sqrt{q_i}} \sim n(0,\,1) \tag{4.389}$$

At a significant level of 5% the critical value is 1.96. The critical value for this test is not a function of the number of observations in the adjustment. The statistic (4.389) uses the a priori value σ_0 and not the a posteriori estimate $\hat{\sigma}_0$.

The statistics (4.384) and (4.389) are functions of the individual redundancy numbers. For a given size of residuals, the respective observations are more likely to be rejected the smaller the redundancy numbers. Because the absorption of blunders increases with decreasing redundancy numbers, the heightened sensitivity of the critical value to small redundancy numbers is a desirable property. Both the τ and the data snooping procedures work best for iterative solutions. At each iteration the observation with the largest blunder should be removed. Since least squares attempts to distribute blunders, several correct observations might receive large residuals and might be flagged mistakenly.

4.11.3 Changing Weights of Observations

This method, although not based on rigorous statistical theory, is an automated method whereby blunders are detected and their effects on the adjustment are minimized or even eliminated. The advantage that this method has over the other two methods is that it (potentially) locates and eliminates the blunders automatically. Basically, this method examines the residuals of each observation during each iteration. If the magnitude of the residual is outside a certain range, the weight of the corresponding observation is reduced. This process of reweighting and readjusting is continued until the solution converges. In the end, the effects of the blunders should have been minimized or even eliminated. This method robustizes the residuals by means of a reweighting strategy. The criteria for judging when a residual is "too large" and the function for reweighting the observations are somewhat arbitrary. One simple technique for selection and weight after each iteration is

$$p^{i+1} = p^i \begin{cases} p^i e^{|v|/3\sigma_i} & \text{if } |v| > 3\sigma_i \\ p^i & \text{if } |v| \leq 3\sigma_i \end{cases} \qquad (4.390)$$

where σ_i denotes the standard deviations of the observation. Other functions may be used for the reweighting of observations, but they must be rapidly declining, positive functions. See Eeg (1986) for the underlying theory of the reweighting technique.

The method works efficiently when the redundancy number is large. One possible danger with this method occurs when the initial approximations of the parameters are fairly inaccurate. This can lead to the weight reduction of many correct observations after the first iteration, because of large residuals caused by the nonlinearity of the adjustment. To avoid unnecessary rejection and reweighting, one might not change the weights during the first iteration. If the residuals do not exceed three times the standard deviation, the weights should not be changed; that is, the original weights should be used. Use of this method requires some experience. All observations whose weights are changed must be investigated, and the cause for change in weight must be established.

CHAPTER 5

LEAST-SQUARES ADJUSTMENT EXAMPLES

In the previous chapter the theoretical foundation of basic least-squares adjustment was developed. As in many mathematical developments, it is devoid of intuitive interpretation by the reader unfamiliar with the adjustment techniques or the mathematical apparatus used in their definition. This intuitive "feel" for what is actually happening mathematically can best be imparted by detailed examples of the mechanics of the least-squares adjustment process and the interpretation of the results.

To this end a series of detailed adjustments are presented with the goal of reinforcing the concepts underlying the adjustment process and of providing enough information to give useful examples for the testing least-squares adjustment computer programs that the reader may choose to implement.

5.1 INTRODUCTION

All adjustment examples discussed here are based on the set of simulated observations performed on a 15-station network shown in Appendix F. The complete data set consisting of station coordinates, observations, and observation standard deviations is also listed in this appendix. Five types of observations are represented in the sample adjustments: horizontal angles, vertical angles, distances, height differences, and GPS baseline components. The angular observations, horizontal and vertical angles, are given in degrees, minutes, and seconds of arc. Distances, height differences, and GPS baseline components are given in meters. All observations except the GPS observations must be reduced to the conformal mapping plane in order to obtain the model observations for the plane adjustments discussed in this chapter. In the following

175

examples the adjustments are performed on the transverse Mercator using the GRS 80 reference ellipsoid and the following specifications:

$$\text{Longitude of the central meridian} = 291° \ 0' \ 0''$$

$$\text{Central meridian scale factor} = 0.9996$$

Appropriate direct and inverse transformation equations and expressions for reductions of observations can be found in Chapter 12. The reduced observations are not listed explicitly in Appendix F.

To keep the tutorial examples manageable, only observations on the quadrangle consisting of stations Six Mile (1), Trav-01 (6), Trav-09 (9), and Trav-14 (15) are presented in complete detail here. The entire network, exclusive of the GPS baseline data, is employed in later examples to emphasize specific points in the various adjustments.

Consider Figure 5.1, showing the four-station quadrangle on which distance and horizontal angle observations have been performed. In this figure stations Six Mile (1) and Trav-14 (15) are assumed to have known positions and the object of the adjustment is to estimate station positions for stations Trav-01 (6) and Trav-09 (9) based on the given observations.

In performing a least-squares adjustment using the observation equation model, two matrices must be formed and updated during the iteration procedure. These are the design matrix A and the misclosure matrix L. Note that the weight matrix need be specified only once and remains unchanged during the adjustment; that is, automatic blunder detection using the method of variation of weights is not considered in these examples. Each row of the design matrix represents a single observation, such as a distance or an angle. Each column of the design matrix represents the observational contribution of all observations to a single parameter to be estimated during the adjustment. The intersection of a row and a column, then, according to the definition of the A matrix, represents the partial derivative of the particular observation equation with respect to the unknown parameter. In forming the A matrix, the sequence in which the parameters are ordered in the columns is not important. What is critical is that the order be preserved throughout the adjustment procedure. Likewise, the order in which the observations are treated is not important, but the sequence of observations must be the same for the design matrix, the observation matrix, and the weight matrix. The tutorial adjustments assume that all observations are uncorrelated; thus, the weight matrix P is diagonal.

The four-station example consists of six observations (four distances and two horizontal angles) and four parameters (the coordinates of stations 6 and 9). Recall that the coordinates of Six Mile (1) and Trav-14 (15) stations are assumed to be known. Therefore, the matrices have the following dimensions (row, column): $A(6, 4)$, $L(6, 1)$, $P(6, 6)$, and $X(4, 1)$. Returning to Figure 5.1, note the ordering of the parameters listed as column headings and the

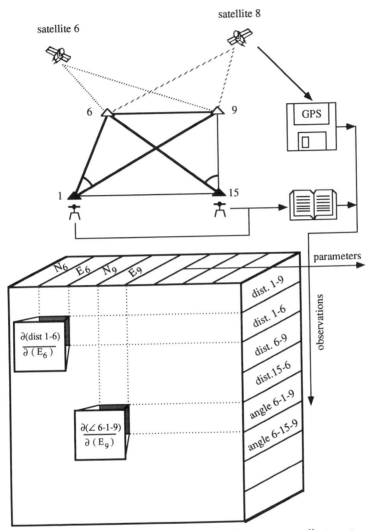

Figure 5.1 Design matrix population for least-squares adjustments.

ordering of the observation listed as row labels. Two elements of the design matrix have been highlighted. The first concerns the distance observation from station 1 to station 6 and the easting coordinate of station 6. This element, (2, 2), of the design matrix contains the partial derivative of the distance observation equation with respect to the easting coordinate of station 6. The second highlighted element relates an angle observation with its parameters. The (5, 4) element of the design matrix contains the partial derivative of the angle observation equation for the angle at station 1 from station 6 to station 9 with respect to the easting coordinate of station 9. The remaining elements are filled in precisely the same manner. Care must be exercised in assigning the correct algebraic signs to the partial derivatives.

Before beginning the adjustment, we must specify the approximate coordinates for the unknown stations. These approximate coordinates can be computed from the observations or scaled from a map or plat. The partial derivatives must be evaluated using the latest approximate coordinates for the unknown station parameters.

The vector L can be formed concurrently with the design matrix because approximate coordinates are normally required to compute the partial derivatives. The computed value of the observations is found from the appropriate inverse equations and the approximate coordinates. The L vector consists of the difference between the approximate observation as computed using the latest estimates for station coordinates and the actual observation recorded from the particular instrument (reduction to the mapping plane is, of course, required in this case). The actual matrices for this simple example are presented in the first sample adjustment).

The operations performed in a typical least-squares adjustment are summarized in Table 5.1. The items listed in the inset box are the core operations of the least-squares adjustment, that is, those that must be repeated for each iteration; the remaining operations are performed only once. A summary of the appropriate observation equation model equations is given in Table 5.2.

5.2 ADJUSTMENT OF A SIMPLE NETWORK ON THE MAPPING PLANE

The adjustment initially presented in the discussion of the design matrix is now adjusted to estimate the coordinates of stations 6 and 9. The summary of observations and unknowns is shown in Table 5.3. Unless otherwise noted, both

TABLE 5.1   Least-Squares Adjustment Operations

Populate P, X_0
Populate A, L Compute N, U, X Update parameters Test $V^T P V$
Compute adjustment statistics Compute adjusted quantities

TABLE 5.2 Observation Equation Model Summary

P weight matrix
observation equations
$V = AX + L$ $L = L_0 - L_b$
normal equations
$NX = -U$ $N = A^T PA$ $U = A^T PL$
adjustment statistics
$\hat{\sigma}_0^2 = \dfrac{V^T PV}{n - u}$ $Q_X = N^{-1}$
redundancy number
$r = \text{diag}\left(I - AN^{-1}A^T P\right)$

the northing and easting coordinates are estimated for the stations in the parameter list.

The approximate coordinates for the unknown stations X_0 were taken directly from Table F.3 of Appendix F, the latitude and longitude being converted to map coordinates. The weight matrix for the observation data, assuming uncorrelated observation, is

$$P_{\text{diag}} = \left[\frac{1}{0.540^2} \; \frac{1}{0.217^2} \; \frac{1}{0.342^2} \; \frac{1}{0.330^2} \; \frac{1}{\left(\dfrac{3}{3600}\right)^2} \; \frac{1}{\left(\dfrac{3}{3600}\right)^2} \right]$$

TABLE 5.3 Example 1: Adjustment Summary

Observations:	as shown in Fig. 5.1
Parameters	6,9
Constraint :	stations 1,15 fixed

The a priori variance of unit weight is assumed to be unity. The validity of this assumption is tested following the adjustment. The design and observation matrices for each iteration are listed along with the adjusted station coordinates and observations.

For the first iteration, the design matrix A and the observation vector L are

$$A_{X_0} = \frac{\partial(\text{observation equation})}{(\partial N_6, \partial E_6, \partial N_9, \partial E_9)}$$

$$= \begin{bmatrix} 0.0000 & 0.0000 & 0.2887 & 0.9574 \\ 0.5750 & 0.8181 & 0.0000 & 0.0000 \\ -0.0925 & -0.9957 & 0.0925 & 0.9957 \\ 0.9270 & -0.2349 & 0.0000 & 0.0000 \\ 0.0109 & -0.0076 & -0.0051 & 0.0015 \\ -0.0020 & -0.0085 & -0.0039 & 0.0052 \end{bmatrix}$$

$$L = F(X_0) - L_b = \begin{bmatrix} 0.3425 \\ 0.1203 \\ 0.1757 \\ 0.0018 \\ -0.0002 \\ -0.0006 \end{bmatrix}$$

By forming the normal equations and solving for X, we obtain the corrections to the initial estimates of the station coordinates for the first iteration, indicated by X_1:

$$X_1 = \begin{bmatrix} -0.1083 \\ -0.0829 \\ -0.2316 \\ -0.2351 \end{bmatrix}$$

The initial estimates X_0 are updated using the computed corrections, and the iteration procedure is repeated. Thus, after the first iteration the station coordinates for station 6 and 9 are

$$X_{01} = X_1 + X_0 = \begin{bmatrix} 4,970,087.701 \\ 16,760.151 \\ 4,970,717.161 \\ 23,537.144 \end{bmatrix}$$

The design matrix and observation vector for the second iteration are now evaluated using the latest estimates for the unknown station coordinates, i.e., X_{01}. Thus,

$$A_{X_{01}} = \begin{bmatrix} 0.0000 & 0.0000 & 0.2887 & 0.9574 \\ 0.5750 & 0.8181 & 0.0000 & 0.0000 \\ -0.0925 & -0.9957 & 0.0925 & 0.9957 \\ 0.9270 & -0.2349 & 0.0000 & 0.0000 \\ 0.0109 & -0.0076 & -0.0051 & 0.0015 \\ -0.0020 & -0.0085 & -0.0039 & 0.0052 \end{bmatrix}$$

$$L = F(X_{01}) - L_b = \begin{bmatrix} 0.0505 \\ -0.0098 \\ 0.0128 \\ -0.0840 \\ 0.0000 \\ -0.0001 \end{bmatrix}$$

By solving the normal equations for X, we obtain the corrections to the latest estimates for the parameters:

$$X_2 = \begin{bmatrix} -0.9161 \times 10^{-6} \\ 0.1272 \times 10^{-6} \\ -0.6295 \times 10^{-6} \\ -0.2272 \times 10^{-6} \end{bmatrix}$$

and

$$X_{02} = X_{01} + X_2 = \begin{bmatrix} 49,70,087.701 \\ 16,760.151 \\ 4,970,717.161 \\ 23,537.144 \end{bmatrix}$$

After iteration 2, the change in V^TPV from the previous iteration can be computed. This quantity is tested to determine when the adjustment has converged. The difference after the second iteration is

$$|V^TPV_2 - V^TPV_1| = 0.08$$

Performing a third iteration, we obtain

$$A_{X_{02}} = \begin{bmatrix} 0.0000 & 0.0000 & 0.2287 & 0.9574 \\ 0.5750 & 0.8181 & 0.0000 & 0.0000 \\ -0.0925 & -0.9957 & 0.0925 & 0.9957 \\ 0.9270 & -0.2349 & 0.0000 & 0.0000 \\ 0.0109 & -0.0076 & -0.0051 & 0.0015 \\ -0.0020 & -0.0085 & -0.0039 & 0.0052 \end{bmatrix}$$

$$L = F(X_{02}) - L_0 = \begin{bmatrix} 0.0505 \\ -0.0098 \\ 0.0128 \\ -0.0840 \\ 0.0000 \\ -0.0001 \end{bmatrix}$$

$$X_3 = \begin{bmatrix} 0.3302 \times 10^{-09} \\ -0.8335 \times 10^{-10} \\ 0.4161 \times 10^{-10} \\ -0.1103 \times 10^{-10} \end{bmatrix}$$

and

$$X_{03} = X_{02} + X_3 = \begin{bmatrix} 4,970,087.701 \\ 16,760.151 \\ 4,970,717.161 \\ 23,537.144 \end{bmatrix}$$

After iteration 3,

$$|V^TPV_3 - V^TPV_2| = 0.3 \times 10^{-7}$$

Continued iteration reveals that the magnitude of the change in V^TPV has stabilized and fluctuates only in algebraic sign. This indicates that no further improvement in the adjustment will be obtained from further iteration, and thus the adjustment is said to have converged. At this point the corrections to the approximate unknown station coordinates are usually quite small, further indicating that adjustment has reached a satisfactory solution. So far, only the condition that the residual sum of squares be a minimum has been imposed and adjusted station coordinates have been determined. The statistics for the adjustment can now be computed. The first item to be computed is the a posteriori variance of unit weight $\hat{\sigma}_0^2$:

$$\hat{\sigma}_0^2 = \frac{V^TPV}{n - u}$$

Recall from (4.271) that the a priori and a posteriori variances are tested using the chi-square test as a measure of the success of the adjustment. The a posteriori variance is computed using the final value for V^TPV. The computation for this example are summarized in Table 5.4.

Next, the adjusted observations are computed using

$$L_a = L_b + V$$

The adjusted observations for this example are shown in Table 5.5 (recall that all observations listed in the tables of Chapter 5 refer to the mapping plane). In addition to the adjusted observations, the redundancy number r for each observation has been included. These redundancy numbers will be useful in pointing deficiencies in some of the following adjustments. Note that the sum of the redundancy numbers for the network observations equals the degree of freedom of the adjustment as expected.

The standard deviations for the estimated station coordinates can now be computed. They are obtained from the variance–covariance matrix of the ad-

TABLE 5.4 Example 1: Adjustment Statistics

Number of observations	6
Number of parameters	4
Number of observed parameters	0
Number of conditions	0
Degrees of freedom	2
V^TPV	0.0835
A-posteriori variance of unit weight	0.0417

TABLE 5.5 Example 1: Adjustment Observations

Stations	Observation	Residual	Adj. observations	r
1 9	10759.585	0.050	10759.635	0.6629
6 1	4307.814	-0.010	4307.804	0.0895
6 9	6806.150	0.013	6806.163	0.3404
15 6	6568.369	-0.084	6568.285	0.8237
6 1 9	18 19 06.6	0.2	18 19 06.8	0.0374
6 15 9	50 19 07.2	-0.2	50 19 07.0	0.0459
				$\Sigma = 2.00$

justed parameters. The lower triangle of the symmetric cofactor matrix Q_X of the adjusted parameters is

$$
Q_X = \begin{bmatrix}
0.0150 & & & \\
-0.0347 & 0.0615 & & \\
0.0339 & -0.0824 & 0.1990 & \\
0.0165 & 0.0422 & 0.0031 & 0.0872
\end{bmatrix}
$$

Finally, the adjusted station coordinate standard deviations obtained from the diagonal of the parameter cofactor matrix and the a posteriori variance of unit weight are given in Table 5.6.

To assess the success of the adjustment, we can compare the a posteriori variance of unit weight to the a priori variance to test whether they are statistically different. The null hypothesis is that they are not different. The chi-square test with two degrees of freedom is used with a significance level of 5%. A failure of the chi-square test and, in turn, the rejection of the null hypothesis may indicate errors in the stochastic model, errors in the functional model, incomplete reduction of observations, or some other systematic but

TABLE 5.6 Example 1: Adjusted Parameters

Station	Northing	Std. dev.	Easting	Std. dev.
6	4970087.701	.025	16760.148	.051
9	4970717.161	.091	23537.144	.060

TABLE 5.7 Example 1: Chi-Square Test

χ^2 statistic with $DF = 2$, $\alpha=0.05$		0.0835
Tabulated values	0.05	7.38
Do not reject H_0 at $\alpha=0.05$		

unmodeled error. The results of the chi-square test are given in Table 5.7. For the specified significance level, $\alpha = .05$, the chi-square has passed. This specific test is not meaningful, because simulated observations were used, but is presented to demonstrate the procedure. In practice, the significance level α is usually taken to 0.05 or 0.10. The significance level α, which is also equal to the probability of a type 1 error, cannot be made arbitrarily small because the probability of a type II error increases as α decreases.

The statistical interdependency of the adjusted parameters is expressed in the correlation matrix, which is derived from the cofactor matrix of the adjusted parameters. The correlation between parameters ranges from $-1 \leq \rho \leq 1$. Parameters may be negatively correlated as well as positively correlated. The correlation coefficient is given in equation (4.22). The correlation matrix is a matrix of all possible correlation coefficients for the given parameter set. By definition, the correlation matrix is symmetric with unit diagonal elements. The correlation matrix for this adjustment is

$$C = \begin{bmatrix} 1 & & & \\ -0.11 & 1 & & \\ 0.62 & -0.74 & 1 & \\ 0.45 & 0.58 & 0.02 & 1 \end{bmatrix}$$

The interpretation and evaluation of the numerical results of a least-squares adjustment requires some experience. A simple yet useful graphical tool can be used to represent the statistics of the adjusted station coordinates. This tool is the construction of standard ellipses for the adjusted station coordinates. The dimensions and orientation of the standard ellipses are obtained from (4.313) through (4.317) using the a posteriori variance of unit weight. Note that care must be exercised to assign the proper algebraic sign to the orientation angle of the ellipse. The shape and orientation of the standard ellipses are influenced by the choice of a minimal constraint. This point is emphasized in the next example. To uniquely define the coordinate system in this simple example, we need to hold as fixed only three parameters. In this example four parameters were in fact held fixed rather than the minimum of three. Such an adjustment problem is referred to as an overconstrained adjustment. The standard ellipses of Example 1 are shown in Figure 5.2. Note that only stations where both

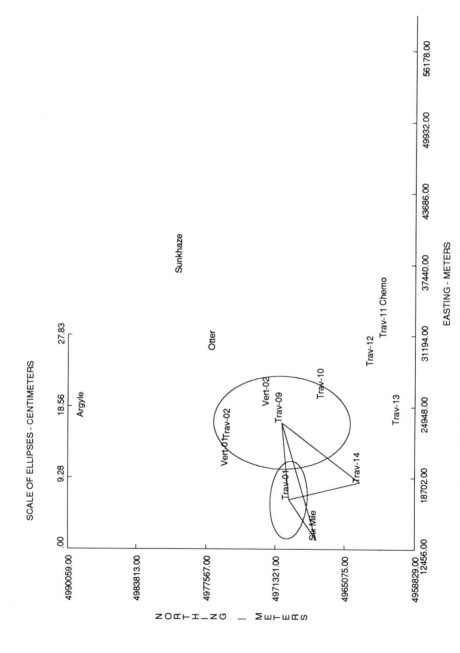

Figure 5.2 Example 1: Adjusted network with standard ellipses.

186

coordinates were estimated have ellipses plotted. For stations having a single parameter estimated, the ellipse collapses to a line.

In the example to follow, we are primarily concerned with estimating the quality of observations. Therefore, the adjusted parameters are not presented.

5.3 MINIMAL CONSTRAINT FOR NETWORK ADJUSTMENT ON THE MAPPING PLANE

Consider the same network adjustment as in Section 5.2. Now, rather than overconstraining the adjustment, the minimal constraint of fixing three parameters is imposed. Station 1 is completely fixed, whereas only the easting coordinate is held fixed for station 15. In the first example a distance observation was included between stations 1 and 15. However, because both stations were held fixed, this distance did not contribute to the adjustment and was removed. The following minimal and inner constraint examples include this distance observation. The adjustment summary, adjustment statistics, and adjusted observations for Example 2a are given in Tables 5.8–5.10. The observations are listed by number according to Tables F.1 and F.2. A plot of the adjusted network with the computed standard ellipses is shown in Figure 5.3. A different set of adjusted observations is estimated as a result of using only the min-

TABLE 5.8 Example 2a: Minimal Constraint Adjustment Summary

Observations:	1,2,3,5,14,30,42-44
Parameters	6,9, 15(N)
Minimal constraint:	stations 1,15(E) fixed

TABLE 5.9 Example 2a: Adjustment Statistics

Number of observations	9
Number of parameters	5
Number of observed parameters	0
Number of conditions	0
Degrees of freedom	4
$V^T PV$	1.5780
A-posteriori variance of unit weight	0.3946

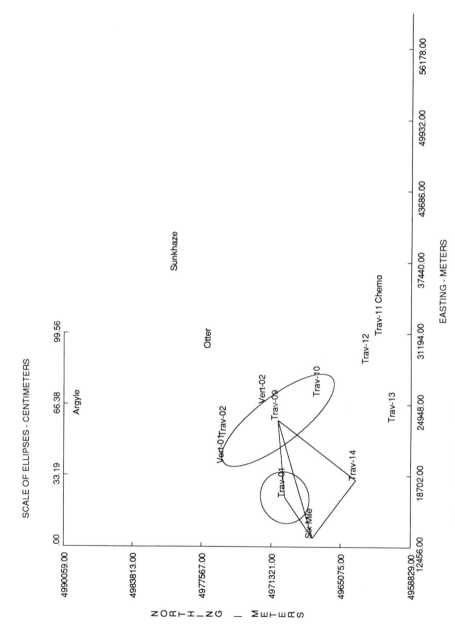

Figure 5.3 Example 2a: Minimal constraint adjustment with standard ellipses.

TABLE 5.10 Example 2a: Adjusted Observations

Stations	Observation	Residual	Adj. observations	r
6 1	4307.814	0.038	4307.852	0.1814
9 1	10759.585	0.278	10759.863	0.7940
9 6	6806.150	0.191	6806.341	0.4857
15 9	8751.961	-0.181	8751.780	0.7640
15 1	6399.312	-0.269	6399.043	0.6706
6 1 15	72 44 07.4	-0.5	72 44 06.9	0.2787
1 6 9	209 47 32.9	0.1	209 47 33.0	0.2537
15 9 6	47 57 39.6	0.2	47 57 39.8	0.3051
1 15 9	89 05 47.1	-0.8	89 05 46.3	0.2666

imal constraint. Note also the change in shape and orientation of the standard ellipses as compared to those in the overconstrained adjustment.

The choice a particular minimal constraint is arbitrary. Consider the same set of observations. In this case the minimal constraint holds fixed station 9 and the northing coordinate of station 6. The adjustment results of Example 2b in Tables 5.11–5.13 follow from this specification for the minimal constraint. The network plot with the standard ellipses is shown in Figure 5.4.

Note that the adjusted observations, the residuals, and the a posteriori variance of unit weight have not changed from the previous adjustment. In general, all estimated quantities *except* for the adjusted parameters $\hat{X}$ and their variance–covariance matrix Σ_X are invariant with respect to the choice of minimal constraint. That is, the quantities of V, L_a, $\hat{\sigma}_0^2$, Σ_{L_a}, and Σ_V will remain unchanged under different specifications of minimal constraint. Note also that the redundancy numbers have not changed, indicating that varying minimal constraint does not change the internal network geometry.

5.4 INNER CONSTRAINT FOR NETWORK ADJUSTMENT ON THE MAPPING PLANE

A final example of a minimal condition least-squares adjustment is that of the inner constraint solution presented in Section 4.8. Using the inner constraint technique, we adjust all stations parameters, and thus all stations have standard ellipses. The inner constraint solution more accurately represents the internal geometry of the network because the choice of the coordinate system is no longer reflected in the standard ellipses. This feature makes the inner constraint approach extremely useful for interpreting standard ellipses.

TABLE 5.11   Example 2b: Alternative Minimal Constraint

Observations:	1,2,3,5,14,30,42-44
Parameters	6(E),1,15
Minimal constraint:	stations 9,6(N) fixed

TABLE 5.12   Example 2b: Adjustment Statistics

Number of observations	9
Number of parameters	5
Number of observed parameters	0
Number of conditions	0
Degrees of freedom	4
$V^T P V$	1.5780
A-posteriori variance of unit weight	0.3946

TABLE 5.13   Example 2b: Adjusted Observations

Stations	Observation	Residual	Adj. observations	r
6 1	4307.814	0.038	4307.852	0.1814
9 1	10759.585	0.278	10759.863	0.7940
9 6	6806.150	0.191	6806.341	0.4857
15 9	8751.961	-0.181	8751.780	0.7640
15 1	6399.312	-0.269	6399.043	0.6706
6 1 15	72 44 07.4	-0.5	72 44 06.9	0.2787
1 6 9	209 47 32.9	0.1	209 47 33.0	0.2537
15 9 6	47 57 39.6	0.2	47 57 39.8	0.3051
1 15 9	89 05 47.1	-0.8	89 05 46.3	0.2666

The inner constraint is simply an alternative minimal constraint. The operational details involve replacing the standard normal matrix with a generalized inverse, called a pseudoinverse (see Section 4.8). To compute the pseudoinverse, we must specify the matrix E. Recall from (4.181) that

$$AE^T = 0$$

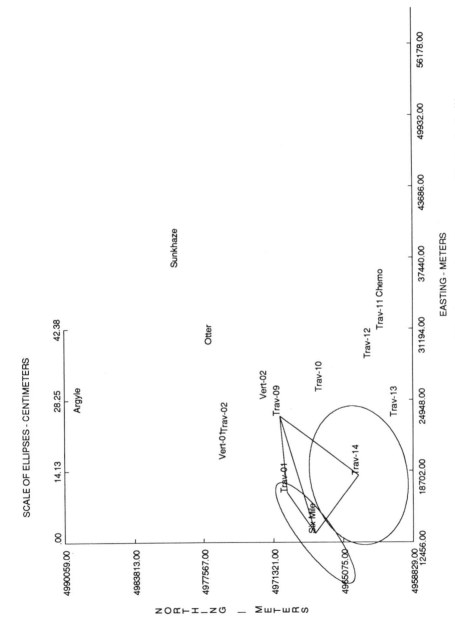

Figure 5.4 Example 2b: Alternative minimal constraint adjustment with standard ellipses.

For many adjustments the elements of the E matrix can be specified by inspection. The E matrix will have the same number of rows as the rank defect of the design matrix A. Consider the example of a trilateration network consisting of k stations. A row of the design matrix for the distance from station i to station k might look like

$$A_i = \left[\frac{-(y_k - y_i)}{s_{ik}} \quad \frac{-(x_k - x_i)}{s_{ik}} \quad \cdots \quad \frac{y_k - y_i}{s_{ik}} \quad \frac{x_k - x_i}{s_{ik}} \right]$$

The inner constraint matrix E would be specified as

$$E = \begin{bmatrix} 1 & 0 & \cdots & 1 & 0 \\ 0 & 1 & \cdots & 0 & 1 \\ -x_i & y_i & \cdots & -x_k & y_k \end{bmatrix}$$

In the expressions above the coordinates y and x are associated with northing and easting, respectively. The reader is left to verify that (4.181) holds. The case of triangulation networks is no more complicated. Consider a sample row of the A matrix for an angle at station i measured from station j to station k:

$$A_i = \left[\frac{x_i - x_j}{s_{ij}^2} - \frac{y_i - y_j}{s_{ij}^2} \quad \cdots \quad \frac{x_k - x_j}{s_{kj}^2} - \frac{x_i - x_j}{s_{ij}^2} \right.$$
$$\left. - \frac{y_k - y_j}{s_{kj}^2} + \frac{y_i - y_j}{s_{ij}^2} \quad \cdots \quad - \frac{x_k - x_j}{s_{kj}^2} \quad \frac{y_k - y_j}{s_{kj}^2} \right]$$

In this case the E matrix must have four rows because triangulation networks are invariant with respect to rotation, translations, and scale. Thus, E would be specified as

$$E = \begin{bmatrix} 1 & 0 & \cdots & 1 & 0 & \cdots & 1 & 0 \\ 0 & 1 & \cdots & 0 & 1 & \cdots & 0 & 1 \\ -x_i & y_i & \cdots & -x_j & y_j & \cdots & -x_k & y_k \\ y_i & x_i & \cdots & y_j & x_j & \cdots & y_k & x_k \end{bmatrix}$$

The reader is again invited to verify (4.181) for the triangulation network. For mixed observation networks, the E matrix is that of the trilateration network.

The adjustment from Section 5.3 is repeated using the inner constraint technique. The results from Example 2c are shown in Table 5.14–5.16 and Figure 5.5. The plane coordinate system of Figure 5.5 has been translated; this has no effect on the ellipses.

TABLE 5.14 Example 2c: Inner Constraint Adjustment Summary

Observations:	1,2,3,5,14,30,42-44
Parameters	1,6,9,15
Minimal constraint:	three inner constraints

TABLE 5.15 Example 2c: Adjustment Statistics

Number of observations	9
Number of parameters	8
Number of observed parameters	0
Number of conditions	3
Degrees of freedom	4
$V^T PV$	1.5780
A-posteriori variance of unit weight	0.3946

TABLE 5.16 Example 2c: Inner Constraint Adjustment Observations

Stations	Observation	Residual	Adj. observations	r
6 1	4307.814	0.038	4307.852	0.1814
9 1	10759.585	0.278	10759.863	0.7940
9 6	6806.150	0.191	6806.341	0.4857
15 9	8751.961	-0.181	8751.780	0.7640
15 1	6399.312	-0.269	6399.043	0.6706
6 1 15	72 44 07.4	-0.5	72 44 06.9	0.2787
1 6 9	209 47 32.9	0.1	209 47 33.0	0.2537
15 9 6	47 57 39.6	0.2	47 57 39.8	0.3051
1 15 9	89 05 47.1	-0.8	89 05 46.3	0.2666

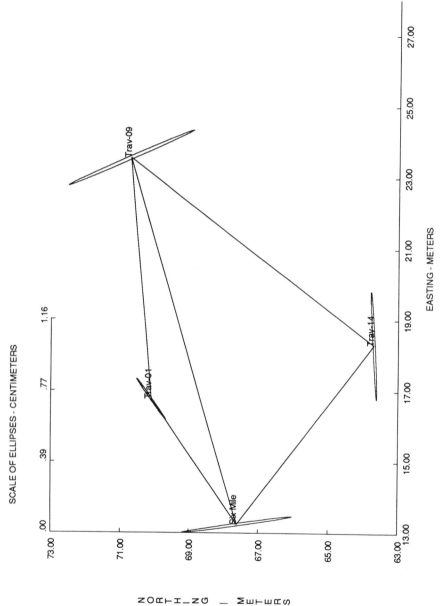

SCALE OF ELLIPSES - CENTIMETERS

Figure 5.5 Example 2c: Inner constraint adjustment with standard ellipses.

194

5.5   ADJUSTMENT OF THE COMPLETE NETWORK ON THE MAPPING PLANE

In the previous three examples, a simple four-station network was adjusted under varying minimal constraints. Consider now the entire network incorporating all distance and horizontal angle observations. This is the network size that is often encountered in day-to-day practice. The network consists of a set of braced quadrilaterals with a circumferential traverse. This pattern may arise when linking existing networks. The adjustment results are presented in Tables 5.17–5.19. Stations 2 (Argyle) and 4 (Sunkhaze) are not included because they are determined by GPS vectors only.

The adjusted network of Example 3 is shown in Figure 5.6. One immediately notices the large disparity in the size of the standard ellipses at the adjusted stations. A closer look at the observation table, however, reveals that the coordinates of station Otter (3) are controlled by only two distances. The intersection of two distances defines the location of the station in the plane. Looking at the standard ellipse at station 3, we can see that the station is relatively well determined in the east–west direction but poorly controlled in the north–south direction. The scale of the standard ellipse at station 3 is so large as to obscure the ellipses at the remaining stations. This situation dramatically points out poor network geometry in the determination of station 3. Notice that the two distance observations into station 3 have zero residuals

TABLE 5.17   Example 3: Summary

Observations:	1-44
Parameters	3,5,6(E),7,8,9,10
	11,12,13,14,15
Minimal constraint:	stations 1,6(N) fixed

TABLE 5.18   Example 3: Adjustment Statistics

Number of observations	44
Number of parameters	23
Number of observed parameters	0
Number of conditions	0
Degrees of freedom	21
$V^T P V$	9.1787
A-posteriori variance of unit weight	0.4371

TABLE 5.19   Example 3: Adjusted Observations

Stations	Observation	Residual	Adj. observations	r
6 1	4307.814	0.039	4307.853	0.3544
9 1	10759.585	0.348	10759.933	0.8927
15 1	6399.312	-0.201	6399.111	0.8198
14 6	11885.803	0.123	11885.926	0.9021
9 6	6806.150	0.258	6806.408	0.8390
8 6	7534.978	0.193	7535.171	0.8343
7 6	6238.000	-0.125	6237.875	0.7233
8 7	2385.420	-0.011	2385.409	0.2084
10 7	6434.498	-0.167	6434.331	0.8645
3 7	10208.799	0.000	10208.799	0.0000
9 8	4895.411	-0.116	4895.295	0.7093
10 8	4548.906	0.124	4549.030	0.7061
3 8	7882.288	0.000	7882.288	0.0000
15 9	8751.961	-0.138	8751.823	0.8595
14 9	10469.437	0.002	10469.439	0.9211
13 9	9530.297	-0.390	9529.907	0.8859
11 9	4321.513	0.126	4321.639	0.6076
10 9	1876.256	0.009	1876.265	0.0498
13 10	9936.289	-0.271	9936.018	0.8973
11 10	4986.565	-0.121	4986.444	0.6529
13 11	5215.127	0.217	5215.344	0.6335
12 11	7741.877	-0.154	7741.723	0.7714
5 11	9930.112	-0.377	9929.735	0.8411
13 5	5506.146	0.295	5506.441	0.6922
12 5	2848.943	-0.060	2848.883	0.2131
13 12	2797.991	-0.028	2797.963	0.2404
14 13	5628.181	0.012	5628.193	0.3195
15 13	10294.564	0.036	10294.600	0.8386
15 14	6180.573	0.156	6180.729	0.5864
6 1 15	72 44 07.4	-0.9	72 44 06.5	0.3640
1 15 14	176 20 53.9	-0.4	176 20 53.5	0.1649
15 14 13	121 15 29.7	-0.6	121 15 29.1	0.1337
14 13 12	229 40 25.3	-1.1	229 40 24.2	0.1235
13 12 5	154 23 32.0	-1.4	154 23 30.6	0.1660
12 5 11	34 13 22.5	-1.5	34 13 21.1	0.2244
5 11 10	229 01 27.7	-1.2	229 01 26.5	0.0929
11 10 9	58 44 45.2	-1.4	58 44 43.8	0.1259
10 9 8	291 41 42.4	-1.1	291 41 41.3	0.1276
9 8 7	111 14 57.7	-1.3	111 14 56.4	0.1489
8 7 6	114 09 16.8	-1.4	114 09 15.4	0.1488
7 6 1	206 30 12.6	-0.9	206 30 11.7	0.1937
1 6 9	209 47 32.9	-0.2	209 47 32.7	0.3418
15 9 6	47 57 39.6	0.5	47 57 40.1	0.3491
1 15 9	89 05 47.1	-0.9	89 05 46.2	0.3527

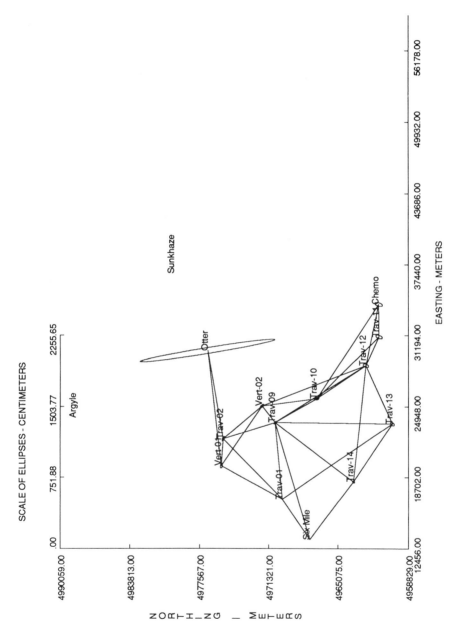

Figure 5.6 Example 3: Full network adjustment with standard ellipses.

resulting from the adjustment. These two observations are completely uncontrolled. No matter what the observed values for these distances, they would remain unchanged during the adjustment. This is even more evident when examining the redundancy number for these two observations. Both are zero! This is not a desirable situation; however, it serves to emphasize the point that the results of any least-squares adjustment should be examined carefully for zero residuals, small redundancy numbers, or unexpectedly large standard ellipses. The deficiency in this example adjustment can be corrected by introducing an additional observation to station 3. This is done in the next example.

5.6   ADJUSTMENT OF THE COMPLETE NETWORK WITH ADDITIONAL INFORMATION

We have seen that the construction of standard ellipses and the inspection of redundancy numbers can be of use in pointing out deficiencies in the observations. In Example 4 one additional observation has been included, a distance observation between stations 3 and 5 (Chemo) to provide some redundancy in the determination of the coordinates of station 3. The adjustment results incorporating the additional observation are given in Tables 5.20–5.22 and Fig-

TABLE 5.20   Example 4: Summary

Observations:	1-44, distance 3-5
Parameters	3,5,6(N),7,8,9,10
	11,12,13,14,15
Minimal constraint:	stations 1,6(E) fixed

TABLE 5.21   Example 4: Adjustment Statistics

Number of observations	45
Number of parameters	23
Number of observed parameters	0
Number of conditions	0
Degrees of freedom	22

$V^T PV$	9.1787
A-posteriori variance of unit weight	0.4172

TABLE 5.22   Example 4: Adjusted Observations

			Adjusted observations	
Stations	Observation	Residual	Adj. observations	
...	...	...	...	...
7 6	6238.000	-.125	6237.875	0.7235
8 7	2385.420	-.011	2385.409	0.2282
10 7	6434.498	-.167	6434.331	0.8656
3 7	10208.799	0.000	10208.799	0.6103
9 8	4895.411	-.116	4895.295	0.7097
10 8	4548.906	.124	4549.030	0.7064
3 8	7882.288	0.000	7882.288	0.3590
15 9	8751.961	-.138	8751.823	0.8595
14 9	10469.437	.002	10469.439	0.9211
13 9	9530.297	-.390	9529.907	0.8859
3 5	15876.770	0.000	15876.770	0.0051
6 1 15	72 44 07.4	-0.9	72 44 06.5	0.3700
1 15 14	176 20 53.9	-0.4	176 20 53.5	0.1650
...	...	...	...	...

ure 5.7. Only a portion of the adjusted observations is given in Table 5.22. In the previous adjustment the standard ellipses for station 3 were of the order of 22 m in the semimajor axis. Notice the improvement in the determination of station 3 by better than one order of magnitude. The residuals in the observations to station 3 are still small (recall that the observations are simulated), but the redundancy numbers have changed considerably. Although there has been a substantial improvement, additional observations are required to reduce the uncertainty in the estimated coordinates for station 3.

The improvement in the adjustment by the addition of a single distance is dramatic. The scale of the standard ellipses has been reduced by an order of magnitude. Although station 3 still has a large standard ellipse compared to the remaining stations, the problem is not solely due to a deficiency in the observations. We can plainly see the systematic increase in the sizes of the standard ellipses with increasing distance from the fixed stations. This is the effect of the choice of the particular minimal constraint. This is supported by the intuition that, as a survey proceeds further from control points, the stations are more weakly determined with respect to the fixed stations. The effects of the choice of minimal constraint somewhat obscure the internal geometry of the network by itself. To evaluate the internal geometry of the network, one could perform an inner constraint solution.

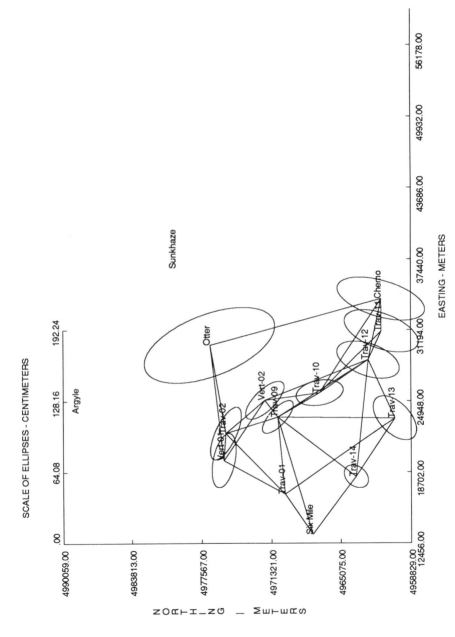

SCALE OF ELLIPSES - CENTIMETERS

Figure 5.7 Example 4: Full network adjustment with additional observation and standard ellipses.

5.7 DETECTING POOR NETWORK GEOMETRY: THE DANGEROUS CIRCLE

To emphasize the utility of the standard ellipses and redundancy numbers in indicating poor geometry, consider the classical resection problem. In this example the coordinates of a set of stations are to be estimated by observing horizontal angles between four known stations. The angle observations are observed with the equal precision.

The four known stations nearly lie on an imaginary circle, the so-called dangerous circle. In the adjustment the unknown stations vary in location from near the center of the imaginary circle to outside the imaginary circle to demonstrate the variation in the size and shape of the standard ellipses and the redundancy numbers.

Because the observations from each of the unknown stations, Resect-1 to Resect-6 (5–10), are uncorrelated, all the resections can be computed in a single adjustment. The summary of the resection adjustment (Example 5) is given in Table 5.23. The redundancy numbers for the observations to each station are given in Table 5.24. These bear close examination along with the network plot (Figure 5.8). The station numbers listed in the first column of Table 5.24 are the station numbers for angle observations, with n denoting the particular unknown station at which the observations were made. The first observation at station 5 is the angle 4–5–3; the second is the angle 4–5–2, etc.

From column 2 of Table 5.24 we see that the redundancy numbers for the

TABLE 5.23 Example 5: Resection Adjustment Summary

Observations:	three interior angles at each resection station
Parameters	5,6,7,8,9,10
constraint:	stations 1,2,3,4 fixed

TABLE 5.24 Example 5: Resection Adjustment Redundancy Numbers

Stations	5	6	7	8	9	10
4 n 3	0.2753	0.7276	0.9055	0.9269	0.9207	0.9157
4 n 2	0.4056	0.1142	0.0206	0.0166	0.0143	0.0197
4 n 1	0.3191	0.1581	0.0739	0.0566	0.0650	0.0645

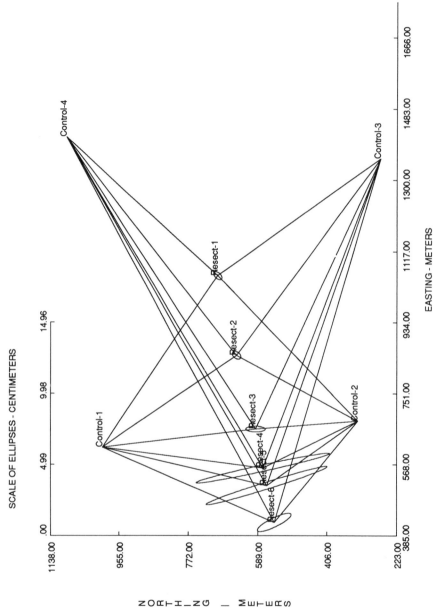

Figure 5.8 Example 5: Dangerous circle resection adjustment with standard ellipses.

angle observations at station 5 (Resect-1) are similar, indicating a strong geometry for the resection at that station. From Figure 5.8 we see that station 5 lies near the center of the imaginary circle. The redundancy numbers for stations 6–8 show a very different situation. As the unknown station proceeds from the center of the imaginary circle to the edge of the circle, the redundancy numbers decrease dramatically. If the unknown station lay directly on the circle, the redundancy numbers for two of the three observations would in fact be zero! In the final two columns the unknown station has moved outside the circle, and redundancy numbers begin to show improvement. The size and shape of the standard ellipses also indicate this improvement in the network geometry. Thus, the combination of the standard ellipses and redundancy numbers in an effective tool for detecting poor network geometry in a least-squares adjustment. Recall also the relationship between the redundancy number and internal reliability.

5.8 CLOSED TRAVERSE ADJUSTMENT ON THE MAPPING PLANE

A more traditional traverse adjustment can be performed by utilizing the observations on the circumference of the 15-station network. In this example all interior angles and distances in the enclosing traverse are observed. This is a closed traverse with minimal constraint imposed to define the coordinate system. As described earlier, the effect of the minimal constraint is evident in the size and shape of the standard ellipses. The degree of this kind of traverse is always the rank defect of the normal matrix or the number of minimal constraints that must be imposed, which is equal to 3 for horizontal networks. This low degree of freedom is reflected in the relatively weak determination of the station coordinates. Thus, when feasible, additional observations should be included in the observational scheme of traditional traverses. The results of Example 6a are given in Tables 5.25–5.27 and Figure 5.9.

The resulting redundancy numbers are quite small because of the less than optimal network geometry. This is a common problem with traditional closed

TABLE 5.25   Example 6a: Closed Traverse Adjustment Summary

Observations:	1,3,7,8,11,18,20,23,24
	25,27,29-41
Parameters	5,6(N),7,8,9,10,11,12
	13,14,15
Minimal constraint:	stations 1,6(E) fixed

TABLE 5.26   Example 6a: Closed Traverse Adjustment Statistics

Number of observations	24
Number of parameters	21
Number of observed parameters	0
Number of conditions	0
Degrees of freedom	3
$V^T P V$	2.0095
A-posteriori variance of unit weight	0.6698

TABLE 5.27   Example 6a: Closed Traverse Adjusted Observations

	Adjusted observations			
Stations	Observation	Residual	Adj. observations	r
⋯	⋯	⋯	⋯	⋯
12 5	2848.943	0.009	2848.952	0.0397
13 12	2797.991	0.014	2798.005	0.0325
14 13	5628.181	0.009	5628.190	0.1847
15 14	6180.573	0.073	6180.646	0.1539
6 1 15	72 44 07.4	-0.8	72 44 06.6	0.1343
1 15 14	176 20 53.9	-0.9	176 20 53.0	0.0996
15 14 13	121 15 29.7	-1.0	121 15 28.7	0.1031
14 13 12	229 40 25.3	-1.2	229 40 24.1	0.1092
⋯	⋯	⋯	⋯	⋯

traverses. Additional observations would improve the overall geometry of the network.

This same closed traverse is computed now with a different minimal constraint. As noted previously, the only effect of varying the minimal constraint is a different adjusted parameter vector and a different parameter covariance matrix. The summary for Example 6b is given in Table 5.28, with the network plot and the standard ellipses in Figures 5.10. Compare this figure with Figure 5.9

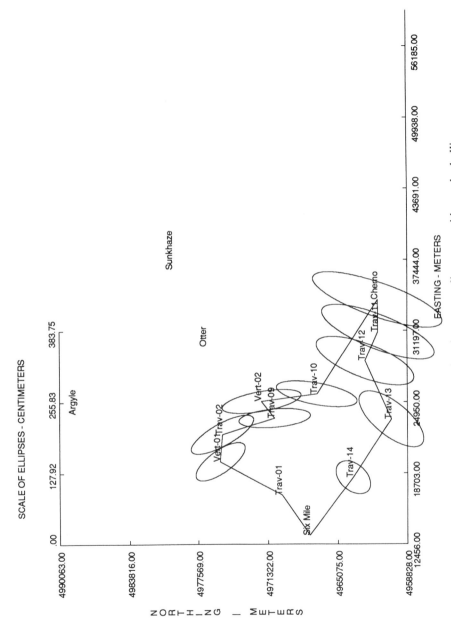

Figure 5.9 Example 6a: Closed traverse adjustment with standard ellipses.

TABLE 5.28   Example 6b: Closed Traverse with Alternative Minimal Constraint Summary

Observations:	1,3,7,8,11,18,20,23,24
	25,27,29-41
Parameters	5,6(N),7,8,9,10,11,12
	13,14,15
Minimal constraint:	stations 1,5(N) fixed

5.9   OPEN TRAVERSE ADJUSTMENT ON THE MAPPING PLANE

The final adjustment example is that of an open or cantilever traverse starting from two fixed stations but not closing on the starting stations or on other control stations. The simplest open traverse, consisting of single distance and interior angle observations, has no degree of freedom. Thus, a least-squares adjustment cannot be performed. However, to examine the geometric strength of an open traverse, we have inverted the normal matrix and used the respective elements to draw ellipses. The network geometry of Example 7 is seen in Figure 5.11. This figure also supports the intuition that, as one moves farther from the control stations, the station coordinate determinations degrade quickly. All the adjusted observations have zero residuals and zero redundancy numbers (except for the duplicated observation) because this open traverse is uniquely determined. The point to be made here is that, whenever possible, open traverses should be closed onto control.

5.10   ADJUSTMENT OF LEVELING OBSERVATIONS

Up to this point we have concentrated on two-dimensional networks. The same techniques can be applied to the adjustment of one-dimensional networks, such as spirit leveling observations. The mathematical model for leveling is

$$h_{ij} = H_j - H_i$$

where h_{ij} is the height difference from station i to j and H_i, H_j are the elevations of stations i and j, respectively. This mathematical model is linear and therefore need not be linearized or iterated. A single solution of the normal equations is all that is required. The design matrix is formed just as with two-dimensional cases. However, the coefficients of the design matrix will simply be -1, 0, or 1. The misclosure vector for a linear model is the negative of the observed height differences, that is, $L_0 = 0$. Because each observation equation involves at most two stations, there can be no more than two nonzero elements

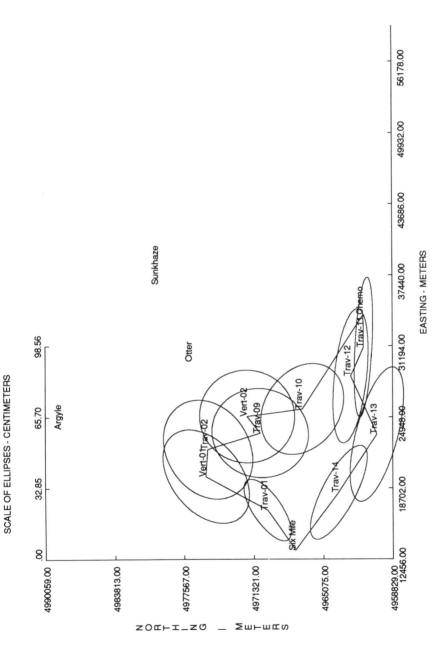

Figure 5.10 Example 6b: Closed traverse adjustment with alternative constraint and standard ellipses.

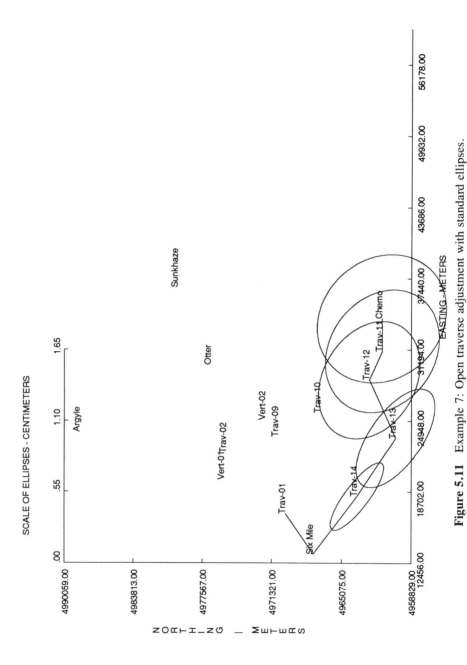

Figure 5.11 Example 7: Open traverse adjustment with standard ellipses.

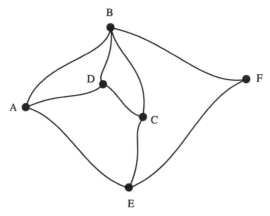

Figure 5.12 Leveling network with four connecting level loops.

per row of the design matrix. Furthermore, there must be one positive term and one negative term per row. Consider the leveling network of Figure 5.12, consisting of four interconnected closed level loops. Nine observation equations can be written, and there are six unknown station heights. Using the observation equation model

$$L_a = F(X_a)$$

we obtain the observation equations for the observed height differences:

$$h_{ab} = H_b - H_a$$

$$h_{bd} = H_d - H_b$$

$$h_{da} = H_a - H_d$$

$$h_{bc} = H_c - H_b$$

$$h_{cd} = H_d - H_c$$

$$h_{ce} = H_e - H_c$$

$$h_{ea} = H_a - H_e$$

$$h_{bf} = H_f - H_b$$

$$h_{fe} = H_e - H_f$$

Because the model is already linear, the elements of the design matrix can be specified directly. The resulting design matrix, assuming the parameters are ordered alphabetically, i.e., $H_a, H_b, \ldots$, is

$$A = \begin{bmatrix} -1 & 1 & 0 & 0 & 0 & 0 \\ 0 & -1 & 0 & 1 & 0 & 0 \\ 1 & 0 & 0 & -1 & 0 & 0 \\ 0 & -1 & 1 & 0 & 0 & 0 \\ 0 & 0 & -1 & 1 & 0 & 0 \\ 0 & 0 & -1 & 0 & 1 & 0 \\ 1 & 0 & 0 & 0 & -1 & 0 \\ 0 & -1 & 0 & 0 & 0 & 1 \\ 0 & 0 & 0 & 0 & 1 & -1 \end{bmatrix}$$

The rank of A is 5, because the sum of columns 1 through 5 equals the negative of column 6. Thus, a single constraint must be specified to solve the normal equations. This single constraint is usually imposed by holding one station height fixed. It should be pointed out that heights are not estimable from geometric leveling; rather, height differences are observable.

The stochastic model for leveling is most often expressed as a function of the length of the leveling leg. Thus,

$$\sigma^2_{h_{ij}} = cs_{ij}$$

The constant term c may be known from experience or estimated from the adjustment.

We have seen that the one-dimensional adjustment of leveling networks require a single constraint to remove the rank deficiency of the normal matrix. To avoid the rank deficiency in the leveling adjustment, we can perform the inner constraint solution without arbitrarily fixing a single station. In this case the E matrix is

$$E = [1 \quad 1 \quad \cdots \quad 1 \quad \cdots \quad 1]$$

5.11 STOCHASTIC MODELS FOR DISTANCES AND ANGLES

The design matrix represents the geometry of the network. In addition to the functional model, statistical information about the observations and possibly the parameters must be incorporated into the adjustment process. This information is called the stochastic model. The stochastic model for each type of observation is based on the instrument used and the procedure involved in taking an observation. The stochastic model also affects the size and shape of standard ellipses.

To characterize a stochastic model for an EDM distance, we must consider systematic errors of an EDM, such as the electronic center corrections, non-linearity of optical paths, reflector errors, and external factors, such as temperature and humidity on the observation. Instrumental errors are determined using repeated measurements over baselines specifically established for EDM calibration. The length observations of the baseline and precise temperature, humidity, and slope are combined to generate a correction constant for an EDM–reflector pair. A correction constant must be determined for each reflector used in normal operations. The frequency of calibration depends on the accuracy requirements of the observations and the age of the instrument. The standard errors of EDM observations are most often stated as a constant term and a term proportional to the length of the line. Short-range (<5 km) EDM instruments may have standard errors in the range of $\pm(2{-}10 \text{ mm} + 1{-}5 \text{ ppm})$. Medium- to long-range (<60 km) instruments are capable of observations to $\pm(5 \text{ mm} + 2 \text{ ppm})$. The determination of the calibration constants of an EDM is itself an application of least-squares adjustments.

As with EDM observations, angle observations are subject to instrumental errors and to environmental effects. Instrumental errors include eccentricity of the circle, index errors, graduation errors, and crosshair misalignment. The stochastic model for a theodolite can be estimated using sequences of directions distributed about the reading circle to a group of stations (Figure 5.13). The number of sequences and stations is arbitrary. The estimation of the sto-

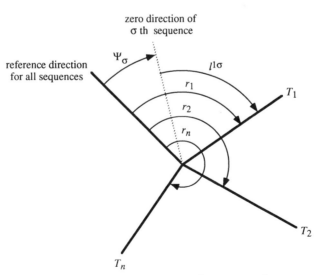

Figure 5.13 Direction observations at a station.

chastic model is in itself a simple least-squares adjustment. The observed directions are assumed to be uncorrelated and have equal weight ($P = I$).

Suppose that n targets are observed in s sequences. The observations are

$$l_b^{v\sigma} = \text{observed direction to target } v \text{ in the } \sigma\text{th sequence}$$

The mathematical model for the adjusted observations in this case is linear:

$$l_a^{v\sigma} = r_{va} - \psi_{\sigma a}$$

$$l_b^{v\sigma} + v_{v\sigma} = r_v - \psi_\sigma$$

with

$$v = 1, \ldots, n$$

$$\sigma = 1, \ldots, s$$

There are a total of ns observation equations, with each sequence having its own orientation unknown ψ_σ for the zero direction. The total vector of parameters for the adjustment is

$$X = (r_1, r_2, \cdots, r_n, \psi_1, \psi_2, \cdots, \psi_s)$$

Finally, the computed observations L_0 are zero because the mathematical model is linear. The degrees of freedom of the adjustment are given by

$$DF = \text{number of observations} + \text{number of conditions}$$

$$- \text{number of parameters}$$

$$= (ns) + 1 - (n + s)$$

$$= (n - 1)(s - 1)$$

The resulting normal matrix has a rank deficiency of 1. Thus, a single condition must be imposed to remove this singularity. An intuitive condition in the measurement of directions is that the sum of the orientation unknowns ψ_σ be zero, or

$$\sum_{\sigma = 1}^{s} \psi_\sigma = 0$$

Consider Table 5.29, which shows such an observational scheme with four targets observed over three sequences. The first column is the target sighted; the second and third columns are the readings with the telescope in the direct

TABLE 5.29 Station Adjustment Observations

T	Direct	Inverse	Mean	Reduced mean l_b	Adjusted direction $\hat{r}_i$	d	v	v^2
	° ' "	° ' "	° ' "	° ' "	° ' "	"	"	
1	000 34 34	180 34 44	000 34 39.0	000 00 00.0	000 00 00.0	0	2	4
2	055 37 05	235 37 23	055 37 14.0	055 02 35.0	055 02 31.5	-4	-2	4
3	095 20 28	275 20 42	095 20 35.0	094 45 56.0	094 45 50.7	-5	-3	9
4	266 29 20	086 29 28	266 29 24.0	265 54 45.0	265 54 46.3	1	1	1
					$\Sigma(d)$ =	-8	-2	26
					$\psi = \Sigma(d)/n$=	-2		
	067 30 22	247 30 47	067 30 34.5	000 00 00.0		0	0	0
	122 32 53	302 33 11	122 33 02.0	055 02 27.0		5	5	25
	162 16 12	347 16 44	162 16 28.0	094 45 53.0		-2	-2	4
	333 25 19	153 25 26	333 25 22.5	265 54 47.5		-1	-1	1
					$\Sigma(d)$ =	2	2	29
					$\psi = \Sigma(d)/n$=	0		
	134 09 00	314 09 29	134 09 14.5	000 00 00.0		0	-2	4
	189 11 38	009 11 56	189 11 47.0	055 02 32.5		-1	-3	9
	228 54 47	048 55 08	228 54 57.5	094 45 43.0		8	6	36
	039 03 54	219 04 08	039 04 01.0	265 54 46.5		0	-2	4
					$\Sigma(d)$ =	7	-1	54
					$\psi = \Sigma(d)/n$=	2		
							Σv^2 =	109

$$\hat{\sigma}_0 = \pm\sqrt{\frac{109}{(s-1)(n-1)}} = \pm 4.3'' \qquad \hat{\sigma}_r = \pm\sqrt{\frac{109}{s(s-1)(n-1)}} = \pm 2.5''$$

s = # of sequences = 3 n = # of observations / seq. = 4

and inverse positions (face left and face right); the fourth column is the mean direction; the fifth column is the reduced mean reading; and, finally, the sixth column is the adjusted direction to each of the targets. The adjusted directions are obtained from the mean of the reduced mean observations for each target:

$$\hat{r}_\nu = \frac{\sum\limits_{1}^{s} (\text{reduced means})_\sigma}{s}$$

The variable $\hat{r}_\nu$ in this context denotes an adjusted direction observation and should not be confused with the redundancy number used previously. The deviation from the mean for each observation is given by

$$d_v = \hat{r}_v - l_b$$

The orientation parameters ψ_σ are given by

$$\psi_\sigma = \frac{\sum\limits_{1}^{n} d_v}{n}$$

The residuals, $v = d_v - \psi_\sigma$, and the squares of the residuals are given in the last two columns. The variance of unit weight, which equals the variance of a single direction observation, is

$$\hat{\sigma}_0^2 = \frac{V^T V}{(n-1)(s-1)}$$

The direction observations were assumed to be uncorrelated, as a consequence of this and of the minimal condition, the adjusted directions are also uncorrelated. Therefore the variance–covariance of the adjusted directions is diagonal with $1/s$ on the diagonal. Thus, the variance of an adjusted direction is

$$\hat{\sigma}_r^2 = \frac{V^T V}{s(n-1)(s-1)}$$

We can see that for direction observations made with this particular instrument, a standard error of 2.5″ of arc can be used as the stochastic model for each observed direction. If angles are to be used as observations in an adjustment, variance–covariance propagation can be used to derive the stochastic model for the angle.

This station adjustment problem can easily be expressed in terms of the adjustment notation from Chapter 4. This is left as an exercise.

CHAPTER 6

LINKS TO PHYSICAL OBSERVATIONS

Observations refer to well-defined physical reference elements on the earth. A typical example is astronomic latitude, longitude, and azimuth measured with a theodolite from star observations. These measurements refer to the instantaneous rotation axis, the instantaneous equator of the earth, and the local astronomic horizon, which is the plane perpendicular to the local plumb line. Vertical angle and horizontal angle observations are referenced with respect to the plumb line and the local astronomic horizon. Even though astronomic latitude, longitude, and azimuth determinations are less frequent today than in the pre-satellite era, these quantities are still conceptually important when defining the geodetic reference frame as a basis for processing all observations in surveying. *From the surveyor's point of view, the geodetic reference frame is simply an ellipsoid of revolution of given size and shape and an accompanying set of geoid undulations and deflection of the vertical angles.* It is acknowledged that for the geodesist or the scientist the term ''geodetic reference frame'' might have an entirely different meaning. It is further recognized that the typical surveyor is unlikely to carry out a survey for the specific goal of improving deflections of the vertical and geoid undulations beyond what is available from the geodetic community. The principle observations of today's surveyor are horizontal directions, angles, slant distances, zenith angles, and, of course, GPS observations. In this chapter, I will show how the angular observations are reduced to corresponding ellipsoidal observations referring to the ellipsoidal normal and the local geodetic horizon. These reduced observations (model observations) can be used in the three-dimensional adjustment model of Chapter 7. Detail on links to physical observations and other elements from physical geodesy such as geoid, equipotential surfaces, and deflections of the vertical can be found in Heiskanen and Moritz (1967) and Vaníček and Krakiwsky (1982).

6.1 GEOID

The geoid is a fundamental physical reference surface. Its shape reflects the distribution of mass inside the earth. Consider two point masses of mass m_1 and m_2 separated by a distance s that attract each other with the force

$$F = \frac{km_1m_2}{s^2} \tag{6.1}$$

where k denotes the constant of gravitation (previously denoted by k^2 in Section 2.3.2). This is Newton's law of gravitation. The attraction between the two point masses is symmetric and opposite in direction. As a matter of convenience, we consider one mass to be the ''attracting'' mass and the other to be the ''attracted'' mass. Furthermore, we assign to the attracted mass the unit mass ($m_2 = 1$) and denote the attracting mass with m. The force equation then becomes

$$F = \frac{km}{s^2} \tag{6.2}$$

In this form the force is that between an attracting mass and a unit attracted mass. Introducing an arbitrary coordinate system, as seen in Figure 6.1, we can decompose the force vector into components parallel to the coordinate axes. Thus,

$$\mathbf{F} = \begin{bmatrix} F_x \\ F_y \\ F_z \end{bmatrix} = -F \begin{bmatrix} \cos \alpha \\ \cos \beta \\ \cos \gamma \end{bmatrix} = -\frac{km}{s^2} \begin{bmatrix} \dfrac{x - \xi}{s} \\ \dfrac{y - \eta}{s} \\ \dfrac{z - \zeta}{s} \end{bmatrix} \tag{6.3}$$

where

$$s = \sqrt{(x - \xi)^2 + (y - \eta)^2 + (z - \zeta)^2} \tag{6.4}$$

The negative sign in the decomposition indicates the convention of the sense of the force vector, here implying that the force vector points from the attracted mass toward the attracting mass. The coordinates (x, y, z) identify the location of the attracted mass in the specified coordinate system, and (ξ, η, ζ) denote the location of the attracting mass. The expression

$$V = \frac{km}{s} \tag{6.5}$$

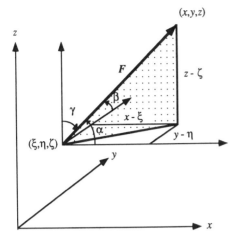

Figure 6.1 Components of the gravity vector.

is called the potential of gravitation. It is a measure of the amount of work required to transport the unit mass from its initial position, a distance s from the attracting mass, to infinity. For example, integrating the force equation (6.2) gives

$$V = \int_s^\infty F \, ds = \int_s^\infty \frac{km}{s^2} \, ds = -\frac{km}{s}\Big|_s^\infty = \frac{km}{s} \tag{6.6}$$

In vector notation the potential of gravitation V and the gravitational force vector **F** are related by

$$F_x = \frac{\partial V}{\partial x} = km \frac{\partial}{\partial x}\left(\frac{1}{s}\right) = -\frac{km}{s^2}\frac{\partial s}{\partial x} = -\frac{km}{s^2}\frac{x-\xi}{s} \tag{6.7}$$

Similar expressions can be written for F_y and F_z. Thus,

$$\text{grad } V \equiv \left(\frac{\partial V}{\partial x}, \frac{\partial V}{\partial y}, \frac{\partial V}{\partial z}\right) = [F_x, F_y, F_z] \tag{6.8}$$

Grad V is read as the gradient V. From (6.5) it is apparent that the gravitational potential is a function only of the separation of the masses and is independent of any coordinate system used to describe the position of the attracting mass and the direction of the force vector **F**. The gravitational potential, however, completely characterizes the gravitational force at any point by use of (6.8). The gravitational potential can thus be used to express properties that are invariant with respect to coordinate systems.

The simple analogy of two masses can be expanded to the general case of several attracting masses acting on a point mass simultaneously. Because the potential is a scalar parameter, the potential at a point is the sum of the individual potentials:

$$V = \sum V_i = \sum \frac{km_i}{s_i} \qquad (6.9)$$

Considering a solid body M rather than individual masses. The discrete summation is replaced by a volume integral over the body:

$$V(x, y, z) = k \iiint_M \frac{dm}{s} = k \iiint_v \frac{\rho \, dv}{s} \qquad (6.10)$$

where ρ denotes a density that can vary throughout the body and v denotes the mass volume.

In deriving (6.10) for the potential of a body, we assumed that the body was at rest. To represent the earth's potential realistically, we must take into consideration the earth's rotation. Consider Figure 6.2. The vector $\mathbf{f}$ denotes the centrifugal force acting on the unit mass. If the angular velocity of the earth's rotation is denoted by ω, then the centrifugal force vector can be written

$$\mathbf{f} = \omega^2 \mathbf{p} = [\omega^2 x, \omega^2 y, 0]^T \qquad (6.11)$$

The centrifugal force acts parallel to the equatorial plane and is directed away from the axis of rotation. The vector $\mathbf{p}$ is the distance from the rotation axis. By using the definition of the potential and by having the z axis coincide with

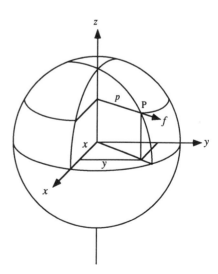

Figure 6.2 Centrifugal force vector.

the rotation axis, we obtain the centrifugal potential:

$$\Phi = \tfrac{1}{2}\omega^2(x^2 + y^2) \tag{6.12}$$

Equation (6.12) can be verified by taking the gradient to get (6.11). Note again that the potential is a function only of the distance from the rotation axis and is not affected by a particular coordinate system definition. The potential of gravity W is the sum of the gravitational and centrifugal potentials:

$$W(x, y, z) = V + \Phi = k \int\!\!\int\!\!\int_v \frac{\rho\, dv}{s} + \frac{1}{2}\omega^2(x^2 + y^2) \tag{6.13}$$

The gravity force vector **g** is then

$$\mathbf{g}(x, y, z) = \text{grad } W = \left[\frac{\partial W}{\partial x}, \frac{\partial W}{\partial y}, \frac{\partial W}{\partial z}\right]^T \tag{6.14}$$

which is the total force acting at a point resulting from the gravitational and centrifugal forces. The magnitude $|\mathbf{g}| = g$ is called gravity. It is traditionally measured in units of gals where 1 gal = 1 cm/sec^2. The gravity increases as one traverses from the equator to the poles because of the decrease in centrifugal force. Approximate values for gravity are $g_{\text{equator}} \cong 978$ gal and $g_{\text{poles}} \cong 983$ gal. Note that the units of gravity are those of acceleration, implying the equivalence of force per unit mass and acceleration. Because of this, the gravity vector **g** is often termed gravity acceleration. The direction of **g** at a point is the direction of the plumb line or the direction of the vertical.

Surfaces for which $W(x, y, z)$ is a constant are called equipotential surfaces or level surfaces. These surfaces can principally be determined by evaluating (6.13) if the density distribution and angular velocity are known. Of course, the density distribution of the earth is not precisely known. Geodetic theories are available to determine the equipotential surface without explicit knowledge of the density distribution. The geoid is defined to be a specific equipotential surface having a gravity potential

$$W(x, y, z) = W_0 \tag{6.15}$$

In practice, this reference gravity potential is chosen such that on the average it coincides with the global ocean surface. Note that this is a purely arbitrary specification chosen for ease of the physical interpretation of the approximate location of the geoid. The geoid is an equipotential surface, not an ideal ocean surface.

There is an important relationship between the direction of the gravity force and equipotential surfaces, demonstrated by Figure 6.3. The total differential

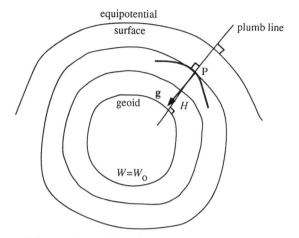

Figure 6.3   Equipotential surfaces and the gravity force vector.

of the gravity potential at a point is

$$dW = \frac{\partial W}{\partial x}\,dx + \frac{\partial W}{\partial y}\,dy + \frac{\partial W}{\partial z}\,dz$$

$$= \left[\frac{\partial W}{\partial x}, \frac{\partial W}{\partial y}, \frac{\partial W}{\partial z}\right]\begin{bmatrix} dx \\ dy \\ dz \end{bmatrix}$$

$$= \operatorname{grad} W \cdot d\mathbf{x}$$

$$= \mathbf{g} \cdot d\mathbf{x} \qquad\qquad (6.16)$$

The quantity dW is the change in potential between two differentially separated points $P(x, y, z)$ and $P'(x + dx, y + dy, z + dz)$. If the vector, $d\mathbf{x}$ is chosen such that P and P' occupy the same equipotential surface, then

$$dW = 0$$

and

$$\mathbf{g} \cdot d\mathbf{x} = 0$$

This last expression implies that the direction of the gravity force vector at a point is normal or perpendicular to the equipotential surface passing through the point.

The shapes of equipotential surfaces, which are related to the mass distribution within the earth through (6.13), have no simple analytic expressions.

The plumb lines or normals to the equipotential surfaces are space curves with finite radii of curvature and torsion. The distance along a plumb line from the geoid to a point is called the orthometric height H. The orthometric height is often misidentified as the "height above sea level." This confusion stems from the specification that the geoid closely approximates the global ocean surface. As stated, the geoid does not precisely coincide with the sea level. One of the goals of current research in geodesy is to determine the separation of the ocean surface and the geoid.

Consider a differential line element $d\mathbf{x}$ along the plumb line, $|d\mathbf{x}| = dH$. By noting that H is reckoned positive upward and $\mathbf{g}$ points downward, we can rewrite (6.16) as

$$
\begin{aligned}
dW &= \mathbf{g} \cdot d\mathbf{x} \\
&= g \, dH \cos (\mathbf{g}, d\mathbf{x}) \\
&= g \, dH \cos 180° \\
&= -g \, dH
\end{aligned}
\tag{6.17}
$$

This expression relates the change in potential to a change in the orthometric height. This equation is central in the development of the theory of geometric leveling. Rewriting (6.17) as

$$
g = -\frac{dW}{dH}
\tag{6.18}
$$

it is obvious that the gravity g cannot be constant on the same equipotential surface because the equipotential surfaces are neither regular nor concentric with respect to the center of mass of the earth. This is illustrated in Figure 6.4, which shows two differentially separate equipotential surfaces. It is observed that

$$
g_1 = \frac{dW}{dH_1} \ne g_2 = \frac{dW}{dH_2}
\tag{6.19}
$$

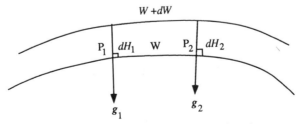

Figure 6.4 Gravity on the equipotential surface.

Now that the definitions of the geoid and the plumb lines are complete, a coordinate system for natural coordinates based on these equipotential surfaces and the plumb lines can be established. Natural coordinates have the important property of being directly observable because they arise directly from the physical phenomena that also describe the level surfaces and the direction of the plumb lines, namely, the earth's potential. Figure 6.5 depicts an equipotential surface through a surface point P and the instantaneous rotation axis and equator. The astronomic normal at point P, also called the local vertical, is identical with the direction of the gravity force vector at point P, which is tangent to the plumb line at P. The astronomic latitude Φ at P is defined as the angle subtended on the instantaneous equator by the astronomic normal. The astronomic meridian plane at point P is described by the astronomic normal and the parallel to the instantaneous rotation axis through the point P. Note that the instantaneous rotation axis and the astronomic normal may or may not intersect. The astronomic longitude Λ is the angle subtended in the instantaneous equatorial plane between this astronomic meridian and a reference meridian, nominally the Greenwich meridian. The third coordinate is the orthometric height of point P, that is, the length measured along the curving plumb line from the geoid to the point P. Thus, $P(\Phi, \Lambda, H)$ is fixed in position by means of natural coordinates (Φ, Λ, H).

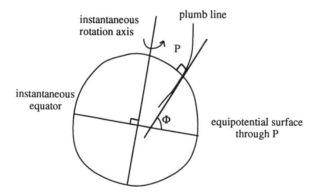

Figure 6.5 Astronomic latitude.

Alternatively, one can describe the third coordinate in terms of potential, using geopotential numbers. The geopotential number C is simply the algebraic difference between the potential at the geoid and the point P:

$$C = W_0 - W \qquad (6.20)$$

The position of P can be described by $P(\Phi, \Lambda, C)$ or $P(\Phi, \Lambda, W)$. From (6.17) it follows that

$$W = W_0 - \int_0^H g \, dH \qquad (6.21)$$

or

$$C = W_0 - W = \int_0^H g \, dH \tag{6.22}$$

or

$$H = -\int_{W_0}^W \frac{dW}{g} = \int_0^C \frac{dC}{g} \tag{6.23}$$

For the sake of completeness, let me point out that, according to (6.22), gravity observations and leveling height differences yield potential differences. The increment dH can be thought of as being obtained from spirit leveling, and the gravity g is measured along the leveling path. Consider the example of a loop being leveled. Because one would return to the same point, that is, to the same equipotential surface, (6.22) implies that the integral (or the sum) of the products $g \, dH$ should add up to zero. Because the gravity g varies along the loop, the sum over the leveled differences dH does not necessarily add up to zero for the loop. In general, the difference between the orthometric height difference and the leveled heights is called the orthometric correction. Expressions for computing the orthometric correction are found in the geodetic literature, for example, van Hees (1992).

6.2 ELLIPSOID OF REVOLUTION

If we recognize that the geoid has no simple mathematical expression, we will find it convenient to define a relatively simple mathematical surface that closely approximates the actual geoid, thus simplifying the computations. This approximation can be made for a local area or on a global scale. Because it rotates, the earth assumes the shape of a sphere that is flattened at the poles and bulging at the equator. This figure is represented very well by an ellipsoid of revolution formed by rotating an ellipse around its minor axis. The geometry of the ellipse is shown in Figure 6.6. The ellipsoid is defined by two parameters, for example, the length of the semimajor axis a and the flattening f. The flattening is related to the semiminor axis b by

$$f = \frac{a - b}{a} \tag{6.24}$$

Figure 6.7 shows an ellipsoid of revolution and the associated coordinates. The Cartesian coordinates (u, v, w) have their origin at the center of the ellipsoid; the w axis coincides with the semiminor axis; the u and v axes are located in the equatorial plane of the ellipsoid, and the u axis is generally directed toward the Greenwich meridian. The ellipsoidal normal through a point

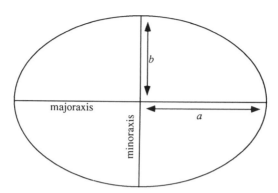

Figure 6.6   Geometry of the ellipse.

P intersects the w axis because of the rotational symmetry of the ellipsoid, but it does not pass through the origin of the Cartesian coordinate system because of the flattening of the ellipsoid. The length of the ellipsoidal normal from the ellipsoid to the point is called the ellipsoidal or geodetic height h. The angle between the ellipsoidal normal and the equatorial plane is the geodetic latitude φ. Any intersection of the ellipsoid (of revolution) with a plane containing the w axis is an ellipse called the ellipsoidal meridian. The geodetic longitude λ is the angle between two meridional planes counted in a clockwise sense from the u axis. Thus the geodetic coordinates (φ, λ, h) completely describe the position of a point in space. The plane at the earth's surface point P, which is perpendicular to the ellipsoidal normal, defines the local geodetic horizon. Notice the distinction between the local geodetic horizon and the local astronomic horizon (the latter is perpendicular to the plumb line at P).

The relationship between the Cartesian coordinates (u, v, w) and the geo-

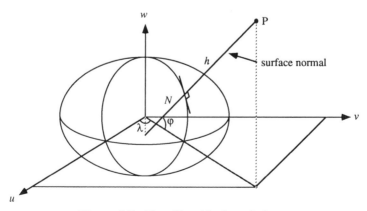

Figure 6.7   The ellipsoid of revolution.

detic coordinates (φ, λ, h) is

$$u = (N + h) \cos \varphi \cos \lambda \tag{6.25}$$

$$v = (N + h) \cos \varphi \sin \lambda \tag{6.26}$$

$$w = [N(1 - e^2) + h] \sin \varphi \tag{6.27}$$

where the auxiliary quantities N and e are

$$N = \frac{a}{\sqrt{1 - e^2 \sin^2 \varphi}} \tag{6.28}$$

$$e^2 = 2f - f^2 \tag{6.29}$$

A derivation of these classical equations can be found readily in the literature, for example, Leick (1980a). The inverse solution of (6.25) to (6.27) can be computed as follows. The longitude λ follows immediately from (6.25) and (6.26):

$$\tan \lambda = \frac{v}{u} \tag{6.30}$$

An expression of the latitude is obtained by noting that N is the radius of curvature of the prime vertical section and corresponds geometrically to the length of the ellipsoidal normal from the ellipsoidal surface to its intersection with the w axis. The prime vertical is the intersection of a plane (which contains the surface normal and which is perpendicular to the meridional plane) with the ellipsoid. In the terminology of differential geometry, any plane containing the normal of the surface is called a normal plane; the intersection of the normal plane with the surface is called a normal section. With this geometric interpretation of N, we can readily see from Figure 6.7 that

$$\tan \varphi = \frac{(N + h) \sin \varphi}{\sqrt{u^2 + v^2}} \tag{6.31}$$

This is a nonlinear equation for the geodetic latitude φ. Rewriting (6.27) as

$$(N + h) \sin \varphi = w + e^2 N \sin \varphi \tag{6.32}$$

and substituting it into (6.31) gives

$$\tan \varphi = \frac{w}{\sqrt{u^2 + v^2}} \left(1 + \frac{e^2 N \sin \varphi}{w} \right) \tag{6.33}$$

The second term in (6.33) can be modified using an approximation for the w coordinate. As a first approximation, one takes a zero ellipsoidal height $h = 0$ and obtains from (6.27) for w

$$w_{h=0} = N(1 - e^2) \sin \varphi \qquad (6.34)$$

Substituting (6.34) in (6.33) gives a first solution for the geodetic latitude:

$$\tan \varphi_1 = \frac{w}{(1 - e^2)\sqrt{u^2 + v^2}} \qquad (6.35)$$

Once φ_1 is computed from (6.35), it is substituted into the second term of (6.33) and a new and improved value for φ is computed. The iteration stops after successive solutions yield negligible changes in geodetic latitude. The ellipsoidal height follows from (6.31):

$$h = \frac{\sqrt{u^2 + v^2}}{\cos \varphi} - N \qquad (6.36)$$

It should be noted that there are alternative procedures for solving the nonlinear inverse solution. Several solutions in closed form exist.

6.3 LOCATION OF THE ELLIPSOID AND REDUCTIONS OF OBSERVATIONS

What remains to be done is to establish a relationship between the ellipsoid and the geoid. This is necessary to reduce physical observations to appropriate simpler mathematical model observations, which are then used in practical computations. The astronomic latitude, longitude, and azimuth must be reduced to the corresponding geodetic latitude, longitude, and azimuth, provided that such astronomic observations are still made in today's era of GPS. Measured horizontal angles or directions must be reduced to angles between normal planes.

One of the first tasks is to specify the direction of the semiminor axis of the ellipsoid. A natural choice is to have this axis coincide with the instantaneous rotation axis. However, the instantaneous rotation axis moves with respect to the earth's crust by what is called polar motion. The consequence is that the astronomic latitude, longitude, and azimuths become a function of time. Time independence is achieved by reducing these astronomic observations to a conventional pole, called the conventional terrestrial pole (CTP). (See the discussion on polar motion given in Chapter 2.) Equations (2.19) to (2.21) are used to carry out the polar motion reduction. Therefore, it makes sense to select the w axis parallel to the direction of the CTP.

The translatory position of the ellipsoid must also be specified. Intuitively one would like the center of the ellipsoid to coincide with the center of mass of the earth. By using techniques of dynamical satellite geodesy, we can determine the geocentric station coordinates, whose origin coincides with the center of mass. These coordinates can be used mathematically to force the center of the ellipsoid to coincide with the center of mass, resulting in a geocentric ellipsoid.

The size and shape of the ellipsoid are defined by just two parameters, for example, the semimajor axis and the flattening. Recall that the reason for using the ellipsoid at all is to obtain a simple surface that approximates the geoid; it is obvious that one should choose the parameters that define the ellipsoid such that the resulting ellipsoid coincides as much as possible with the geoid. Figure 6.8 shows both the geoid and the ellipsoid. The relationship between the ellipsoidal height and the orthometric height is

$$h = H + N \tag{6.37}$$

where N is the geoid undulation with respect to the specific ellipsoid. [There should be no cause for confusion by using the same symbol for the geoid undulations in (6.37) and the radius of curvature of the prime vertical in (6.28); both notations are traditional in the geodetic literature]. The geoid undulation is positive if the geoid is above the ellipsoid. A desirable condition for the best-fitting ellipsoid is that the magnitudes of the geoid undulations be minimized. Thus,

$$\sum_{i=1}^{n} N_i^2 = \text{minimum} \tag{6.38}$$

Minimizing the sum of the squares of the geoid undulations distributed over the whole earth ensures a balance of positive and negative undulations. The resulting ellipsoid is called a globally best-fitting ellipsoid. In special cases the

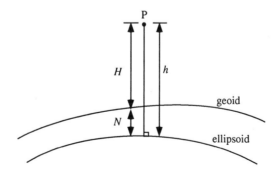

Figure 6.8 Orthometric versus ellipsoidal heights.

minimization (6.38) can be applied to undulations of a small area the size of a country or continent. This would yield a best-fitting local ellipsoid.

Parallelity of the semiminor axis of the ellipsoid and the direction of the CTP leads to important relationships between the reduced astronomic quantities (Φ_{CTP}, Λ_{CTP}, A_{CTP}) and the corresponding ellipsoidal or geodetic quantities (φ, λ, α). The geometric relationships are shown in Figures 6.9 and 6.10. The following symbols are used:

Z_a	astronomic zenith (= intersection of local vertical with the sphere direction)
CTP	position of the conventional terrestrial pole
Z_e	ellipsoidal zenith (= intersection of the ellipsoidal normal through P_1 with the sphere of direction)
T	target point to which the azimuth is measured
A_{CTP}	reduced astronomic azimuth
Φ_{CTP}	reduced astronomic latitude
Λ_{CTP}	reduced astronomic longitude
z'	observed zenith angle
φ	ellipsoidal (geodetic) latitude
λ	ellipsoidal (geodetic) longitude
α	ellipsoidal (geodetic) azimuth between two normal planes
z	ellipsoidal geodetic zenith angle
θ	total deflection of the vertical
ϵ	deflection of the vertical in the direction of azimuth
ξ	deflection of the vertical along the meridian
η	deflection of the vertical along the prime vertical

By applying spherical trigonometry to the various triangles in Figure 6.10, we can eventually derive the following relations:

$$A_{CTP} - \alpha = (\Lambda_{CTP} - \lambda) \sin \varphi + (\xi \sin \alpha - \eta \cos \alpha) \cot z \qquad (6.39)$$

$$\xi = \Phi_{CTP} - \varphi \qquad (6.40)$$

$$\eta = (\Lambda_{CTP} - \lambda) \cos \varphi \qquad (6.41)$$

$$z = z' + \xi \cos \alpha + \eta \sin \alpha \qquad (6.42)$$

The derivations of these classical equations can be found in most of the geodetic literature, for example, Leick (1980a). Equation (6.39) is called the Laplace equation. It relates the reduced astronomic azimuth and the geodetic azimuth of the normal section containing the target point. Equations (6.40) and (6.41) define the deflection of the vertical components. The deflection of the vertical is simply the angle between the direction of the plumb line and the ellipsoidal normal through the same point. By convention, the deflection of the vertical is decomposed into two components, one lying in the meridian and the other lying in the prime vertical, or orthogonal to the meridian. The deflection components depend directly on the shape of the geoid in the region. Because the deflections of the vertical are merely another manifestation of the irregularity of the gravity field or the geoid, they are mathematically related to the geoid undulation. Equation (6.42) relates the ellipsoidal and observed zenith angle (refraction not considered).

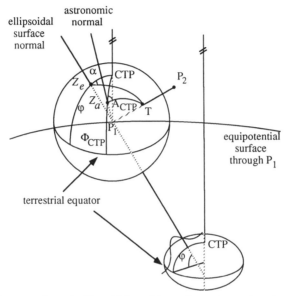

Figure 6.9 Astronomic and ellipsoidal zenith on the topocentric sphere of direction. The astronomical vertical is perpendicular to the equipotential surface at P_1.

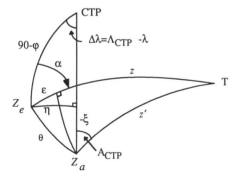

Figure 6.10 Deflection of the vertical components. This figure gives an exaggerated geometry for the deflections; the magnitude of θ is only a few seconds of arc.

Equations (6.39) to (6.41) can be used to correct the observed astronomic latitude, longitude, and azimuth and thus to obtain the ellipsoidal latitude, longitude, and azimuth. It is important to note that the reduction of a horizontal angle due to deflection of the vertical is obtained from the difference of (6.39) as applied to both legs of the angle. If the zenith angle to the endpoints of both legs is close to 90°, then the corrections are small and can possibly be neglected. In the days of classical geodesy, equation (6.39) was also used as a condition between the observed astronomic azimuth and the computed geodetic azimuth. Today systematic errors can be controlled more efficiently and better with GPS. However, if surveyors were to check the orientation of a GPS vector with the astronomic azimuth from the sun or polaris, they must expect a discrepancy, indicated by (6.39).

Equations (6.39) to (6.41) also show how to specify a local ellipsoid that is tangent to the geoid at some centrally located station, called the initial point, and whose semiminor axis is still parallel to the CTP. Such a local ellipsoid is shown in Figure 6.11. If we specify that, for the initial point the reduced astronomic latitude, longitude, and azimuth equal the ellipsoidal latitude, longitude, and azimuth, respectively, then we ensure parallelity of the semimajor axis and the direction of the CTP and the geoidal normal and the ellipsoidal normal coincide at that initial point. If, in addition, we set the undulation at the initial point to zero, then the ellipsoid touches the geoid tangentially at the initial point. Thus the local ellipsoid is specified by

$$\varphi = \Phi_{\text{CTP}} \tag{6.43}$$

$$\lambda = \Lambda_{\text{CTP}} \tag{6.44}$$

$$\alpha = A_{\text{CTP}} \tag{6.45}$$

$$N = 0 \tag{6.46}$$

at the initial point. Other specifications are also possible.

The local ellipsoid is suitable for serving as a computation reference for least-squares adjustments of networks encountered typically in local and re-

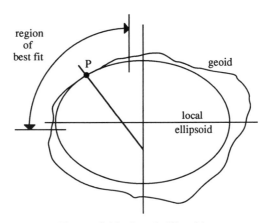

Figure 6.11 Local ellipsoid.

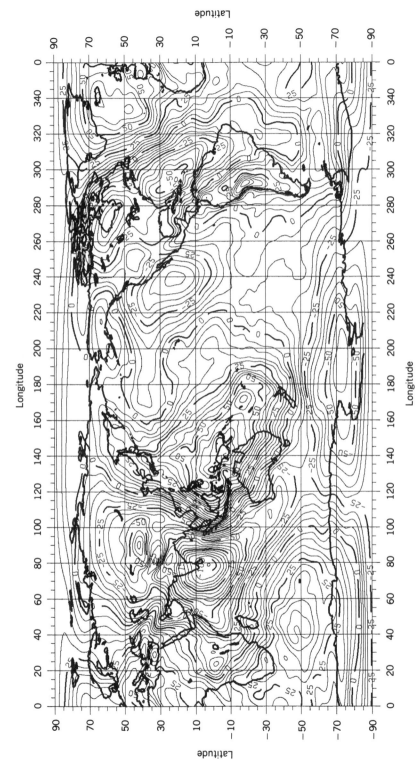

Figure 6.12 Geoid undulation map of the world. (Courtesy of R. Rapp, Ohio State University.)

231

gional surveying. In these cases it is not necessary to determine the size and shape of the locally best-fitting ellipsoid by minimizing the square of undulations [as indicated by (6.38)]. It is sufficient to adopt the size and shape of any of the currently valid geocentric ellipsoids. Because the deflection of the vertical will be small in the region around the initial point, it can often be neglected completely. This is especially true for the reduction of angles. The local ellipsoid is even more useful than it appears at first sight. So long as typical observations, such as horizontal directions, angles, and slant distances, are adjusted, the position of the initial point in (6.43) and (6.44) may not be accurately known. In fact, if the (local) undulation variation is negligible, the coordinate values for the position of the initial point are arbitrary. The same is true for the azimuth condition (6.45). These simplifications make it attractive to use an ellipsoid as a reference for the adjustment of even the smallest survey, thus providing a unified adjustment approach for surveys of large and small areas.

Figure 6.12 shows a geoid undulation map of the world. It is based on the OSU91A potential coefficient model taken to degree 50 only (Rapp et al. 1991). The contour interval is 5 m and the reference ellipsoid is geocentric with a flattening of 1/298.257. This geoid model is part of a more general solution with a geopotential model of degree 360 and a sea surface topography model of degree 15. In addition to terrestrial gravity data, the solution includes altimeter observations made by the *Geosat* satellite.

CHAPTER 7

THE THREE-DIMENSIONAL GEODETIC MODEL

Once the angular observations are reduced to the ellipsoid, it is a simple matter to develop the mathematical model for the adjustment. It is assumed that measured horizontal directions, horizontal angles, and zenith angles have been reduced to the geodetic horizon and the ellipsoidal normal. The reduced horizontal angles are angles between two normal planes containing the target points and having the ellipsoidal normal at the observing stations in common. The reduced zenith angles are the angles between the ellipsoidal normals and the line of sight to the targets. Measured slant distances require no model reductions.

Traditionally the three-dimensional geodetic models were applied to the measurements themselves; that is, the angular measurements were not reduced first to the ellipsoidal normal and the geodetic horizon. This more expanded three-dimensional model allows for the estimation of astronomic and geodetic positions. The deflections of the vertical are not necessarily considered known, but attempts are occasionally made to estimate them. Much effort has gone into applying the general three-dimensional model, particularly in mountainous regions. However, the results have been marginal compared to what can be achieved with GPS today. One of the prime limits of the classical three-dimensional approach is the uncertainty in the vertical refraction when measuring zenith angles (and, of course, the need for visibility between stations). There was also the burdensome procedure of astronomic position and azimuth determination. Much of the work on three-dimensional geodesy was done by Bruns (1878), Wolf (1963), Hotine (1969), Vincenty (1979), and Hradilek (1984). The three-dimensional approach can be generalized even more by requiring that gravity observations be processed together with all the other types of observations. Thus, position, gravity potential, and functionals of the gravity potential, are all processed in one step in a unified model. This approach also allows a theoretically sound inclusion of spirit leveling observation. In simple terms, one may envision that the "classical" three-dimensional model, the spirit leveling model, and the undulation computation procedure explained in Chapter 11 are combined in one unified model capable of dealing with all types of observations, whether dependent on gravity or not. This modern approach was born largely because of the need and desire to provide geoid undulation with an accuracy comparable to the GPS-derived vertical component. This new approach is called *integrated geodesy* (Hein 1986).

The mathematical model presented here is a simplification of a more general three-dimensional geodetic model. Thanks to GPS, we no longer depend on zenith angle to provide the vertical dimension, and thanks to the progress of geodetic satellite techniques over the past several decades, we can now compute deflection of the vertical and geoid undulation whenever necessary. Highly accurate geoid undulation differences at the 2- to 4-cm level can be computed with the help of local gravity data. Details on this technique are given in Chapter 11. Thus, if specifications call for a vertical accuracy of 2–4 cm, even spirit leveling can be replaced by GPS. Therefore, the types of observations most likely to be used by the modern surveyors are horizontal angles, slant distances, and GPS vectors. Thanks to the readjustment of the North American Datum, now referred to as the NAD 83, high-quality geodetic positions are available that can be incorporated into a GPS survey. On the other side, quality control on the observations in a local or regional area can be carried out with a local reference ellipsoid; the local reference ellipsoid can be readily specified by the minimal constraints during the adjustment.

7.1 THE PARTIAL DERIVATIVES

Figure 7.1 shows the local geodetic coordinate system, which plays a central role in the development of the mathematical model. The axes n and e span the local geodetic horizon (plane perpendicular to the ellipsoidal normal through the surface point P_1). The n axis points north; the e axis points east, and the h axis coincides with the ellipsoidal normal with the positive end outward of the ellipsoid. Thus, the local geodetic coordinates are (n, e, h). The spatial orientation of the local geodetic coordinate system is completely specified by the geodetic latitude φ and the geodetic longitude λ. Recall that the w axis coincides with the direction of the CTP.

Figure 7.2 shows how the ellipsoidal (geodetic) azimuth and vertical angle (or zenith angle) between points P_1 and P_2 relate to the local geodetic coordinate system. The same symbol a is used for the altitude angle and the semimajor axis of the ellipsoid. Also, the symbol h denotes the height of the second station P_2 in the local geodetic coordinate system (of P_1). Unless used in the context of local geodetic coordinates, the symbol h denotes the ellipsoidal height. Often the height in the local geodetic coordinate system is denoted by

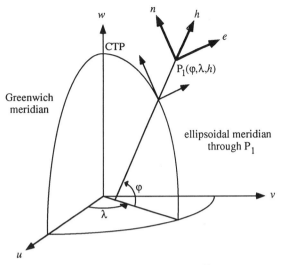

Figure 7.1 The local geodetic coordinate system.

u (up). However, this notation is not without conflict either, since u denotes one of the three geocentric Cartesian coordinates (u, v, w). These explanations should avoid confusion about the notation. It follows that

$$n = s \cos a \cos \alpha \tag{7.1}$$

$$e = s \cos a \sin \alpha \tag{7.2}$$

$$h = s \sin a \tag{7.3}$$

The inverses of (7.1) to (7.3) are

$$\alpha = \tan^{-1}\left(\frac{e}{n}\right) \tag{7.4}$$

$$a = \sin^{-1}\left(\frac{h}{s}\right) \tag{7.5}$$

$$s = \sqrt{n^2 + e^2 + h^2} \tag{7.6}$$

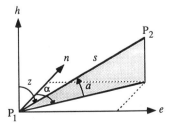

Figure 7.2 Reduced observations in the local geodetic coordinate system.

The relationship between the local geodetic coordinate system and the geocentric Cartesian system $(U) = (u, v, w)$ is seen in Figure 7.1:

$$
\begin{bmatrix} n \\ -e \\ h \end{bmatrix} = R_2(\varphi - 90°)R_3(\lambda - 180°) \begin{bmatrix} \Delta u \\ \Delta v \\ \Delta w \end{bmatrix}
\tag{7.7}
$$

where R_2 and R_3 denote the usual rotation matrices given in Appendix D, and

$$
\begin{bmatrix} \Delta u \\ \Delta v \\ \Delta w \end{bmatrix} = \begin{bmatrix} u_2 - u_1 \\ v_2 - v_1 \\ w_2 - w_1 \end{bmatrix}
\tag{7.8}
$$

Changing the algebraic sign of e in (7.7) and combining the rotation matrices R_2 and R_3 results in

$$
\begin{bmatrix} n \\ e \\ h \end{bmatrix} = R(\varphi, \lambda) \begin{bmatrix} \Delta u \\ \Delta v \\ \Delta w \end{bmatrix}
\tag{7.9}
$$

where

$$
R = \begin{bmatrix} -\sin\varphi\cos\lambda & -\sin\varphi\sin\lambda & \cos\varphi \\ -\sin\lambda & \cos\lambda & 0 \\ \cos\varphi\cos\lambda & \cos\varphi\sin\lambda & \sin\varphi \end{bmatrix}
\tag{7.10}
$$

Substituting (7.9) and (7.10) into (7.4) to (7.6) gives expressions for the reduced observations as functions of the geocentric Cartesian coordinate differences and the geodetic position of P_1:

$$
\alpha_1 = \tan^{-1}\left(\frac{-\sin\lambda_1\Delta u + \cos\lambda_1\Delta v}{-\sin\varphi_1\cos\lambda_1\Delta u - \sin\varphi_1\sin\lambda_1\Delta v + \cos\varphi_1\Delta w}\right)
\tag{7.11}
$$

$$
a_1 = \sin^{-1}\left(\frac{\cos_1\cos\lambda_1\Delta u + \cos\varphi_1\sin\lambda_1\Delta v + \sin\varphi_1\Delta w}{\sqrt{\Delta u^2 + \Delta v^2 + \Delta w^2}}\right)
\tag{7.12}
$$

$$
s = \sqrt{\Delta u^2 + \Delta v^2 + \Delta w^2}
\tag{7.13}
$$

It is important to note the subscripts in (7.11) and (7.12); the geodetic latitude and longitude in these expressions refer to point P_1.

Equations (7.11) to (7.13) constitute the adjustment model of the observation equation type $L_a = F(X_a)$ where the observations consist of the reduced azimuth (horizontal angle), the reduced vertical angle, and the slant distance.

The parameters to be estimated by least-squares adjustment are $(\Delta u, \Delta v, \Delta w)$. The nonlinear model has the general form

$$\alpha_1 = f(u_1, v_1, w_1, u_2, v_2, w_2) \tag{7.14}$$

$$a_1 = f(u_1, v_1, w_1, u_2, v_2, w_2) \tag{7.15}$$

$$s = f(u_1, v_1, w_1, u_2, v_2, w_2) \tag{7.16}$$

To find the elements of the design matrix, we require the total partial derivatives with respect to the parameters. The general form is

$$\begin{bmatrix} d\alpha_1 \\ da_1 \\ ds \end{bmatrix} = \begin{bmatrix} g_{11} & g_{12} & g_{13} & \vdots & g_{14} & g_{15} & g_{16} \\ g_{21} & g_{22} & g_{23} & \vdots & g_{24} & g_{25} & g_{26} \\ g_{31} & g_{32} & g_{33} & \vdots & g_{34} & g_{35} & g_{36} \end{bmatrix} \begin{bmatrix} du_1 \\ dv_1 \\ dw_1 \\ du_2 \\ dv_2 \\ dw_2 \end{bmatrix}$$

$$= [G_1 : G_2] \begin{bmatrix} dU_1 \\ dU_2 \end{bmatrix} \tag{7.17}$$

The partial derivatives are listed in Table 7.1. This particular form of the partial derivatives follows after some algebraic manipulations (Wolf 1963).

Often it is preferable to take differences in geodetic latitude, longitude, and height as parameters instead of the Cartesian components $(\Delta u, \Delta v, \Delta w)$. One of the reasons for such a reparameterization is that changes in latitude, longitude, and height can be visualized more readily than the values in geocentric coordinates. The required transformation is given by (6.25) through (6.28). Differentiating these equations yields

$$\begin{bmatrix} du \\ dv \\ dw \end{bmatrix} = \begin{bmatrix} -(M + h) \cos \lambda \sin \varphi & -(N + h) \cos \varphi \sin \lambda & \cos \varphi \cos \lambda \\ -(M + h) \sin \lambda \sin \varphi & (N + h) \cos \varphi \cos \lambda & \cos \varphi \sin \lambda \\ (M + h) \cos \varphi & 0 & \sin \varphi \end{bmatrix}$$

$$\cdot \begin{bmatrix} d\varphi \\ d\lambda \\ dh \end{bmatrix} \tag{7.18}$$

$$= J(\varphi, \lambda, h) \begin{bmatrix} d\varphi \\ d\lambda \\ dh \end{bmatrix}$$

TABLE 7.1 Partial Derivatives of the Three-Dimensional Geodetic Model in Terms of Cartesian Coordinates

$$g_{11} = \frac{\partial \alpha_1}{\partial u_1} = -g_{14} = \frac{-\sin\varphi_1 \cos\lambda_1 \sin\alpha_1 + \sin\lambda_1 \cos\alpha_1}{s\cos a_1} \tag{a}$$

$$g_{12} = \frac{\partial \alpha_1}{\partial v_1} = -g_{15} = \frac{-\sin\varphi_1 \sin\lambda_1 \sin\alpha_1 - \cos\lambda_1 \cos\alpha_1}{s\cos a_1} \tag{b}$$

$$g_{13} = \frac{\partial \alpha_1}{\partial w_1} = -g_{16} = \frac{\cos\varphi_1 \sin\alpha_1}{s\cos a_1} \tag{c}$$

$$g_{21} = \frac{\partial a_1}{\partial u_1} = -g_{24} = \frac{-s\cos\varphi_1 \cos\lambda_1 + \sin a_1 \Delta u}{s^2 \cos a_1} \tag{d}$$

$$g_{22} = \frac{\partial a_1}{\partial v_1} = -g_{25} = \frac{-s\cos\varphi_1 \sin\lambda_1 + \sin a_1 \Delta u}{s^2 \cos a_1} \tag{e}$$

$$g_{23} = \frac{\partial a_1}{\partial w_1} = -g_{26} = \frac{-s\sin\varphi_1 + \sin a_1 \Delta w}{s^2 \cos a_1} \tag{f}$$

$$g_{31} = \frac{\partial s_1}{\partial u_1} = -g_{34} = -\frac{\Delta u}{s} \tag{g}$$

$$g_{32} = \frac{\partial s_1}{\partial v_1} = -g_{35} = -\frac{\Delta v}{s} \tag{h}$$

$$g_{33} = \frac{\partial s_1}{\partial w_1} = -g_{36} = -\frac{\Delta w}{s} \tag{i}$$

where

$$M = \frac{a(1 - e^2)}{(1 - e^2 \sin^2 \varphi)^{3/2}} \tag{7.19}$$

is the radius of curvature of the ellipsoidal meridian. The matrix $J(\varphi, \lambda, h)$ must be evaluated for the geodetic latitude and longitude of the point under consideration; thus, $J_1(\varphi_1, \lambda_1, h_1)$ and $J_2(\varphi_2, \lambda_2, h_2)$ denote the transformation matrices for points P_1 and P_2, respectively. Substituting (7.18) into (7.17), we obtain the parameterization in terms of geodetic latitude, longitude and height:

$$
\begin{bmatrix} d\alpha_1 \\ da_1 \\ ds \end{bmatrix} = [G_1 J_1 : G_2 J_2]
\begin{bmatrix} d\varphi_1 \\ d\lambda_1 \\ dh_1 \\ d\varphi_2 \\ d\lambda_2 \\ dh_2 \end{bmatrix}
\tag{7.20}
$$

To achieve a parameterization that is even easier to interpret, we transform the geodetic latitude and longitude parameters $(d\varphi, d\lambda)$ into differentials of the local geodetic coordinates (dn, de). The latter coordinate changes can be interpreted as linear shifts at the station in the direction of north and east. Figure 7.3 shows how this transformation is simply related to the radii of curvature of the prime vertical and of the meridian. Because the units of (dn, de) are in lengths, say meters, the standard deviations of the adjusted coordinates can be visualized more readily than those of geocentric angles regardless of whether the angles are given in degrees, seconds, or fractions of seconds. The figure shows that

$$
\begin{bmatrix} dn \\ de \\ dh \end{bmatrix} = \begin{bmatrix} M + h & 0 & 0 \\ 0 & (N + h)\cos\varphi & 0 \\ 0 & 0 & 1 \end{bmatrix} \begin{bmatrix} d\varphi \\ d\lambda \\ dh \end{bmatrix} = H(\varphi, h) \begin{bmatrix} d\varphi \\ d\lambda \\ dh \end{bmatrix} \quad (7.21)
$$

The transformation (7.21) does not change the differential ellipsoidal height dh. The matrix H is evaluated for the station under consideration. Because H is a diagonal matrix, the inverse H^{-1} is readily computed. Thus, the final parameterization is

$$
\begin{bmatrix} d\alpha_1 \\ da_1 \\ ds \end{bmatrix} = [G_1 J_1 H_1^{-1} \; : \; G_2 J_2 H_2^{-1}] \begin{bmatrix} dn_1 \\ de_1 \\ dh_1 \\ dn_2 \\ de_2 \\ dh_2 \end{bmatrix}
$$

$$
= \begin{bmatrix} a_{11} & a_{12} & a_{13} & : & a_{14} & a_{15} & a_{16} \\ a_{21} & a_{22} & a_{23} & : & a_{24} & a_{25} & a_{26} \\ a_{31} & a_{32} & a_{33} & : & a_{34} & a_{35} & a_{36} \end{bmatrix} \begin{bmatrix} dn_1 \\ de_1 \\ dh_1 \\ dn_2 \\ de_2 \\ dh_2 \end{bmatrix} \quad (7.22)
$$

The partial derivatives are given in Table 7.2 (Wolf 1963; Heiskanen and Moritz 1967; Vincenty 1979). Note that some of the partial derivatives have been expressed in terms of the back azimuth α_2 and the back vertical angle a_2, meaning azimuth and vertical angle from station 2 to station 1.

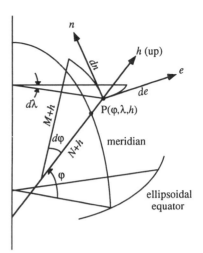

Figure 7.3 Relating changes in geodetic latitude and longitude to changes in local geodetic coordinates.

TABLE 7.2   Partial Derivatives for the Local Geodetic Coordinates

$$a_{11} = \frac{\partial \alpha_1}{\partial n_1} = \frac{\sin \alpha_1}{s \cos a_1} \quad \text{(a)} \qquad a_{12} = \frac{\partial \alpha_1}{\partial e_1} = -\frac{\cos \alpha_1}{s \cos a_1} \quad \text{(b)} \qquad a_{13} = \frac{\partial \alpha_1}{\partial h_1} = 0 \quad \text{(c)}$$

$$a_{14} = \frac{\partial \alpha_1}{\partial n_2} = -\frac{\sin \alpha_1}{s \cos a_1} \Big[\cos(\varphi_2 - \varphi_1) + \sin \varphi_2 \sin(\lambda_2 - \lambda_1) \cot \alpha_1 \Big] \quad \text{(d)}$$

$$a_{15} = \frac{\partial \alpha_1}{\partial e_2} = \frac{\cos \alpha_1}{s \cos a_1} \Big[\cos(\lambda_2 - \lambda_1) - \sin \varphi_1 \sin(\lambda_2 - \lambda_1) \tan \alpha_1 \Big] \quad \text{(e)}$$

$$a_{16} = \frac{\partial \alpha_1}{\partial h_2} = \frac{\cos \alpha_1 \cos \varphi_2}{s \cos a_1} \Big[\sin(\lambda_2 - \lambda_1) + \big(\sin \varphi_1 \cos(\lambda_2 - \lambda_1) - \cos \varphi_1 \tan \varphi_2 \big) \tan \alpha_1 \Big] \quad \text{(f)}$$

$$a_{21} = \frac{\partial a_1}{\partial n_1} = \frac{\sin a_1 \cos \alpha_1}{s} \quad \text{(g)}$$

$$a_{22} = \frac{\partial a_1}{\partial e_1} = \frac{\sin a_1 \sin \alpha_1}{s} \quad \text{(h)}$$

$$a_{23} = \frac{\partial a_1}{\partial h_1} = -\frac{\cos a_1}{s} \quad \text{(i)}$$

$$a_{24} = \frac{\partial a_1}{\partial n_2} = \frac{-\cos \varphi_1 \sin \varphi_2 \cos(\lambda_2 - \lambda_1) + \sin \varphi_1 \cos \varphi_2 + \sin a_1 \cos a_2 \cos \alpha_2}{s \cos a_1} \quad \text{(j)}$$

$$a_{25} = \frac{\partial a_1}{\partial e_2} = \frac{-\cos \varphi_1 \sin(\lambda_2 - \lambda_1) + \sin a_1 \cos a_2 \sin \alpha_2}{s \cos a_1} \quad \text{(k)}$$

$$a_{26} = \frac{\partial a_1}{\partial h_2} = \frac{\sin a_1 \sin a_2 + \sin \varphi_1 \sin \varphi_2 + \cos \varphi_1 \cos \varphi_2 \cos(\lambda_2 - \lambda_1)}{s \cos a_1} \quad \text{(l)}$$

$$a_{31} = \frac{\partial s}{\partial n_1} = -\cos a_1 \cos \alpha_1 \quad \text{(m)} \qquad\qquad a_{32} = \frac{\partial s}{\partial e_1} = -\cos a_1 \sin \alpha_1 \quad \text{(n)}$$

$$a_{33} = \frac{\partial s}{\partial h_1} = -\sin a_1 \quad \text{(o)} \qquad\qquad a_{34} = \frac{\partial s}{\partial n_2} = -\cos a_2 \cos \alpha_2 \quad \text{(p)}$$

$$a_{35} = \frac{\partial s}{\partial e_2} = -\cos a_2 \sin \alpha_2 \quad \text{(q)} \qquad\qquad a_{36} = \frac{\partial s}{\partial h_2} = -\sin a_2 \quad \text{(r)}$$

7.2 ADDITIONAL REMARKS

As in any adjustment, the partial derivatives must be evaluated at the current point of expansion; that is, the adjusted positions of the previous iteration must be used. This is true whether the partials have been expressed in terms of Cartesian coordinates, geodetic latitudes, longitudes, and heights, or azimuths and vertical angles.

There is no need to reduce the observations to the marks on the ground. This is, in fact, one of the advantages of the three-dimensional adjustment. During the course of the adjustment, the vector $L = L_0 - L_b$ must be computed. When computing L_0 from (7.11) to (7.13) use $h' = h + \Delta h$ instead of h for the station heights. The symbol Δh denotes the height of the instrument or the height of the target above the mark on the ground. As mentioned, L_b denotes the measured value (which is not reduced). After the adjustment is completed, the adjusted observations L_a with respect to the marks on the ground can be computed from the adjusted positions using h in (7.11) to (7.13).

The (φ, λ, h) or (n, e, h) parameterizations are particularly useful for introducing observed heights or height differences. Minimal constraints are conveniently introduced by fixing individual coordinates. The set of minimal constraints depends on the available type of observations and on where the observations are located within the network. One choice for the minimal constraints might be to fix the coordinates (φ, λ, h) of one station (translatory component), the azimuth or the longitude or another station (rotation on azimuth), and the heights of two additional stations. If the adjustment has been carried out in the (u, v, w) parameterization and it is deemed desirable that the result be transformed into the (φ, λ, h) or (n, e, h) coordinates, then the transformations (7.18) and (7.21) can be used accordingly; that is,

$$\begin{bmatrix} dn \\ de \\ dh \end{bmatrix} = H(\varphi, h) J^{-1}(\varphi, \lambda, h) \begin{bmatrix} du \\ dv \\ dw \end{bmatrix} \qquad (7.23)$$

$$\begin{bmatrix} du \\ dv \\ dw \end{bmatrix} = J(\varphi, \lambda, h) H^{-1}(\varphi, h) \begin{bmatrix} dn \\ de \\ dh \end{bmatrix} \qquad (7.24)$$

where

$$J^{-1}(\varphi, \lambda, h) = \begin{bmatrix} \dfrac{-\sin \varphi \cos \lambda}{M + h} & \dfrac{-\sin \varphi \sin \lambda}{M + h} & \dfrac{\cos \varphi}{M + h} \\[2ex] \dfrac{-\sin \lambda}{(N + h) \cos \varphi} & \dfrac{\cos \lambda}{(N + h) \cos \varphi} & 0 \\[2ex] \cos \varphi \cos \lambda & \cos \varphi \sin \lambda & \sin \varphi \end{bmatrix} \qquad (7.25)$$

Using the law of variance–covariance propagation, we obtain the transformation of the 3×3 covariance submatrices:

$$\Sigma_{n,e,h} = (HJ^{-1}) \, \Sigma_{u,v,w} \, (HJ^{-1})^{\mathrm{T}} \qquad (7.26)$$

$$\Sigma_{\varphi,\lambda,h} = (J^{-1}) \, \Sigma_{u,v,w} \, (J^{-1})^{\mathrm{T}} \qquad (7.27)$$

Note that the product HJ^{-1} equals the matrix R of (7.10).

By noting that the horizontal angle is the difference of two azimuths, we can obtain the observation equations for the horizontal angle through differencing the azimuth observation equation. The angles might be correlated, depending on the observation procedure. If geoid undulation differences are available (Chapter 11), then leveled height differences can be corrected for the undulation differences to yield ellipsoidal height differences. The respective elements of the design matrix are 1 and -1. The accuracy of incorporating leveling data in this manner is limited by our ability to compute undulation differences. If vertical angles are observed, it may or may not be necessary to introduce vertical refraction parameters into the adjustment. If this is done, we must be careful to avoid overparameterization, that is, introducing too many refraction parameters that could potentially absorb other systematic effects not caused by refraction. Optionally, the vertical angles can be corrected for refraction using a standard model for the atmosphere. In view of GPS capability, high-precision vertical angle measurement is not likely to be important to the modern surveyor. The primary purpose of vertical angles is to give sufficient height information to process the slant distances. A so-called horizontal distance, which is measured in (or reduced to) the local geodetic horizon, can readily be introduced in the three-dimensional adjustment as a new type of observation. Equation (7.6) gives

$$l_{12} = \sqrt{n_1^2 + e_1^2} \qquad (7.28)$$

and

$$l_{21} = \sqrt{n_2^2 + e_2^2} \qquad (7.29)$$

For short distances the approximations $l = l_{12} \approx l_{21}$, $n = n_1 \approx -n_2$, $e = e_1 \approx -e_2$ are valid. Thus, the observation equation expressed in the local geodetic coordinate system is

$$dl = \frac{n}{l} \, dn_1 + \frac{e}{l} \, de_1 - \frac{n}{l} \, dn_2 - \frac{n}{l} \, de_2 \qquad (7.30)$$

Equation (7.21) relates the local geodetic coordinates for short distances to the differences of ellipsoidal latitude and longitude, as follows:

$$n \approx (\varphi_2 - \varphi_1)M_1 \tag{7.31}$$

$$e \approx (\lambda_2 - \lambda_1)N_1 \cos \varphi_1 \tag{7.32}$$

Substituting (7.31) and (7.32) in (7.30) gives

$$dl = \frac{\varphi_2 - \varphi_1}{l} M_1(dn_1 - dn_2) + \frac{\lambda_2 - \lambda_1}{l} N_2 \cos \varphi_1(de_1 - de_2) \tag{7.33}$$

The components of L_0 that are required in the adjustment are computed from (7.28), (7.31), and (7.32) using the approximate latitude and longitude. Note that a distance in the local geodetic horizon can always be treated as a slant distance with zero vertical angle.

If the observations contain little or no vertical information, that is, if zenith angles and leveling data are not available, it is still possible to adjust the network in three dimensions. The height parameters h can be weighted using reasonable estimates for their a priori variances. This is the so-called height-controlled two-dimensional adjustment. In the extreme case the height parameters can even be eliminated from the list of parameters altogether. Even a strictly planar network can be adjusted in three dimensions by using artificial height parameters and weights or by fixing all height parameters. It is clear that the geodetic latitude and longitude parameters can also be assigned a priori weights; this is also true for the local geodetic coordinates n and e. The latter option is preferable, because the units of n and e are units of length rather than degrees of angles. Weighting parameters is a convenient method for the incorporation of existing control stations into the adjustment.

7.3 AN EXAMPLE

The purpose of this adjustment is to demonstrate the three-dimensional model for testing the consistency of the terrestrial observations (angles, distances, leveled height differences) of Table F.1. These observations are already reduced to the ellipsoidal normal, and their stochastic model is given. The leveled height differences are combined with the geoid undulation differences to yield ellipsoidal height differences. These height differences are the model observations to be used in the three-dimensional model described in this chapter. The mathematical model for the angles and distances is given by (7.11) through (7.13); the partial derivatives are listed in Table 7.2.

The minimal constraint adjustment is best suited for the consistency check (or quality control) of the observations; the observations freely determine the polyhedron of stations without any external constraint caused by existing network stations. In the most detailed procedure, the observations could be adjusted in components. For example, those network components that are completely determined by angles or by distances could be adjusted separately to

TABLE 7.3   Summary of Terrestrial Observations

2 9	Slant distances (EDM)
0	Azimuths
2 8	Altitudes (trigonometric)
1 5	Horizontal angles
0	Coordinate differences (relative positions)
0	Horizontal distances
1 1	Leveled height differences
0	GPS vector observations

No. of observations	8 3
No. of parameters	3 6
No. of observed parameters	6
Degree of freedom	5 3
A-posteriori variance of unit weight	0.723
$V^T PV$	38.3

investigate the internal consistency of each type of observation. Alternatively, one could adjust subgroups of stations using all observations of the subgroup. The subgroup could be as small as three stations forming a triangle.

Table 7.3 gives a summary of the observations. Because no azimuths are observed, the rank defect of the normal matrix is 4. The respective minimal constraints for defining the coordinate system are implemented by fixing the three approximate coordinates of one station (to fix translation) and the approximate latitude or longitude of another station (to fix rotation in azimuth). Alternatively, these three parameters could be weighted using the formalism of Table 4.3. In the most general case all parameters could be weighted. However, one must ensure that the standard deviations assigned to the coordinates are large enough to allow adjustment of the coordinates; that is, the approximate coordinates must be allowed to move to the correct value despite the weights. One way to avoid this difficulty is to set $U_2 = 0$ in Table 4.3; this, in effect, results in a weighting of the current approximate positions for each iteration. If we assign weights to each parameter, we can proceed as indicated by Table 4.4; that is, we can add one observation at a time to the adjustment. Analyzing the sequence of $V^T PV$ for jumps will provide helpful hints for detecting blunders. The adjustment specified in Table 7.3 includes all stations except stations 2 (Argyle) and 4 (Sunkhaze), which are determined by GPS vectors only. One station is held fixed and six coordinates are weighted. The ability to specify which coordinates are adjusted freely and which coordinates are assigned weights provides much flexibility and power to the least-squares technique.

Table 7.4 lists some of the results. The observations in this table refer to the surveying marks on the ground and not to the centers of the instrument and the target. The reduction of the observations to the surveying mark is done as

TABLE 7.4   Residuals and Adjusted Observations

St	St	St	Code	Observation mark to mark	Residual	Adj. obs
6	1		1	4309.531	0.012	4309.543
9	1		1	10763.943	0.339	10764.282
15	1		1	6402.034	0.071	6402.105
14	6		1	11890.478	0.154	11890.632
9	6		1	6808.932	0.248	6809.180
6	1	15	4	.72 44 7.2	2.9	72 44 10.1
1	15	14	4	176 20 53.5	-2.6	176 20 50.9
15	14	13	4	121 15 29.7	-1.6	121 15 28.1
14	13	12	4	229 40 25.4	-1.2	229 40 24.2
13	12	5	4	154 23 31.9	-1.1	154 23 30.8
6	1		3	7 37.374	10.889	7 48.263
15	1		3	23 47.950	0.573	23 48.523
1	15		3	-27 21.253	5.779	-27 15.474
14	15		3	-28 30.973	3.608	-28 27.366
9	15		3	-3 34.693	9.277	-3 25.416
7	6		7	1.439	-0.007	1.432
6	15		7	-36.309	-0.003	-36.312
15	14		7	48.185	-0.000	48.185
14	13		7	-39.119	0.001	-39.118
14	11		7	2.781	0.001	2.782

the last step in the adjustment. Because the observations are simulated, the residuals do not reveal much and require no further analysis. The table is presented primarily to "optically document" the outcome of a least-squares solution of a three-dimensional network. The value of the a priori variance of unit weight is also artificial because of the simulation. For real data the fundamental chi-square test (4.267) is used to find out whether or not the minimal constraint adjustment is distorted. If the test fails, then a detailed analysis is required, such as searching for blunders, checking the stochastic model, studying unmodeled effects or neglected reductions (refraction, deflection of the vertical), recomputing geoid undulations, and carrying out more iterations. It may also be instructive to compute the internal and external reliability of the network to find the size of the blunders that are still marginally detectable and to study the impact of undetectable blunders on the parameters (or functions thereof). See Section 4.10 for details on internal and external reliability. As an important practical matter it should be noted that the normal matrix should be inverted only when required. Not inverting the normal matrix saves substantial computer time if the system of normal equations is large.

If there is a need to compute a function of the adjusted coordinates, this should be done with the minimal constraint solution as well. Examples are distances or angles not measured (because of accessibility problems), areas, and volumes. Applying the law of covariance propagation (4.34) gives the

TABLE 7.5 Functions of the Adjusted Parameters and their Statistics

Station	Function	Estimate	St. dev.
13 9	distance	9533.8703 m	0.123 m
8 7 6	horizontal angle	114° 9' 14".5233	2".31310
7 3	ellip. height diff.	8.4500 m	0.021 m

variance of the function. Table 7.5 contains a few selected functions. Note that the value of these functions and their variances are independent of the specific choice of the minimal constraints. The adjusted coordinates are not important in this context, because they do depend on the choice of the minimal constraints as do their variance and covariance.

CHAPTER 8

GPS OBSERVABLES

There are two types of GPS observations (observables): pseudoranges and carrier phases. Pseudoranges are often used in navigation. The carrier phases are sometimes also called carrier beat phases in order to stress the measurement process of correlating the incoming carrier with the receiver-generated reference frequency. In high-precision surveying the carrier phase is favored. Increasingly, however, combination solutions of pseudoranges and carrier phases are becoming more common. Although the (undifferenced) phases can be used directly, it has become common practice to take advantage of various linear combinations of the original carrier phase observation, such as double differences and triple differences.

Measuring pseudoranges and carrier phases involves advanced techniques in electronics. Some of these were discussed briefly in Chapter 3. In this chapter the equations are developed to use pseudoranges and carrier phases, as downloaded from the receiver, to determine geocentric positions (point positioning) or relative positions between co-observing stations (differential or relative positioning). Examples of early literature on the subject are Counselman and Shapiro (1979), Bossler et al. (1980), Counselman and Gourevitch (1981), Remondi (1984, 1985a), Vanicek et al. (1984), King et al. (1985), and Schaffrin and Grafarend (1986). Recent additions to the GPS literature include Hofmann-Wellenhof et al. (1992) and Seeber (1993).

In addition to deriving and discussing the basic pseudorange and carrier phase equations, and the double- and triple-difference functions, frequently asked questions of novice GPS users are addressed. Examples include simultaneity of observations, singularities, a priori knowledge of initial station and ephemeris, and selective availability. The implications of relativity to GPS observables have widely been addressed in the literature, for example, Grafarend (1992), Hatch (1992b), Ashby (1993), and Schwarze et al. (1993). In this chapter remarks on relativity are limited to Section 8.3.1 where a clock correction term is given to account for satellite orbital eccentricity. This correction and the adjustment to the fundamental frequency of 10.23 MHz mentioned in Chapter 3 are the only references on relativity and GPS in this book. In relative positioning most of the relativistic effects cancel or become negligible during differencing.

Throughout this and the following chapters, a superscript identifies the satellite, and a subscript identifies the receiver when used with pseudorange or carrier phase symbols. Lower case letters p and q generally label the satellites, whereas k and m refer to receivers. In many cases the superscripts and subscripts also indicate specific functions of the undifferenced observables. The L1 and L2 carriers are indicated by the subscripts 1 and 2, respectively. Terms whose numerical values depend on the pseudorange or the carrier phase are identified with the subscripts P and φ, respectively. The subscript Φ is used for scaled (metric) carrier phases.

8.1   PSEUDORANGES

The pseudorange is a measure of the distance between the satellite and the receiver's antenna, referring to the epochs of emission and reception of the codes (Figure 8.1). The transmission (travel) time of the signals is measured by correlating identical pseudorandom noise (PRN) codes generated by the satellite with those generated internally by the receiver. The code-tracking loop within the receiver shifts the internal replica of the PRN code in time, until maximum correlation occurs. The receiver codes are derived from the receiver's own clock and the codes of the satellite transmissions are generated by the satellite system of clocks. Unavoidable timing errors in both the satellite and the receiver clock will cause the measured pseudorange to differ from the geo-

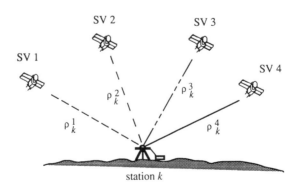

Figure 8.1   Pseudoranging to four satellites.

metric distance corresponding to the epochs of emission and reception. Pseudoranging is applicable to P-codes and/or C/A-codes.

Let the symbol t_k denote the nominal time of the receiver clock k (receiver time frame) at the instant of reception of a signal and t^p denote the nominal time of the satellite clock p at emission. Nominal time refers to the time actually indicated by the respective clocks. In this terminology the "satellite clock" indicates space vehicle time as determined by onboard atomic clocks which guide the code generation. The nominal times are related to true times, such as GPS time, via the clock error terms as follows:

$$t_{r,k} = t_k + dt_k \tag{8.1}$$

$$t_r^p = t^p + dt^p \tag{8.2}$$

The subscript r denotes true time. The pseudorange measurement between satellite p and receiver k, $P_k^p(t_k)$ is the scaled difference of the nominal times such that

$$P_k^p(t_k) = (t_k - t^p)c \tag{8.3}$$

where c denotes the speed of light. The code correlation process yields the nominal time difference $t_k - t^p$ for a specific code sequence. Because t_k is known, the nominal emission time t^p is readily inferred. In this sense, measuring the reception time and the pseudorange at the receiver is conceptually identical to measuring the nominal emission time in the satellite time frame. The pseudorange (8.3) would equal the geometric distance that the signal traveled (from the instant of emission at the satellite to reception at the receiver), if the propagation medium were a vacuum and if there were no clock errors and no other biases. Taking these errors and biases into account, the complete expression for the pseudorange becomes

$$P_k^p(t_k) = (t_{r,k} - t_r^p)c - c(dt_k - dt^p) + I_{k,P}^p(t_k) + T_k^p(t_k)$$
$$+ d_{k,P}(t_k) + d_{k,P}^p(t_k) + d_P^p(t_k) + \epsilon_P$$
$$= \rho_k^p(t_{r,k}) - c(dt_k - dt^p) + I_{k,P}^p(t_k) + T_k^p(t_k)$$
$$+ d_{k,P}(t_k) + d_{k,P}^p(t_k) + d_P^p(t_k) + \epsilon_P \tag{8.4}$$

The signs in (8.4) are such that the left-hand side, $P_k^p(t_k)$, represents the actual measurement. The scaled difference of the true times corresponds to the vacuum distance that the signal traveled. Subtracting the clock corrections is in agreement with the understanding that the left side of the equation represents the actual measurement. Because the signal travels through the ionosphere and the troposphere, propagation delays of the codes must be taken into account. As is explained in Chapter 9, the measured range is longer than the geometric distance. Thus, the numerical values of $I_{k,P}^p(t_k)$ and $T_k^p(t_k)$ are always positive and are a function of time depending on the varying conditions along the path of the signal. The ionospheric correction is a function of the carrier frequency while the tropospheric correction is not. However, no carrier frequency identifier is used in (8.4) in order to emphasize the general form of the pseudorange equation which is applicable to both carriers. The frequency-dependent notation will be introduced later as necessary. The symbols $d_{k,P}(t_k)$ and $d_P^p(t_k)$ denote the receiver hardware and satellite hardware code delays. To be general, these delays are marked as time-dependent, although in practice they are, hopefully, stable. The symbol $d_{k,P}^p(t_k)$ denotes the multipath of the codes which depends on the geometry of the antenna and satellite with respect to surrounding reflective surfaces. Multipath is discussed in detail in Section 9.5. The term ϵ_P denotes the random measurement noise. The relativity correction is not listed explicitly. As is explained in Section 8.3.2.1, the satellite clock will be corrected to account for relativity.

The subscript P identifies all terms whose numerical values are specific to pseudoranges. Note that the tropospheric term has no such subscript. The units of all terms in (8.4) are meters. To avoid overloading expressions for the subsequent development, the ionospheric and tropospheric terms, the hardware delays, the multipath term, and the measurement noise will be listed only when relevant to the discussion.

The topocentric distance $\rho_k^p(t_{r,k})$ equals the geometric distance traveled by the signal between the instants of signal emission and reception. Because distances between the receiver and the satellites differ, signals arriving simultaneously must have been emitted at different times. Since we do not know the true time $t_{r,k}$, the topocentric distance is linearized around the known nominal receiver time t_k by

$$\rho_k^p(t_{r,k}) = \rho_k^p(t_k) + \dot{\rho}_k^p(t_k)\, dt_k \tag{8.5}$$

where higher-order terms are neglected. Substituting (8.5) in (8.4) gives the fully developed expression for the pseudorange:

$$P_k^p(t_k) = \rho_k^p(t_k) - c\left[1 - \frac{\dot{\rho}_k^p(t_k)}{c}\right] dt_k + c\, dt^p + \cdots \tag{8.6}$$

The topocentric distance $\rho_k^p(t_k)$ at the nominal receiver epoch t_k is given by

$$\rho_k^p(t_k) = \sqrt{(u^p - u_k)^2 + (v^p - v_k)^2 + (w^p - w_k)^2} \tag{8.7}$$

The Cartesian coordinates refer to the same reference system, e.g., the WGS 84 system. The triplet (u_k, v_k, w_k) denotes the approximate position of the receiver whose accurate position is sought. The coordinates (u^p, v^p, w^p) denote the position of the satellite at the instant of emission. Computation of the distance $\rho_k^p(t_k)$ is required for both pseudorange and carrier phase processing.

The solution is carried out iteratively, given a satellite ephemeris and approximate position of the receiver. One method begins with the nominal reception time, whereas the other starts with the nominal emission time as computed from the pseudoranges.

In the first approach, the iteration begins by assuming an average value for the travel time τ_k^p. Next, the satellite position is interpolated for the epoch $(t_k - \tau_k^p)$ and the topocentric distance is computed, which is then used to recompute the travel time by dividing the distance by the velocity of light. If the discrepancy between the first and second approximation of τ_k^p is greater than a specified criterion, the iteration is repeated, i.e., a new satellite position is interpolated, a new distance is computed, etc. Usually a couple of iterations are sufficient.

In the second method, which requires pseudoranges, the nominal emission time is obtained from equation (8.3) as

$$t^p = t_k - \frac{P_k^p(t_k)}{c} \tag{8.8}$$

This computation is not affected by the receiver clock error dt_k and the satellite clock error dt^p. The computed nominal emission time, however, is corrupted by the ionospheric, tropospheric, hardware, and multipath delays, as seen from equation (8.4). All of these effects are negligible in the present context. For example, if all of these delays together equaled 300 m, light would transverse that distance in just 1 μsec, thus falsifying t^p by that amount. However, during 1 μsec the topocentric distance changes less than 1 mm using the estimate $|\dot{\rho}_k^p| < 800$ m/sec. Even the difference between the nominal and true satellite time, which is less than 1 μsec, is negligible as far as computing the topocentric distance is concerned. Having the nominal emission time, the satellite is interpolated from the ephemeris for that particular instant, giving (u^p, v^p, w^p). Computation of the topocentric distance requires that the earth's rotation be taken into account during the signal travel time. The process starts by picking a good estimate for the received time, giving a first estimate of the travel time. Next, the earth is rotated by an amount corresponding to this initial estimate of the travel time, i.e., the coordinates of the receiver change. For this new coordinate position the topocentric distance is computed and divided by the velocity of light to give a better estimate of the travel time. The earth is now rotated to correspond to this better estimate of the travel time. The loop is continued until the topocentric range, or the travel time, has converged.

Finally, the topocentric range rate $\dot{\rho}_k^p(t)$, which is required in (8.6), can be derived from changes in the topocentric distance over time. If the station clock error is 1 μsec, the topocentric range rate term contributes less than 1 mm to the range, since the absolute value of $\dot{\rho}$ is less than 800 m/sec. Because of this small contribution, the $\dot{\rho}$ term is often neglected and is simply not listed in all equations explicitly.

The notation is now simplified by omitting the subscript k to identify the nominal reception time of the station clock k. The exception is dt_k. The expression (8.6) is valid for all code observations. For pseudoranges on the L1 carrier, the complete expression is

$$P_{k,1}^P(t) = \rho_k^P(t) - c\left[1 - \frac{\dot{\rho}_k^P(t)}{c}\right] dt_k + c \, dt^P + I_{k,1,P}^P(t) + T_k^P(t)$$

$$+ \, d_{k,1,P}(t_k) + d_{k,1,P}^P(t_k) + d_{1,P}^P(t_k) + \epsilon_{1,P} \qquad (8.9)$$

All bias terms, except the tropospheric delay $T_k^P(t)$, depend on the carrier frequency and are identified accordingly with the subscript 1. Pseudoranges on L2 are identified with a subscript 2. A subscript C/A will identify C/A-code pseudoranges.

8.1.1 The Navigation Solution

In applications where low accuracy but instantaneous positions are required, the pseudorange is the preferred observable. Given the satellite ephemeris (that is, the position of the satellite at the epoch of emission), the receiver clock error and the receiver location are the only unknowns considered in equation (8.9). The effects of residual satellite clock errors are negligible for the typical navigation solution. This is particularly true because the time errors are indistinguishable from common ionospheric and tropospheric delays. The satellite clocks are constantly monitored with respect to GPS time as maintained by the control center. Actual offsets of the satellite clocks are approximated by polynomials in time, and are transmitted to the user as part of the navigation message. See Section 8.3.1 for additional details on receiver and satellite clock errors. Ionospheric and tropospheric delays can be computed approximately from ionospheric and tropospheric models, as explained in Chapter 9. Hardware delays and multipath are neglected in the case of the navigation solution. The $\dot{\rho}$ term is usually omitted as well. Thus, there are four unknowns left in (8.9). These can be computed using four pseudoranges measured simultaneously to four GPS satellites, giving the following system of equations:

$$P_k^1(t) = \sqrt{(u^1 - u_k)^2 + (v^1 - v_k)^2 + (w^1 - w_k)^2} - c \, dt_k \qquad (8.10)$$

$$P_k^2(t) = \sqrt{(u^2 - u_k)^2 + (v^2 - v_k)^2 + (w^2 - w_k)^2} - c \, dt_k \qquad (8.11)$$

$$P_k^3(t) = \sqrt{(u^3 - u_k)^2 + (v^3 - v_k)^2 + (w^3 - w_k)^2} - c \, dt_k \qquad (8.12)$$

$$P_k^4(t) = \sqrt{(u^4 - u_k)^2 + (v^4 - v_k)^2 + (w^4 - w_k)^2} - c \, dt_k \qquad (8.13)$$

The receiver clock error dt_k is solved together with the position of the receiver's antenna for every epoch. Thus, relatively inexpensive quartz crystal clocks can be used in the receiver rather than expensive atomic clocks. The basic requirement, however, is that there be four satellites visible at a given epoch. This visibility requirement was taken into consideration when designing the basic GPS satellite constellation for universal coverage.

If more than four satellites are visible, a least-squares solution can actually be performed. Modifications of the basic navigation solutions are readily made. For example, for a ship on the ocean, the orthometric heights of the receiver can be accurately derived from the height of the receiver above the water surface and the geoid undulations. The geoid undulations are computed from spherical harmonic expressions like (11.31). Equations (8.10) through (8.13) can be expressed in terms of ellipsoidal latitude, longitude, and height using transformations (6.25) through (6.27). In these transformed equations the ellipsoidal height can be entered as a known quantity. Thus, in principle, pseudoranging to three satellites is sufficient to determine positions at sea. Other variations of the basic navigation solution are possible. However, having the complete satellite constellation available, one would naturally observe all satellites in view.

The pseudorange positioning technique depends on the accuracy of the satellite ephemeris, because the (u^i, v^i, w^i) coordinates are considered known in (8.10) through (8.13). The best real-time source of ephemeris data is the navigation message read by the receiver. The accuracy of the navigation solution is typically about 5–20 m if the P-code is used and less accurate if the C/A-code is used. However, the achievable position accuracy with codes is subject to selective availability degradation, changes in receiver technology, e.g., narrow correlation for C/A-codes, and the quality of satellite transmissions which might improve in the future with replacement satellites.

8.1.2 DOP Factors

It has become common practice to use dilution of precision (DOP) factors to describe the effect of the geometric satellite distribution on the accuracy of the navigation solution. The DOP factors are simple functions of the diagonal elements of the covariance matrix of the adjusted parameters (to follow adjustment terminology). In general,

$$\sigma = \sigma_0 \text{ DOP} \tag{8.14}$$

where σ_0 denotes the standard deviation of the observed pseudoranges, and σ is the standard deviation of, for example, the horizontal or vertical position. For DOP computations the pseudorange observations are considered uncorrelated and of equal accuracy; that is, the weight matrix is

$$P = \sigma_0^2 I \tag{8.15}$$

If the ordered set of parameters is

$$X = \{du_k, dv_k, dw_k, dt_k\} \tag{8.16}$$

then the design matrix follows from (8.10) through (8.13) upon linearization around the nominal station location:

$$
_4A_4 = -
\begin{bmatrix}
\dfrac{u^1 - u_k}{\rho_k^1} & \dfrac{v^1 - v_k}{\rho_k^1} & \dfrac{w^1 - w_k}{\rho_k^1} & c \\[2ex]
\dfrac{u^2 - u_k}{\rho_k^2} & \dfrac{v^2 - v_k}{\rho_k^2} & \dfrac{w^2 - w_k}{\rho_k^2} & c \\[2ex]
\dfrac{u^3 - u_k}{\rho_k^3} & \dfrac{v^3 - v_k}{\rho_k^3} & \dfrac{w^3 - w_k}{\rho_k^3} & c \\[2ex]
\dfrac{u^4 - u_k}{\rho_k^4} & \dfrac{v^4 - v_k}{\rho_k^4} & \dfrac{w^4 - w_k}{\rho_k^4} & c
\end{bmatrix}
$$

$$
= -
\begin{bmatrix}
(\mathbf{e}_k^1)^{\mathrm{T}} & c \\[1ex]
(\mathbf{e}_k^2)^{\mathrm{T}} & c \\[1ex]
(\mathbf{e}_k^3)^{\mathrm{T}} & c \\[1ex]
(\mathbf{e}_k^4)^{\mathrm{T}} & c
\end{bmatrix}
\tag{8.17}
$$

In the alternative form of A the symbol $\mathbf{e}_k^i$ denotes the unit vector from the nominal station location toward satellite i. The components of this unit vector are the direction cosines. The cofactor matrix for the adjusted position and receiver clock is

$$
Q_X = (A^{\mathrm{T}}A)^{-1} =
\begin{bmatrix}
\sigma_u^2 & \sigma_{uv} & \sigma_{uw} & \sigma_{ut} \\[1ex]
\sigma_{vu} & \sigma_v^2 & \sigma_{vw} & \sigma_{vt} \\[1ex]
\sigma_{wu} & \sigma_{wv} & \sigma_w^2 & \sigma_{wt} \\[1ex]
\sigma_{tu} & \sigma_{tv} & \sigma_{tw} & \sigma_t^2
\end{bmatrix}
\tag{8.18}
$$

Usually it is more convenient to view results in the local geodetic coordinate system. The cofactor matrix (8.18) is readily transformed into such a coordinate system using (7.26) and (7.27) of Section 7.2. The result is

$$
Q_{\mathrm{LG}} =
\begin{bmatrix}
\sigma_n^2 & \sigma_{ne} & \sigma_{nh} \\[1ex]
\sigma_{en} & \sigma_e^2 & \sigma_{eh} \\[1ex]
\sigma_{hn} & \sigma_{he} & \sigma_h^2
\end{bmatrix}
\tag{8.19}
$$

The DOP factors are functions of the diagonal elements of (8.18) or (8.19). Table 8.1 shows the various dilution factors: VDOP, dilution of precision for height; HDOP, dilution of precision for horizontal positions; PDOP, dilution

TABLE 8.1 DOP Expressions

$$VDOP = \sigma_h \quad\quad\quad (a)$$

$$HDOP = \sqrt{\sigma_n^2 + \sigma_e^2} \quad\quad\quad (b)$$

$$PDOP = \sqrt{\sigma_n^2 + \sigma_e^2 + \sigma_h^2}$$
$$= \sqrt{\sigma_u^2 + \sigma_v^2 + \sigma_w^2} \quad\quad\quad (c)$$

$$TDOP = \sigma_t \quad\quad\quad (d)$$

$$GDOP = \sqrt{\sigma_n^2 + \sigma_e^2 + \sigma_h^2 + \sigma_t^2 c^2} \quad\quad\quad (e)$$

of precision of position; TDOP, dilution of precision in time; and GDOP, geometric dilution of precision. GDOP is a composite measure reflecting the geometry on the position and time estimates.

The more usable satellites that are available, the better. However, the DOP criteria can also be used to find the four best satellites out of a geometric constellation consisting of more than four satellites (particularly desirable if only a four-channel receiver is available). The average value for HDOP and VDOP is about 2 for the best possible constellation of 4 satellites. The DOPs can readily be computed in advance given the approximate receiver location and the predicted satellite ephemeris. The DOPs change from epoch to epoch with changing satellite geometry. Because most receivers now observe all satellites in view, the DOPs are primarily used to identify temporal weakness in the geometric strength for rapid static and kinematic applications.

8.2 CARRIER PHASES

The phase observable is the difference between the received satellite carrier phase (as sensed by the receiver's antenna) and the phase of the internal receiver oscillator. The measurements are recorded at equally spaced nominal receiver clock epochs t. The measurement process cannot account for the number of whole carrier waves between the receiver and the satellite. The antenna does not "sense" the exact number of carrier waves to the satellite. The mathematical development uses the fact that the received phase was emitted at some earlier instant of time. The variation of the carrier phase observable in time relates to changes in the topocentric distance in the same way as the integrated Doppler. Sometimes the term interferometry is used to describe carrier phase techniques.

Measuring waves to subcentimeter accuracy demands that utmost attention be given to the treatment of time. After all, the time for light to travel a distance of 3 mm is only 0.01 nsec! A typical pseudorange solution allows us to synchronize the clocks, or to determine the receiver clock errors, to (only) 0.1-μsec accuracy. The remaining, unavoidable clock synchronization errors will be neutralized through appropriate differencing of the carrier phase observables.

8.2.1 Undifferenced Carrier Phase Observable

The carrier phase observable $\varphi_k^p(t)$ for station k and satellite p is written as

$$\varphi_k^p(t) = \varphi_k(t) - \varphi^P(t) + N_k^p(1) + I_{k,\varphi}^p(t) + \frac{f}{c} T_k^p(t)$$

$$+ d_{k,\varphi}(t) + d_{k,\varphi}^p(t) + d_\varphi^p(t) + \epsilon_\varphi \qquad (8.20)$$

The symbols $\varphi_k(t)$ and $\varphi^P(t)$ denote the receiver's phase and the received satellite phase at the nominal reception time t. The subscript k is not used in connection with the nominal time t in order to simplify the notation. The symbol $N_k^p(1)$ denotes the initial integer ambiguity, sometimes also called the lane identifier. It represents the arbitrary counter setting of the tracking register at the start of observations (phase lock). The terms $I_{k,\varphi}^p(t)$ and $T_k^p(t)$ denote the ionospheric and tropospheric effects. The ionospheric term has a negative value because the carrier phase advances due to the ionosphere. See equation (9.28) of Chapter 9 for additional details on the ionosphere. The terms $d_{k,\varphi}(t)$ and $d_\varphi^p(t)$ refer to the hardware delays of the receiver and the satellite, $d_{k,\varphi}^p(t)$ is the multipath, and ϵ_φ denotes the random carrier phase measurement noise. The carrier identification subscript is not used in (8.20) in order to keep the expression general. Terms having a subscript φ are expressed in units of cycles. The tropospheric delay is converted to cycles using the factor f/c, where f is the nominal carrier frequency.

The receiver's phase lock loop measures the time-varying function $\varphi_k(t) - \varphi^P(t)$ by shifting the receiver-generated $\varphi_k(t)$ to track (lock) the received $\varphi^P(t)$. Thus, a phase difference is measured rather than the time difference which matches identical codes. As the phase difference increases or decreases by one cycle (2π), the carrier phase observable $\varphi_k^p(t)$ changes by one cycle accordingly. If there is a temporary blockage of the transmitted signal $\varphi^P(t)$, the receiver loses lock on the signal and there is the possibility that the receiver will miss some of the whole cycle change in $\varphi_k(t) - \varphi^P(t)$. Once phase lock is regained, the fractional part of $\varphi_k(t) - \varphi^P(t)$ is again measured correctly, but the counter register might show an incorrect value. In this case a cycle slip is said to have occurred. Plotting (8.20) as a function of time will typically show a step function. Often receivers try to resolve cycle slips internally using extrapolation. They usually write flags into the data stream to warn users that a slip might have occurred.

To further the understanding of $\varphi_k(t) - \varphi^P(t)$ consider the following simplified situation: Let the receiver and the satellite be located on nearby fixed locations, both operating with perfect clocks. The satellite continuously transmits carrier phases and the receiver internally generates waves at the same frequency. For further simplification assume that the receiver has been tuned such that at some instant, the receiver-generated phase and the received phase are both zero. Since neither the receiver nor the satellite move in this experiment, the measured difference $\varphi_k(t) - \varphi^P(t)$ will remain zero. Next, let the satellite be moved one wavelength closer to the receiver over a period of 1 sec. Over that same period the receiver will register one extra incoming wave, and the measurement of $\varphi_k(t) - \varphi^P(t)$ will then be one. Thus, the change in the phase observable reflects the change in receiver–satellite separation. It does not matter whether the satellite or the receiver moves, or whether both move at the same time. Nor has the number of waves between the receiver and the satellite had any impact on the measured value. As the satellite moves closer by another wavelength, the counter will increase again by one. The observable $\varphi_k^P(t)$, when considered over time, is often called the accumulated carrier phase observable.

The difference $\varphi_k(t) - \varphi^P(t)$ is developed for a vacuum, because the contribution from the propagation media along with other biases and delays are added separately. See equation (8.20). The idea in the development of the carrier phase equation is the equivalence of the received carrier phase and the emitted phase at the satellite, exactly τ_k^P seconds earlier. This fact is expressed by the following equation:

$$\varphi^P(t) = \varphi_T^P(t - \tau_k^P) \tag{8.21}$$

where τ_k^P is vacuum travel time. Signals received at the same time have different emission times because of the different distances to the satellites. The subscript T denotes transmission at the satellite, that is, the phase emitted by satellite p. The phases for nominal time t and true time t_r are related to the clock errors as follows

$$\varphi_k(t_r) = \varphi_k(t) + f\, dt_k \tag{8.22}$$

$$\varphi_T^P(t_r - \tau_k^P) = \varphi_T^P(t - \tau_k^P) + f\, dt^P \tag{8.23}$$

These equations make use of the fact that the frequency of an oscillator is constant with respect to its own time system. Therefore, it is permissible to take the oscillator frequency as a constant and, per definition, to have the clock error absorb any variation. Solving (8.22) for $\varphi_k(t)$ and combining (8.23) with (8.21) gives

$$\varphi_k(t) = \varphi_k(t_r) - f\, dt_k \tag{8.24}$$

$$\varphi^P(t) \equiv \varphi_T^P(t - \tau_k^P) = \varphi_T^P(t_r - \tau_k^P) - f\, dt^P \tag{8.25}$$

Substituting these two equations into (8.20), and ignoring the propagation biases and the other delay terms, we obtain

$$\varphi_k^p(t) = \varphi_k(t_r) - \varphi_T^p(t_r - \tau_k^p) - f\,dt_k + f\,dt^p + N_k^p(1) \qquad (8.26)$$

Since τ_k^p is the signal travel time, which is about 70 msec, the expansion of $\varphi_T^p(t_r - \tau_k^p)$ requires that the satellite frequency be modeled. Here we use the simple model

$$\dot{\varphi}_T^p(t_r) = f + a^p + b^p t \qquad (8.27)$$

where a^p and b^p denote the satellite frequency offset and drift at emission time. Thus,

$$\begin{aligned}
\varphi_T^p(t_r - \tau_k^p) &= \varphi_T^p(t_r) - \int_\tau \dot{\varphi}_T^p(t)\,dt \\
&= \varphi_T^p(t_r) - [f + a^p + \tfrac{1}{2}b^p\tau_k^p]\,\tau_k^p
\end{aligned} \qquad (8.28)$$

Substituting (8.28) into (8.26) gives

$$\begin{aligned}
\varphi_k^p(t) = \varphi_k(t_r) &- \varphi_T^p(t_r) - f\,dt_k + f\,dt^p \\
&+ (f + a^p + \tfrac{1}{2}b^p\tau_k^p)\,\tau_k^p + N_k^p(1)
\end{aligned} \qquad (8.29)$$

The phase terms $\varphi_k(t_r)$ and $\varphi_T^p(t_r)$ can be ignored altogether because they cannot be separated from the clock errors $f\,dt_k$ and $f\,dt^p$, and because these terms will cancel in the double difference expressions which will be developed in Section 8.2.4. Equation (8.29) shows that the satellite frequency offset and drift affect the received phase by the amount

$$\Delta\varphi^p = (a^p + \tfrac{1}{2}b^p\tau_k^p)\tau_k^p \qquad (8.30)$$

The term $\Delta\varphi^p$ represents the impact of the change in satellite clock error during the travel time of the signal. This impact on the received phase is, of course, a function of the signal travel time, and is the manner in which selective availability manifests itself in the carrier phase observable. See Section 8.3.1.4 for additional remarks on the impact of selective availability. Since the travel time τ_k^p is short, the impact of the frequency drift term $b^p\tau_k^p$ is usually negligible. Deleting the respective terms, the expression (8.29) becomes

$$\varphi_k^p(t) = -f\,dt_k + f\,dt^p + (f + a^p)\tau_k^p + N_k^p(1) \qquad (8.31)$$

The last step in the development is to relate the signal travel time to the topocentric range as

$$\tau_k^p = \frac{\rho_k^p(t) + \dot{\rho}_k^p(t)\, dt_k}{c} \tag{8.32}$$

Substituting (8.32) into (8.31) and including the ionospheric and tropospheric terms and the other delay terms gives

$$\varphi_k^p(t) = \frac{f}{c}\rho_k^p(t) - f\left[1 - \frac{\dot{\rho}_k^p(t)}{c}\right] dt_k + f\, dt^p + N_k^p(1) + \frac{a^p}{c}\rho_k^p(t)$$
$$+ I_{k,\varphi}^p(t) + \frac{f}{c} T_k^p(t) + d_{k,\varphi}(t) + d_{k,\varphi}^p(t) + d_\varphi^p(t) + \epsilon_\varphi \tag{8.33}$$

This expression shows that the station clock error enters in two ways: the large term consisting of the product $f\, dt_k$ and the smaller term, which is a function of the topocentric range rate $\dot{\rho}_k^p$. For a dt_k of merely 1 nsec, the term $f\, dt_k$ already contributes 1.5 cycles. This is about 150 times the expected carrier phase measurement accuracy. For a phase measurement accuracy of 0.01 cycles the required receiver clock accuracy is about 0.01 nsec. A station clock error of 1 μsec contributes 0.004 cycles via Doppler term $f\dot{\rho}_k^p(t)/c$ assuming $|\dot{\rho}_k^p(t)|$ < 800 m/sec. This geometry-dependent term is zero if the Doppler is zero and changes sign for approaching and departing satellites. If the station clock error does not exceed 0.1 μsec then the $\dot{\rho}$ term is negligible.

Satellite clock errors affect the phase observable through the large term $f\, dt^p$ and the small frequency-offset term $a^p\rho_k^p/c$. The latter term depends on the travel time of the signal to traverse the topocentric distance.

Equation (8.33) is the fully developed expression for the undifferenced carrier phase observation. Because of the linear relationship between the clock errors and the other unmodeled effects, it would probably be necessary to model the combined effect by one parameter per epoch per observation. Fortunately, most of these error terms are either eliminated or their impact is significantly reduced employing techniques discussed in following sections.

It is instructive to compare the carrier phase expression (8.33) with that of the pseudorange (8.9). Apart from the general scaling factor f/c these two expressions differ only because of the satellite frequency offset term a^p. Therefore, the carrier phase equation is sometimes developed in the same way as the pseudorange was derived. The small a^p term is often not listed explicitly or it is simply added to account for the change in received phase due to the satellite frequency offset. Following that line of thought, there is no need to introduce terms like $\varphi_k(t_r)$ and $\varphi_T^p(t_r)$ found in (8.29).

8.2.2 Single Differences

If two receivers k and m observe the same satellite p at the same preset nominal receiver epoch, one can write two equations of the form (8.33). The *between-receiver single-difference* phase observable, more commonly called the *single-*

difference phase observable, is defined by

$$\varphi^P_{km}(t) \equiv \varphi^P_k(t) - \varphi^P_m(t)$$

$$= \frac{f}{c} [\rho^P_k(t) - \rho^P_m(t)] + \frac{a^P}{c} [\rho^P_k(t) - \rho^P_m(t)]$$

$$+ \frac{f}{c} [\dot\rho^P_k(t)\, dt_k - \dot\rho^P_m(t)\, dt_m] + N^P_{km}(1) - f(dt_k - dt_m)$$

$$+ I^P_{km,\varphi}(t) + \frac{f}{c} T^P_{km}(t) + d_{km,\varphi}(t) + d^P_{km,\varphi}(t) + \epsilon^P_{km,\varphi}$$

(8.34)

Note how the subscripts k and m in $\varphi^P_{km}(t)$ are used to indicate the difference operation of $\varphi^P_k(t)$ minus $\varphi^P_m(t)$. Following the same convention, we have

$$N^P_{km}(1) = N^P_k(1) - N^P_m(1) \tag{8.35}$$

$$I^P_{km,\varphi}(t) = I^P_{k,\varphi}(t) - I^P_{m,\varphi}(t) \tag{8.36}$$

$$T^P_{km}(t) = T^P_k(t) - T^P_m(t) \tag{8.37}$$

$$d_{km,\varphi}(t) = d_{k,\varphi}(t) - d_{m,\varphi}(t) \tag{8.38}$$

$$d^P_{km,\varphi}(t) = d^P_{k,\varphi}(t) - d^P_{m,\varphi}(t) \tag{8.39}$$

$$\epsilon^P_{km,\varphi}(t) = \epsilon^P_{k,\varphi}(t) - \epsilon^P_{m,\varphi}(t) \tag{8.40}$$

Figure 8.2 shows a conceptual view of the single-difference approach. The principal advantage of the single-difference observation is that most of the errors common to the satellite cancel. For example, the large satellite clock term dt^P has canceled. The remaining small term of the satellite frequency offset converges toward zero as the separation of the receivers decreases. The satellite hardware delay also cancels. All of these errors cancel as long as they remain constant between satellite emissions. Recall that even though the nom-

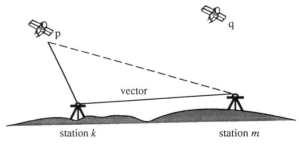

Figure 8.2 The single difference. Two receivers observe the same satellite at the same epoch.

inal reception time is the same, the emission times differ slightly because of different distances between the satellite and the two stations k and m. The single-difference observations, however, remain sensitive to both receiver clock errors dt_k and dt_m.

8.2.3 Double Differences

If two receivers k and m observe two satellites p and q at the same nominal time, the *double-difference* phase observable is

$$
\begin{aligned}
\varphi_{km}^{pq}(t) &\equiv \varphi_{km}^{p}(t) - \varphi_{km}^{q}(t) \\[4pt]
&= \frac{a^p}{c} \left[\rho_k^p(t) - \rho_m^p(t) \right] - \frac{a^q}{c} \left[\rho_k^q(t) - \rho_m^q(t) \right] \\[4pt]
&\quad + \frac{f}{c} \left[\rho_k^p(t) - \rho_m^p(t) \right] - \frac{f}{c} \left[\rho_k^q(t) - \rho_m^q(t) \right] \\[4pt]
&\quad + \frac{f}{c} \left[\dot{\rho}_k^p(t)\, dt_k - \dot{\rho}_m^p(t)\, dt_m \right] - \frac{f}{c} \left[\dot{\rho}_k^q(t)\, dt_k - \dot{\rho}_m^q(t)\, dt_m \right] \\[4pt]
&\quad + N_{km}^{pq}(1) + I_{km,\varphi}^{pq}(t) + \frac{f}{c} T_{km}^{pq}(t) + d_{km,\varphi}^{pq}(t) + \epsilon_{km,\varphi}^{pq}
\end{aligned}
\tag{8.41}
$$

where

$$
N_{km}^{pq}(1) = N_{km}^{p}(1) - N_{km}^{q}(1) \tag{8.42}
$$

$$
I_{km,\varphi}^{pq}(t) = I_{km,\varphi}^{p}(t) - I_{km,\varphi}^{q}(t) \tag{8.43}
$$

$$
T_{km}^{pq}(t) = T_{km}^{p}(t) - T_{km}^{q}(t) \tag{8.44}
$$

$$
d_{km,\varphi}^{pq}(t) = d_{km,\varphi}^{p}(t) - d_{km,\varphi}^{q}(t) \tag{8.45}
$$

$$
\epsilon_{km,\varphi}^{pq} = \epsilon_{km,\varphi}^{p} - \epsilon_{km,\varphi}^{q} \tag{8.46}
$$

See Figure 8.3 for a conceptual representation. The most important feature of the double-difference observation is the cancellation of the large receiver clock errors dt_k and dt_m (in addition to the cancellation of the large satellite clock errors). These receiver clock errors cancel completely as long as observations to satellites p and q are taken at the same time, or the receiver clock drifts between the observation epochs are negligible. The small clock terms, which are a function of the topocentric range rate, remain in (8.41). Because multipath is a function of the specific receiver–satellite–reflector geometry, it does not cancel in the double-difference observable.

The integer ambiguity plays an important role in double differencing. If it is possible during the least-squares estimation to fix the integer, that is, to

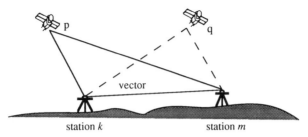

Figure 8.3 The double difference. Two receivers observe two satellites at the same epoch.

constrain the estimates $\hat{N}^{pq}_{km}$ to integers, then the fixed solution is the preferred one. Because of residual model errors the estimated ambiguities will, at best, be close to integers. Imposing the integer constraints adds strength to the solution because the number of parameters is reduced and the correlations are reduced as well. Much effort has gone into extending the baseline length over which ambiguities can be fixed. At the same time much research has been carried out in order to develop algorithms which allow the ambiguities to be fixed over short baselines using increasingly shorter observation spans.

8.2.4 Triple Differences

The *triple difference* is the difference of two double differences between different epochs:

$$\varphi^{pq}_{km}(t_2,\, t_1) = \varphi^{pq}_{km}(t_2) - \varphi^{pq}_{km}(t_1) \tag{8.47}$$

See Figure 8.4 for a conceptual representation. Assuming phase lock, the initial integer ambiguity $N^{pq}_{km}(t)$ has canceled in equation (8.47). Notice that the

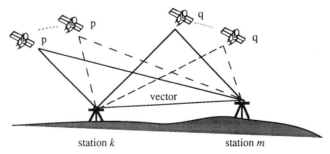

Figure 8.4 The triple difference. Two receivers observe two satellites from one epoch to the next.

triple and double differences have the same sensitivity with regard to clock errors and satellite frequency offset.

The triple-difference observable is probably the easiest to deal with because the ambiguity cancels. The triple-difference solution is often considered a pre-processing technique to get good approximate positions for the double difference solution. Triple differences have a major advantage in that cycle slips are mapped as individual outliers in the computed residuals. Individual outliers can usually be detected and removed or corrected. The resulting cycle slip free (or nearly so) observations can be used in the double difference solution. Because of the additional differencing over time, the triple differences lose some geometric strength. The triple difference can be computed in any order. Consider the following:

$$
\begin{aligned}
\varphi_{km}^{pq}(t_2, t_1) &= [\varphi_{km}^{p}(t_2) - \varphi_{km}^{q}(t_2)] - [\varphi_{km}^{p}(t_1) - \varphi_{km}^{q}(t_1)] \\
&= [\varphi_{km}^{p}(t_2) - \varphi_{km}^{p}(t_1)] - [\varphi_{km}^{q}(t_2) - \varphi_{km}^{q}(t_1)] \\
&= \varphi_{km}^{p}(t_2, t_1) - \varphi_{km}^{q}(t_2, t_1) \\
&= [\varphi_{k}^{p}(t_2) - \varphi_{m}^{p}(t_2)] - [\varphi_{k}^{p}(t_1) - \varphi_{m}^{p}(t_1)] \\
&\quad - [\varphi_{k}^{q}(t_2) - \varphi_{m}^{q}(t_2)] + [\varphi_{k}^{q}(t_1) - \varphi_{m}^{q}(t_1)] \\
&= [\varphi_{k}^{pq}(t_2) - \varphi_{m}^{pq}(t_2)] - [\varphi_{k}^{pq}(t_1) - \varphi_{m}^{pq}(t_1)] \\
&= [\varphi_{k}^{p}(t_2) - \varphi_{k}^{p}(t_1)] - [\varphi_{m}^{p}(t_2) - \varphi_{m}^{p}(t_1)] \\
&\quad - [\varphi_{k}^{q}(t_2) - \varphi_{k}^{q}(t_1)] + [\varphi_{m}^{q}(t_2) - \varphi_{m}^{q}(t_1)] \\
&= \varphi_{k}^{p}(t_2, t_1) - \varphi_{m}^{p}(t_2, t_1) - [\varphi_{k}^{q}(t_2, t_1) - \varphi_{m}^{q}(t_2, t_1)] \\
&= \varphi_{k}^{pq}(t_2, t_1) - \varphi_{m}^{pq}(t_2, t_1)
\end{aligned}
\tag{8.48}
$$

Whenever a new observable is formed by differencing over time, the epoch identifiers t_1 and t_2 become part of the symbol for the new observable. For example, the symbol $\varphi_{km}^{p}(t_2, t_1)$ denotes the difference over time of two between-receiver single differences involving the same satellite, whereas $\varphi_{k}^{pq}(t)$ is the *between-satellite single difference*. The symbol $\varphi_{k}^{pq}(t_2, t_1)$ denotes the difference over time of two between-satellite single differences.

8.2.5 Delta Ranges

The differences in time of carrier phase observables involving one station and one or several satellites are often called the *delta range observables*. For example,

$$
\varphi_{k}^{p}(t_2, t_1) = \varphi_{k}^{p}(t_2) - \varphi_{k}^{p}(t_1)
\tag{8.49}
$$

relates two undifferenced observations over time, while

$$
\begin{aligned}
\varphi_k^{pq}(t_2, t_1) &\equiv \varphi_k^p(t_2, t_1) - \varphi_k^q(t_2, t_1) \\
&= [\varphi_k^p(t_2) - \varphi_k^p(t_1)] - [\varphi_k^q(t_2) - \varphi_k^q(t_1)] \\
&= [\varphi_k^p(t_2) - \varphi_k^q(t_2)] - [\varphi_k^p(t_1) - \varphi_k^q(t_1)] \\
&= \varphi_k^{pq}(t_2) - \varphi_k^{pq}(t_1)
\end{aligned}
\tag{8.50}
$$

involves one station and two satellites. These delta ranges are a function of the change in topocentric distance between the station and the satellite(s) provided there is no cycle slip between the epochs t_1 and t_2. They do not depend on the initial ambiguity because of the differencing over time. The delta range (8.49) depends on the change of the receiver and satellite clock errors from epoch t_1 to epoch t_2. If the delta range observable is used for point positioning, that is, determining the geocentric location of an individual station as opposed to the relative position between a pair of stations, the stability of the clocks becomes of critical concern. The receiver clock errors must be carefully modeled and estimated. Alternatively, one might want to connect the receiver to a stable rubidium or cesium clock. Fortunately, satellite time is already derived from atomic frequency standards. However, with selective availability, an unknown but significant variation in time is imposed on the satellite clocks (in addition to possible degradation of the broadcast ephemeris). The differenced delta range (8.50) does not depend on the station clock term dt_k. See Hermann (1992) and Malys et al. (1992) for additional details on point positioning and results. Generally speaking, the geocentric point positioning accuracy is around 1 m or slightly better for observation spans of several hours. Up until now there has not been a widespread effort by industry to develop techniques for geocentric point positioning with carrier phase observations, probably because relative positioning works so well, and because of the fear of adverse effects from selective availability on the satellite clocks. But this trend is changing due to the requirements of major users, such as the exploration industry.

8.3 EVALUATION

The expressions given above are valid for both carriers. Even though the difference observables are formulated in the previous section in terms of the carrier phase observables, differencing is applicable to pseudoranges as well. The corresponding functions could be labeled $P_{km}^p(t)$, $P_{km}^{pq}(t)$, $P_k^{pq}(t_2, t_1)$, etc.

Let R, S, T denote the number of receivers, satellites, and epochs, respectively. If all receivers observe all satellites in all epochs, there is a total of RST undifferenced observations. Such a set yields $(R - 1)ST$ independent single-difference observations, $(R - 1)(S - 1)T$ double-difference observations, or $(R - 1)(S - 1)(T - 1)$ triple-difference observations. When processing the differenced observations, it is necessary to consider fully the mathematical correlations between observations.

The carrier phase solutions require initial approximate coordinates for the receiver locations because of linearization. Conveniently, the code solutions yield such geocentric positions with sufficient accuracy for typical relative positioning over short distances. See Section 8.1 regarding the navigation solution and Section 8.3.4 on the required a priori position accuracy. In principle one can attempt to estimate the coordinates of both stations in the double-difference and triple-difference solutions, and can avoid the receiver clock error terms in the undifferenced and single-differenced observables. Unfortunately, the station–satellite geometry of the double-difference solutions has little "geocentric" positional strength. Therefore, usually one station is held fixed, while the other is adjusted.

Double differencing is, undoubtedly, the most popular technique in use today for achieving relative positioning, primarily because the clock errors cancel. Since differencing reduces the geometric strength of the newly formed observable, it might be preferable to work with single differences in specialized applications. Especially good solutions might be achieved with between-receiver single differences if the receiver clock terms are not explicitly estimated, but are somehow eliminated by means of the physical design of the application. Schreiner (1990) discusses a case for orbital accuracy improvements where the receivers are linked by a common oscillator. The receivers and oscillators are located at a common site and widely separated antennas are connected to the receivers via fiber optic cables. Santerre and Beutler (1993) propose to connect two or more antennas with the same receiver. They are interested in improving height determination and foresee applications in dam deformation monitoring where the receiver-antenna connections are permanently installed.

Much can be learned about positioning with GPS from the expressions developed in the previous sections. Although these mathematical formulations are important, their study alone is no substitute for predicting GPS performance. Implementation of these expressions varies considerably among manufacturers. There are still plenty of opportunities for innovative approaches, addition of various quality checks, and other enhancements. However, several frequently asked questions about GPS positioning can be explained best with the help of mathematics.

8.3.1 The Clock Error Terms

Both pseudorange and carrier phase observables are affected by receiver and satellite clock errors as evidenced from the many clock terms encountered in the previous derivations. One of the primary purposes of differencing is to reduce or even to eliminate residual clock errors. Table 8.2 summarizes the relevant receiver and satellite clock error terms. The numerical values on which this table is based are listed in the last row.

TABLE 8.2 Clock Error Magnitudes

Term	Magnitude [L1-cycles]
$a^p \rho_k^p(t)/c$	10^{-4}
$a^p \left[\rho_k^p(t) - \rho_m^p(t)\right]/c$	$< 10^{-4}$
$f \dot{\rho}_k^p dt_k /c$	$< 4 \times 10^{-3}$
$f\, dt_k$	1500

$a^p / f = 10^{-12}$	$\dot{\rho}_k^p(t) \leq 800$ m/sec	$\rho_k^p(t) - \rho_m^p(t) \leq 1000$ km
$c = 3 \times 10^5$ km	$\rho_k^p = 20,000$ km	$dt_k = 1\mu$ sec

8.3.1.1 Receiver Clocks An important element in this discussion is that the navigation solution provides station clock accuracy to 1 μsec and better. The limit in accuracy results from the combined effects of code measurement resolution, unmodeled errors in troposphere and ionosphere, ephemeris errors, and selective availability. The navigation solution estimates the receiver location and the receiver clock at the same time.

If good initial positions are available for the receiver, an alternative for computing the receiver clock error is to use pseudorange observations individually. The process begins with computing the topocentric distance $\rho_k^p(t)$ from the signal emission times as outlined in Section 8.1. Since the satellite clocks are carefully controlled, as is explained further below, the residual satellite clock errors are small, such as in the range of 1 μsec (falsifying the computed topocentric distance by less than 1 mm). The computed topocentric distance is substituted into equation (8.4). This equation, in turn, is solved for the receiver clock error dt_k. The accuracy of this receiver clock determination is, again, limited by the satellite clock error, the ionospheric and tropospheric delays, multipath, and hardware delays. In this manner, we can compute a value for the receiver clock error for each pseudorange observation. These solutions will differ slightly because the individual satellite clock errors differ and because of direction-dependent effects of ephemeris errors, ionospheric, tropospheric, and multipath delays. The error in the assumed receiver position used to compute the topocentric distance $\rho_k^p(t)$ also causes the individual clock estimates to differ. Figure 8.5 demonstrates the result of receiver clock determinations at every epoch. The single line plot refers to the case when accurate WGS 84 coordinates were used for the receiver. The receiver clock drifted about 200 μsec during 20 min. In the other case, approximate coordinates were used for the receiver position. The individual clock solutions do not coincide. The separation of the lines reflects the projection of the position error in the direction of the satellite.

8.3.1.2 The Question of Simultaneity Because of differing distances between the receiver and the satellites, in principle at least, one must chose between simultaneity in reception time or emission time. Signals that are received at the same time were emitted at different times, and signals that are emitted

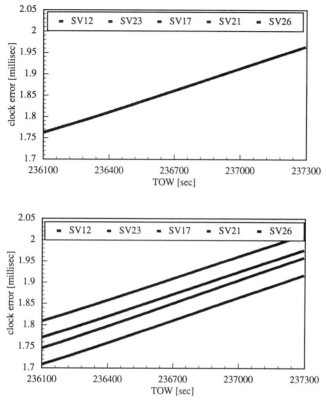

Figure 8.5 Receiver clock errors computed from pseudoranges (Experiment 1, Orono, DOY 356, 1992).

at the same time will be received at different times. Prior to selective availability it was natural to think in terms of simultaneity of signal reception at the receiver, because atomic clocks in the satellites are superior to crystal clocks in the receivers. This understanding is not necessarily true with selective availability being active. The impact of selective availability is discussed in more detail in Section 8.3.1.4.

When processing the double-difference equation (8.41) what really matters is the computation of the topocentric distance $\rho_k^p(t)$. This distance is required for computing the partial derivatives, although the computational accuracy is not so critical in this case. More importantly, this distance is used directly in the misclosure $L_0 - L_b$ (adjustment notation). The computational accuracy of L_0 must correspond to the accuracy of the observation. As was mentioned previously, this distance can be computed with millimeter accuracy using the signal emission time (the error caused by the approximate receiver location is not relevant because its position is estimated as part of the solution). The actual received time need not ever be used again. The primary requirement for eliminating $f\,dt_k$ is that all observations at one receiver are simultaneous. Note that the double differences can be written as the difference of two between-satellite single differences. If the two receivers observe at different times and the respective times are known to 1 μsec, then the topocentric ranges can always be accurately computed.

As the time offset between both receivers increases, the double-difference misclosure $L_0 - L_b$ might not be accurate enough because the satellite clock errors dt^p and dt^q will not cancel in the single differences. If the emission times are substantially different, the satellite clocks might have drifted due to selective availability. Allowing a synchronization offset of 1 msec between both receivers and assuming that changes in dt^p and dt^q are negligible is roughly equivalent to assuming that the terms a^p and a^q cancel for a 300-km baseline in the single differences.

Often the observations are corrected to accommodate the known receiver clock offset. The receiver clock errors are used to shift the observations in time using the relation

$$\Delta \varphi_k^p = \frac{f}{c} \dot{\rho}_k^p \, dt_k \tag{8.51}$$

The undifferenced observations could also be corrected for the large term $f\,dt_k$. However, this is not necessary because this term cancels during double differencing. Correcting the observations in this manner is identical to the mathematical operation of moving the known station clock term from the right side to the left side in the carrier phase equations, and adding it to the observed values. What remains on the right side are the effects of residual clock errors, which usually do not exceed 1 μsec and which, consequently, have negligible effects.

It follows that the term "simultaneity of double observations" can be understood rather loosely: Accurate results are obtained as long as the observations at the individual receiver are simultaneous, the receiver drift is known to about 1 μsec, and the drift between the receivers does not exceed 1 msec. When the distances $\rho_k^p(t)$ are computed for the corrected observation times, the $\dot{\rho}$ terms in (8.51) can be ignored. Alternatively, the $\dot{\rho}$ terms (8.51) are applied to the observations and $\rho_k^p(t)$ is computed for the common nominal receiver time. The actual receiver clock drift should be small enough so that $\dot{\rho}$ in (8.51) can be treated constant over the time of the receiver clock offset.

8.3.1.3 Satellite Clock Corrections
The control segment maintains GPS time to within 1 μsec of UTC(USNO) according to the Interface Control Document (ICD-GPS-200 1991). However, GPS time does not participate in UTC's leap-second jumps. The necessary data for relating GPS time and the individual satellite time are included in the navigation message. The accuracy of these data is such that, during the transmission interval, the space vehicle time can be related to UTC(USNO) within 90 nsec (1 standard deviation). In the nota-

tion and sign convention as used by the interface control document, the time correction of the space vehicle is

$$\Delta t_{SV} = a_{f0} + a_{f1}(t_{SV} - t_{oc}) + a_{f2}(t_{SV} - t_{oc})^2 + \Delta t_R \quad (8.52)$$

with

$$t_{GPS} = t_{SV} - \Delta t_{SV} \quad (8.53)$$

and

$$\Delta t_R = -\frac{2}{c^2} \sqrt{a\mu} \, e \sin E = -\frac{2}{c^2} \mathbf{X} \cdot \dot{\mathbf{X}} \quad (8.54)$$

The second part of this equation follows from (2.96) of Chapter 2. The polynomial coefficients are transmitted in units of sec, sec/sec, and sec/sec^2. The clock data reference time t_{oc} is also broadcast in seconds. As is required when using the ephemeris expressions, the value of t_{SV} must account for the beginning or end-of-week crossovers. That is, if $(t_{SV} - t_{oc})$ is greater than 302,400, subtract 604,800 from t_{SV}. If $(t_{SV} - t_{oc})$ is less than $-302,400$, add 604,800 to t_{SV}. The expression (8.54) is a small relativistic clock correction caused by the orbital eccentricity e. The symbol μ denotes the gravitational constant, a is the semimajor axis of the orbit, and E is the eccentric anomaly. See Chapter 2 for details on these elements. Using $a \approx 26,000$ km we have

$$\Delta t_{R[\mu sec]} \approx -2e \sin E \quad (8.55)$$

Figure 8.6 shows several samples of satellite clock polynomials. Because the SV time is derived from atomic frequency standards, a polynomial representation is valid over a time interval of 4 to 6 hours. Since clock coefficients

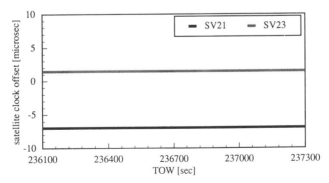

Figure 8.6 Example of satellite clock polynomials (DOY 356, 1992).

and the elements for the relativistic corrections are all part of the navigation message, the respective corrections can be applied inside the receiver.

8.3.1.4 Residual Satellite Clock Error

Assuming that the nominal satellite emission times are always corrected according to (8.53), the satellite clock term dt^k in the pseudorange equation (8.9) and carrier phase equations (8.33) denotes the small residual clock error between GPS time and UTC that "defies the grip of the control segment" and is about 1 μsec or less. The correction (8.52) directly impacts the navigation solution, whereas it cancels in the difference observations. Fully developed single- or double-difference phase expressions like (8.34) and (8.41) depend on the satellite frequency offset. This impact is a function of the station–satellite geometry and the baseline length as follows:

$$\Delta\varphi_{km}^p(t) = \frac{a^p}{c} [\rho_k^p(t) - \rho_m^p(t)] \tag{8.56}$$

Denoting the topocentric range difference by d yields

$$\Delta\varphi_{km[\text{cyc}]}^p = 0.3 \times 10^{-5} \, a_{[\text{cyc/sec}]}^p d_{[\text{km}]} \tag{8.57}$$

For a relative frequency offset of $a^p/f = 10^{-12}$, which is typical for cesium standards, the offset equals $a^p = 0.0015$ cyc/sec. Such offsets are negligible even for long baselines. However, selective availability entails dithering the satellite clock and frequency; the broadcast coefficients in (8.52) do not reflect these intentional variations. Fortunately, in relative positioning the impact of the clock error is reduced according to (8.57). In Section 8.3.7 an example of selective availability is discussed.

8.3.2 Summary of Undifferenced Expressions

The undifferenced expressions for pseudoranges and carrier phases, equations (8.9) and (8.33), are summarized in Table 8.3. The first row of the table refers to the pseudoranges, the second to the carrier phases expressed in cycles, and the third row contains the carrier phase expressions scaled to distance. The metric scaling is done according to

$$\Phi_{k,i}^p = \lambda_i \varphi_{k,i}^p = \frac{c}{f_i} \varphi_{k,i}^p \tag{8.58}$$

The scaled carrier phase is denoted by $\Phi_{k,i}^p$. Scaling the carrier phase is merely a matter of convenience in occasionally achieving simpler-looking expressions. The receiver clock terms $\dot{\rho}_k^p \, dt_k$ and the satellite frequency offset terms $a^p \rho_k^p/c$ are not listed. It is assumed that the station clock term has been determined with sufficient accuracy and that the satellite frequency offset term is

TABLE 8.3   Undifferenced Pseudorange and Carrier Phase Expressions

$$P_{k,1}^p(t) = \rho_k^p(t) - cdt_k + cdt^p + I_{k,1,P}^p(t) + T_k^p(t) \tag{a}$$
$$+ d_{k,1,P}(t) + d_{k,1,P}^p(t) + d_{1,P}^p(t) + \varepsilon_{1,P}$$

$$P_{k,2}^p(t) = \rho_k^p(t) - cdt_k + cdt^p + \frac{f_1^2}{f_2^2} I_{k,1,P}^p(t) + T_k^p(t) \tag{b}$$
$$+ d_{k,2,P}(t) + d_{k,2,P}^p(t) + d_{2,P}^p(t) + \varepsilon_{2,P}$$

$$\varphi_{k,1}^p(t) = \frac{f_1}{c} \rho_k^p(t) - f_1 dt_k + f_1 dt^p + N_{k,1}^p(1) - \frac{f_1}{c} I_{k,1,P}^p(t) + \frac{f_1}{c} T_k^p(t) \tag{c}$$
$$+ d_{k,1,\varphi}(t) + d_{k,1,\varphi}^p(t) + d_{1,\varphi}^p(t) + \varepsilon_{1,\varphi}$$

$$\varphi_{k,2}^p(t) = \frac{f_2}{c} \rho_k^p(t) - f_2 dt_k + f_2 dt^p + N_{k,2}^p(1) - \frac{f_1^2}{cf_2} I_{k,1,P}^p(t) + \frac{f_2}{c} T_k^p(t) \tag{d}$$
$$+ d_{k,2,\varphi}(t) + d_{k,2,\varphi}^p(t) + d_{2,\varphi}^p(t) + \varepsilon_{2,\varphi}$$

$$\Phi_{k,1}^p(t) = \rho_k^p(t) - cdt_k + cdt^p + \frac{c}{f_1} N_{k,1}^p(1) - I_{k,1,P}^p(t) + T_k^p(t) \tag{e}$$
$$+ d_{k,1,\Phi}(t) + d_{k,1,\Phi}^p(t) + d_{1,\Phi}^p(t) + \varepsilon_{1,\Phi}$$

$$\Phi_{k,2}^p(t) = \rho_k^p(t) - cdt_k + cdt^p + \frac{c}{f_2} N_{k,2}^p(1) - \frac{f_1^2}{f_2^2} I_{k,1,P}^p(t) + T_k^p(t) \tag{f}$$
$$+ d_{k,2,\Phi}(t) + d_{k,2,\Phi}^p(t) + d_{2,\Phi}^p(t) + \varepsilon_{2,\Phi}$$

negligible. The ionospheric terms are expressed as a function of the ionospheric delay for P1-codes. Such a distinction is not necessary for the tropospheric delay. Chapter 9 gives details on the impact of the propagation media. Separate hardware delays and multipath terms are listed for codes and carrier phases.

Figure 8.7 shows an example of raw accumulated carrier phases. The carrier phases diverge because of different wavelengths for the L1 and L2 carriers. Both observation series are shifted to coincide at the first epoch. Cycle slips typically cause jumps in these graphs. However, because of the large scale it would be difficult to notice small slips of one or two cycles. Small cycle slips can readily be discovered from the misclosure $L_0 - L_b$ (adjustment notation) using good coordinates for the receiver. Because the topocentric range rate to the satellites is at most about 800 m/sec, the slope of the accumulated carrier phases does not exceed 4000 cycles per second. Similar graphs can be given for the double-difference and triple-difference observables. These would show a smaller slope.

Figure 8.8 shows the difference of the P1 and P2 pseudoranges. Antispoofing was not activated at that time. The trends and the variations are primarily

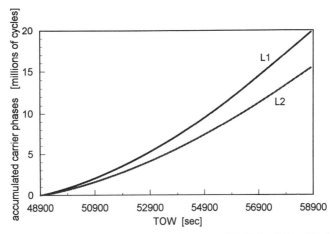

Figure 8.7 Accumulated carrier phases (DOY 356, 1992, SV 1).

caused by the ionosphere and multipath. The variations increase as the altitude of the satellite decreases. That is typical behavior.

8.3.3 Singularities

A case of a critical configuration for terrestrial observations is discussed at length in Chapter 5. For example, Figure 5.8 shows how ellipses of standard deviation display changes in the solution geometry as the critical configuration is reached for the plane resections. Blaha (1971) gives a general discussion of critical configurations for range networks. As a general rule, for critical configurations the columns of the design matrix are linearly dependent, and the

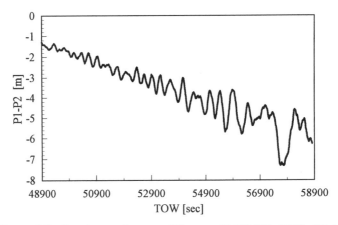

Figure 8.8 P-code pseudorange difference (DOY 356, 1992, SV 1).

normal matrix is singular. Even a quick look at dilution of precision (DOP) plots often shows a dramatic variation over a brief period of time. In such cases the satellite constellation approaches a critical (dangerous) configuration resulting in either a singularity or an ill-conditioned solution. This happens most often when a minimum number of four satellites is used.

For the navigation solution we must analyze the design matrix (8.17). The first three elements in each row of this matrix represent the direction cosines of the unit vector pointing from the station to the satellite. For station k and satellite p, the general form of this vector is

$$\mathbf{e}_k^p = \frac{1}{\rho_k^p} \, \boldsymbol{\rho}_k^p \tag{8.59}$$

The change in the pseudorange observable due to differential changes in the station coordinates is given by

$$dP_k^p(d\mathbf{X}) = -\mathbf{e}_k^p \cdot d\mathbf{X}_k = -\left[\frac{u^p - u_k}{\rho_k^p} \quad \frac{v^p - v_k}{\rho_k^p} \quad \frac{w^p - w_k}{\rho_k^p} \right] \begin{bmatrix} du_k \\ dv_k \\ dw_k \end{bmatrix} \tag{8.60}$$

Figure 8.9 shows a situation where all satellites are located on a circular cone. This is obviously a special situation. The vertex of the cone is at the receiver. The unit vector $\mathbf{e}_{\text{axis}}$ specifies the axis of the cone. For all those satellites that are located on the cone, the dot product

$$\mathbf{e}_k^i \cdot \mathbf{e}_{\text{axis}} = \cos \theta = \text{constant} \tag{8.61}$$

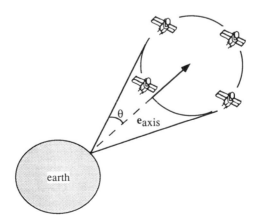

Figure 8.9 Critical configuration on a circular cone.

is a constant. The dot product (8.61) is applicable to each row of the design matrix. Thus, there is a perfect linear dependency of the first three columns of the design matrix (8.17). Such critical configurations usually do not last long because of the continuous motion of the satellites. The critical configurations present a problem only in continuous kinematic or supershort rapid static applications. Furthermore, the more satellites that are observed, the less likely it is that a critical configuration will occur. Another critical configuration occurs when the satellites and the receiver happen to be located on the same plane. In this case the first three columns of the design matrix fulfill the cross product vector function

$$\mathbf{e}_k^i \times \mathbf{e}_k^j = 0 \tag{8.62}$$

In relative positioning one can encounter critical configurations as well. Clearly, the satellites cannot be located on a perfectly circular cone as viewed from each of the stations. However, for short baselines the satellites could be located approximately on circular cones. Consider the relevant portion of the double-difference phase equation (8.41), scaled to distances,

$$P_{km}^{pq}(t) = \rho_k^p(t) - \rho_m^p(t) - [\rho_k^q(t) - \rho_m^q(t)] + \cdots \tag{8.63}$$

The total differential

$$\begin{aligned} dP_{km}^{pq}(d\mathbf{X}_k, d\mathbf{X}_m) &= -\mathbf{e}_k^p \cdot d\mathbf{X}_k + \mathbf{e}_m^p \cdot d\mathbf{X}_m + \mathbf{e}_k^q \cdot d\mathbf{X}_k - \mathbf{e}_m^q \cdot d\mathbf{X}_m \\ &= [\mathbf{e}_k^q - \mathbf{e}_k^p] \cdot d\mathbf{X}_k + [\mathbf{e}_m^p - \mathbf{e}_m^q] \cdot d\mathbf{X}_m \end{aligned} \tag{8.64}$$

expresses the change in the double-difference observable in terms of differential changes in station coordinates. The coefficients in the brackets represent the differences in the direction cosines from one station and two satellites. For short baselines these differences approach zero. It can readily be seen that the direction vectors are related to the vector of directions from the center of the baseline to the same satellite as

$$\mathbf{e}_k^p = \mathbf{e}_c^p + \boldsymbol{\varepsilon}_k^p \tag{8.65}$$

where the components of the vector $\boldsymbol{\varepsilon}_k^p$ are of the order $O(b/\rho_k^p)$ and b denotes the length of the baseline. When the other vectors are also referenced to the center of the baseline, equation (8.64) becomes

$$\begin{aligned} dP_{km}^{pq}(d\mathbf{X}_k, d\mathbf{X}_m) &= [\mathbf{e}_c^q - \mathbf{e}_c^p + \boldsymbol{\varepsilon}_k^q - \boldsymbol{\varepsilon}_k^p] \cdot d\mathbf{X}_k \\ &\quad + [\mathbf{e}_c^p - \mathbf{e}_c^q + \boldsymbol{\varepsilon}_m^p - \boldsymbol{\varepsilon}_m^q] \cdot d\mathbf{X}_m \end{aligned} \tag{8.66}$$

Considering the special case that the vertex of the circular cone is at the center of the baseline, the condition

$$\mathbf{e}_c^i \cdot \mathbf{e}_{\text{axis}} = \cos \theta \tag{8.67}$$

is valid for all satellites on the cone. This means that the dot products

$$[\mathbf{e}_c^q - \mathbf{e}_c^p + \boldsymbol{\varepsilon}_k^q - \boldsymbol{\varepsilon}_k^p] \cdot \mathbf{e}_{\mathrm{axis}} = [\boldsymbol{\varepsilon}_k^q - \boldsymbol{\varepsilon}_k^p] \cdot \mathbf{e}_{\mathrm{axis}} = O\left(\frac{b}{\rho_k^p}\right) \quad (8.68)$$

in (8.66) are of the order $O(b/\rho_k^p)$. These products become smaller the shorter the baseline. A product like (8.68) applies to every double-difference observation. Therefore, we are dealing with a near-singular situation since the columns of the double-difference design matrix are nearly dependent. The shorter the baseline, the more likely it is that the near-singularity damages the baseline solution.

8.3.4 A Priori Position Errors

A frequent concern in GPS applications is the a priori knowledge of geocentric station positions and the effects of ephemeris errors on the relative positions. The answer to these concerns lies again in the linearized double-difference equations. Without loss of generality, it is sufficient to investigate the difference between one satellite and two ground stations. Scaled to distances, the relevant portion of the double-difference equations is

$$P_{km}^{pq}(t) = \rho_k^p(t) - \rho_m^p(t) + \cdots \quad (8.69)$$

The linearized form is

$$dP_{km}^{pq}(d\mathbf{X}_k, d\mathbf{X}_m, d\mathbf{X}^p) = -\mathbf{e}_k^p \cdot d\mathbf{X}_k + \mathbf{e}_m^p \cdot d\mathbf{X}_m + [\mathbf{e}_k^p - \mathbf{e}_m^p] \cdot d\mathbf{X}^p$$

$$(8.70)$$

It is advantageous to transform the coordinate corrections into their differences and sums. This is accomplished by

$$d\mathbf{X}_k - d\mathbf{X}_m = d(\mathbf{X}_k - \mathbf{X}_m) = d\mathbf{b} \quad (8.71)$$

$$\frac{d\mathbf{X}_k + d\mathbf{X}_m}{2} = d\left(\frac{\mathbf{X}_k + \mathbf{X}_m}{2}\right) = d\mathbf{X}_c \quad (8.72)$$

The difference (8.71) represents the change in the baseline vector, i.e., the change in length and orientation of the baseline, and (8.72) represents the change in the geocentric location of the baseline center. The latter can be interpreted as the translatory uncertainty of the baseline, or the uncertainty of the fixed baseline station. Transforming (8.70) to the difference and sum gives

$$dP_{km}^{pq}(d\mathbf{X}_k, d\mathbf{X}_m, d\mathbf{X}^p)$$

$$= -\tfrac{1}{2}[\mathbf{e}_k^p + \mathbf{e}_m^p] \cdot d\mathbf{b} - [\mathbf{e}_k^p - \mathbf{e}_m^p] \cdot d\mathbf{X}_c + [\mathbf{e}_k^p - \mathbf{e}_m^p] \cdot d\mathbf{X}^p \quad (8.73)$$

There is a characteristic difference in magnitude between the first bracket and the other brackets. Allowing an error of the order $O(b/\rho_k^p)$, the first bracket simplifies to $2e_m^p$ or $2e_k^p$. The second and the third brackets have opposite signs but have the same magnitude. It is readily seen that the terms in the latter two brackets are of the order $O(b/\rho_k^p)$. When the baseline vector is defined by

$$\mathbf{b} \equiv \boldsymbol{\rho}_m^p - \boldsymbol{\rho}_k^p \qquad (8.74)$$

equation (8.73) becomes, after neglecting the usual small terms,

$$dP_{km}^{pq}(d\mathbf{X}_k, d\mathbf{X}_m, d\mathbf{X}^p) = -\mathbf{e}_m^p \cdot d\mathbf{b} + \frac{\mathbf{b}}{\rho_m^p} \cdot d\mathbf{X}_c - \frac{\mathbf{b}}{\rho_m^p} \cdot d\mathbf{X}^p \qquad (8.75)$$

The orders of magnitude for the coefficients in this equation will not change even if double-difference expressions are fully considered. The relative impact of changes in the baseline and the translatory position of the baseline follow from

$$-\boldsymbol{\rho}_m^p \cdot d\mathbf{b} = \mathbf{b} \cdot d\mathbf{X}_c \qquad (8.76)$$

Similarly, changes in the baseline vector and ephemeris position are related by

$$\boldsymbol{\rho}_m^p \cdot d\mathbf{b} = \mathbf{b} \cdot d\mathbf{X}^p \qquad (8.77)$$

These relations are usually quoted in terms of absolute values, thereby neglecting the cosine terms of the dot product. In this sense, a rule of thumb for relating baseline accuracy, a priori geocentric position accuracy, and ephemeris accuracy is

$$\frac{|d\mathbf{b}|}{b} = \frac{|d\mathbf{X}_c|}{\rho_m^p} = \frac{|d\mathbf{X}^p|}{\rho_m^p} \qquad (8.78)$$

It is seen that the same accuracy is required for the a priori geocentric station coordinates as that for the satellite orbital positions.

Some of the pertinent values are given in Table 8.4, which is based on a topocentric satellite distance of 20,000 km. It is seen that an orbital uncertainty of 20 m will cause about 1-mm errors in baselines of 1 km. Because the best antennas available today define the electronic center no better than 1 mm, there is no need for higher orbital accuracy for such a small baseline. The GPS broadcast ephemerides available today are of such quality that 1–2 ppm can be achieved routinely. If higher accuracy is required over longer distances, it becomes necessary to estimate some of the GPS orbital parameters as part of the carrier phase solution. See Chapter 10.6 for additional details on orbital relaxation. Note that a 1000-km baseline can be measured to 1 cm if the

TABLE 8.4 Approximate Effects of Ephemeris Errors and Geocentric Station Location on Baselines

b [km]	dr [m]	$\dfrac{db}{b}$ [ppm]	db [cm]
1	200	10	1
	20	1	0.1
	2	0.1	0.01
10	200	10	10
	20	1	1
	2	0.1	0.1
100	200	10	100
	20	1	10
	2	0.1	1
1000	200	10	1000
	20	1	100
	2	0.1	10
	0.2	0.01	1

ephemeris errors can be reduced to 0.2 m according to the rule of thumb given above. This case corresponds to a relative positioning accuracy of 0.01 ppm.

The simplified derivation given in this section neglects the impact of the satellite constellation on the geometry of the solution. The only elements that enter the derivations are the baseline length and the receiver–satellite distance. More complete studies are found in the literature, e.g., Santerre (1991).

8.3.5 Cancellation of Common Mode Errors

GPS positioning benefits considerably from absorption or cancellation of common-mode errors. It has been pointed out in great detail how single and double differences reduce the effects of clock errors. Additional detail is provided here for both the navigation solution and relative positioning.

8.3.5.1 Navigation Solution
Whereas the propagation media effects on satellite signals, indeed, show a variation with azimuth and elevation angle, there is a station average that can be considered common to all signals. For example, the ionospheric delay can be split into a common component and a correction, as in

$$I^p_{k,P}(t) = I_{k,P}(t) + \Delta I^p_{k,P}(t) \tag{8.79}$$

The tropospheric delay can be split in a similar manner. It is even imaginable that the satellite clocks contain a common offset, e.g., caused by an incomplete correction due to relativity. The receiver hardware delay can also be a common source of errors. Consider once again the expressions of the navigation solution as listed in equation (8.9) and combine the common errors with the station clock error as follows:

$$B_k \equiv dt_k - \Delta_k = dt_k - c(dt^p)_k - \frac{I_{k,P}(t)}{c} - \frac{T_k(t)}{c} - \frac{d_{k,P}}{c} \qquad (8.80)$$

The subscript k, in the case of the satellite clock, should indicate that portion which is common to station k. The symbols for the common ionosphere and the troposphere have no superscript p in (8.80). The symbol B_k represents a new unknown, which, in addition to the station clock error, also contains the common components of all the other errors, denoted by Δ_k. The relevant portion of the pseudorange equations can now be written as

$$P_k^p(t) = \rho_k^p(t) - cB_k + c\Delta(dt^p) + \Delta I_{k,P}^p(t) + \Delta T_k^p(t) \qquad (8.81)$$

Introducing this newly defined parameter B_k does not change the design matrix (8.17) at all. The coefficients for the parameter B_k are still equal to c. In adjustments notation the nonredundant navigation solution is

$$\begin{bmatrix} d\mathbf{X}_k \\ B_k \end{bmatrix} = -A^{-1}(L_0 - L_b) \qquad (8.82)$$

where L_b denotes the four pseudorange observations, and a component of L_0, $l_{k,0}^p$, is given by

$$l_{k,0}^p \equiv l_{k,0,1}^p = \rho_k^p(t)|_{X_0} + c\Delta(dt^p) + \Delta I_{k,P}^p(t) + \Delta T_k^p(t) \qquad (8.83)$$

if B_k is intended to be estimated, or

$$l_{k,0}^p \equiv l_{k,0,2}^p = \rho_k^p(t)|_{X_0} + c\,dt^p + I_{k,P}^p(t) + T_k^p(t) \qquad (8.84)$$

if dt_k is intended to be estimated. Note that both misclosures differ by a constant and that this constant is the same for each pseudorange. In practice this common station error is not known, therefore the distinction between (8.83) and (8.84) cannot be realized. Consequently, it is of interest to find out what difference, if any, the constants $l_{k,0,1}^p$ and $l_{k,0,2}^p$ will cause to the estimated positions.

Let q_{ij} denote the element (i, j) of the inverse matrix A^{-1}; then the interesting relation

$$\sum_{j=1}^{4} q_{ij} = 0 \qquad i = 1, 2, 3 \qquad (8.85)$$

holds, as was pointed out by Greenspan et al. (1991). The property (8.85) assures that any common constant term in the components of L_0 does not affect the position, but merely changes the estimated clock parameter.

It is concluded that unmodeled errors that are common to all observations at a particular station do not affect the estimated position. Thus, modeling of, for example, the ionosphere and troposphere is useful only if it reduces the variability with respect to the common portion. This example also demonstrates why the requirements for positioning and timing with GPS are quite different. If the goal is to determine time, then modeling or controlling the common station errors is of critical importance.

8.3.5.2 *Relative Positioning* In relative positioning the common portion for both stations is automatically eliminated during the double-differencing operation. For example, the tropospheric correction can be decomposed into the common part and the satellite-dependent part as follows:

$$\Delta T_{km}^{pq} = [(T_{km} + \Delta T_k^p) - (T_{km} + \Delta T_m^p)] - [(T_{km} + \Delta T_k^q) - (T_{km} + \Delta T_m^q)]$$

$$= (\Delta T_k^p - \Delta T_m^p) - (\Delta T_k^q - \Delta T_m^q) \tag{8.86}$$

In relative positioning it is useful to apply tropospheric and ionospheric corrections if the differential correction between the stations can be determined accurately. If this is not the case, because, say, the meteorological data are not representative of the actual tropospheric conditions, it might be better not to apply the correction at all and to rely on the common-mode elimination. Because the ionosphere and the troposphere are highly correlated over short distances, most of their delays are common to both stations. In terms of the tropospheric effect, an exception to this rule might apply to nearby stations that are located at significantly different elevations. Additional details are given in the next chapter.

Table 8.5 summarizes the most important features of the single-, double-, and triple-difference observations as they apply to carrier phases and pseudoranges. It is because of the cancellation of most of the effects of the propagation media and the clock errors, that relative positioning has become so popular and useful in surveying. Although the presence of the ambiguity parameters in the double differences might initially be perceived as a nuisance, they provide a unique vehicle to improve the solution if they can be successfully constrained to integers.

8.3.6 Connection to Other Systems

The development of GPS positioning techniques has benefited very much from concepts developed earlier for the TRANSIT Doppler satellite system, and very long baseline interferometry (VLBI). Not only could certain algorithms be directly transferred to GPS, but, most importantly, there was a large pool of experienced researchers ready to jump on the GPS bandwagon, causing a very rapid pace in the development of positioning using GPS.

TABLE 8.5 Common-mode Cancellations

Observation	Effects eliminated	Effects reduced	Option
Single difference	first order satellite clock	orbit errors geometric	constrain ambiguity
Double difference	first order satellite & station clocks	position error ionosphere	constrain ambiguity
Triple difference	first order satellite & station clocks	troposphere	ambiguity eliminated

8.3.6.1 The Doppler Connection

Occasionally questions arise as to the connection between the carrier phase approach and the familiar Doppler approach. The difference between these two approaches is not in the physics but in the mathematical treatment of the observations. Consider the Doppler equation as it is found in the standard literature:

$$\frac{f^p}{f^p_T} = \frac{1 - (v \cos \beta)/c}{\sqrt{1 - v^2/c^2}} = \left(1 - \frac{v}{c} \cos \beta\right)\left(1 + \frac{v^2}{2c^2} + \frac{v^4}{8c^4} + \cdots\right)$$

(8.87)

The symbol f^p denotes the received satellite frequency as detected at the receiver, and f^p_T is the stable frequency emitted by the satellite. The tangential velocity of the satellite is v, and β is the angle to the receiver as indicated in Figure 8.10. The figure shows that the topocentric range rate is related to the tangential velocity by

$$\dot{\rho} = \frac{d\rho}{dt} = -v \cos \beta$$

(8.88)

Because $v \ll c$ the higher order terms in (8.87) are usually deleted. With (8.88) the ratio of the received and emitted frequency becomes

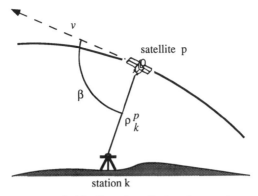

Figure 8.10 Doppler effect and geometry.

$$\frac{f^P}{f^P_T} = 1 - \frac{\dot{\rho}^P_k}{c} \tag{8.89}$$

and the difference is

$$f^P_T - f^P = \frac{f^P_T}{c} \dot{\rho}^P_k \tag{8.90}$$

This is the Doppler shift; it is directly related to the topocentric range rate of the satellite. In practice, a receiver runs at a stable frequency f_k, and the receiver frequency and the received frequency are compared and integrated over time. Since the integration of frequency over time gives phase, we obtain

$$
\begin{aligned}
\varphi^P_k(t_2, t_1) &= \int_{t_1}^{t_2} (f_k - f^P)\, dt = \int_{t_1}^{t_2} f_k\, dt - \int_{t_1}^{t_2} f^P\, dt \\
&= [\varphi_k(t_2) - \varphi_k(t_1)] - [\varphi^P(t_2) - \varphi^P(t_1)] \\
&= [\varphi_k(t_2) - \varphi^P(t_2)] - [\varphi_k(t_1) - \varphi^P(t_1)]
\end{aligned}
\tag{8.91}
$$

This expression is identical to the delta range (8.49) and agrees with the basic form (8.20) for the carrier phase observable. The second integral in (8.91) is often written as

$$\int_{t_1}^{t_2} f^P\, dt = \int_{t_1 - \tau_1}^{t_2 - \tau_2} f^P_T\, dt \tag{8.92}$$

where τ_i denotes the travel time at the respective epochs. Equation (8.92) implies conservation of cycles; that is, there are as many cycles received at the receiver during the time interval $[t_2 - t_1]$ as there are transmitted during $[(t_2 - \tau_2) - (t_1 - \tau_1)]$. The equivalent statement for the carrier phase approach is

$$\varphi^P(t_2) = \varphi_T^P(t_2 - \tau_2) \qquad (8.93)$$

$$\varphi^P(t_1) = \varphi_T^P(t_1 - \tau_1) \qquad (8.94)$$

which also implies that the phase is a continuous function, and that all phases are accumulated at the receiver. See equation (8.21). Equations (8.93) and (8.94) still allow the addition of an initial register counter N_k^p as long as it is constant in time.

The difference between the Doppler and the carrier phase approach, therefore, lies in the processing of the data and not in the physics. Positioning with the Doppler approach focuses automatically on the integrated Doppler, which encompasses two epochs. This approach leads directly to the delta range equations given above, which were discussed in Section 8.2.5. The carrier phase formulation focuses on one epoch and explicitly introduces the ambiguity parameter. The advantage of the ambiguity parameter is that we have a priori knowledge that it is an integer. If its estimate is statistically close to an integer, we may be able to constrain it to an integer and thus introduce strength into the solution. If the estimated ambiguity differs statistically from an integer value, then this indicates the presence of unmodeled effects which, unfortunately, cannot always be avoided.

8.3.6.2 The VLBI Connection

A primary purpose of very long baseline interferometry is to map extragalactic radio sources and, in particular, quasars. The radiation from these celestial objects often looks like broadband Gaussian noise as opposed to transmission generated from stable frequencies, as is the case with GPS satellites. Radiation sources often do not look like a point source but have an extended shape with varying intensity. But non-point sources are not used for geodetic VLBI. Because these cosmologically distant extragalactic radio sources represent inertially stable direction in the sky, they are used extensively to solve geodetic problems such as the definition of reference coordinate systems. Because of their long distance from the earth, the same radio sources can be observed from widely separated stations for accurate determination of long baselines. Because the arriving signals from these extragalactic radio sources are weaker than those of GPS satellites, it is necessary to use large, directional antennas.

The observations are recorded separately at the observatories and are accurately time-tagged using atomic clocks. Avoiding a discussion here of the complexities of processing VLBI observations, it suffices to say that both data streams are cross-correlated, yielding a quantity closely resembling the single-difference observable $\varphi_{km}^p(t)$. In principle, at least, this VLBI function also suffers from the ambiguity problem. In the VLBI literature this observable is called the fringe phase. In addition, the delay,

$$\tau(t) = \frac{\partial \varphi^{p}_{km}(t)}{\partial f^{p}} \tag{8.95}$$

and the fringe rate

$$\nu(t) = \frac{d[\varphi^{p}_{km}(t)]}{dt} \tag{8.96}$$

are used. Since the extragalactic radio sources are stable in space, the fringe rate is extrasensitive to the earth's rotation.

Because the observable is primarily of the single-difference type, time tagging the observations requires very accurate atomic clocks. However, VLBI researchers discovered the utility of double-difference observables long before GPS became available. An opportunity to explore double differencing was provided by the signals from the Apollo Lunar Surface Experiments Packages (ALSEPs), which were deposited by the various Apollo missions. Double differencing was also used for the successful and accurate tracking of the lunar rover (Salzberg 1973). Counselman (1973) addresses applications of VLBI for geodesy, geophysics, planetary sciences, etc. It is amazing to read about the concerns addressed in that paper and to compare these with similar concerns still facing GPS positioning today, such as the ionosphere, the troposphere, and multipath.

8.3.7 Selective Availability

The intentional degradation of the GPS signals has become known as selective availability (SA). This degradation is accomplished through manipulation of the broadcast ephemeris data and dithering of the satellite clocks. A faulty broadcast ephemeris affects the single-receiver navigation solution directly. According to the discussion of Section 8.3.4 the computed ranges are in error by $\mathbf{e}^{p}_{k} \cdot d\mathbf{X}^{p}$, whereas in relative positioning the impact on the range difference is much less, i.e., $(\mathbf{e}^{p}_{k} - \mathbf{e}^{p}_{m}) \cdot d\mathbf{X}^{p}$. Analysis shows that the induced ephemeris error appears to be a nonperiodic trend covering the duration of the satellite pass. Dithering of the satellite clock results in erroneous pseudorange and carrier phase observations. This error appears to be of a (pseudo) random nature with periods in the range of minutes. Again, unknown satellite clock errors affect the navigation solution directly, whereas the effect potentially cancels as a common-mode error in relative positioning. Thus, SA is primarily of concern to nondifferential users.

Research on the implications of SA is still in progress (Lear et al. 1992; Braasch 1990; Braasch et al. 1993). Because the formula for SA is classified, not much is known about the statistical properties of SA and whether or not these properties change with time. Researchers attempt to model SA-induced time dithering with stochastic models or analytical models with random coef-

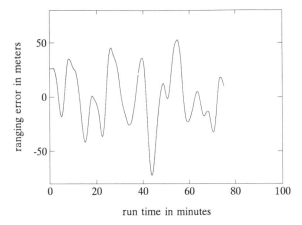

Figure 8.11 SA induced ranging error (DOY 104, 1992, SV 28). (Courtesy of M. S. Braasch, Ohio University, Athens, Ohio.)

ficients. SA can be investigated on pseudoranges, on carrier phases in undifferenced form, or in the form of delta ranges. Investigations usually proceed on the assumption that the ephemeris errors can be removed because they change slowly compared to clock dithering. Figure 8.11 shows an example of ranging errors which result from SA. The amplitude varies to 70 m and oscillations have periods between 5 to 10 min. The generation of these data requires a number of steps. In this particular case, the true topocentric ranges were computed using a precise ephemeris. These ranges were then subtracted from the carrier phase observations. The remainder reflects SA, receiver clock drift, ionospheric and tropospheric delays, multipath, and a bias due to the initial

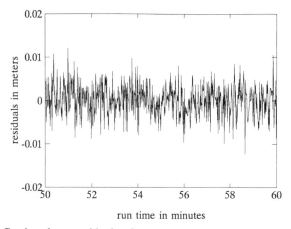

Figure 8.12 Carrier phase residuals after removing SA (DOY 104, 1992, SV 28). (Courtesy of M. S. Braasch, Ohio University, Athens, Ohio.)

carrier phase ambiguity. The receiver had been controlled by an external rubidium clock during the observations. Because of the high stability of such an atomic clock, the clock drift can be modeled sufficiently by a first-order polynomial. Assuming further that atmospheric delays typically introduce long-term trends, a second-order polynomial was eventually fitted to the data to remove these effects. This procedure, unfortunately, also removes any long-term trend that SA might have generated. The success of modeling SA in this case is demonstrated by Figure 8.12, which shows the residuals after removing the modeled SA. Only 10 min of data is shown. The standard deviation of the remaining residuals is 4.12 mm, which is slightly above the measurement noise.

CHAPTER 9

PROPAGATION MEDIA, MULTIPATH, AND PHASE CENTER

The propagation media affect electromagnetic wave propagation at all frequencies, resulting, among other things, in a bending of the signal path, time delays of arriving modulations, advancement of carrier phases, and scintillation. In GPS positioning one is not concerned with angular changes in the zenith angle but with the arriving times of carrier modulations and carrier phases. Geometric bending of the signal path causes a small delay which is negligible for elevation angles above 5°. The local propagation speed in the atmosphere differs from that in a vacuum and varies with location and time. The most relevant propagation regions are the troposphere, tropopause, stratosphere, and the ionosphere.

The ionosphere covers the region between approximately 50 and 1500 km above the earth and is characterized by the presence of free (negatively charged) electrons. Carrier frequencies below 30 MHz are reflected by the ionosphere; only higher frequencies penetrate the ionosphere. The ionosphere is a dispersive medium, meaning that the modulations on the carrier and the carrier phases are affected differently and that this effect is a function of the carrier frequency (the impact decreases with increased frequency).

The combined effect of the electronically neutral atmosphere, consisting of the troposphere, tropopause, and the stratosphere, is labeled tropospheric refraction. The effective height of the atmosphere in terms of tropospheric refraction is about 40 km. For frequencies below 30 GHz the troposphere behaves essentially like a nondispersive medium; that is, the refraction is independent of the frequency of the signals transmitted through it.

The propagation of electromagnetic waves through the various atmospheric regions is a complex matter and is still the subject of much research. Whereas positioning with GPS requires careful consideration of the impacts of the propagation media, GPS, in turn, has become a tool for studying the atmosphere, in particular the ionosphere. The subject of propagation is treated in this chapter only to the extent required for standard applications using GPS positioning. Additional details on the ionosphere and the upper atmosphere in general are found in Hargreaves (1992) and Davies (1990). Many publications about tropospheric refraction are available in the standard geodetic literature.

Once the satellite signals have reached the earth's surface, ideally they enter the antenna directly. However, the signals may be received as a result of being reflected off objects in the receiver's surrounding area. This multipath of the signals causes unwanted variations in the code phase and carrier phase observations. Multipath is currently one of the dominating, if not the dominant, sources of error in GPS positioning. The electronic center and the geometric center of the antenna do not always exactly coincide and, in fact, tend to vary slightly with time. This variation is a function of the geometry of the satellites. Since the magnitude of this variation can be in the centimeter range, the phenomenon of phase center variation is to be taken seriously for accurate positioning. Tranquilla and Colpitts (1989) give an overview of GPS antenna design characteristics for high-precision applications.

9.1 THE ATMOSPHERE IN BRIEF

The earth's atmosphere is usually subdivided into regions. These regions are based on common physical properties and appearances such as temperatures, composition, state of mixing, and ionization. In this scheme, the regions are referred to as "spheres" and the boundary between two spheres is called a "pause." Figure 9.1 shows some of the subdivisions of the atmosphere.

In the nomenclature of temperature, the lowest portion of the atmosphere is called the troposphere. The temperature in this region decreases about 10 kelvins per kilometer. The first boundary, the tropopause, is reached at 10–12 km. In the following stratosphere, the temperature increases slightly until the maximum is reached at the stratopause around 50 km. In the mesopause the temperature decreases again in height until the minimum is reached in the mesopause at about 80–85 km. At the beginning of the thermopause the temperature increases rapidly, followed by a long stretch of constant or slightly increasing temperature.

Most of the mass of the atmosphere is located in the troposphere and the tropopause. In regard to propagation classification, the term tropospheric refraction includes the combined effect of the neutral, gaseous atmosphere. The effective height of this layer is about 40 km. The density of atoms and molecules in higher regions is too small to have a measurable effect. The tropospheric propagation is usually expressed as a function of temperature, pressure, and relative humidity.

There are several other spheres and pauses, but they are of interest primarily to atmospheric researchers. For example, below 100 km the atmosphere is well

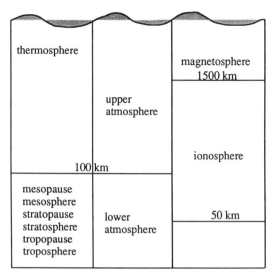

Figure 9.1   Atmospheric regions.

mixed and is often referred to as the turbosphere. The region above the turbosphere is the heterosphere, in which the components of the atmosphere may separate under the force of gravity. The lower portion of the heterosphere (where helium is dominant) is called the heliosphere. The higher region of the heterosphere is called the protonosphere, where hydrogen is dominant.

In the nomenclature of ionization, there are two distinct regions: the ionosphere and the magnetosphere. The magnetosphere refers to the outermost region where the geomagnetic field controls particle motion. The ionosphere covers the region between 50 and 1500 km. This region is characterized by a significant number of free (negatively charged) electrons and positively charged atoms and molecules called ions. The free electrons affect the propagation of radio waves, and thus affect the signals from GPS satellites. The ionization, i.e., the breaking away of electrons from the atoms of atmospheric gases, is caused primarily by the impact of solar ultraviolet radiation. After ionization, the electrons and ions tend to recombine and/or to react with other gaseous components to form new ions. Thus the amount of ultraviolet radiation from the sun determines the state of the ionosphere. The number of free electrons (referred to as the electron density) is a key parameter for formulating the propagation characteristics of radio waves. Once the radiation source is removed, the electron density decreases. It follows that the electron density variation strongly depends on the relative position of the sun. For example, on the dark side of the earth the electron density should be low and fairly stable. This dependency of the ionospheric state on the solar position can be effectively used for temporal and spatial modeling of ionospheric effects on GPS signals.

Figure 9.2 indicates the complexity in the spatial distribution of an average

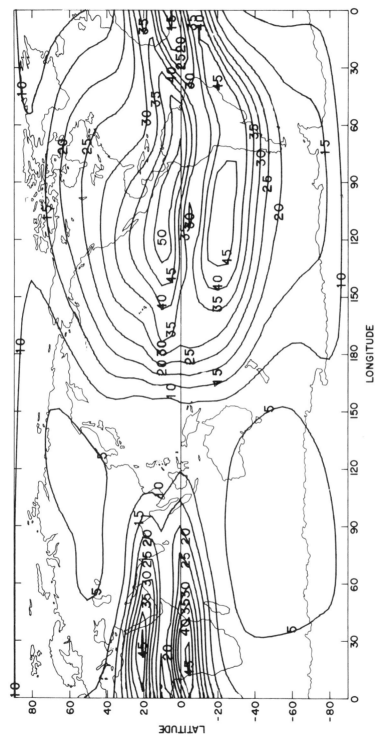

Figure 9.2 Worldwide distribution of the ionospheric time delay for a moderately high solar maximum. (Courtesy of J. A. Klobuchar, Air Force Geophysics Laboratory, Hanscom AFB, Massachusetts.)

monthly ionospheric group delay on the L1 frequency. The group delay represents the time by which a P-code sequence on L1 is delayed by the ionosphere as compared to travel time in a vacuum. The contours are in units of nanoseconds (three nanoseconds corresponds roughly to 0.9 m distance). Rather than reaching a maximum on the equator, there is a minimum in the electron density along the magnetic equator, with two maxima 10 to 20° north and south of it.

The short-term spectrum of ionospheric variations is a complicated function of the elevation angle of the satellite, the time of day, the latitude of the observer, the season, and solar flare activities. Indeed, several natural or artificial phenomena, such as a solar eclipse or the launch of a satellite, can leave a footprint in the electron density profile. At regions between midlatitude and high latitude there are often troughs, which are bands of depleted ionosphere. This phenomenon is generally attributed to the mechanism underlying the polar ionosphere. Significant ionization can occur in the polar cap area from auroral activities and magnetic storms, causing significant spatial and temporal variability. Ionospheric disturbances of wave-like structures have been observed and appear to travel rapidly through space. These phenomena have appropriately been called traveling ionospheric disturbances (TIDs).

Since ionization is driven largely by the magnitude and frequency of solar flare activities, it is a function of the sunspot cycle. Sunspots are seen as dark areas in the solar disk. They appear dark because of a difference in temperature with surrounding regions. Sunspots tend to group and have an approximate lifetime of from a few days to a month. The systematic recording of these events began in 1848 when the Swiss astronomer Johann Wolf introduced the sunspot number. This number captures the total number of spots seen, the number of disturbed regions, and the sensitivity of the observing instrument. Wolf searched observatory records to tabulate past sunspot activities. He apparently traced the activities to 1610, the year Galileo Galilei first observed sunspots through his telescope (McKinnon 1987). Sunspot activities follow a periodic variation, with the principal period being 11 years. The cycles are usually not symmetric. The time from minimum to maximum is shorter than the time from maximum to minimum. A sample of the most recent cycles is seen in Figure 9.3. The strongest cycle ever recorded was cycle 19, which peaked in 1958. Details on the current cycle, cycle 22, which began in September 1986, are found in Kunches and Hirman (1990).

9.2 ELEMENTS OF WAVE PROPAGATION

In the context of positioning with GPS we are primarily concerned with the advancement of the carrier phase and the group delay of the codes. To gain an intuitive understanding of these phenomena, it is sufficient to consider a simplified situation of wave propagation in a homogeneous and isotropic medium. In a homogeneous medium the index of refraction is constant and the isotropic

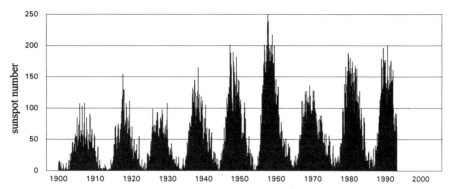

Figure 9.3 Yearly mean sunspot numbers. (Courtesy of the National Geophysical Data Center, Boulder, Colorado.)

property implies that the propagation velocity at any given point in the medium is independent of the direction of the propagation.

In a homogeneous and isotropic medium a harmonic wave with unit amplitude is described by

$$\varphi = \cos \omega \left(t - \frac{x}{c_\varphi} \right) \tag{9.1}$$

The symbol t denotes the time, c_φ [m/sec] is the phase velocity or the propagation speed of the wave, and x is the distance from the transmitting source. The angular frequency ω [rad/sec], the frequency f [Hz], the wavelength λ_φ [m], and the wave number or phase propagation constant k [rad/m] are related by

$$\omega = 2\pi f \tag{9.2}$$

$$\lambda_\varphi = \frac{c_\varphi}{f} \tag{9.3}$$

$$k = \frac{2\pi}{\lambda_\varphi} \tag{9.4}$$

Using relations (9.2) to (9.4) the wave equation (9.1) can be written as

$$\varphi_1 = \cos (\omega t - kx) \tag{9.5}$$

Let us consider another wave which has a slightly different frequency and wave number:

$$\varphi_2 = \cos [(\omega + \Delta\omega)t - (k + \Delta k)x] \tag{9.6}$$

These two harmonic waves can be superimposed as follows:

$$\varphi_s = \varphi_1 + \varphi_2$$

$$= 2 \cos \frac{\Delta\omega t - \Delta k x}{2} \cos \left[\left(\omega + \frac{\Delta\omega}{2} \right) t - \left(k + \frac{\Delta k}{2} \right) x \right] \quad (9.7)$$

This wave is displayed in Figure 9.4a. The combined signal shows two component waves of significantly different frequency. The envelope wave, or the beat signal, is given by

$$\Psi = 2 \cos \tfrac{1}{2} (\Delta\omega t - \Delta k x) \quad (9.8)$$

Analogously to equation (9.1) the expression (9.8) implies a propagation velocity of this beat signal of

$$c_g = \frac{\Delta\omega}{\Delta k} \quad (9.9)$$

At the limit, $\Delta\omega \to 0$ and $\Delta k \to 0$ we obtain the differential form

$$c_g = \frac{d\omega}{dk} \quad (9.10)$$

The quantity c_g is called the group velocity. It is the velocity by which the modulation or the energy travels. Since modulation (P-codes and C/A-codes) is used in GPS positioning in addition to carrier phase observables, it is important to establish a relationship between the group velocity and the phase velocity. The second wave component in (9.7) is the carrier. The frequency of this component approaches ω at the limit. In the case of GPS, the carrier is binary phase modulated.

In general, the product of phase refractive index n_φ and phase velocity equal the velocity of light in a vacuum:

$$n_\varphi c_\varphi = c \quad (9.11)$$

An analogous expression holds for the group refractive index n_g and the group velocity:

$$n_g c_g = c \quad (9.12)$$

Using the various relationships identified above, the group refractive index can

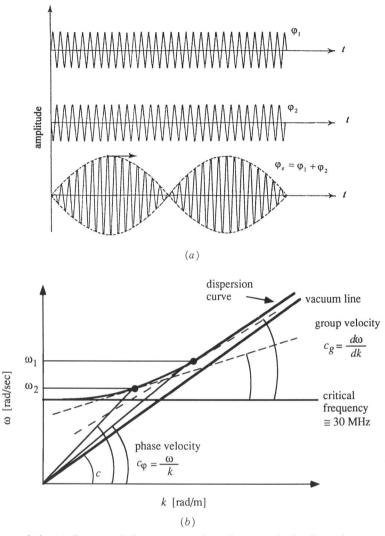

Figure 9.4 (a) Group and phase propagation; (b) Ionospheric dispersion curve.

be expressed as

$$n_g = \frac{c}{c_g} = c\,\frac{dk}{d\omega} = \frac{d}{d\omega}\left(\frac{2\pi c}{\lambda_\varphi}\right) = \frac{d}{d\omega}\left(n_\varphi\,\frac{c_\varphi}{c}\,\frac{2\pi c}{\lambda_\varphi}\right)$$

$$= \frac{d}{d\omega}\left(n_\varphi\,\frac{c_\varphi}{c}\,\frac{2\pi c}{\lambda_\varphi}\,\frac{\omega}{2\pi f}\right) \tag{9.13}$$

$$= \frac{d}{d\omega}\,(n_\varphi\omega) = n_\varphi + \omega\,\frac{dn_\varphi}{d\omega} = n_\varphi + f\,\frac{dn_\varphi}{df}$$

If the group refractive index depends on the frequency of the carrier, i.e., if the derivative

$$\frac{dn_\varphi}{df} \neq 0 \tag{9.14}$$

is not zero, then the medium is called dispersive. Equations (9.10) to (9.13) imply that the phase velocity and the group velocity differ in a dispersive medium. At the L1 and L2 frequencies the ionosphere is dispersive, whereas the troposphere is nondispersive. The vacuum, of course, is a special case for which both indices of refraction are one. The superimposition (9.7) can result in a phase refractive index being smaller than one, which implies a phase velocity greater than the vacuum speed according to (9.11). The fact that $c_\varphi >$ c does not conflict with relativity theory, because the energy, i.e., the modulation, travels with a speed $c_g < c$. Figure 9.4b displays the angular velocity as a function of the phase propagation constant for an ionized medium such as the ionosphere. The dispersion curve varies with time because of ionospheric variations. The phase velocity is always greater than the vacuum velocity in such a medium. The group velocity equals the slope of the dispersion curve.

9.3 IONOSPHERIC EFFECTS ON GPS OBSERVABLES

The primary purpose of the second frequency in GPS satellites is to neutralize the effect of the ionosphere on signal propagation. Functions of the dual-frequency observables are readily available that do not depend on the ionosphere. Single-frequency users typically depend on the elimination of ionospheric effects as common-mode errors in double differencing, as discussed in Chapter 8. Efforts are being made to measure the ionospheric effects with just a single-frequency receiver using code and carrier phase observations together with a physically plausible model of the ionosphere.

9.3.1 Code Delays and Phase Advances

The phase refractive index of the ionosphere is directly tied to the number of free electrons along the path of the signal in addition to being a function of several fundamental constants (Hargreaves 1992, p. 67). The developed expression for the phase refractive index is usually given in the following form:

$$n_\varphi = \sqrt{1 - 2N_I} = 1 - N_I + O(N_I^2) \tag{9.15}$$

where ionospheric refractivity N_I is

$$N_I = \frac{40.30}{f^2} N_e \ll 1 \tag{9.16}$$

The symbol f denotes the carrier frequency. The local electron density is N_e and it is given in units of electrons per cubic meter [el/m^3]. The density varies along the path of the signal; the highest electron densities are in the ionospheric region. The total electron content (TEC) along the path follows from the integral

$$\text{TEC} = \int_{\text{path}} N_e \, ds \qquad (9.17)$$

The TEC represents the number of free electrons in a 1-square meter column along the path and is given in units of [el/m^2]. Substituting the phase refractivity (9.16) and phase refractive index (9.15) into equation (9.13) and carrying out the respective differentiation gives the expression for the group refractive index:

$$n_g = n_\varphi + f \frac{dn_\varphi}{df} = 1 + N_I \qquad (9.18)$$

Neglecting terms of the order $O(N_I^2)$ and smaller, the expressions for the phase and group velocities become

$$c_\varphi = \frac{c}{n_\varphi} = \frac{c}{1 - N_I} = c(1 + N_I) \qquad (9.19)$$

$$c_g = \frac{c}{n_g} = \frac{c}{1 + N_I} = c(1 - N_I) \qquad (9.20)$$

Since N_I is a positive number, the phase velocity is larger than vacuum speed and the group velocity is smaller than vacuum speed by the same amount Δc; that is,

$$\Delta c = cN_I = \frac{40.30c}{f^2} N_e \qquad (9.21)$$

The time delay of a code sequence or the phase advancement that is registered by the GPS receiver is directly related to the velocity difference Δc and its variations along the path. The time delay and the phase advancement are often transformed to the corresponding distance. Integrating (9.21) over time gives the ionospheric distance

$$I^p_{k,f,P} = \frac{40.30c}{f^2} \int_{\text{path}} N_e \, dt \qquad (9.22)$$

in meters. In (9.22) we have introduced the complete notation that is necessary in identifying the ionospheric effect. The first subscript and the superscript identify the path by means of station k and satellite p, the second subscript denotes the frequency of the carrier, and the third subscript identifies the sign and the unit. If the third subscript is P, the distance is positive and is given in units of meters. Other subscripts will be introduced below. Along the path of the signal there exists the following differential relationship between traveled distance and time:

$$ds = c(1 \pm N_I)\, dt \tag{9.23}$$

Neglecting again the small and insignificant terms, in a similar manner as that shown above, the inverse of this relation is

$$dt = \frac{1}{c}(1 \mp N_I)\, ds \tag{9.24}$$

Substituting (9.24) in (9.22) gives the distance

$$I^p_{k,f,P} = \frac{40.30}{f^2} \int_{\text{path}} N_e(1 \mp N_I)\, ds = \frac{40.30}{f^2} \int_{\text{path}} N_e\, ds \tag{9.25}$$

The integral equals the TEC according to (9.17). Thus,

$$I^p_{k,f,P} = \frac{40.30}{f^2}\, \text{TEC} \tag{9.26}$$

The corresponding ionospheric time delay or time advance follows as

$$\nu_f = \frac{I^p_{k,f,P}}{c} = \frac{40.30\ \text{TEC}}{cf^2} \tag{9.27}$$

The time delay is proportional to the inverse of the frequency squared. Consequently, higher frequencies are less affected by the ionosphere; that is, the time delay is shorter. It is because of this relationship that the high-frequency GPS signals pass through the ionosphere relatively well and can be measured accurately on arrival at the receiver.

With GPS signals we must distinguish between the delays of the P1- and P2-codes and the advances of the L1 and L2 carrier phases. The unit for the ionospheric code delay is in meters, whereas it is typically described in cycles for the carrier phases, unless the carrier phases have explicitly been scaled to distance. The following notation identifies the units and the signs:

$$I^p_{k,1,P} = -I^p_{k,1,\Phi} = -\frac{c}{f_1} I^p_{k,1,\varphi} \tag{9.28}$$

$$I_{k,2,P}^p = -I_{k,2,\Phi}^p = -\frac{c}{f_2} I_{k,2,\varphi}^p \tag{9.29}$$

$$\frac{I_{k,1,P}^p}{I_{k,2,P}^p} = \frac{f_2^2}{f_1^2} \tag{9.30}$$

$$\frac{I_{k,1,\varphi}^p}{I_{k,2,\varphi}^p} = \frac{f_2}{f_1} \tag{9.31}$$

Table 9.1 shows the ionospheric delays for several frequencies and TEC values. The major impact of the ionosphere on the L1 and L2 frequency ranges is readily seen. Typically the TEC values range from 10^{16} to 10^{18}. Often the total electron content is expressed in terms of TEC units, with one TEC unit being 10^{16} electrons per 1-square meter column.

Although very important, phase advancement and group delay are not the only manifestations of the ionosphere on the signal propagation. Some of the phase variations are converted to amplitude variation by means of diffraction. The result is an irregular but rapid variation in amplitude and phase, called scintillation. The signal can experience short-term fading, i.e., it loses strength. These scintillations might occasionally cause phase-lock problems to occur in receivers. The term scintillation is analogous to the intensity variation of luminous stars when seen through a turbulent atmosphere.

The receiver bandwidth must be sufficiently wide as to accommodate the normal rate of change of the geometric Doppler shift. The frequency shift due to the Doppler can be 5000 Hz, whereas the rate of change can be up to 1 Hz per second. For example, if the receiver bandwidth is set to 1 Hz to deal with the rate of change of the geometric Doppler shift, and if the ionosphere causes an additional 1-Hz shift, the receiver might lose phase lock. Assuming a maximum TEC of 10^{18} el/m², a change of 1.12% TEC causes a single cycle change in L1. Scintillations are particularly severe in tropical regions during evening hours at the time of equinoxes, and in the auroral zones. See Klobuchar (1991)

TABLE 9.1 Ionospheric Range Correction in Meters

Frequency	TEC=10^{16} [el/m²]	TEC=10^{18} [el/m²]
100 MHz	40.	4000.
400 MHz	2.5	250.
f_2	0.26	26.
f_1	0.16	16.
2 GHz	0.1	10.
10 GHz	0.004	0.4

and Wanninger (1993) for more details on the temporal and regional distribution of scintillation.

9.3.2 Single-Frequency Solutions

Single-frequency users cannot take advantage of the full capability of all GPS signals to eliminate the ionospheric effects. To support single-frequency users, the broadcast message contains ionospheric model data which allow the computation of the approximate ionospheric group delay. Several attempts have been made to determine the TEC from single-frequency observations. Such efforts might eventually lead to practical implementations, especially as the C/A-code measurement resolution improves. Single-frequency ionospheric estimation might eventually become useful in long-baseline applications where the ionosphere becomes uncorrelated. This section summarizes some of the options available to single frequency users, starting with the code and carrier phase equations summarized in Table 8.3:

$$P_{k,1}^p(t) = \rho_k^p(t) - c\,dt_k + c\,dt^p + I_{k,1,P}^p(t) + T_k^p(t)$$
$$+ d_{k,1,P}(t) + d_{k,1,P}^p(t) + d_{1,P}^p(t) + \epsilon_{1,P} \tag{9.32}$$

$$\Phi_{k,1}^p(t) = \rho_k^p(t) - c\,dt_k + c\,dt^p - I_{k,1,P}^p(t) + T_k^p(t) + \frac{c}{f_1} N_{k,1}^p(1)$$
$$+ d_{k,1,\Phi}(t) + d_{k,1,\Phi}^p(t) + d_{1,\Phi}^p(t) + \epsilon_{1,\Phi} \tag{9.33}$$

Note that the carrier phase equation is expressed in distance units and that the ionospheric term is added to the pseudorange and subtracted from the carrier phase expression in accordance with the understanding that the left-hand sides of (9.32) and (9.33) denote the observed values and $\rho_k^p(t)$ denotes the geometric straight-line distance between the receiver's antenna and the satellite. Differencing both equations at epoch t gives

$$P_{k,1}^p(t) - \Phi_{k,1}^p(t)$$

$$= 2I_{k,1,P}^p(t) - \frac{c}{f_1} N_{k,1}^p(1) + d_{k,1,P}(t) + d_{k,1,P}^p(t) + d_{1,P}^p(t) \tag{9.34}$$

$$- d_{k,1,\Phi}(t) - d_{k,1,\Phi}^p(t) - d_{1,\Phi}^p(t) + \epsilon_{1,P} - \epsilon_{1,\Phi}$$

which is a function of the group delay, the initial ambiguity, the receiver and satellite hardware delays, and the multipath. A similar expression can be given for the L2 carrier. The many terms in equation (9.34) clearly demonstrate the difficulty of accurately solving for the ionospheric delay. Even if we assume that the receiver and hardware delays are constant over time, the multipath terms are difficult to model as they vary with satellite direction and receiver environment. Even if we ignore all delay and multipath terms, the group delay and the ambiguity cannot be estimated separately, because each epoch adds a new ionospheric unknown. Several solutions have been proposed in the literature for modeling some of the primary features of the ionosphere. The more sophisticated models make use of the fact that ionization strongly depends on the relative position of the sun. All implementations of (9.34) are limited by the low measurement accuracy of the codes.

9.3.2.1 Ionospheric Plate Model
In the simplest model the ionosphere is approximated at a particular receiver location by a flat plate of equal thickness having a homogeneous distribution of free electrons. This model does not incorporate the curvature of the earth. For such a simple model, the vertical group delay, $I_{k,1,P}$, and the group delay along the path are related as

$$I^p_{k,1,P}(t) = \frac{I_{k,1,P}}{\sin a^p_k} \tag{9.35}$$

The symbol a^p_k denotes the elevation angle of the satellite as seen from the receiver. Notice that the vertical group delay symbol does not include the superscript p. The vertical group delay is treated as a constant in this model, having no dependency on the azimuth of the satellite. Combining (9.35) and (9.34), and neglecting the hardware delay and multipath terms, we get

$$P^p_{k,1}(t) - \Phi^p_{k,1}(t) = 2I_{k,1,P}\frac{1}{\sin a^p_k} - \frac{c}{f_1}N^p_{k,1}(1) \tag{9.36}$$

At least conceptually the vertical group delay and the ambiguity can be estimated from a series of code and phase observations. In a graphical representation with the horizontal axis given in units of $1/\sin a^p_k$, the function (9.36) is linear, with the slope indicating the vertical group delay, and the intercept denoting the ambiguity (Xia 1992). Actually, any hardware delays in the satellite and/or in the receiver will be absorbed by the ambiguity estimate as long as these delays are constant. This is not necessarily a concern because the emphasis here is on determining the ionosphere and not the ambiguity.

9.3.2.2 Daily Cosine Model
A slightly more advanced ionospheric model takes into consideration the earth rotation and the daily motion of the sun with respect to the receiver location. For example,

$$I_{k,1,P}(t) = I_{k,1,P,\max}\cos(h_s - 2^h) \tag{9.37}$$

The symbol $I_{k,1,P,\max}$ denotes the maximum vertical group delay, which occurs around 14:00 local time (Klobuchar 1987). The hour angle of the sun is h_s. The model (9.37) can be substituted in (9.36).

9.3.2.3 *Ionospheric Point Model* A further improvement in modeling the ionosphere is achieved by recognizing that the ionosphere begins at a height of about 50 km rather than at the earth's surface. Usually a mean ionospheric height of 350 km is assumed. The simple model (9.35) is replaced with

$$I_{k,1,P}^{p}(t) = \frac{I_{k,1,P}}{\sin \theta_{k}^{p}}$$ (9.38)

where θ_{k}^{p} is the elevation angle of the satellite at the ionospheric point. See Figure 9.5 to view the geometry. The ionospheric point is located along the line of sight to the satellite and is at a height of 350 km. The projection of the ionospheric point on the earth is called the subionospheric point. The factor F,

$$F = \frac{1}{\sin \theta_{k}^{p}}$$ (9.39)

is called the slant factor or the obliquity (not to be confused with the angle between the equator and the ecliptic). Klobuchar (1987) relates the slant factor to the elevation angle of the satellites as viewed from the receiver, by the following expressions:

$$F = \frac{1}{\sin \theta_{k}^{p}} = \frac{1}{\cos \left[\sin^{-1} (0.94792 \cos a_{k}^{p}) \right]} \approx 1 + 2 \left(\frac{96 - a_{k}^{p}}{90} \right)^{3}$$ (9.40)

As is seen from Figure 9.6, the slant factor varies from 1 to slightly above 3.

9.3.2.4 *Generalization in Azimuth and Altitude* The ionospheric models discussed thus far do not incorporate an azimuth dependency of the vertical group delay. The variation in the vertical angle is modeled by the sine function. More sophisticated models can readily be envisioned. Cohen et al. (1992) have experimented with the function

$$I_{k,1,p} = J_{1} + J_{1} \sin \varphi^{*} + C_{11} \cos \varphi^{*} \cos \lambda^{*} + S_{11} \cos \varphi^{*} \sin \lambda^{*}$$ (9.41)

where φ^{*} and λ^{*} denote the latitude and longitude of the subionospheric point in a corotating coordinate system attached to the sun. Coco et al. (1991) used the polynomial

$$I_{k,1,P} = c_{1} + c_{2}\varphi^{*} + c_{3}\lambda^{*} + c_{4}(\varphi^{*})^{2} + c_{5}(\lambda^{*})^{2} + c_{6}\varphi^{*}\lambda^{*}$$ (9.42)

Caution must be exercised to avoid overparameterization to prevent weak solutions and also to prevent effects other than ionospheric delay from being absorbed in the estimated parameters. It is quite conceivable that in the near future single-frequency receivers will measure the ionospheric delay more ac-

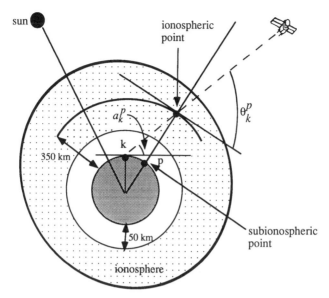

Figure 9.5 Ionospheric point geometry.

curately than can be computed from the standard ionospheric models and the data transmitted via the navigation message. Dual-frequency receivers present an attractive alternative, since the ionospheric effects can be eliminated altogether. Unfortunately, dual-frequency receivers are more expensive than their single-frequency counterparts.

9.3.2.5 *Broadcast Message and Ionospheric Model*   The satellite message contains eight coefficients for computing the group delay. The algorithm is published in Klobuchar (1987) and is also listed in ICD-GPS-200 (1991). It

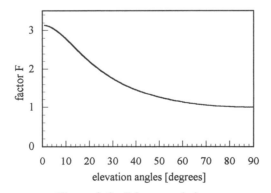

Figure 9.6 *F*-factor variation.

uses the cosine model for the daily variation of the ionosphere, with the maximum being at 14:00 local time, similar to expression (9.37). The amplitude and the period of the cosine term are functions of geomagnetic latitude and are represented by third-degree polynomials. The coefficients of these polynomials are transmitted as part of the navigation message.

The algorithm is summarized in Table 9.2. Note that several angular arguments are given in semicircles (SC). No attempt has been made to convert these values to other (more popular) units. The input parameters are the location of the receiver, the azimuth and elevation angle of the satellite, time, and broadcast coefficients. All auxiliary quantities in the middle portion of Table 9.2 can be computed one at a time from the top down. The symbol F denotes, again, the slant factor or the obliquity. The angle ψ denotes the earth's central angle between the user location and the ionospheric point, and φ_{IP} and λ_{IP}

TABLE 9.2   The Broadcast Ionospheric Model

φ , λ geodetic lat. and lon. of receiver (SC)	T = GPS time (TOW)
α_k^p, a_k^p azimuth and altitude of satellite (SC)	α_n, β_n broadcast coefficients

$$F = 1 + 16(0.53 - a_k^p)^3 \quad \text{(a)} \qquad \psi = \frac{0.0137}{a_k^p + 0.11} - 0.022 \qquad \text{(b)}$$

$$\varphi_{IP} = \begin{cases} \varphi + \psi \cos\alpha_k^p & if \; |\varphi_{IP}| \le 0.416 \\ 0.416 & if \; \varphi_{IP} > 0.416 \\ -0.416 & if \; \varphi_{IP} < -0.416 \end{cases} \quad \text{(c)} \qquad \lambda_{IP} = \lambda + \frac{\psi \sin\alpha_k^p}{\cos\varphi_{IP}} \qquad \text{(d)}$$

$$\phi = \varphi_{IP} + 0.064 \cos(\lambda_{IP} - 1.617) \qquad \text{(e)}$$

$$t = \begin{cases} \lambda_{IP} \, 4.32 \times 10^4 + T & if \; 0 \le t < 86400 \\ \lambda_{IP} \, 4.32 \times 10^4 + T - 86400 & if \; t \ge 86400 \\ \lambda_{IP} \, 4.32 \times 10^4 + T + 86400 & if \; t < 0 \end{cases} \qquad \text{(f)}$$

$$x = \frac{2\pi(t - 50400)}{P} \qquad \text{(g)}$$

$$P = \begin{cases} \sum_{n=0}^{3} \beta_n \, \phi^n & if \; P \ge 72000 \\ 72000 & if \; P < 72000 \end{cases} \quad \text{(h)} \qquad A = \begin{cases} \sum_{n=0}^{3} \alpha_n \phi_m^n & if \; A \ge 0 \\ A = 0 & if \; A < 0 \end{cases} \qquad \text{(i)}$$

$$I_{k,1,P}^p = \begin{cases} cF\left[5 \times 10^{-9} + A\left(1 - \frac{x^2}{2} + \frac{x^4}{24}\right)\right] & if \, |x| < 1.57 \\ cF(5 \times 10^{-9}) & if \, |x| > 1.57 \end{cases} \qquad \text{(j)}$$

Conversion of unit : 1 SC=180 degrees

denote the geodetic latitude and longitude of the ionospheric point. The geomagnetic latitude of the ionospheric point is ϕ. Implicit in these expressions is an ionospheric height of 350 km and the position of 78.3°N and 291°E for the geodetic latitude and longitude of the magnetic pole. Furthermore, t denotes the local time, P is the period in seconds, x is the phase in radians, and A denotes the amplitude in seconds. The function in the third part of the table has been multiplied with the velocity of light in order to yield the group delay directly in meters. Note that the group delay along the actual path is computed, whereas the vertical group delay is not. The algorithm presented here compensates for 50–60% of the actual group delay. Additional insight on the effects of ionospheric errors and this algorithm is given in Greenspan et al. (1991).

9.3.3 Dual-Frequency Ionospheric-free Solutions

Because both ionospheric delays and advances are frequency dependent, it is possible to eliminate the ionospheric effects with dual-frequency receivers, at least mathematically. Terms neglected in (9.26) are only of the order of about 5 cm (Spilker 1980). The remaining ionospheric errors are further reduced during double differencing. There are slight complications to be taken into consideration if the code phases are used to eliminate the ionosphere or to determine the absolute TEC. The difficulty arises from a modulation offset between L1 and L2 at the satellite, and possibly at the receiver. According to the ICD-GPS-200 (1991) the code phases are offset by the amount T_{GD}^p. The code phase emission times of both frequencies are related to this offset as follows:

$$t_{1,P}^p - t_{2,P}^p = T_{\mathrm{GD}}^p \left(1 - \frac{f_1^2}{f_2^2} \right) \tag{9.43}$$

The offset T_{GD}^p is determined by the control segment based on measurements made by the satellite manufacturer during factory testing and is part of the broadcast message. This offset T_{GD}^p can be treated as an additional clock correction for the code phases (not to be confused with the symbol for the tropospheric correction). To be general, we also allow a code offset for the receiver, $T_{k,\mathrm{GD}}$, and assume that a relation similar to (9.43) also applies to the receiver. The pseudorange equations now become

$$P_{k,1}^p(t) = \rho_k^p(t) - c(dt_k + T_{k,\mathrm{GD}}) + c(dt^p + T_{\mathrm{GD}}^p) + I_{k,1,P}^p(t)$$
$$+ T_k^p(t) + d_{k,1,P}(t) + d_{k,1,P}^p(t) + d_{1,P}^p(t) + \epsilon_{1,P} \tag{9.44}$$

$$P_{k,2}^p(t) = \rho_k^p(t) - c\left(dt_k + \frac{f_1^2}{f_2^2} T_{k,\mathrm{GD}}\right) + c\left(dt^p + \frac{f_1^2}{f_2^2} T_{\mathrm{GD}}^p\right) + \frac{f_1^2}{f_2^2} I_{k,1,P}^p(t)$$
$$+ T_k^p(t) + d_{k,2,P}(t) + d_{k,2,P}^p(t) + d_{2,P}^p(t) + \epsilon_{2,P} \tag{9.45}$$

These equations are most general in the sense that hardware delay terms have been included in addition to those following the specifications (9.43). For a navigation solution of at least four satellites, the receiver code phase offsets, and other error components common to the station, are absorbed by the receiver clock estimate. See the respective discussion in Section 8.3.5. Some of the satellite code phase offsets, on the other hand, can be corrected with the broadcast value for T_{GD}^P. This correction is usually included with the satellite clock correction as indicated by (9.44) and (9.45).

The objective is to find a function of the codes that does not depend on the ionosphere. As is readily verified, the ionospheric-free function $P_{k,IF}^P$,

$$
\begin{aligned}
P_{k,IF}^P(t) &\equiv \frac{f_1^2}{f_1^2 - f_2^2} P_{k,1}^P(t) - \frac{f_2^2}{f_1^2 - f_2^2} P_{k,2}^P(t) \\
&\equiv \frac{1}{(1 - f_1^2/f_2^2)} \left\{ P_{k,2}^P(t) - \frac{f_1^2}{f_2^2} P_{k,1}^P(t) \right\} \\
&= \rho_k^P(t) - c\, dt_k + c\, dt^P + T_k^P(t)
\end{aligned}
\tag{9.46}
$$

serves this purpose. Equation (9.46) is not a function of the ionospheric term $I_{k,1,P}^P$. The satellite and receiver code phase offsets $T_{k,GD}$ and T_{GD}^P have canceled, whereas the other hardware delays and multipath terms do not cancel (but are not listed explicitly in (9.46) in order to keep the expression simple). The coefficients have the following values:

$$
\frac{f_1^2}{f_1^2 - f_2^2} = 2.546
\tag{9.47}
$$

$$
\frac{f_2^2}{f_1^2 - f_2^2} = 1.546
\tag{9.48}
$$

The dual-frequency carrier phase expression in units of cycles is available in Table 8.3 as follows:

$$
\begin{aligned}
\varphi_{k,1}^P(t) = \frac{f_1}{c} \rho_k^P(t) - f_1\, dt_k + f_1\, dt^P - \frac{f_1}{c} I_{k,1,P}^P(t) + \frac{f_1}{c} T_k^P(t) + N_{k,1}^P(1) \\
+ d_{k,1,\varphi}(t) + d_{k,1,\varphi}^P(t) + d_{1,\varphi}^P(t)
\end{aligned}
\tag{9.49}
$$

$$
\begin{aligned}
\varphi_{k,2}^P(t) = \frac{f_2}{c} \rho_k^P(t) - f_2\, dt_k + f_2\, dt^P - \frac{f_2}{c} I_{k,2,P}^P(t) + \frac{f_2}{c} T_k^P(t) + N_{k,2}^P(1) \\
+ d_{k,2,\varphi}(t) + d_{k,2,\varphi}^P(t) + d_{2,\varphi}^P(t)
\end{aligned}
\tag{9.50}
$$

The ionospheric-free carrier phase function $\varphi^p_{k,IF}$ is

$$\varphi^p_{k,IF}(t) \equiv \frac{f_1^2}{f_1^2 - f_2^2}\, \varphi^p_{k,1}(t) - \frac{f_1 f_2}{f_1^2 - f_2^2}\, \varphi^p_{k,2}(t)$$

$$= \frac{f_1}{c}\, \rho^p_k(t) - f_1\, dt_k + f_1\, dt^p + \frac{f_1^2}{f_1^2 - f_2^2}\, N^p_{k,1}(1) \qquad (9.51)$$

$$- \frac{f_1 f_2}{f_1^2 - f_2^2}\, N^p_{k,2}(1) + T^p_k(t)$$

where again the general hardware and multipath terms are not listed explicitly. The multiplier for the L1 carrier phase is given in (9.47) and the new multiplier for the L2 carrier phase is

$$\frac{f_1 f_2}{f_1^2 - f_2^2} = 1.984 \qquad (9.52)$$

This ionospheric-free phase function does not contain the ionospheric term. Unfortunately, the integer nature of the initial ambiguities $N^p_{k,1}$ and $N^p_{k,2}$ on this new function is lost because of the multipliers (9.47) and (9.52). If (9.51) is used alone, either in undifferenced or double differencing form, only the combined L1 and L2 ambiguity are estimable. Any hardware delays in the receiver and/or the satellite that are constant in time are absorbed by the ambiguities. However, if the hardware phase delays follow the same laws of physics as the code delays, the analogous expression to (9.43) for the carrier phases is

$$t^p_{1,\varphi} - t^p_{2,\varphi} = T^p_{GD}\left(1 - \frac{f_1}{f_2}\right) \qquad (9.53)$$

with a similar expression for the hardware delay in the receiver. It can readily be seen that delays following (9.53) cancel in the ionospheric-free function (9.51).

9.3.4 Dual-Frequency Ionospheric Solutions

Because of code phase delays and multipath, the determination of the absolute ionosphere is not at all a straightforward matter even for dual-frequency observations. The code difference of (9.44) and (9.45) is

$$P^p_{k,I}(t) \equiv P^p_{k,1}(t) - P^p_{k,2}(t)$$

$$= \left(1 - \frac{f_1^2}{f_2^2}\right)[I^p_{k,1,P}(t)] + c\left(1 - \frac{f_1^2}{f_2^2}\right)(T_{k,GD} - T^p_{GD})$$

$$+ d_{k,1,P}(t) + d^p_{k,1,P}(t) + d^p_{1,P}(t) \qquad (9.54)$$

$$- d_{k,2,P}(t) - d^p_{k,2,P}(t) - d^p_{2,P}(t)$$

This function readily shows the difficulties encountered when measuring the total ionosphere, or the TEC, with dual-frequency receivers. The system specifications for the stability of the satellite offset T_{GD}^p is ± 3 nsec (2 sigma) level. This poses a limit in accuracy to determine the absolute TEC, because 3 nsec correspond to 0.9 m delay or 8.5 TEC units. Mannucci et al. (1993) give additional details on potential limitations for absolute TEC measurements with dual-frequency receivers and review results accomplished thus far. They report on multi-site solutions which allows solving for receiver and satellite biases simultaneously with the TEC. The separation of the hardware delays and the TEC becomes, in principle, possible because of overlapping geometry (lines of sight from neighboring stations).

The ionospheric function for the carrier phases follows readily from (9.49) and (9.50) as

$$\varphi_{k,I}^p(t) \equiv \varphi_{k,1}^p(t) - \frac{f_1}{f_2}\, \varphi_{k,2}^p(t)$$

$$= N_{k,1}^p(t) - \frac{f_1}{f_2} N_{k,2}^p(t) - \frac{f_1}{c}\left(1 - \frac{f_1^2}{f_2^2}\right) I_{k,1,P}^p(t) \qquad (9.55)$$

The ionospheric function (9.55) reflects the time variation of the TEC. The hardware delays are not listed explicitly. This variation can be measured accurately because of high carrier phase resolution and small multipath (as compared to the code measurements). Unfortunately, expression (9.55) does not permit the estimation of the absolute TEC, because the initial ambiguities $N_{k,1}^p$ and $N_{k,2}^p$ are not known. Carrier phase hardware delays that follow (9.53) do not cancel in (9.55); these delays will be absorbed by the ambiguity estimate.

From the single-frequency solution (9.34), it is clear that the code and phase offsets cancel as long as the relations (9.43) and (9.53) are valid. The complete form is

$$R_{k,1}^p(t) \equiv P_{k,1}^p(t) - \Phi_{k,1}^p(t)$$

$$= 2I_{k,1,P}^p(t) - \left[c(T_{k,GD}^p - T_{GD}^p) \right.$$

$$\left. - c(T_{k,GD}^p - T_{GD}^p) + \frac{c}{f_1} N_{k,1}^p(1) \right] \qquad (9.56)$$

A similar equation can be written for the L2 carrier. Even though the relation (9.53) might not be valid for all receivers, single-frequency estimation of the ionosphere, at least in principle, is not affected by the code and carrier phase offsets of the receiver and the satellite. These offsets are absorbed with the ambiguity estimate. The most crucial element in single-frequency estimation of the ionosphere, therefore, is the quality of the ionospheric model in capturing the ionospheric variations with azimuth and altitude. The initial ambiguity and the code phase offsets cancel completely when differenced over time,

$$R^p_{k,1}(t_2, t_1) \equiv 2[I^p_{k,1,P}(t_2) - I^p_{k,1,P}(t_1)]$$

$$= \frac{2 \times 40.30}{f_1}[\text{TEC }(t_2) - \text{TEC }(t_1)] \qquad (9.57)$$

as long as the offsets are constant. The function (9.57) is often called the range difference equation. It is useful for cycle slip detection since slips cause jumps or outliers. Equations (9.56) and (9.57) apply to both carriers and all codes, including the C/A-code.

It is important to realize that both offsets, $T_{k,\text{GD}}$ and T^p_{GD}, and other constant hardware delays are eliminated in double differences. Because double differencing and relative positioning are generally anticipated, the hardware offset terms are usually not explicitly listed in the undifferenced code and carrier phase equations. They are, however, relevant to ionospheric researchers and to the single-frequency navigation solution. In the latter case they are part of the satellite clock correction.

9.4 TROPOSPHERIC EFFECTS ON GPS OBSERVABLES

The neutral atmosphere (troposphere, tropopause, and stratosphere) is a nondispersive medium. Its impact, here referred to as tropospheric refraction, does not depend on the frequency and, consequently, affects both the code modulation and the carrier phases in the same way. The effect is a delay (same sign as the ionosphere has on the codes) that reaches 2.0–2.5 m in the zenith direction and increases approximately with the cosecant of the elevation angle, yielding about a 20–28 m delay at a 5° angle. The measured ranges are longer than the geometric distance between the receiver and the satellite. The delay depends on the temperature, humidity and pressure, varies with the height of the user, and the type of terrain below the signal path. Because tropospheric refraction does not depend on the carrier frequency, it cannot be eliminated with a dual-frequency observation, unlike the ionospheric refraction.

Analogous to the ionospheric refraction expressions (9.20) and (9.22), the tropospheric refraction delay is written as

$$T^p_k = \int_{\text{path}} cN_T \, dt \qquad (9.58)$$

where N_T denotes the tropospheric refractivity. Since $ds = c(1 - N_T) \, dt$ and $N_T \ll 1$ equation (9.58) reduces to

$$T^p_k = \int_{\text{path}} N_T \, ds \qquad (9.59)$$

The expression for the tropospheric refractivity is based on the following empirical formula that includes meteorological parameters:

$$N_T = \left(77.624 \, \frac{p}{T} - 12.92 \, \frac{e}{T} + 371900 \, \frac{e}{T^2} \right) 10^{-6} \qquad (9.60)$$

where p denotes the total atmospheric pressure in mbar, e is the partial pressure of water vapor in mbar, and T is the surface temperature in kelvins. The numerical coefficients are those adopted from Essen and Froome (1951). The total atmospheric pressure p is the sum of the dry air pressure p_d and the water vapor pressure e. In terms of refractivity, therefore, equation (9.60) allows an unambiguous identification of "dry" refractivity (depending only on p_d) and "wet" refractivity (depending only on e). One can define the tropospheric delay by substituting (9.60) in (9.59) and integrating; the contribution of the first term in (9.60), which depends on the total pressure, is sometimes labeled the "dry" or hydrostatic component.

The dry atmosphere contributes about 90% of the total tropospheric refraction. It can be accurately modeled to about 2–5% using surface pressure and temperature. The various models of the dry atmosphere are based on the laws of ideal gases; they assume spherical layers, symmetrical refractivity with respect to the zenith direction, no temporal change in refractivity (which is not a function of pressure and temperature), and an effective height for the dry layer of about 40 km. The wet component is much more difficult to quantify, because water vapor cannot be accurately predicted and modeled. Even under normal conditions there are localized sources of water vapor, often in the form of liquid water. Therefore, these water vapor sources along with turbulence in the lower atmosphere cause variations in the concentration of water vapor that cannot be correlated over time or space. These variations cannot be accurately predicted from surface measurements. Fortunately, the wet contribution is only about 10% of the total tropospheric refraction. Despite the variability of water vapor, one way to model it is by an exponential vertical profile. The height of the wet layer is typically about 12 km. The wet delay is about 5–30 cm in continental midlatitudes. It can be modeled to about 2–5 cm. The combined models for dry and wet layers together predict the time delay caused by the neutral atmosphere.

Much research has gone into the creation and testing of tropospheric refraction models to compute N_T along the path of signal travel: Saastamoinen (1972, 1973), Hopfield (1969), Goad and Goodman (1974), Black (1978), and Robinson (1986). The various tropospheric models differ primarily with respect to the assumptions made on the vertical refractivity profiles and the mapping of the vertical delay with elevation angle. Summaries are found in Hofmann-Wellenhof et al. (1992) and Seeber (1993). Janes et al. (1991) compares several well-known tropospheric models. Yan et al. (1992) focus on ray-bending effects to improve model performance at low elevation angles. The expressions of Table 9.3 are taken from Goad and Goodman (1974). In this table the sub-

TABLE 9.3 Tropospheric Correction

$i=1$ dry component $(h_1 \equiv h_d)$	$i=2$ wet component $(h_2 \equiv h_w)$
p pressure in mb at height (h_p)	h_t height of temperature meas. in km
T temperature in degree Kelvin	h_h height of humidity meas. in km
h_k height of station. in km	h humidity (%) at height h_h
h_p height of pressure meas. in km	β elevation angle
Δ_r range correction (meter)	

$$a_e = 6378.137 \qquad x = T - 6.5(h_h - h_t) \qquad y = \frac{7.5(x - 273.15)}{237.3 + x - 273.15} \qquad e_0 = 0.0611 * h * 10^y$$

$$t_k = T + 6.5 h_t \qquad e_m = 5.2459587 \qquad e_s = e_0 * \left(\frac{t_k}{x}\right)^{4 e_m} \qquad t_e = T + 6.5(h_t - h_p)$$

$$p_{sea} = p(t_k / t_e)^{e_m} \qquad r_{sea} = (77.624 * 10^{-6} / t_k) * p_{sea}$$

$$\alpha_{1i} = 1 \qquad \alpha_{2i} = 4 a_i \qquad \alpha_{3i} = 6 a_i^2 + 4 b_i \qquad \alpha_{4i} = 4 a_i \left(a_i^2 + 3 b_i\right)$$

$$\alpha_{5i} = a_i^4 + 12 a_i^2 b_i + 6 b_i^2 \qquad \alpha_{6i} = 4 a_i b_i \left(a_i^2 + 3 b_i\right) \qquad \alpha_{7i} = b_i^2 \left(6 a_i^2 + 4 b_i\right)$$

$$\alpha_{8i} = 4 a_i b_i^3 \; ; \quad \alpha_{9i} = b_i^4 \; ; \quad a_i = -\left(\frac{\sin\beta}{h_i - h_k}\right) \; ; \quad b_i = -\frac{1}{2 a_e} * \frac{\cos^2\beta}{(h_i - h_k)}$$

$$R_i = \sqrt{(a_e + h_i)^2 - (a_e + h_k)^2 \cos^2\beta} \; - (a_e + h_k)\sin\beta$$

$$h_d = (1.1385 * 10^{-5} / r_{sea}) \qquad h_w = 1.1385 * 10^{-5} [(1255 / t_k + 0.05) / r_{sea}]$$

$$N_1 = r_{sea}\left(\frac{h_d - h_k}{h_d}\right)^4 \qquad N_2 = \frac{\left(\dfrac{0.371900}{t_k} - 12.92 * 10^{-6}\right)}{t_k} * e_s * \left(\frac{h_w - h_k}{h_w}\right)^4$$

$$\Delta_r = 10^3 \left[N_1 \sum_{j=1}^{9} \frac{\alpha_{j1}}{j} R_1^j + N_2 \sum_{j=1}^{9} \frac{\alpha_{j2}}{j} R_2^j \right]$$

scripts 1 and 2 refer to the dry and wet components, respectively. A graphical display for typical tropospheric delay values is given in Figure 9.7. The zenith delay is obtained by substituting a vertical angle of 90°. These atmospheric models should not be used at vertical angles below 5–10°.

Actually, one can attempt to measure the delay caused by water vapor along the line of transmission. Such measurements will probably be necessary if the highest possible accuracy in GPS measurements is to be achieved. The instru-

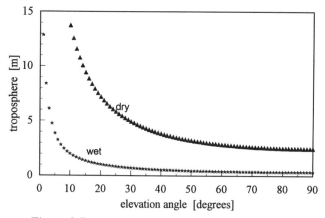

Figure 9.7 Dry and wet tropospheric model delays.

ments capable of measuring water vapor and water liquid content are called water vapor radiometers (WVR). Davis (1986) describes in detail the physical principles involved in water vapor radiometry. The WVRs usually operate with two frequencies: near 22 GHz to measure water vapor and near 31 GHz to measure liquid water. With WVR, the wet atmospheric delay can be computed to 1–2 cm accuracy. WVRs are very expensive and are unlikely to become a standard in practical GPS surveying. Alternatively, radiosondes measure pressure, temperature, and humidity along their vertical path. Water vapor pressure can be computed from these values. Unfortunately, radiosondes drift with varying wind conditions and require considerable logistics to operate. Like WVRs, radiosondes are expensive and are unlikely to be used in routine GPS surveying.

Because of the difficulty in measuring water vapor, one may try to estimate the tropospheric scale bias as part of the station position estimation. Generally, the solutions for tropospheric delay are expressed as a product of the vertical delay $T_k(t)$ and a mapping function $m(a_k^p)$. For example,

$$T_k^p(t) = T_k(t)m(a_k^p) \tag{9.61}$$

The mapping function relates the vertical delay to the delay of other elevation angles. Such mapping functions typically do not depend on azimuth, because atmospheric layers are assumed to be spherically symmetric. A simple model for the mapping function is $1/\sin a_k^p$, which is similar to the one considered for the ionosphere. Because the refractivity changes with height, and because the earth is only approximately spherical, more elaborate mapping functions have been devised. An example is the Marini model (Davis 1986):

$$T_k^p(t) = (1 + \delta_k)T_k(t) \frac{1 + \kappa}{\sin a_k^p + \dfrac{\kappa/(1 + \kappa)}{\sin a_k^p + 0.015}} \tag{9.62}$$

where $\kappa = B(\varphi, h)/T_k(t)$, and B being a small factor which varies as a function of station latitude and height. The a priori value of the vertical delay is computed from any of the tropospheric models. The tropospheric scale bias δ_k is estimated. When computing the partial derivatives, the ratio κ is treated as a constant. Caution must be exercised against overparameterization when estimating the tropospheric scale and the station coordinates. There will always be a high correlation between the estimated tropospheric scale bias and the vertical station coordinate (height). Longer satellite arcs reduce this correlation.

The expression (9.62) represents a constant delay model where a single estimate of the vertical delay is obtained per arc per station. Because of the spatial and temporal variation of water vapor, much attention has been given to modeling the vertical delay by stochastic processes. For example, Tralli and Lichten (1990) use first-order Gauss–Markov and random walk temporal models of tropospheric path delay and compare the results with WVR measurements. Their conclusion is that GPS measurements alone have sufficient strength to resolve centimeter-level zenith path delay fluctuations over periods of a few minutes. Tranquilla and Al-Rizzo (1993) have developed a model to study relative carrier phase positioning under the condition of extremely high ice cloud volume concentrations and snowfall rates.

Finally, surface meteorological data should be used with caution. Typical field observations can be influenced by ''surface layer biases'' introduced by micrometeorological effects. The measurements at the earth's surface are not necessarily representative of adjacent layers along the line-of-sight to the satellites. Temperature inversion can occur during nighttime when the air layers close to the ground are cooler than the higher air layers caused by the ground's radiation loss. Convection can occur during noontime when the sun heats the air layers near the ground. Occasionally, it might be better to use standard atmospheric values rather than locally measured values.

9.5 MULTIPATH AND PHASE CENTER

Multipath is a major source of error to be concerned with in GPS positioning. Satellite signals can arrive at the receiver via multiple paths due to reflection. Multipath distorts the C/A-code and P-code modulations and the carrier phase observations. Although multipath signals have common emission time at the satellite, they arrive with code and carrier phase offsets due to reflection differences in their path length. Multipath signals are always delayed compared to line-of-sight signals because of the longer travel paths caused by the reflection. The composition signal arriving at the antenna is processed by the receiver. Since the geometry between the GPS satellites and a specific receiver-reflector location repeats every sidereal day, multipath shows the same pattern between consecutive days. This repetition can be useful for verifying the presence of multipath by analyzing observations from different days. The impact of multipath signals on the pseudorange or the carrier phase observables depends on a variety of factors, such as the strength and the delay of the reflected signal as compared to the line-of-sight signal, the attenuation characteristics of the antenna, and the sophistication of the receiver's measuring and processing techniques. Often a local reflector imposes its dominant signature on the multipath. In general, however, multipath exhibits random low-frequency and high-frequency features.

Signals can be reflected at the satellite (satellite multipath) or in the surroundings of the receiver (receiver multipath). Satellite multipath is likely to cancel in the single-difference observables for short baselines. The reflected signal is always weaker because of attenuation at the reflector. This attenuation depends on the material of the reflector, the incident angle of the reflection and the polarization. In general, reflections with very low incident angle have practically no attenuation. This explains why satellites at a low elevation angle tend to generate strong multipath interference. Reflective objects for ground receivers can be the earth's surface itself (ground and water), buildings, trees, hills, etc. Rooftops are bad multipath environments since there are often many vents and other reflective objects within the GPS antenna field of view. For a moving antenna in kinematic applications, the effect of multipath changes rapidly because of changing geometry between the antenna and the surrounding reflective objects. Airborne receivers may experience additional multipath generated from reflection off the airframe. Highly reflective surfaces change the polarization from right-hand to left-hand circular. GPS antennas that are designed for right-hand polarized signals will attenuate signals of opposite polarization.

Antennas do not receive signals equally well from all directions. Partial multipath rejection can be built into the antenna by shaping the gain pattern. Since most multipath arrives from angles near the horizon, multipath may be sharply reduced and eventually rejected by shaping the pattern to have low gain in these directions. Such antennas may be acceptable for ground receivers but may become a problem for airborne applications. Airborne antennas must continue to receive the satellite signals while undergoing dynamics, i.e., loss of lock must be avoided while the aircraft is banking. Multipath arriving from below the antenna can be significant. Depending on the type of antenna used, a ground plane may be required. Energy originating from below the ground plane enters the antenna via edge diffraction. Usually the ground plane is a metallic surface of circular or rectangular form. Improved multipath resistance is achieved with choke rings. These are metallic circular fins at a spacing of about one-half of the carrier wavelength.

The impact of multipath on the carrier phase observable can be highlighted with the help of a very simple example. The following case serves to demonstrate multipath on the L1 and L2 carrier phases and the correlation between multipath and ionosphere. See Georgiadou and Kleusberg (1988) and Bishop et al. (1985) for further details. Figure 9.8 shows the situation where the signal is reflected on a planar surface. To emphasize the role of signal strength, the carrier phase observable is written as

$$S_{k,D}^p = V_k^p \cos \varphi_k^p \tag{9.63}$$

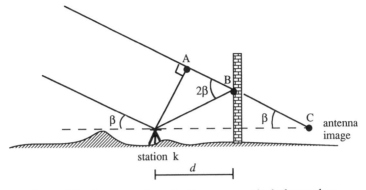

Figure 9.8 Geometry for reflection on a vertical planar plane.

where $S_{k,D}^P$ is the direct line-of-sight carrier signal. The symbols V_k^P and φ_k^P denote the signal voltage and the phase, respectively. The reflected signal is written as

$$S_{k,R}^P = \alpha_k^P V_k^P \cos(\varphi_k^P + \theta_k^P) \qquad (9.64)$$

The amplitude reduction factor α_k^P fulfills the inequality $0 \le \alpha_k^P \le 1$. The phase shift caused by multipath is θ_k^P. The composite signal at the antenna consists of the superposition

$$S_k^P = S_{k,D}^P + S_{k,R}^P = R_k^P \cos(\varphi_k^P + \psi_k^P) \qquad (9.65)$$

where the resultant carrier phase voltage $R_k^P(V_k^P, \alpha_k^P, \theta_k^P)$ is

$$R_k^P(V_k^P, \alpha_k^P, \theta_k^P) = V_k^P[1 + 2\alpha_k^P \cos \theta_k^P + (\alpha_k^P)^2]^{1/2} \qquad (9.66)$$

and the carrier phase multipath delay $\psi_k^P(\alpha_k^P, \theta_k^P)$ is

$$\psi_k^P(\alpha_k^P, \theta_k^P) = \tan^{-1}\left(\frac{\alpha_k^P \sin \theta_k^P}{1 + \alpha_k^P \cos \theta_k^P}\right) \qquad (9.67)$$

Considering the case of constant reflectivity, $\alpha_k^P = $ constant, the maximum path delay fulfills the condition $\partial \psi_k^P / \partial \theta_k^P = 0$. The maximum value is $\psi_{max} = \pm \sin^{-1} \alpha_k^P$, which occurs at $\theta_k^P(\psi_{max}) = \pm \cos^{-1}(-\alpha_k^P)$. The maximum multipath carrier phase error is thus only a function of the signal strength ratio α_k^P. The maximum phase error is $\pm 90°$ and occurs for $\alpha_k^P = 1$, which is the case of no attenuation for the reflected signal. This maximum error corresponds to $\lambda_1/4 \approx 4.8$ cm for the L1 carrier.

The multipath delay shown in Figure 9.8 is the sum of the distances AB and BC. Converting this sum into L1 cycles and converting to radians gives

$$\theta_{k,1}^P = \frac{4\pi d}{\lambda_1} \cos \beta_k^P \qquad (9.68)$$

The multipath error frequency f_ψ is the same as that of the phase delay θ_k^P, as can be seen from (9.67). Differentiating (9.68) gives the expression for the multipath error frequency for L1 as

$$f_{\psi,1} \equiv f_{\theta,1} = \frac{1}{2\pi} \frac{d\theta_{k,1}^P}{dt} = \frac{2d}{\lambda_1} \sin \beta_k^P |\dot{\beta}_k^P| \tag{9.69}$$

It is seen that the multipath frequency is proportional to the perpendicular distance of the reflector, is inversely proportional to the wavelength, and is a function of the satellite elevation angle. Since the satellite is continuously moving, the multipath frequency is a function of time. It is helpful to evaluate expression (9.69) for some typical values. For example, if $d = 10$ m, $\beta_k^P = 45°$, and $\dot{\beta}_k^P = 0.07$ mrad/sec (= one-half of the satellite's mean motion), then the period of the multipath error is about 5 min. If $d = 1$ m, the respective period is 50 min. According to (9.69) the ratio of the multipath frequencies for L1 and L2 equals the ratio of the carrier frequencies, $f_{\psi,1}/f_{\psi,2} = f_1/f_2$.

The ionospheric carrier phase observable is given by (9.55). The corresponding effect of the multipath is

$$\begin{aligned}
\varphi_{k,MP}^P &\equiv \psi_{k,1}^P - \frac{f_1}{f_2} \psi_{k,2}^P \\
&= \frac{1}{2\pi} \left[\tan^{-1} \left(\frac{\alpha_k^P \sin \theta_{k,1}^P}{1 + \alpha_k^P \cos \theta_{k,1}^P} \right) \right. \\
&\quad \left. - \frac{f_1}{f_2} \tan^{-1} \left(\frac{\alpha_k^P \sin \theta_{k,2}^P}{1 + \alpha_k^P \cos \theta_{k,2}^P} \right) \right]
\end{aligned} \tag{9.70}$$

Figure 9.9 demonstrates that the multipath $\varphi_{k,MP}^P$ might impact the ionospheric observable in a complicated manner. The figure refers to $d = 10$ m and to three different power ratios α_k^P. The amplitude of the cyclic phase variation is nearly proportional to the power ratio. It is readily verified that the frequency decreases with decreasing distance d. In practical situations one expects several reflections of the signal to occur and power ratios to be time dependent. The receiver will attempt to relate the internal receiver-generated phase to the average of the received phases. All of this makes multipath of the carrier phase observables a difficult error source to deal with. When analyzing the ionospheric–free observable in order to map the time variation of the ionospheric delay, the multipath signatures of (9.70) cannot be ignored. The variation of (9.67) will make it occasionally impossible to fix the integer ambiguities, even for short baselines.

Pseudorange multipath behaves very much like that of carrier phases, except that the variation is several orders of magnitude larger. Similar to the carrier phase, in which the maximum multipath is a fraction of the wavelength, multipath for pseudoranges is related to the chipping rate, and is a function of the

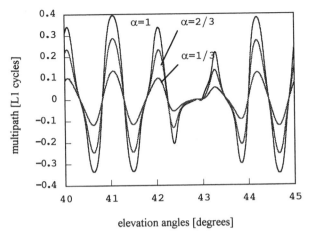

Figure 9.9 Example of multipath effects on the ionospheric carrier phase observable from reflection on a planar surface.

length of the codes. The chipping rate sets a natural limit for the maximum multipath. The higher the chipping rate, the lower the maximum multipath. According to this general rule, the expected multipath on the P-code pseudoranges (chipping rate 10.23 MHz) is smaller than for C/A-code pseudoranges (chipping rate 1.023 MHz). The same dependency is true for the pseudorange measurement accuracy. Because of the large size of the multipath on pseudoranges, much effort has been made to refine receiver processing algorithms both to reduce the threshold for multipath detection and rejection and simultaneously to improve the pseudorange measurement accuracy. Meehan and Young (1992) discuss four on-receiver processing methods to reduce carrier phase and pseudorange multipath. Similar efforts have resulted in a new generation of C/A-code receivers that make use of narrow correlator spacing techniques. See van Dierendonck et al. (1992) for details on the theory and performance of narrow correlator technology. Fenton et al. (1991) discuss one of the first low-cost receivers employing narrow correlation and provide additional background on the theory behind achievable pseudorange accuracy.

Finally, in addition to multipath interference, the antenna may show phase center variation and may be affected by antenna imaging. Phase center variation is primarily a result of the nonspherical phase response pattern of the antenna. The measured phase of the same incoming signal depends on the direction of the incidence angle. Antenna imaging occurs when another conducting body in the immediate vicinity becomes electrically a part of the antenna. Imaging can cause rapid and complex phase variations. The metal structure of a vessel or airplane might contribute to imaging. Furthermore, the L1 and L2 frequencies may show slightly different phase variation properties, and consequently have no common phase center, at least not at the subcentimeter to millimeter level. The difficulty in accurately relating the phase center with the geometric center still prevents GPS from being a truly universal millimeter measurement tool even over short distances, this despite the high carrier phase measurement resolution. To minimize the phase variation and phase centering problem, manufacturers usually recommended that the same type of antenna be used and that the antenna be oriented consistently in the same direction. There is still much need for improved antenna designs; some of the required antenna characteristics depend on the application, for example, of airborne kinematic versus ground-based static. Whereas a uniform phase response, i.e., a spherical phase pattern, is very desirable, it conflicts with the need to "blind" the antenna for low elevation angles in order to desensitize it to multipath signals.

CHAPTER 10

PROCESSING GPS CARRIER PHASES

This chapter deals primarily with double- and triple-difference observations, which are used to determine the vector between simultaneously observing receivers. The chapter begins with a general discussion of dual-frequency code and carrier phase observations and their geometric content per epoch. It is assumed that full wavelength dual-frequency carrier phases and pseudoranges are available. It is up to the reader to modify the results, for example, for the case of half-wavelength L2 carrier phases. For example, Goad (1992) discusses solutions for the case of two carrier phases and one pseudorange.

The second section gives the observation equations for double and triple difference and addresses methods for coping with cycle slips. A large portion of the section is devoted to various ambiguity search and fixing schemes and the discernibility of solutions. The discussion includes the methods that utilize primary and secondary satellites, normal points, and the ambiguity function. The section does not elaborate much on the distinction between static GPS, rapid static GPS, and pseudostatic GPS. Rather, the view is taken that all three terms describe the same physics of GPS surveying, with the difference being merely algorithmic implementation and refinement, and the availability of suitable observations.

The section on kinematic GPS makes heavy use of developments given in the static GPS section. This is done to emphasize the unified approach to GPS surveying and navigation which diminish the distinction between static and kinematic approaches. The kinematic section, which deals with the search volume and propagation of trial positions, concludes with a general Kalman filtering formulation.

Special consideration to long baselines is given in the fourth section, where the base receiver and base satellite scheme is generalized, weighting of the

317

residual double-differenced ionosphere is discussed, and orbital relaxation is presented in order to achieve highest accuracy for long baselines.

If more than two receivers observe at the same time, we speak of a session network consisting of all co-observing stations. The result is a set of correlated vectors between the stations. This is a generalization of the two-receiver arrangement, which yields separate vectors. For economic reasons, it might be preferable to use more than two receivers at the same time. A complete survey network may thus consist of sets of correlated vectors from various sessions.

10.1   GETTING ACQUAINTED WITH THE OBSERVABLES

The observables and the functions that were derived in Chapters 8 and 9 are at the base of any GPS positioning algorithm. This section focuses on the geometric content of some of these observables and functions. The epoch solution, which uses observations from just one epoch, is analyzed for dual-frequency codes and carrier phases. Several figures are used to illustrate the behavior of the observables and functions with time, in order to identify the geometric strength and weakness and to demonstrate adverse effects, such as hardware delays and multipath. Brief remarks are made regarding OTF (on-the-fly) ambiguity fixing and the possibility of using OTF as a unifying approach for positioning with GPS in surveying and navigation. The illustrations use two data sets collected by students for classroom demonstration at the University of Maine at Orono. In Experiment 1 (DOY 356, 1992, stations LAND and TIDE) the stationary receivers collected data every second for at least three hours, and in Experiment 2 (DOY 014, 1993, stations HOME and AWAY) an OFT solution was carried out over a short baseline in the static and the kinematic modes.

10.1.1   Dual-Frequency Epoch Solutions

Early GPS surveying of a decade ago made use of codeless receiver technology; e.g., Collins and Leick (1985) describe an early geodetic network densification with GPS using codeless technology. Codeless technology was favored for these first-generation receivers because of the fear that antispoofing (AS) would render the codes unusable for civilians. Active antispoofing converts the P-code to the unknown Y-code. The following generations of GPS receivers use C/A-codes (which are not subject to the antispoofing measures). It has long been understood that full implementation of P-code technology will yield the best receiver performance, simply because all possible observables are available at the same time. Because of the fear of AS, the P-code technology has developed slowly. Only in recent years has major progress been reported with P-code technology under AS conditions. See Chapter 3 for details on solutions using the encrypted P-code. Every major receiver manufacturer has developed techniques to deal satisfactorily with AS.

10.1.1.1 Geometry of Epoch Solutions The undifferenced pseudorange and carrier phase equations of Table 8.3 together constitute the epoch solution. The epoch solution for station k and satellite p at epoch t is written in the following form:

$$
\begin{bmatrix} P^p_{k,1}(t) \\ P^p_{k,2}(t) \\ \Phi^p_{k,1}(t) \\ \Phi^p_{k,2}(t) \end{bmatrix} = \begin{bmatrix} 1 & 1 & 0 & 0 \\ 1 & \dfrac{f_1^2}{f_2^2} & 0 & 0 \\ 1 & -1 & \lambda_1 & 0 \\ 1 & \dfrac{-f_1^2}{f_2^2} & 0 & \lambda_2 \end{bmatrix} \begin{bmatrix} \rho^p_k(t) \\ I^p_{k,1,P}(t) \\ N^p_{k,1}(1) \\ N^p_{k,2}(1) \end{bmatrix} + \begin{bmatrix} 1 \\ 1 \\ 1 \\ 1 \end{bmatrix} \Delta^p_k(t)
$$

$$
+ \begin{bmatrix} d_{k,1,P}(t) + d^p_{k,1,P}(t) + d^p_{1,P}(t) + \epsilon_{1,P} \\ d_{k,2,P}(t) + d^p_{k,2,P}(t) + d^p_{2,P}(t) + \epsilon_{2,P} \\ d_{k,1,\Phi}(t) + d^p_{k,1,\Phi}(t) + d^p_{1,\Phi}(t) + \epsilon_{1,\Phi} \\ d_{k,2,\Phi}(t) + d^p_{k,2,\Phi}(t) + d^p_{2,\Phi}(t) + \epsilon_{2,\Phi} \end{bmatrix} \tag{10.1}
$$

where the symbol $\Delta^p_k(t)$ consists of the clock terms and the tropospheric terms as follows

$$
\Delta^p_k(t) = \dot{\rho}^p_k(t)\, dt_k - c\, dt_k + c\, dt^p + T^p_k(t) \tag{10.2}
$$

Following the standard convention, the carriers L1 or L2 are distinguished by the subscripts 1 or 2. The unknowns are the topocentric distance $\rho^p_k(t)$, the ionospheric delay for P1-codes $I^p_{k,1,P}(t)$, and the ambiguities $N^p_{k,1}(1)$ and $N^p_{k,2}(1)$. The carrier phases are expressed in terms of the scaled values $\Phi^p_{k,1}(t)$ and $\Phi^p_{k,2}(t)$. The unknown correction $\Delta^p_k(t)$ is the same for all four observables. Inclusion of the range-rate term $\dot{\rho}^p_k(t)$ is not of particular significance other than keeping (10.2) more general.

The ionosphere is a parameter in the epoch solution. It automatically includes the effects of the satellite and receiver code and phase hardware offsets as long as these follow equations (9.43) and (9.53). The ionosphere parameter $I^p_{k,1,P}(t)$ can readily be replaced with

$$
\tilde{I}^p_{k,1,P}(t) = I^p_{k,1,P}(t) + cT_{k,\text{GD}} - cT^p_{\text{GD}} \tag{10.3}
$$

One needs only to replace the ionosphere $I^p_{k,1,P}(t)$ with the more general function $\tilde{I}^p_{k,1,P}(t)$ in all expressions of this section. Such a simple procedure is, unfortunately, not possible with the other hardware delay terms and multipath terms listed in (10.1).

Equation (10.1) is valid regardless of whether or not the receiver is in motion. The topocentric distance and the ionosphere change with time, whereas the ambiguity parameters are constant as long as there is no cycle slip. The correction term $\Delta_k^p(t)$ is, of course, a function of time. In terms of matrix notation, equation (10.1) can be written as

$$L_b = AX + D + D' \tag{10.4}$$

The association between the terms of (10.1) and (10.4) is readily apparent. The elements D and D' can be viewed as real-world disturbances in an otherwise simple mathematical relationship. Since the A matrix is not singular the solution

$$X = A^{-1}L_b - A^{-1}D - A^{-1}D' \tag{10.5}$$

exists for every epoch. The general form of the inverse matrix is

$$A^{-1} = \frac{1}{f_1^2 - f_2^2}$$

$$\cdot \begin{bmatrix} f_1^2 & -f_2^2 & 0 & 0 \\ -f_2^2 & f_2^2 & 0 & 0 \\ -\lambda_1^{-1}(f_1^2 + f_2^2) & 2\lambda_1^{-1}f_2^2 & \lambda_1^{-1}(f_1^2 - f_2^2) & 0 \\ -2\lambda_2^{-1}f_1^2 & \lambda_2^{-1}(f_1^2 + f_2^2) & 0 & \lambda_2^{-1}(f_1^2 - f_2^2) \end{bmatrix}$$

$$\tag{10.6}$$

as can readily be verified by computing $AA^{-1} = I$. An interesting feature of the epoch solution is that the product

$$A^{-1}D = [\Delta_k^p(t), 0, 0, 0]^T \tag{10.7}$$

contains three zero components. The implication is that the computed ionosphere and the two ambiguities do not depend on the receiver and satellite clock errors or on the tropospheric delay. However, the clock errors and the troposphere impact the computed topocentric distance in full. Because the topocentric distance could eventually be parameterized in terms of receiver coordinates, it follows that the estimated receiver positions will reflect these dependencies. There are no zero elements in $A^{-1}D'$.

The geometry of the epoch solution is implicit in the covariance matrix. As usual the observations L_b are assumed to be uncorrelated. It is further assumed that the ratio of the standard deviations of the pseudorange and the carrier phase for the same carrier are a constant, denoted here by k. The third assumption regarding the stochastic model is that the carrier phases have the same phase resolution, i.e., the relation $\sigma_{2,\Phi} = (f_1/f_2)\sigma_{1,\Phi}$ is valid for the standard deviations of the scaled carrier phases. With these assumptions the covariance matrix of the observations is

$$\Sigma_{L_b} = \sigma_{1,\Phi}^2 \begin{bmatrix} k^2 & 0 & 0 & 0 \\ & \dfrac{f_1^2}{f_2^2} k^2 & 0 & 0 \\ & & 1 & 0 \\ \text{sym} & & & \dfrac{f_1^2}{f_2^2} \end{bmatrix} \qquad (10.8)$$

Applying the law of variance–covariance propagation (4.34), the covariance matrix of the parameters becomes

$$\Sigma_X = A^{-1} \Sigma_{L_b}(A^{-1})^{\mathrm{T}} \qquad (10.9)$$

The elements of matrix Σ_X are listed in Table 10.1. Each of the elements must be multiplied with the factor F listed in element $(4, 1)$. The specification of the numerical elements in (10.8) entails some degree of arbitrariness. The element k, which is simply the ratio of the standard deviations for the codes and the carrier phases,

$$k = \frac{\sigma_{i,P}}{\sigma_{i,\Phi}} \qquad (10.10)$$

TABLE 10.1   Variance–Covariance Matrix of Epoch Solution

$f_1^2\left(f_1^2 + f_2^2\right)$	$-2f_1^2 f_2^2$	$-\lambda_1^{-1} f_1^2\left(f_1^2 + 3f_2^2\right)$	$-\lambda_1^{-1} f_1 f_2\left(3f_1^2 + f_2^2\right)$
	$f_2^2\left(f_1^2 + f_2^2\right)$	$f_2^2\left(f_1^2 + f_2^2\right)$	$\lambda_1^{-1} f_1 f_2\left(f_1^2 + 3f_2^2\right)$
	symmetric	$\lambda_1^{-2}\left(f_1^4 + 6f_1^2 f_2^2 + f_2^4\right)$ $+\lambda_1^{-2} k^{-2}\left(f_1^2 - f_2^2\right)^2$	$4\lambda_1^{-2} f_1 f_2\left(f_1^2 + f_2^2\right)$
$F = \dfrac{\sigma_{1,\Phi}^2 k^2}{\left(f_1^2 - f_2^2\right)^2}$	F is a multiplication factor		$\lambda_1^{-2}\left(f_1^4 + 6f_1^2 f_2^2 + f_2^4\right)$ $+\lambda_1^{-2} k^{-2}\left(f_1^2 - f_2^2\right)^2$

is set to $k = 154$, which corresponds to the ratio of the L1 frequency and the P-code chipping rate. Using $\sigma_{1,\Phi} = 0.002$ m and k as specified, the standard deviations of the parameters are

$$(\sigma_p, \sigma_I, \sigma_{N,1}, \sigma_{N,2}) = (0.99 \text{ m}, 0.77\text{m}, 9.22 \text{ cycL}_1, 9.22 \text{ cycL}_2) \quad (10.11)$$

and the correlation matrix is

$$C_X = \begin{bmatrix} 1 & -0.9697 & -0.9942 & -0.9904 \\ & 1 & 0.9904 & 0.9942 \\ & & 1 & 0.9995 \\ & \text{sym} & & 1 \end{bmatrix} \quad (10.12)$$

Striking features of the epoch solution are the equality of the standard deviation for both ambiguities and the high correlation between all parameters. Of particular interest here is the shape and orientation of the ellipse of standard deviation for the ambiguities. The general expressions (4.312) to (4.317) from adjustment theory also apply in this case. The computations show that the ellipse of the ambiguities $N^p_{k,1}$ and $N^p_{k,2}$ almost degenerates into a straight line. The semiminor and semimajor axes are 0.20 and 13.04, respectively, and the orientation is exactly $45°$ (counted positive in a clockwise direction from the $N^p_{k,1}$ axis.) The $N^p_{k,1}$ and $N^p_{k,2}$ axes, which carry the units L1 cycles and L2 cycles, are perpendicular. This degenerated shape of the ellipse of standard deviation was also pointed out by Melbourne (1985).

The geometry is the same for every epoch solution, as long as the stochastic model (10.8) remains unchanged. Therefore, all parameters computed from solution (10.5) will display highly correlated variations over time in addition to being corrupted by $A^{-1}D$ and $A^{-1}D'$. A standard procedure for breaking high correlation is reparameterization by means of an appropriate transformation. Usually there are several possibilities for such reparameterization. For example, consider the following transformation:

$$Z = TX = TA^{-1}L_b - TA^{-1}D - TA^{-1}D' \quad (10.13)$$

with

$$T = \begin{bmatrix} 1 & 0 & 0 & 0 \\ 0 & 1 & 0 & 0 \\ 0 & 0 & 1 & -1 \\ 0 & 0 & 1 & 0 \end{bmatrix} \quad (10.14)$$

It is readily verified that the ionosphere and the ambiguity parameters, again, do not depend on the correction term $\Delta_k^p(t)$ because

$$TA^{-1}D = [\Delta_k^p, 0, 0, 0]^T \tag{10.15}$$

The transformed epoch solution has the following parameters and covariance matrix

$$Z = [\rho_k^p, I_{k,1,P}^p, N_{k,1}^p - N_{k,2}^p, N_{k,1}^p]^T \tag{10.16}$$

$$\Sigma_Z = TA^{-1}\Sigma_{L_b}(TA^{-1})^T \tag{10.17}$$

Using again the numerical values $\sigma_{1,\Phi} = 0.002$ m and $k = 154$, the standard deviations and the correlation matrix become

$$(\sigma_\rho, \sigma_I, \sigma_{N,w}, \sigma_{N,1}) = (0.99 \text{ m}, 0.77 \text{ m}, 0.28 \text{ cycL}_w, 9.22 \text{ cycL}_1) \tag{10.18}$$

$$C_Z = \begin{bmatrix} 1 & -0.9697 & -0.1230 & -0.9942 \\ & 1 & 0.1230 & 0.9904 \\ & & 1 & 0.0154 \\ & & & 1 \end{bmatrix} \tag{10.19}$$

The third parameter in (10.16) is the difference of the original ambiguities and is called the wide-lane ambiguity,

$$N_{k,w}^p = N_{k,1}^p - N_{k,2}^p \tag{10.20}$$

The subscript w will indicate "wide lane" henceforth. The standard deviation for $N_{k,1}^p$ is not affected by the transformation (10.14). More importantly, there is little correlation between the wide-lane and L1 ambiguities. Furthermore, the correlations between the wide-lane ambiguity and both the topocentric distance $\rho_k^p(t)$ and the ionosphere $I_{k,1,P}^p(t)$ have been reduced dramatically. Considering the small standard deviation for the wide-lane ambiguity in (10.18) and its low correlations with the other parameters, it should be possible to estimate the wide-lane ambiguity from epoch solutions. The ellipse of standard deviation for the ambiguities has a semimajor and a semiminor axis of 9.22 and 0.28, respectively, and an orientation of 89.97°. The semimajor axis is essentially perpendicular to the direction of the wide-lane ambiguity and is almost aligned with the $N_{k,1}^p$ ambiguity axis. Analysis shows that the shape and orientation of the ellipse have little sensitivity in regard to varying the fundamental frequencies f_1 and f_2. The matrix (10.19) still shows high correlations between $N_{k,1}^p$, the ionosphere, and the topocentric distance.

10.1.1.2 Wide Laning With Undifferenced Observations Proper reparameterization and variance–covariance propagation can reduce the correlation between parameters and, consequently, can decrease the variances of some of the new parameters. This is desirable, although it does not tell the whole story. It is always possible to relate mathematically the various transformations and respective statistics. Of course, the case can arise that the correlations are so high that the numerical stability of the inverse is of concern as the matrix becomes near singular. The advantage of reparameterization is the ability it provides to determine a subset of parameters accurately in order to allow decisions to be made. For example, it might be possible to estimate the wide-lane ambiguity in (10.16) and to use the estimated real value and its statistics to infer the correct integer, and then to apply an integer constraint in subsequent solutions. This procedure makes it possible to introduce into the solution our a priori knowledge that the ambiguity is indeed an integer. The ambiguity estimate is likely to differ from an integer the larger the model errors and the worse the observational accuracy. Constraining some of the ambiguities to integers, therefore, can also be viewed as a way to "push the mathematical model into the correct direction." Constraining (fixing) some ambiguities to integers might result in a "chain reaction," in that subsequent solutions identify the integer values of the remaining ambiguities more clearly. Identifying the fixing integer ambiguities is still a challenge in many applications and remains a fruitful area for experimentation. Success in fixing ambiguities is not guaranteed.

Graphical displays of the observables, or their simple functions, are useful for developing and understanding ambiguity fixing and, at the same time, for assessing the overall quality of the observations and for detecting effects such as multipath. The following figures use three hours of data recorded at a one-second interval. At the beginning of the observation session the satellite was near the zenith, and it was close to the mask angle at the end. The antenna was mounted on a regular tripod on the roof of an office building, no ground plate extension was used to reduce the multipath. All functions discussed and shown below do not depend on the troposphere or the station and satellite clock errors.

The ionospheric delay is readily available for the code and carrier observations. For example, the ionospheric delay for the P1-code follows directly from (10.5) as

$$
\begin{aligned}
I_{k,1,P}^p(t) = -&\frac{f_2^2}{f_1^2 - f_2^2} \left[P_{k,1}^p(t) - P_{k,2}^p(t) \right] \\
-&\frac{f_2^2}{f_1^2 - f_2^2} \left[d_{k,1,P}(t) - d_{k,2,P}(t) + d_{k,1,P}^p(t) \right. \\
&\left. - d_{k,2,P}^p(t) + d_{1,P}^p(t) - d_{2,P}^p(t) + \epsilon_{1,P} - \epsilon_{2,P} \right] \\
\approx -&1.545 \left[P_{k,1}^p(t) - P_{k,2}^p(t) \right] + \cdots
\end{aligned}
\tag{10.21}
$$

The transformation (10.14) does not affect the ionosphere delay expression. An alternative expression follows from the third and fourth equations of (10.1),

$$
\begin{aligned}
I^p_{k,1,P}(t) &= \frac{f_2^2}{f_1^2 - f_2^2} \left[\Phi^p_{k,1}(t) - \Phi^p_{k,2}(t) + \lambda_1 N^p_{k,1}(1) - \lambda_2 N^p_{k,2}(1) \right] \\
&+ \frac{f_2^2}{f_1^2 - f_2^2} \left[d_{k,1,\Phi}(t) - d_{k,2,\Phi}(t) + d^p_{k,1,\Phi}(t) - d^p_{k,2,\Phi}(t) \right. \\
&\left. + d^p_{1,\Phi}(t) - d^p_{2,\Phi}(t) + \epsilon_{1,\Phi} - \epsilon_{2,\Phi} \right] \\
&\approx 1.545 \left[\Phi^p_{k,1}(t) - \Phi^p_{k,2}(t) \right] + 0.294 N^p_{k,1}(1) \\
&- 0.377 N^p_{k,2}(1) + \cdots
\end{aligned}
\tag{10.22}
$$

Note the sense of the observation differences in (10.21) and (10.22), which accounts for the fact that the pseudoranges and the carrier phases are affected by the ionosphere in opposite direction. Figure 10.1 shows both ionospheric functions. The carrier phase difference curve is shifted by an arbitrary constant to account for the unknown ambiguities $N^p_{k,1}$ and $N^p_{k,2}$. Since both curves in Figure 10.1 depicts the same ionospheric variation, and since the ionosphere is expected to vary smoothly, the rapid fluctuations displayed by the code function must have origins other than the ionosphere. Assuming constant receiver and satellite hardware delays over this observation time, the variations are

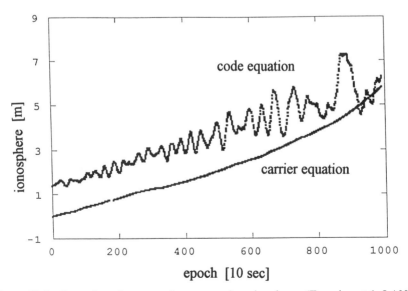

Figure 10.1 Ionosphere from pseudoranges and carrier phases (Experiment 1, LAND, DOY 356, 1992, SV 1).

likely caused by differential multipath on the pseudoranges and are represented by the terms $d^P_{k,1,P}(t)$ and $d^P_{k,2,P}(t)$. The variations of the P-code differences increase as the elevation angle of the satellite decreases. Because the multipath affects the carrier phase much less than codes, we see a smooth variation of the phase curve. A cycle slip in either $N^P_{k,1}$ or $N^P_{k,2}$ will cause a sudden change in the ionospheric function. Some cycle slip detection techniques analyze the carrier phase ionospheric function (10.22). Because of the noninteger factors, an integer change in the ambiguities does not map as an integer jump in the ionospheric function. An additional difficulty is that cycle slips can occur on both carriers independently in terms of times and magnitude. Instead of (10.22) the phase function $I^P_{k,1,\varphi}(t) = -I^P_{k,1,P}(t)/\lambda_1$ is often used to identify cycle slips. See also equation 9.55.

Another function of interest is the L1 ambiguity $N^P_{k,1}(1)$, which can be computed directly from (10.5) or (10.16) as

$$N^P_{k,1}(1) = -\frac{(f_1^2 + f_2^2)}{\lambda_1(f_1^2 - f_2^2)} [P^P_{k,1}(t) + d_{k,1,P}(t) + d^P_{k,1,P}(t) + d^P_{1,P}(t) + \epsilon_{1,P}]$$

$$+ \frac{2f_2^2}{\lambda_1(f_1^2 - f_2^2)} [P^P_{k,2}(t) + d_{k,2,P}(t) + d^P_{k,2,P}(t) + d^P_{2,P}(t) + \epsilon_{2,P}]$$

$$+ \frac{1}{\lambda_1} [\Phi^P_{k,1}(t) + d_{k,1,\Phi}(t) + d^P_{k,1,\Phi}(t) + d^P_{1,\Phi}(t) + \epsilon_{1,\Phi}]$$

$$\approx -21.50\, P^P_{k,1}(t) + 16.26\, P^P_{k,2}(t) + \varphi^P_{k,1}(t) + \cdots \qquad (10.23)$$

This equation readily shows that the computed ambiguity parameter $N^P_{k,1}(1)$ is directly related to the initial ''register setting'' of the carrier phase counter of the receivers; for example, adding an integer value to $\varphi^P_{k,1}(t)$ changes the computed $N^P_{k,1}(1)$ by the same amount. If the initial carrier phase observation is only a fraction of a cycle, i.e., the ''register setting'' is zero or an initial constant is subtracted from the carrier phase observations during preprocessing, then the value of computed $N^P_{k,1}(1)$ approximates the average pseudorange (expressed in L1 wavelengths). In the hypothetical case of no ionospheric and tropospheric delays and the complete absence of all other errors, the thus computed $N^P_{k,1}(1)$ would equal the topocentric distance expressed in units of L1 wavelengths. Considering the phase measurement process by itself, the receiver can only measure the fraction of the phase as it arrives at the receiver, and the receiver has no way of knowing how many waves there are between the antenna and the satellite.

Since receivers track the accumulated change in phase, there is only one initial ambiguity required (as long as there is no cycle slip). Thus, ideally the function (10.23) should be a constant representing the initial ambiguity value.

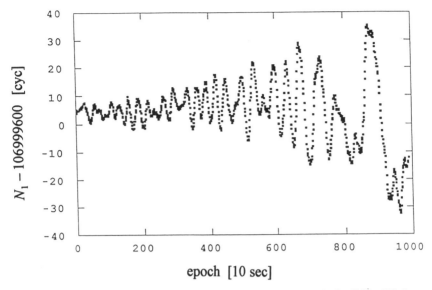

Figure 10.2 Variations of L1 ambiguity (LAND, DOY 356, 1992, SV 1).

Because of the large multiplier of the P-code observations, one should expect a large variation in $N_{k,1}^p(1)$ if the codes are corrupted by multipath. Figure 10.2 shows the L1 ambiguity as a function of time. The high correlation with the P-code ionosphere of Figure 10.1 is expected, and is supported by the high correlation in the matrices (10.12) and (10.19). Given the strong fluctuations in Figure 10.2, one does not expect success in determining the correct $N_{k,1}^p(1)$ ambiguity from these epoch solutions. More accurate pseudoranges are required and, most importantly, a much better control of the multipath is necessary, in particular for satellites at low elevation angles.

The multipath of the P1- and P2-codes is the primary cause for the variations seen in Figure 10.2. Notice the large multiplication factors for the codes and the multipath terms in equation (10.23). The multipath terms on the carrier phases are much smaller than those for pseudoranges. Since the pseudoranges and carrier phases measure the same temporal ionospheric variations, we can use the carrier ionospheric curve in Figure 10.1 to smooth the pseudorange ionospheric curve. As a first step, the unknown ambiguity constant $g[N_{k,1}^p(1), N_{k,2}^p(1)]$ in (10.22) is determined by equating (10.22) and (10.21) at a particular epoch. At other epochs the same equations are used but are solved for the second pseudorange $P_{k,2}^p$ using the known ambiguity constant. The thus computed pseudorange $P_{k,2}^p$ is, in turn, substituted into (10.23). The result of this substitution is shown in Figure 10.3. A considerable fluctuation is still seen at low satellite elevation angles.

As mentioned, the transformation (10.13) generates the wide-lane ambigu-

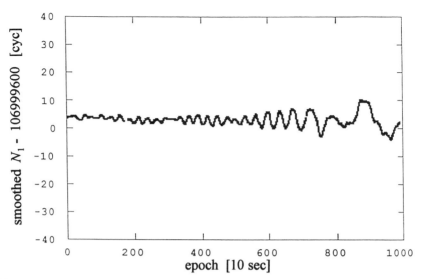

Figure 10.3 Carrier smoothed variations in L1 ambiguity (LAND, DOY 356, 1992, SV 1).

ity. The respective function is

$$N_{k,w}^P = \varphi_{k,2}^P(t) - \frac{f_1 P_{k,1}^P(t) + f_2 P_{k,2}^P(t)}{(f_1 + f_2)\lambda_w}$$

$$+ d_{k,1,\varphi}(t) - d_{k,2,\varphi}(t) + d_{k,1,\varphi}^P(t) - d_{k,2,\varphi}^P(t)$$

$$+ d_{k,1,\varphi}^P(t) - d_{k,2,\varphi}^P(t) + \epsilon_{1,\varphi} - \epsilon_{2,\varphi}$$

$$- \frac{f_1}{(f_1 + f_2)\lambda_w} [d_{k,1,P}(t) + d_{k,1,P}^P(t) + d_{1,P}^P(t) + \epsilon_{1,P}]$$

$$- \frac{f_1}{(f_1 + f_2)\lambda_w} [d_{k,2,P}(t) + d_{k,2,P}^P(t) + d_{2,P}^P(t) + \epsilon_{2,P}]$$

$$\approx \varphi_{k,w}^P - 0.562 P_{k,1}^P(t) - 0.438 P_{k,2}^P(t) + \cdots \tag{10.24}$$

with

$$\varphi_{k,w}^P = \varphi_{k,1}^P - \varphi_{k,2}^P \tag{10.25}$$

$$\lambda_w = \frac{c}{f_w} \approx 0.862 \text{ m} \tag{10.26}$$

and

$$f_w = f_1 - f_2 \tag{10.27}$$

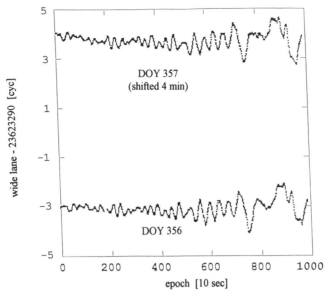

Figure 10.4 Variations in wide-lane ambiguity (LAND, DOY 356 & 357, 1992, SV 1).

The symbol $\varphi^p_{k,w}$ denotes the wide-lane carrier phase function and is expressed in units of wide-lane cycles, and the symbols λ_w and f_w are the wide-lane wavelength and wide-lane frequency, respectively. The wide-lane ambiguity $N^p_{k,w}$ is shown in Figure 10.4. Although there is still some variation in what should be a constant, Figure 10.4 suggests that it should be possible to determine the wide-lane integer value. The fluctuation appears to be about one wide-lane cycle at low elevation angles. The pseudorange coefficients in (10.24) and in (10.23) are literally of a different order of magnitude. The small coefficients in (10.24) reduce the impact of multipath on the wide-lane ambiguity function. The purpose of the wide-laning technique is to determine the wide-lane ambiguity $N^p_{k,w}(1)$ as soon as possible and to accomplish this with the minimum number of observations.

Figure 10.4 shows wide-lane ambiguities for two consecutive days for the same station and the same satellite. Because the same station–satellite geometry occurs on both days, the same pattern should be seen if the variations are caused by multipath. Indeed, both lines appear highly correlated. One of them is shifted by 4 min in order to emphasize that correlation graphically. The same satellite pattern occurs about 4 min earlier each succeeding day.

10.1.1.3 Wide Laning and Extrawide Laning With Double-Difference Carrier Phases
Modern relative positioning algorithms usually use wide laning in one form or another. With dual-frequency P-code receivers the undifferenced wide-lane ambiguities can be determined as was explained above. In fact, the wide-laning idea has been expanded into what is often referred to

as extrawide laning. The extrawide-lane function of the undifferenced L1 and L2 carrier phases is

$$e_k^P(t) \equiv \frac{f_1 \varphi_{k,2}^P(t) - f_2 \varphi_{k,1}^P(t)}{f_1 - f_2} \equiv \varphi_{k,1}^P(t) - \frac{f_1}{f_1 - f_2} \varphi_{k,w}^P(t)$$

$$= N_{k,1}^P(1) - \frac{f_1}{f_1 - f_2} N_{k,w}^P(1) + \frac{f_1 + f_2}{f_2} I_{k,1,\varphi}^P(t)$$

$$+ \frac{f_1}{f_1 - f_2} [d_{k,2,\varphi}(t) + d_{k,2,\varphi}^P(t) + d_{2,\varphi}^P(t) + \epsilon_{2,\varphi}]$$

$$- \frac{f_2}{f_1 - f_2} [d_{k,1,\varphi}(t) + d_{k,1,\varphi}^P(t) + d_{1,\varphi}^P(t) + \epsilon_{1,\varphi}] \qquad (10.28)$$

The undifferenced carrier phase expressions $\varphi_{k,1}^P(t)$ and $\varphi_{k,2}^P(t)$ are listed in Table 8.3. Equation (10.28) uses the wide-lane notation $\varphi_{k,w}^P(t)$ and $N_{k,w}^P(1)$ as defined in (10.25) and (10.20). Assuming that all wide-lane ambiguities for the receiver have been successfully identified with the help of the wide-laning techniques (10.24), it is possible to solve the undifferenced ambiguity $N_{k,1}^P(1)$ directly. Consider the following rearrangement of (10.28):

$$N_{k,1}^P(1) = \varphi_{k,1}^P(t) + \frac{f_1}{f_1 - f_2} [N_{k,w}^P(1) - \varphi_{k,w}^P(t)]$$

$$- \frac{f_1 + f_2}{f_2} I_{k,1,\varphi}^P(t) + \cdots \qquad (10.29)$$

If it were not for the presence of the unknown ionosphere term, it would be possible to compute $N_{k,1}^P(1)$ from the L1 carrier phase, the wide-lane phase, and the wide-lane ambiguity. Unfortunately, the ionospheric term is not known accurately, because it is determined from the P-codes. See equation (10.21). This ionosphere problem can be avoided, at least for short baseline distances for which the ionosphere is highly correlated, by subjecting (10.29) to the double-difference operation, giving

$$N_{km,1}^{pq}(1) = \varphi_{km,1}^{pq}(t) + \frac{f_1}{f_1 - f_2} [N_{km,w}^{pq}(1) - \varphi_{km,w}^{pq}(t)] + \frac{f_1 + f_2}{f_2} I_{km,1,\varphi}^{pq}(t)$$

$$+ \frac{f_1}{f_1 - f_2} d_{km,1,\varphi}^{pq}(t) - \frac{f_2}{f_1 - f_2} d_{km,2,\varphi}^{pq}(t)$$

$$\approx \varphi_{km,1}^{pq}(t) + 4.5 [N_{km,w}^{pq}(1) - \varphi_{km,w}^{pq}(t)] \qquad (10.30)$$

$$+ \frac{f_1 + f_2}{f_2} I_{km,1,\varphi}^{pq}(t) + \cdots$$

Since the double-differenced ionosphere term $I^{pq}_{km,1,\varphi}(t)$ is negligible over short distances, and if the wide-lane ambiguities are known from prior analysis, the double-differenced L1 ambiguity can be computed from the carrier phases. Unfortunately, the double-differenced multipath terms do not necessarily cancel because of the differing geometry in signal reflection between the receivers and satellites. If the computed wide-lane ambiguity happened to be incorrectly identified by one unit, a situation that might occur for satellites at low elevation angles, the computed L1 ambiguity changes by 4.5 cycles. The first decimal of the computed L1 ambiguity would be close to 5. Because of our a priori knowledge that the L1 ambiguity is an integer, this fact can be used in selecting between two candidate wide-lane ambiguities. This procedure is known as extrawide laning. It is a useful technique for identifying wide-lane ambiguities for low-altitude satellites or for reducing the observation span required to identify the wide-lane ambiguity integers.

Figures (10.5) to (10.12) demonstrate wide laning with double differences and the recovery of the L1 double-difference ambiguity. The first 300 observations of Experiment 1 are used. These observations, which were recorded each second, belong to a 100-m baseline with endpoints named LAND and TIDE. Figures 10.5 to 10.10 show the wide-lane ambiguity (10.24), the L1 ambiguity (10.23), and the smoothed L1 ambiguity between station LAND and satellites 1 and 23. The integer values of the undifferenced wide-lane ambiguities can, again, be inferred from Figures 10.5 and 10.8. Figures 10.6 and 10.9 show a variation of the undifferenced L1 ambiguities of about 5–10 cycles. The smoothed L1 ambiguity in Figures 10.7 and 10.10 show a variation of merely 1–3 cycles. Other figures could be given for the station TIDE, but the conclusions would be the same, therefore are not included. Figure 10.11 shows the L1 double-difference wide-lane ambiguity with respect to LAND and TIDE and satellites 1 and 23. There is still evidence of some vari-

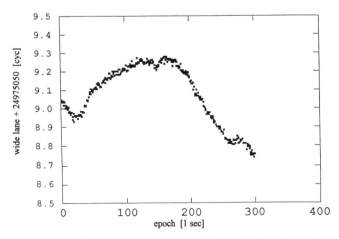

Figure 10.5   Wide-lane ambiguity (LAND, DOY 356, 1992, SV1).

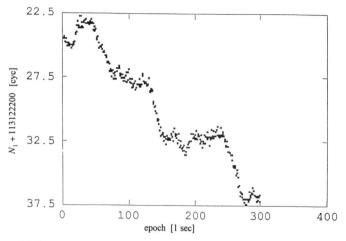

Figure 10.6 Undifferenced L1 ambiguity (LAND, DOY 356, 1992, SV1).

ation due to multipath, but the integer value can readily be identified as 824,293. Substituting this value into equation (10.30) yields the double-differenced L1 ambiguity shown in Figure 10.12. There is no doubt as to what the correct integer value for the L1 ambiguity is. In fact, this ambiguity can, at least in the present case, be determined from just one observation. If the wide-lane ambiguity had been identified only within two wide-lane cycles, then the extrawide-lane technique could have been used to identify the correct L1 double difference ambiguity.

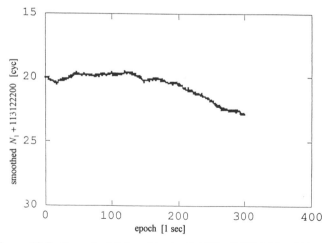

Figure 10.7 Smoothed L1 ambiguity LAND, DOY 356, 1992, SV1).

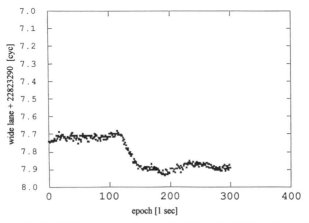

Figure 10.8 Wide-lane ambiguity (LAND, DOY 356, 1992, SV 23).

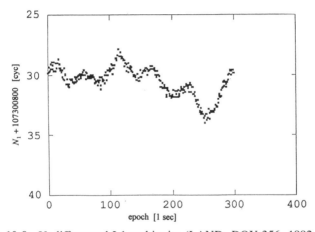

Figure 10.9 Undifferenced L1 ambiguity (LAND, DOY 356, 1992, SV 23).

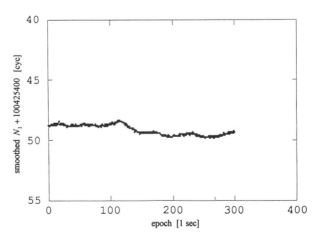

Figure 10.10 Smoothed L1 ambiguity (LAND, DOY 356, 1992, SV 23).

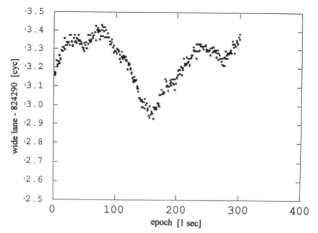

Figure 10.11 Double-differenced wide-lane ambiguity (LAND/TIDE, DOY 356, 1992, SV 1/SV 23).

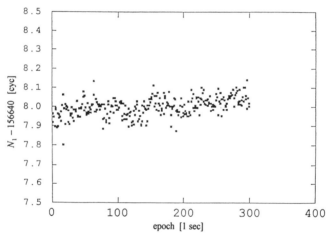

Figure 10.12 Double-differenced L1 ambiguity (LAND/TIDE, DOY 356, 1992, SV 1/SV 23).

10.1.2 Primary Functions of Observables

The expressions of Table 8.3 are often subjected to various linear transformations. The goals usually are to eliminate undesirable effects such as ionospheric delays and advances as well as clock errors. When transforming carrier phases, there is usually the additional goal of extracting a subset of ambiguity parameters with sufficient accuracy as to identify the correct integer value, or of finding a smooth function that is suitable for detecting cycle slips. The question arises as to how the various transformations differ, and as to whether any of them are preferred. The transformation (10.14) is just one example. The general form of a linear combination of the carrier phase observables, denoted by $\varphi^P_{k,m,n}(t)$, is

$$
\begin{aligned}
\varphi^P_{k,m,n}(t) &\equiv m\varphi^P_{k,1}(t) + n\varphi^P_{k,2}(t) \\
&= \frac{mf_1 + nf_2}{c}\,\rho^P_k(t) - (mf_1 + nf_2)\,dt_k + (mf_1 + nf_2)\,dt^P \\
&\quad + mN^P_{k,1}(1) + nN^P_{k,2}(1) + mI^P_{k,1,\varphi}(t) + nI^P_{k,2,\varphi}(t) \\
&\quad + \frac{mf_1 + nf_2}{c}\,T^P_k(t) \\
&\quad + m[d_{k,1,\varphi}(t) + d^P_{k,1,\varphi}(t) + d^P_{1,\varphi}(t) + \epsilon_{1,\varphi}] \\
&\quad + n[d_{k,2,\varphi}(t) + d^P_{k,2,\varphi}(t) + d^P_{2,\varphi}(t) + \epsilon_{2,\varphi}]
\end{aligned}
\tag{10.31}
$$

The factors (m, n) are integer constants. Note that in this notation, the integer m does not denote a station name; equation (10.31) refers to undifferenced carrier phases at the same station. The symbols (m, n) identify the combination, and are used as subscripts when necessary. The frequency, wavelength, ambiguity, and the ionospheric terms for the general function are

$$
f_{m,n} = mf_1 + nf_2
\tag{10.32}
$$

$$
\lambda_{m,n} = \frac{c}{mf_1 + nf_2}
\tag{10.33}
$$

$$
N^P_{k,m,n}(1) = mN^P_{k,1}(1) + nN^P_{k,2}(1)
\tag{10.34}
$$

$$
I^P_{k,m,n,\varphi}(t) = mI^P_{k,1,\varphi}(t) + nI^P_{k,2,\varphi}(t) = \frac{I^P_{k,1,\varphi}(t)}{f_2}(mf_2 + nf_1)
\tag{10.35}
$$

$$
I^P_{k,m,n,\Phi}(t) = \lambda_{m,n}I^P_{k,m,n,\varphi}(t)
\tag{10.36}
$$

With the generalized (m, n) notation the equation (10.31) can be written as

$$
\begin{aligned}
\varphi^P_{k,m,n}(t) &= \frac{f_{m,n}}{c}\,\rho^P_k(t) - f_{m,n}\,dt_k + f_{m,n}\,dt^P + N^P_{k,m,n}(1) \\
&\quad + I^P_{k,m,n,\varphi}(t) + \frac{f_{m,n}}{c}\,T^P_k(t) \\
&= \frac{\rho^P_k(t)}{\lambda_{m,n}} - f_{m,n}\,dt_k + f_{m,n}\,dt^P + N^P_{k,m,n}(1) \\
&\quad + \frac{I^P_{k,m,n,\Phi}(t)}{\lambda_{m,n}} + \frac{T^P_k(t)}{\lambda_{m,n}} + \cdots
\end{aligned}
\tag{10.37}
$$

Expressed in units of length this function becomes

$$\Phi^P_{k,m,n}(t) = \rho^P_k(t) - c\, dt_k + c\, dt^P + \lambda_{m,n} N^P_{k,m,n}(1)$$

$$+ I^P_{k,m,n,\Phi}(t) + T^P_k(t) + \cdots \tag{10.38}$$

The ionospheric time advance of the combination phase is

$$\Delta t[I^P_{k,m,n}(t)] = \frac{I^P_{k,m,n,\Phi}(t)}{c} = \frac{\lambda_{m,n}}{c} I^P_{k,m,n,\varphi}(t)$$

$$= \frac{I^P_{k,1,\varphi}(t)}{f_2} \left(\frac{mf_2 + nf_1}{mf_1 + nf_2}\right) \tag{10.39}$$

The ionospheric ratio with respect to the L1 carrier is

$$\frac{I^P_{k,m,n,\Phi}(t)}{I^P_{k,1,\Phi}(t)} = \frac{f_1}{f_2} \left(\frac{mf_2 + nf_1}{mf_1 + nf_2}\right) \tag{10.40}$$

In (10.37) the distances are expressed in units of the wavelength. A change in $\varphi^P_{k,m,n}(t)$ by one cycle, or an equivalent change of the ambiguity $N^P_{k,m,n}(1)$ by one, is the same as a distance change along the topocentric station–satellite distance by one wavelength of $\lambda_{m,n}$. The distance corresponding to one wavelength is frequently called a lane. Determination of the ambiguity $N^P_{k,m,n}(1)$ thus implies that the topocentric range has been resolved within the unit of $\lambda_{m,n}$. One might, therefore, prefer transformations that give large wavelengths and solve the respective ambiguities. The assumption is that the unmodeled errors are small enough to allow a unique determination of these ambiguities. In a subsequent solution for the original ambiguities, such as $N^P_{k,1}(1)$ and $N^P_{k,2}(1)$, the estimated ambiguity integers of $N^P_{k,m,n}(1)$ are constrained. Unfortunately, the m and n factors, which generate long wavelengths according to (10.33), might also increase the multipath errors and other disturbances according to (10.31) and (10.39).

The combinations for which m and n have different signs are called the wide-lane observables. Because the specific observable $(1, -1)$ is the most important of all the wide-lane observables, it is usually referred to simply as the wide lane (without explicitly mentioning the m and n); the subscript w is also used to identify this lane. Equations (10.20) and (10.24) to (10.27) already made reference to the wide-lane ambiguity, phase, wavelength, and frequency. If the m and n have the same sign, we speak of narrow-lane observables. The particular combination $(1, 1)$ is simply the narrow lane (without explicitly mentioning the m and n). The subscript n denotes this narrow lane. For example,

$$\varphi^P_{k,n} = \varphi^P_{k,1} + \varphi^P_{k,2} \tag{10.41}$$

$$N^P_{k,n} = N^P_{k,1} + N^P_{k,2} \tag{10.42}$$

$$\lambda_n = \frac{c}{f_n} \approx 0.107 \text{ m} \tag{10.43}$$

$$f_n = f_1 + f_2 \tag{10.44}$$

The m and n can, in principle, be positive or negative integers. The results of all linear combinations are equivalent in the statistical sense as merely transforming observables and applying variance–covariance propagation properly, generates results that are mathematically related. Although the propagation of random errors is only part of what impacts the combination solution, let us look at the geometry of the general transformation

$$T = \begin{bmatrix} 1 & 0 & 0 & 0 \\ 0 & 1 & 0 & 0 \\ 0 & 0 & m & n \\ 0 & 0 & 1 & 0 \end{bmatrix} \tag{10.45}$$

which relates the (m, n) lanes and the L1 lane; the respective ambiguity parameters are $N^p_{k,m,n}$ and $N^p_{k,1}$. Table 10.2 summarizes the covariance matrix (10.17) for selected m and n. The geometry of the solution is, again, quantified by the semimajor and semiminor axes of the ellipse of standard deviation for the ambiguities (columns 3 and 4), the azimuth of the semimajor axis in column 5 (counted clockwise from the wide-lane axis), the wide-lane wavelength in column 6, the standard deviations of the wide lane and the L1 lane in columns 7 and 8, and the ionospheric time advance ratio (10.40), listed in the last column.

TABLE 10.2 Wide-Lane Ambiguity Geometry ($|\lambda_{m,n}| > \lambda_w$)

m	n	Major	Minor	Azimuth [deg.]	$\lambda_{m,n}$	σ_{N_w}	σ_{N_1}	Ratio
-10	13	29.33	1.16	18.20	1.47	27.86	9.22	51.46
-8	10	20.74	1.26	26.19	-0.92	18.62	9.22	-23.26
-7	9	20.72	1.14	26.26	14.65	18.59	9.22	350.35
-4	5	13.07	1.00	44.73	-1.83	9.31	9.22	-23.26
-3	4	13.06	0.80	44.84	1.63	9.28	9.22	18.25
-1	1	9.22	0.28	90.03	-0.86	0.28	9.22	-1.28
1	-1	9.22	0.28	89.97	0.86	0.28	9.22	-1.28
3	-4	13.06	0.80	135.16	-1.63	9.28	9.22	18.25
4	-5	13.07	1.00	135.27	1.83	9.31	9.22	-23.26
7	-9	20.72	1.14	153.74	-14.65	18.59	9.22	350.35
8	-10	20.74	1.26	153.81	0.92	18.62	9.22	-23.26
10	-13	29.33	1.16	161.80	-1.47	27.86	9.22	51.46

For the most preferable solution the ambiguity parameters should be uncorrelated and should have small standard deviations. Table 10.2 shows this to be the case for the wide-lane combinations $(1, -1)$ and $(-1, 1)$ for which the semimajor axis almost coincides with the $N_{k,1}^p$ axis. The special case $(1, -1)$ was discussed in the previous section. Both ellipses are located symmetrically with respect to the $N_{k,1}^p$ axis; which is the only difference for both wide-lane solutions. Only in these two cases is the standard deviation of the wide lane smaller than one wavelength (thus giving reasonable expectation that the correct integer might be identifiable). Table 10.2 suggests that $(1, -1)$ and $(-1, 1)$ are the only serious candidates for wide laning. Therefore, it is justifiable to have the term "wide lane" conventionally referred to the $(1, -1)$ combination unless otherwise stated explicitly. The $(-1, 1)$ lanes does not add anything that is not already obtained from $(1, -1)$. Because its wavelength is negative, which is no problem mathematically as such, this symmetrical combination is usually not mentioned.

The rows of Table 10.2 are symmetric with respect to the wide lane. The entries next to the wide lanes have azimuths for the semimajor axis of 45° (135°). The azimuths will systematically decrease toward zero for negative m and will systematically increase to 180° for positive m. For azimuths smaller than 45° and larger than 135° the L1 ambiguity is determined better than the wide-lane ambiguity. Note, however, that the actual standard deviation of the L1 ambiguity remains a constant for these specific transformations. All of the ellipses have approximately the same flattening. The last column in Table 10.2 shows the ionospheric time delay ratio according to equation (10.40). Again, these ratios are a direct result of mathematical transformations yielding negative and positive delays for the wide-lane phases. There is no connection here to the physical fact that the carrier phases are advanced and the codes are delayed by the ionosphere.

Variations to the transformation (10.45) can readily be made. For example, one could set the element $(4, 4)$ in the matrix T to one. This implies a transformation of the wide-lane ambiguities and the narrow lane $(1, 1)$. This specification will change the values in Table 10.2 only slightly. As a matter of curiosity, the azimuth of the ellipse of standard deviation for the wide lane $(1, -1)$ and the narrow lane $(1, 1)$ is exactly 90°. A further generalization of (10.45) would be to set the element $(4, 3)$ to m and $(4, 4)$ to $-n$. This implies a parameterization of terms for general wide lanes and general narrow lanes. The semimajor axis now remains in the vicinity of 90°, but the standard deviations of the wide-lane ambiguities are the same as those in the previous case. Basically, these variations in transformation will not improve upon what is already documented by Table 10.2, which identifies the wide lane $(1, -1)$ as the primary (only) combination worth considering when m and n are integers.

Table 10.3 summarizes the frequently used linear code and phase combinations. These linear functions are not limited to m and n being integers. The first column lists the name of the combination, columns 2–5 contain the coef-

TABLE 10.3   Popular Functions for Pseudoranges and Carrier Phases

	$P^p_{k,1}$	$P^p_{k,2}$	$\varphi^p_{k,1}$	$\varphi^p_{k,2}$	ρ^p_k	dt_k	dt^p	$I^p_{k,1,P}$	T^p_k	$N^p_{k,1}$	$N^p_{k,2}$	$N^p_{k,w}$	$N^p_{k,n}$
$P^p_{k,IF}$	$\dfrac{-\lambda_w\lambda_n}{\lambda_1^2}$	$\dfrac{-\lambda_w\lambda_n}{\lambda_2^2}$	0	0	1	$-c$	c	0	1	0	0	0	0
$P^p_{k,I}$	$\dfrac{-\lambda_w\lambda_n}{\lambda_2^2}$	$\dfrac{\lambda_w\lambda_n}{\lambda_2^2}$	0	0	0	0	0	1	0	0	0	0	0
$P^p_{k,w}$	$\dfrac{\lambda_n}{\lambda_1}$	$\dfrac{\lambda_n}{\lambda_2}$	0	0	1	$-c$	c	$\dfrac{-\lambda_2}{\lambda_1}$	1	0	0	0	0
$\varphi^p_{k,IF}$	0	0	$\dfrac{\lambda_w\lambda_n}{\lambda_1^2}$	$\dfrac{-\lambda_w\lambda_n}{\lambda_1\lambda_2}$	λ_1^{-1}	$\dfrac{-c}{\lambda_1}$	$\dfrac{c}{\lambda_1}$	0	λ_1^{-1}	$\dfrac{\lambda_w\lambda_n}{\lambda_1^2}$	$\dfrac{-\lambda_w\lambda_n}{\lambda_1\lambda_2}$	0	0
	0	0	$\dfrac{\lambda_w\lambda_n}{\lambda_1^2}$	$\dfrac{-\lambda_w\lambda_n}{\lambda_1\lambda_2}$	λ_1^{-1}	$\dfrac{-c}{\lambda_1}$	$\dfrac{c}{\lambda_1}$	0	λ_1^{-1}	0	0	$\dfrac{\lambda_w}{2\lambda_1}$	$\dfrac{\lambda_n}{2\lambda_1}$
	0	0	$\dfrac{\lambda_w\lambda_n}{\lambda_1^2}$	$\dfrac{-\lambda_w\lambda_n}{\lambda_1\lambda_2}$	λ_1^{-1}	$\dfrac{-c}{\lambda_1}$	$\dfrac{c}{\lambda_1}$	0	λ_1^{-1}	$\dfrac{\lambda_n}{\lambda_1}$	0	$\dfrac{-\lambda_w\lambda_n}{\lambda_1\lambda_2}$	0
$\varphi^p_{k,I}$	0	0	1	$\dfrac{-\lambda_2}{\lambda_1}$	0	0	0	$\dfrac{-\lambda_2^2\lambda_1^{-1}}{\lambda_w\lambda_n}$	0	1	$\dfrac{-\lambda_2}{\lambda_1}$	0	0
$\varphi^p_{k,U}$	0	0	$\dfrac{1}{2}$	$\dfrac{\lambda_2}{2\lambda_1}$	λ_1^{-1}	$\dfrac{-c}{\lambda_1}$	$\dfrac{c}{\lambda_1}$	λ_1^{-1}	λ_1^{-1}	$\dfrac{1}{2}$	$\dfrac{\lambda_2}{2\lambda_1}$	0	0
$\varphi^p_{k,w}$	0	0	1	-1	λ_w^{-1}	$\dfrac{-c}{\lambda_w}$	$\dfrac{c}{\lambda_w}$	$\dfrac{-\lambda_2}{\lambda_w\lambda_1}$	λ_w^{-1}	0	0	1	0
$\varphi^p_{k,n}$	0	0	1	1	λ_n^{-1}	$\dfrac{-c}{\lambda_n}$	$\dfrac{c}{\lambda_n}$	$\dfrac{-\lambda_2}{\lambda_n\lambda_1}$	λ_n^{-1}	0	0	0	1
e^p_k	0	0	$\dfrac{-\lambda_w}{\lambda_2}$	$\dfrac{\lambda_w}{\lambda_1}$	0	0	0	$\dfrac{-\lambda_2}{\lambda_n\lambda_1}$	0	1	0	$\dfrac{-\lambda_w}{\lambda_1}$	0
W^p_k	$\dfrac{-\lambda_n}{\lambda_1}$	$\dfrac{-\lambda_n}{\lambda_2}$	λ_w	$-\lambda_w$	0	0	0	0	0	0	0	λ_w	0
$R^p_{k,1}$	1	0	λ_1	0	0	0	0	2	0	λ_1	0	0	0
R^p_k	1	-1	λ_1	$-\lambda_2$	0	0	0	0	0	λ_1	$-\lambda_2$	0	0

ficients of the basic observables, and the remaining columns contain the coefficients of the parameters and biases. The hardware delay terms and the multipath terms are not listed. Two parameterizations are given for the ambiguities. In addition to the undifferenced ambiguities $N^p_{k,1}(1)$ and $N^p_{k,2}(1)$ the coefficients for the wide-lane and narrow-lane ambiguities $N^p_{k,w}(1)$ and $N^p_{k,n}(1)$ are listed in a few cases. The first three rows of the table refer to pseudoranges, the next six functions pertain to carrier phase combinations, and the last three functions represent combinations of pseudoranges and carrier phases.

   Several of these functions do not depend on the receiver location implied by the topocentric range $\rho^p_k(t)$. These functions are $P^p_{k,I}$, e^p_k, W^p_k, $R^p_{k,1}$, R^p_k, and $\varphi^p_{k,I}$. These particular functions are also independent of the receiver clock error and the satellite clock error. The W^p_k and R^p_k functions, in addition, are independent of the ionosphere; they depend only on the ambiguities. Because these functions do not depend on the topocentric distance $\rho^p_k(t)$, they are directly usable in kinematic application when one receiver is in motion. The ionospheric-free functions are $P^p_{k,IF}$, $\varphi^p_{k,IF}$, W^p_k, and R^p_k. The ionospheric function are $P^p_{k,I}$ and $\varphi^p_{k,I}$. The latter still depends on the ambiguities. The preferred functions for analyzing cycle slips are $\varphi^p_{k,I}$, W^p_k, and R^p_k. The carrier phase functions $\varphi^p_{k,IF}$, $\varphi^p_{k,w}$, $\varphi^p_{k,U}$, and $\varphi^p_{k,n}$ can be scaled to their distance equivalents by multiplying with their respective wavelengths. This is not true for the ionospheric observation $\varphi^p_{k,I}$ because its "$f_{m,n}$ frequency" is zero. The function R^p_k is sometimes called the range difference equation. The function e^p_k leads to the wide-lane and extrawide-lane formulation. The wide-lane pseudorange function $P^p_{k,w}$ and the carrier wide lane $\varphi^p_{k,w}$ have the same dependency on the ionosphere, i.e., magnitude and sign. The functions $\varphi^p_{k,IF}$ and $\varphi^p_{k,U}$ are statistically uncorrelated, which is not true for the pair ($\varphi^p_{k,IF}$, $\varphi^p_{k,I}$). There is no standard notation yet in use for the various functions. However, occasionally the following identity in notation is found: L1 $\equiv \varphi^p_{k,1}(t)$, L2 $\equiv \varphi^p_{k,2}(t)$, L3 $\equiv \varphi^p_{k,IF}(t)$, and L4 $\equiv \varphi^p_{k,w}(t)$.

Not all functions listed in Table 10.3 are independent. Only four observables are available, leading to, at most, four independent functions. The following relations between the functions are readily seen:

$$\lambda_w \varphi^p_{k,w}(t) - \frac{\lambda_n}{\lambda_1} P^p_{k,1}(t) - \frac{\lambda_n}{\lambda_2} P^p_{k,2}(t) = W^p_k(t) \qquad (10.46)$$

$$\frac{1}{2\lambda_1} [\lambda_w \varphi^p_{k,w}(t) + \lambda_n \varphi^p_{k,n}(t)] = \varphi^p_{k,IF}(t) \qquad (10.47)$$

$$\frac{\lambda_2}{2} \left[\frac{\varphi^p_{k,w}(t)}{\lambda_n} - \frac{\varphi^p_{k,n}(t)}{\lambda_w} \right] = \varphi^p_{k,I}(t) \qquad (10.48)$$

$$P^p_{k,w}(t) = \lambda_w \varphi^p_{k,w}(t) \qquad (10.49)$$

It is interesting to see how the wide-lane and the narrow-lane functions relate to the ionospheric-free and ionospheric functions. The wide-lane carrier phase function in (10.49), of course, contains the ambiguity term.

The actual implementation of the functions in Table 10.3 varies considerably. Each function can be subjected to the double-difference operation for relative positioning to achieve the usual reduction of common-mode errors. Unfortunately, single-frequency C/A-code and carrier phase users do not have many options in forming useful functions. They can take advantage only of the cancellations of common-mode errors in double differencing.

10.1.3   Description of a Unified Approach

The first successes of surveying with GPS in the early 1980s used static techniques. Typically, the receivers remained stationary and recorded observations over several hours. Primarily, receivers with C/A-code and L1 carrier phase capability were used, which could not observe the P-codes. Thus, algorithm developers could not take advantage of most of the functions listed in Table 10.3. At least one dual-frequency P-code receiver was available in those early days, but because of high costs these receivers were used primarily by government agencies and some research institutions. The predominance of the "classical static" approach began crumbling as kinematic techniques were developed. Kinematic GPS became more attractive with the introduction of the antenna swap procedure, which can determine the ambiguities in minutes (Remondi 1985b). The alternatives to antenna swapping at that time were the occupation of two known stations prior to the kinematic survey, or the carrying out of a separate static survey first. The next step in the development was the introduction of pseudo-static GPS. With this technique, the unknown station is occupied twice for a period of about 10 min each. The recommended time between the two station visits is about one hour. At the heart of the pseudo-static approach is the fact that the geometry of the solution derives its strength primarily from observing the same satellites at significantly different orbital locations.

Over time, receiver technology matured and dual-frequency carrier phases became available. Finally, new dual-frequency P-code receivers were introduced. The attention of algorithm developers then shifted to rapid-static GPS. The goal of the rapid-static technique is to eliminate the second station visit required in pseudo-static GPS, and simultaneously to reduce the observation time during the first setup as much as possible. This goal is to be achieved with refined processing, which fully explores the information content of the dual-frequency carrier phases alone or in combination with the P-codes. This development inevitably results in a merging of kinematic and static GPS. For example, if the occupation time for short baselines is sufficiently short, then there is no need for an antenna swap. These modern techniques also deal with the most general case in which the roving receiver is in motion right from the start of the survey. The integer ambiguities are resolved while the roving receiver is moving. This is called the on-the-fly (OTF) technique. If both receivers can communicate the data and carry out the computations in real time, we have a relative positioning technique which not only makes the distinction between static and kinematic GPS surveying obsolete, but also unifies or merges the traditional fields of surveying with high-precision navigation.

The OTF concept is readily explained by means of Experiment 2. The processing was carried out with commercial software that employs a Kalman filter to produce a position and velocity at each epoch (Qin et al. 1992a). The program is flexible in that code and carrier phase processing can optimally be carried out separately and in combination. Even C/A-code processing is possible. Displays on the screen let you see how the estimates of the ambiguity quickly improve over time, and how first the wide-lane ambiguities and then the L1 ambiguities are resolved and constrained to integers. The dynamic model provides knowledge of the dynamics of the moving receiver, for example, the receiver is moved by plane or car, walked around, or is even stationary.

Figure 10.13 shows the time history for Experiment 2. The experiment began with both antennas in static position, separated by 17 m. After approximately the first 200 sec, one antenna was taken off the tripod and turned upside

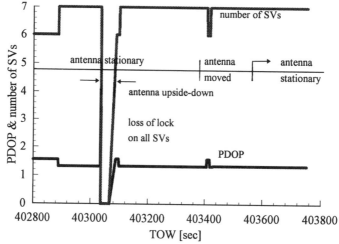

Figure 10.13 PDOP and number of observed satellites for Experiment 2 (DOY 014, 1993).

down. All satellites were lost, of course. Subsequently, the antenna was placed again on the same tripod for 3 min and then was "walked around" for 3 more minutes before it was returned to the tripod. The objective of the walk was to begin a separate solution while the antenna was in motion, and to see whether the same position for the tripod could be obtained if the data were processed continuously from beginning to end.

Figure 10.14 shows a typical convergence for the computed horizontal position. Once the integers have "snapped in," the correct position is obtained with centimeter accuracy. Figure 10.15 shows the variation in the vertical co-

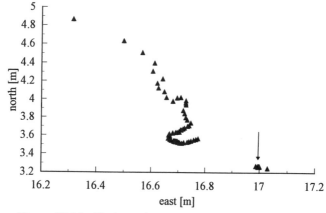

Figure 10.14 Horizontal convergence (DOY 014, 1993).

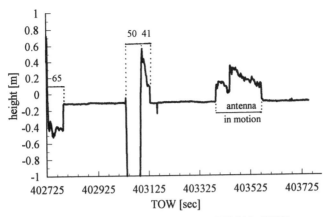

Figure 10.15 Vertical profile (DOY 014, 1993).

ordinate. After only 65 sec the baseline was determined, as seen from the vertical variation in Figure 10.15, showing only system noise after the 65-sec mark. Recall that the antenna was stationary during this part of the experiment. After losing all satellites while the antenna was in the upside-down position, it took merely 41 sec to determine the baseline again. The vertical motion of the antenna during the walk-around in the later part of the experiment is also seen in Figure 10.15.

Figure 10.16 shows the vertical coordinate for the case in which processing began while the antenna was in motion (walked around). The integers are resolved after merely 24 sec. Figure 10.17 shows the difference in height of the current solution (which started while the antenna was walked around) with the previous solution (which started after the antenna had been turned upside

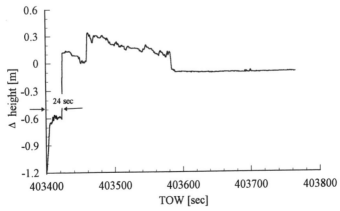

Figure 10.16 Vertical profile for OTF (DOY 014, 1993).

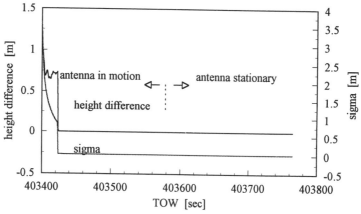

Figure 10.17 Comparison in height for complete and partial solutions (DOY 014, 1993).

down). Once the integers are determined, the discrepancies between both solutions and the standard deviation are in the millimeter range.

This example shows how OTF merges static and kinematic GPS. The specific times quoted for integer resolution depend on software implementations and parameter settings (which are subject to change). Processing backward in time, the positions can even be determined accurately for the time span it took initially to resolve the ambiguities. The only condition now is that cycle slips and loss of lock occur at time separations long enough to allow time for ambiguity resolution. When more than four satellites are being observed, the cycle slip issue becomes even simpler, as will be explained in the following sections. There is no need, at least conceptually, for initial antenna swaps or for returning to known locations. With real-time communication and processing available, OTF merges surveying and navigation to the centimeter level.

10.2 STATIC POSITIONING

In static relative positioning with GPS, the receivers remain stationary until observations have been completed. The observation time varies considerably, depending on variables such as receiver capabilities (single frequency, dual frequency, narrow correlator C/A-codes, P-codes), post processing software features, and the length of the baseline. In the case of rapid static GPS surveying, the station occupation time is very short, in the range of a couple of minutes to about 10–15 min. Again, these occupation times are a function of the state of the development of hardware and software, which are subject to change in the future. The goal in rapid static GPS surveying is always to resolve the integer ambiguities as soon as possible. The term pseudo-static GPS surveying is used when a station is visited twice for a period of about 5–10 min, and with about one hour between visits. Long occupation times are preferred for long baselines and whenever highest accuracy is to be achieved, which tends to better average unmodeled errors such as multipath.

10.2.1 Observation Equations for Double and Triple Differences

If R receivers observe S satellites for T epochs, at most RST carrier phase observations are possible. In many cases one might not have such a complete data set. Algorithms for computing the double and triple difference observations must be able to deal with gaps in the data that can occur as a result of cycle slips, as well as rising and setting satellites during a session. To see and explore the symmetry of the expressions, we assume that the undifferenced phase observations are ordered as shown by the observation vector Ψ:

$$
\begin{aligned}
\Psi = [&\varphi_1^1(1), \varphi_1^2(1), \varphi_1^3(1), \varphi_1^4(1), \varphi_2^1(1), \varphi_2^2(1), \varphi_2^3(1), \varphi_2^4(1), \\
&\varphi_3^1(1), \varphi_3^2(1), \varphi_3^3(1), \varphi_3^4(1), \\
&\varphi_1^1(2), \varphi_1^2(2), \varphi_1^3(2), \varphi_1^4(2), \varphi_2^1(2), \varphi_2^2(2), \varphi_2^3(2), \varphi_2^4(2), \\
&\varphi_3^1(2), \varphi_3^2(2), \varphi_3^3(2), \varphi_3^4(2), \\
&\varphi_1^1(3), \varphi_1^2(3), \varphi_1^3(3), \varphi_1^4(3), \varphi_2^1(3), \varphi_2^2(3), \varphi_2^3(3), \varphi_2^4(3), \\
&\varphi_3^1(3), \varphi_3^2(3), \varphi_3^3(3), \varphi_3^4(3), \ldots]
\end{aligned} \tag{10.50}
$$

In this scheme the phase observations are ordered sequentially first by the epoch of observation, then by the receiver, and then by the satellite. It is usually assumed that all carrier phase observations are uncorrelated and are of the same accuracy. Thus the complete $(RST \times RST)$ cofactor matrix of the undifferenced phase observations is

$$
Q_\varphi = \sigma_\varphi^2 I \tag{10.51}
$$

with σ_φ denoting the standard deviation of the phase measurement expressed in cycles.

As detailed in Section 8.2.3 the double-differencing operations for computing the double differences from the observed undifferenced phases at stations k and m to the satellites p and q is

$$
\varphi_{km}^{pq} = \varphi_k^p - \varphi_m^p - (\varphi_k^q - \varphi_m^q) \tag{10.52}
$$

To establish a simple scheme for identifying independent double-difference observations, we use the ideas of base station and base satellite. The station with label 1 is called the base station, and the satellite with label 1 is called the base satellite. The session network of R stations is now thought of as consisting of $(R - 1)$ baselines emanating from the base station. There are $(S -$

1) independent double differences for each baseline as the superscript q takes the values 2, 3, ... , S. Thus, a total of $(R - 1)(S - 1)$ independent double differences can be computed for the session network. On the basis of the ordered observation vector (10.50), the definition of the double differences (10.52), and the base-station and base-satellite ordering scheme, an independent set of double differences is

$$\Delta = D\Psi \tag{10.53}$$

where

$$\Delta = [\varphi_{12}^{12}(1),\ \varphi_{12}^{13}(1),\ \varphi_{12}^{14}(1),\ \varphi_{13}^{12}(1),\ \varphi_{13}^{13}(1),\ \varphi_{13}^{14}(1),$$
$$\varphi_{12}^{12}(2),\ \varphi_{12}^{13}(2),\ \varphi_{12}^{14}(2),\ \varphi_{13}^{12}(2),\ \varphi_{13}^{13}(2),\ \varphi_{13}^{14}(2), \tag{10.54}$$
$$\varphi_{12}^{12}(3),\ \varphi_{12}^{13}(3),\ \varphi_{12}^{14}(3),\ \varphi_{13}^{12}(3),\ \varphi_{13}^{13}(3),\ \varphi_{13}^{14}(3),\ \ldots]$$

Matrix D is shown in Table 10.4 for the case $R = 3$, $S = 4$, and $T = 3$. The size of D in general is $(R - 1)(S - 1)T \times RST$. The columns of D are labeled with the superscripts and subscripts of the phase observation identifying the satellite and receiver, respectively. According to Section 8.2.4 the triple-difference operator is

$$\varphi_{km}^{pq}(j,\ i) = \varphi_{km}^{pq}(j) - \varphi_{km}^{pq}(i) \tag{10.55}$$

where i and j denote the epochs. The triple differences are computed from the double-difference observations or the undifferenced phase observations using

TABLE 10.4   Specifications of the D Matrix ($R = 3$, $S = 4$, $T = 3$)

<------- epoch 1 phase --------><------- epoch 2 phase --------><------- epoch 3 phase ------->

1	2	3	4	1	2	3	4	1	2	3	4	1	2	3	4	1	2	3	4	1	2	3	4	1	2	3	4	1	2	3	4	1	2	3	4
1	1	1	1	2	2	2	2	3	3	3	3	1	1	1	1	2	2	2	2	3	3	3	3	1	1	1	1	2	2	2	2	3	3	3	3
1	-1			-1	1																														
1		-1		-1		1																													
1			-1	-1			1																												
1	-1							-1	1																										
1		-1						-1		1																									
1			-1					-1			1																								
												1	-1			-1	1																		
												1		-1		-1		1																	
												1			-1	-1			1																
												1	-1							-1	1														
												1		-1						-1		1													
												1			-1					-1			1												
																								1	-1			-1	1						
																								1		-1		-1		1					
																								1			-1	-1			1				
																								1	-1							-1	1		
																								1		-1						-1		1	
																								1			-1					-1			1

the transformation

$$\nabla = T\Delta = TD\Psi \qquad (10.56)$$

For the ordered vector of triple-difference observations

$$\nabla = [\varphi^{12}_{12}(2, 1),\ \varphi^{13}_{12}(2, 1),\ \varphi^{14}_{12}(2, 1),\ \varphi^{12}_{13}(2, 1),\ \varphi^{13}_{13}(2, 1),\ \varphi^{14}_{13}(2, 1),$$
$$\varphi^{12}_{12}(3, 2),\ \varphi^{13}_{12}(3, 2),\ \varphi^{14}_{12}(3, 2),\ \varphi^{12}_{13}(3, 2),\ \varphi^{13}_{13}(3, 2),\ \varphi^{14}_{13}(3, 2) \ldots]$$

$$(10.57)$$

the transformation matrix T is that given in Table 10.5.

Both the double- and the triple-difference observations are linear functions of the observed carrier phases. By applying the law of variance–covariance propagation and by taking the cofactor matrix (10.51) into account, the double-difference and triple-difference cofactor matrices respectively become

$$Q_\Delta = \sigma^2_\varphi DD^{\mathrm{T}} \qquad (10.58)$$

$$Q_\nabla = \sigma^2_\varphi TD(TD)^{\mathrm{T}} \qquad (10.59)$$

These cofactor matrices are shown in Tables 10.6 and 10.7. The double-difference cofactor matrix is block diagonal, implying that double differences of different epochs are uncorrelated. The triple-difference cofactor matrix is band diagonal. The triple differences between consecutive (adjacent) epochs are correlated. Note that the inverse of the triple-difference cofactor matrix, which is required in the least-squares solution, is a full matrix. Eren (1987) gives an algorithm for computing the elements of the cofactor matrices (10.58) and (10.59), requiring no explicit matrix multiplication. He uses rules based on

TABLE 10.5 Specifications of the T Matrix ($R = 3$, $S = 4$, $T = 3$)

	epoch 1 double differences						epoch 2 double differences						epoch 3 double differences					
	$^{12}_{12}$	$^{13}_{12}$	$^{14}_{12}$	$^{12}_{13}$	$^{13}_{13}$	$^{14}_{13}$	$^{12}_{12}$	$^{13}_{13}$	$^{14}_{13}$	$^{12}_{12}$	$^{13}_{12}$	$^{14}_{12}$	$^{12}_{13}$	$^{13}_{13}$	$^{14}_{13}$	$^{12}_{12}$	$^{13}_{13}$	$^{14}_{13}$
	-1						1											
		-1						1										
			-1						1									
				-1						1								
					-1						1							
						-1						1						
							-1						1					
								-1						1				
									-1						1			
										-1						1		
											-1						1	
												-1						1

TABLE 10.6 Matrix DD^T for Double Differences ($R = 3$, $S = 4$, $T = 3$)

4 2 2 2 1 1 2 4 2 1 2 1 2 2 4 1 1 2 2 1 1 4 2 2 1 2 1 2 4 2 1 1 2 2 2 4		
	4 2 2 2 1 1 2 4 2 1 2 1 2 2 4 1 1 2 2 1 1 4 2 2 1 2 1 2 4 2 1 1 2 2 2 4	
		4 2 2 2 1 1 2 4 2 1 2 1 2 2 4 1 1 2 2 1 1 4 2 2 1 2 1 2 4 2 1 1 2 2 2 4

TABLE 10.7 Matrix $TDD^T T^T$ for Triple Differences ($R = 3$, $S = 4$, $T = 3$)

8	4	4	4	2	2	-4	-2	-2	-2	-1	-1
4	8	4	2	4	2	-2	-4	-2	-1	-2	-1
4	4	8	2	2	4	-2	-2	-4	-1	-1	-2
4	2	2	8	4	4	-2	-1	-1	-4	-2	-2
2	4	2	4	8	4	-1	-2	-1	-2	-4	-2
2	2	4	4	4	8	-1	-1	-2	-2	-2	-4
-4	-2	-2	-2	-1	-1	8	4	4	4	2	2
-2	-4	-2	-1	-2	-1	4	8	4	2	4	2
-2	-2	-4	-1	-1	-2	4	4	8	2	2	4
-2	-1	-1	-4	-2	-2	4	2	2	8	4	4
-1	-2	-1	-2	-4	-2	2	4	2	4	8	4
-1	-1	-2	-2	-2	-4	2	2	4	4	4	8

the subscripts and superscripts of the phase observations to compute the elements of the cofactor matrices directly.

Having computed the vectors of double- and triple-difference phase observations and their cofactor matrices, we can now formulate the adjustment. The relevant terms of the double-difference equation (8.41) are

$$\varphi_{km}^{pq}(t) = \frac{f}{c} [\rho_k^p(t) - \rho_m^p(t)] - \frac{f}{c} [\rho_k^q(t) - \rho_m^q(t)] + N_{km}^{pq}(1)$$

$$+ \frac{f}{c} [\dot{\rho}_k^p(t)\, dt_k - \dot{\rho}_m^p(t)\, dt_m] - \frac{f}{c} [\dot{\rho}_k^q(t)\, dt_k - \dot{\rho}_m^q(t)\, dt_m] \quad (10.60)$$

$$+ I_{km,\varphi}^{pq}(t) + \frac{f}{c} T_{km}^{pq}(t) + d_{km,\varphi}^{pq}(t) + \epsilon_{km,\varphi}^{pq}$$

with

$$\rho_k^p(t) = \sqrt{(u^p - u_k)^2 + (v^p - v_k)^2 + (w^p - w_k)^2} \qquad (10.61)$$

$$\rho_m^p(t) = \sqrt{(u^p - u_m)^2 + (v^p - v_m)^2 + (w^p - w_m)^2} \qquad (10.62)$$

The satellite frequency offset terms are neglected because of the short baselines typically encountered in surveying. Furthermore, the satellite ephemeris is assumed to be available, either in the form of the broadcast ephemeris or the precise ephemeris provided by the National Geodetic Survey or other services. These ephemerides typically give the satellite positions in a conventional terrestrial coordinate system, such as the WGS 84 (broadcast ephemeris) or the ITRF 92 (NGS precise ephemeris).

The partial derivatives with respect to the receiver positions are

$$\frac{\partial \varphi_{km}^{pq}}{\partial u_k} = -\frac{f}{c}\frac{u^p - u_k}{\rho_k^p} + \frac{f}{c}\frac{u^q - u_k}{\rho_k^q} \qquad (10.63)$$

$$\frac{\partial \varphi_{km}^{pq}}{\partial u_m} = \frac{f}{c}\frac{u^p - u_m}{\rho_m^p} - \frac{f}{c}\frac{u^q - u_m}{\rho_m^q} \qquad (10.64)$$

Similar expressions can be given for the partial derivatives with respect to second and third coordinates v_k, v_m, w_k, and w_m. These partial derivatives are evaluated for the approximate receiver coordinates and the satellite positions at the instant of signal emission. The topocentric distance is computed by one of the procedures explained in Section 8.1.

The derivatives for the station clocks are

$$\frac{\partial \varphi_{km}^{pq}}{\partial t_k} = \frac{f}{c}\dot{\rho}_k^p - \frac{f}{c}\dot{\rho}_k^q \qquad (10.65)$$

$$\frac{\partial \varphi_{km}^{pq}}{\partial t_m} = -\frac{f}{c}\dot{\rho}_m^p + \frac{f}{c}\dot{\rho}_m^q \qquad (10.66)$$

The topocentric range rate is obtained from the change in the topocentric distance immediately before and after the epoch of observation. Since the station clock errors are available from the navigation solution (Section 8.1), or from the various pseudoranges (Section 8.3.1.2) with an accuracy of about 1 μsec, the station clock parameters are usually not estimated in the double-difference carrier phase solution. If the station clocks are estimated, one might not solve these parameters at every epoch, but might model them by a polynomial in time. In such a case, it could, furthermore, be advantageous to transform the clock errors for two receivers into the difference and the sum such as

$$\eta_{km} = \frac{dt_m - dt_k}{2} \qquad (10.67)$$

$$\xi_{km} = \frac{dt_m + dt_k}{2} \tag{10.68}$$

The coefficients for the transformed clock time parameters follow from

$$\frac{f}{c}(\dot{\rho}_k^p - \dot{\rho}_k^q) \, dt_k - \frac{f}{c}(\dot{\rho}_m^p - \dot{\rho}_m^q) \, dt_m$$

$$= \frac{f}{c}[(\dot{\rho}_k^p - \dot{\rho}_k^q) - (\dot{\rho}_m^p - \dot{\rho}_m^q)]\xi_{km} \tag{10.69}$$

$$- \frac{f}{c}[(\dot{\rho}_k^p - \dot{\rho}_k^q) + (\dot{\rho}_m^p - \dot{\rho}_m^q)]\eta_{km}$$

For small baselines the coefficient of ξ_{km} converges toward zero. It is, therefore, possible in many applications to neglect this parameter or to model it by a constant, and only to model the difference η_{km} by a polynomial in time.

The partial derivatives for the initial ambiguities are 1. There are $(S - 1)$ $(R - 1)$ ambiguities to be estimated. The third line in (10.60) represents the ionospheric and tropospheric delay and the multipath terms. For short baselines the double-differenced ionospheric term $I_{km,\varphi}^{pq}(t)$ is usually not significant. For longer baselines it is best to use dual-frequency carrier phases and the ionospheric-free carrier phase function, such as (9.51). The tropospheric term is usually neglected. In extreme circumstances one can compute the tropospheric effect from an appropriate tropospheric model and meteorological data, or can attempt to estimate corrections to the vertical tropospheric delay. See Section 9.4 for additional details; for example, equation 9.62 could be used to compute the partial derivatives for the vertical tropospheric delay. Finally, the multipath usually remains unmodeled.

The list of parameters for the session with $R = 3$, $S = 4$, and $T = 3$ could possibly look like this:

$$X = (N_{12}^{12}, N_{12}^{13}, N_{12}^{14}, N_{13}^{12}, N_{13}^{13}, N_{13}^{14},$$

$$u_1, v_1, w_1, u_2, v_2, w_2, u_3, v_3, w_3, \text{clock}_{12}, \text{clock}_{13}) \tag{10.70}$$

One usually groups the parameters to conveniently achieve patterns in the design matrix and the normal matrix. The list (10.70) can readily be extended to include parameters for the vertical tropospheric delay. The design matrix corresponding to the sequence of parameters (10.70) and the sequence of double-difference observations (10.57) is shown in Table 10.8. Each baseline contributes one row to that table; each epoch contributes $(R - 1)$ rows. Equation (10.71) explains the shorthand notation used in writing the partial derivatives:

$$
\frac{\partial \varphi_{km}^{1q}(t)}{\partial (u_m, v_m, w_m)} = \begin{bmatrix} \dfrac{\partial \varphi_{km}^{12}(t)}{\partial u_m} & \dfrac{\partial \varphi_{km}^{12}(t)}{\partial v_m} & \dfrac{\partial \varphi_{km}^{12}(t)}{\partial w_m} \\[2.2ex] \dfrac{\partial \varphi_{km}^{13}(t)}{\partial u_m} & \dfrac{\partial \varphi_{km}^{13}(t)}{\partial v_m} & \dfrac{\partial \varphi_{km}^{13}(t)}{\partial w_m} \\[2.2ex] \dfrac{\partial \varphi_{km}^{14}(t)}{\partial u_m} & \dfrac{\partial \varphi_{km}^{14}(t)}{\partial v_m} & \dfrac{\partial \varphi_{km}^{14}(t)}{\partial w_m} \end{bmatrix}
\tag{10.71}
$$

The superscript q takes on the values from 2 to S.

The partial derivatives of the triple differences are related to those of the double differences, as follows:

$$
\frac{\partial \varphi_{km}^{pq}(j, i)}{\partial (\text{parameter})} = \frac{\partial \varphi_{km}^{pq}(j)}{\partial (\text{parameter})} - \frac{\partial \varphi_{km}^{pq}(i)}{\partial (\text{parameter})}
\tag{10.72}
$$

Equation (10.72) is valid because the triple difference is the difference of two double differences. The design matrix of the triple difference contains no columns for the initial ambiguities, because these parameters cancel during the differencing.

The topocentric distances are needed for some of the partial derivatives and for computing the misclosure:

$$
\begin{aligned}
l_{km}^{pq}(t) &= \varphi_{km,0}^{pq}(t) - \varphi_{km,b}^{pq}(t) \\[1.5ex]
&= \frac{f}{c} \rho_{km}^{pq}(t)\big|_{X_0} - \varphi_{km,b}^{p}(t)
\end{aligned}
\tag{10.73}
$$

The topocentric distances are computed as explained in Section 8.1, always using the current point of expansion X_0 as the adjustment is iterated.

The ordering scheme of base station and base satellite used for identifying the set of independent double-difference observations is not the only scheme available. It has been used here because of its simplicity. Other schemes might order baselines by length or number of observations, and then might find the independent set starting with the shortest baseline, etc. Any scheme that iden-

TABLE 10.8 Partial Derivatives for Double Differences

$N_{12}^{12} N_{12}^{13} N_{12}^{14}$	$N_{13}^{12} N_{13}^{13} N_{13}^{14}$	$u_1 \quad v_1 \quad w_1$	$u_2 \quad v_2 \quad w_2$	$u_3 \quad v_3 \quad w_3$
I	0	$\dfrac{\partial \varphi_{12}^{1q}}{\partial(u_1, v_1, w_1)}$	$\dfrac{\partial \varphi_{12}^{1q}}{\partial(u_2, v_2, w_2)}$	0
0	I	$\dfrac{\partial \varphi_{13}^{1q}}{\partial(u_1, v_1, w_1)}$	0	$\dfrac{\partial \varphi_{13}^{1q}}{\partial(u_3, v_3, w_3)}$

tifies a set of independent double-difference observations is suitable. Additional details on this subject are given in Section 10.4. An example where the base-station and base-satellite scheme requires a slight modification occurs when the base station does not observe at a certain epoch due to temporary signal blockage or to some other cause. If station 1 does not observe, then the double difference $\Delta\varphi_{23}^{pq}(t)$ can be computed for this particular epoch. Because of the relationship

$$\varphi_{23}^{pq}(t) = \varphi_{13}^{pq}(t) - \varphi_{12}^{pq}(t) \tag{10.74}$$

the ambiguity N_{23}^{pq} is related to the base-station ambiguities as

$$N_{23}^{pq}(1) = N_{13}^{pq}(1) - N_{12}^{pq}(1) \tag{10.75}$$

Introduction of N_{23}^{pq} as an additional parameter would create a singularity of the normal matrix because of the dependency expressed in (10.75). Instead of adding this new ambiguity, the base-station ambiguities N_{12}^{pq} and N_{13}^{pq} are given the coefficients 1 and -1 respectively in the design matrix. The partial derivatives with respect to the station coordinates can be computed as required by (10.74) and entered directly into the design matrix because the respective columns are already there. A similar situation arises when the base satellite changes. The linear functions in this case are

$$\varphi_{km}^{23}(t) = \varphi_{km}^{13}(t) - \varphi_{km}^{12}(t) \tag{10.76}$$

$$N_{km}^{23}(1) = N_{km}^{13}(1) - N_{km}^{12}(1) \tag{10.77}$$

The respective elements for the base-satellite ambiguities in the design matrix are, again, 1 and -1 in order to avoid a singular normal matrix. In general, if the preprocessing generates double differences in all combinations, i.e., $\varphi_{12}^{pq}(t)$, $\varphi_{13}^{pq}(t)$, and $\varphi_{23}^{pq}(t)$ in the case of three receivers, only two of these should be used because of the linear dependency between them.

Finally, in a session network as described here or in individual baseline processing, it is necessary to fix at least one station. This is most conveniently done by eliminating the three columns that refer to the coordinates of that station. The estimates for the other station coordinates are then automatically obtained relative to the fixed position. Unfortunately, the double differences do not contain sufficient geometric strength to solve for all station positions. See the discussion on a priori station position errors in Section 8.3.4.

10.2.2 Cycle Slip Detection and Correction

A cycle slip is a sudden jump in the carrier phase observable by an integer number of cycles. The fractional portion of the phase is not affected by this discontinuity in the observation sequence. Cycle slips are caused by the loss of lock of the phase lock loops. Loss of lock may occur briefly between two

TABLE 10.9 Effect of Cycle Slip on Carrier Phase Differences

Carrier phase				Double difference	Triple difference
$\varphi_k^p(i-2)$	$\varphi_m^p(i-2)$	$\varphi_k^q(i-2)$	$\varphi_m^q(i-2)$	$\partial_{km}^{pq}(i-2)$	$\partial_{km}^{pq}(i-1,i-2)$
$\varphi_k^p(i-1)$	$\varphi_m^p(i-1)$	$\varphi_k^q(i-1)$	$\varphi_m^q(i-1)$	$\partial_{km}^{pq}(i-1)$	$\partial_{km}^{pq}(i,i-1)-\Delta$
$\varphi_k^p(i)$	$\varphi_m^p(i)$	$\varphi_k^q(i)+\Delta$	$\varphi_m^q(i)$	$\partial_{km}^{pq}(i)-\Delta$	$\partial_{km}^{pq}(i+1,i)$
$\varphi_k^p(i+1)$	$\varphi_m^p(i+1)$	$\varphi_k^q(i+1)+\Delta$	$\varphi_m^q(i+1)$	$\partial_{km}^{pq}(i+1)-\Delta$	$\partial_{km}^{pq}(i+2,i+1)$
$\varphi_k^p(i+2)$	$\varphi_m^p(i+2)$	$\varphi_k^q(i+2)+\Delta$	$\varphi_m^q(i+2)$	$\partial_{km}^{pq}(i+2)-\Delta$	

epochs, or may last several minutes or more if the satellite signals cannot reach the antenna. With the pseudo-static technique this gap might be as much as one hour. It is a characteristic of cyclic slips that all observations taken after the cycle slip are shifted by the same integer amount. This is demonstrated in Table 10.9 where a cycle slip is assumed to have occurred for receiver k while observing satellite q between the epochs $i-1$ and i. The cycle slip is denoted by Δ. Because the double differences are a function of carrier phases at one epoch, all double differences starting with epoch i and above are offset by the amount Δ. Only one of the triple differences is affected by the cycle slip. This is a consequence of the triple difference being a difference of carrier phases at two different epochs. It is quite possible for several cycle slips to occur in one observation sequence. For each slip there is one corrupted triple difference and one additional step in the double-difference sequence. A cycle slip may be limited to one cycle, or could be millions of cycles. There are no unique methods for eliminating cycle slips; available procedures are not guaranteed to work in all cases.

One of the various methods for cycle slip fixing is to carry out a triple-difference solution first. Because only individual triple-difference observations are affected, they can be treated as observations containing blunders and dealt with as explained in Section 4.11. A simple method is to change the weights of those triple-difference observations that have particularly large residuals. Once the least-squares solution has converged, the residuals will indicate the size of the cycle slips. In a good solution the residuals are small fractions of a cycle, and only those triple differences containing cycle slips will have residuals close to one cycle or larger. Not only is triple-difference processing a robust technique for cycle slip detection, it also provides a good solution for the station coordinates, which, in turn, can be used as approximate positions in subsequent double-difference solutions.

Before computing the double-difference solution, the double-difference observations should be corrected for cycle slips identified from the triple-difference solution. Now only small jumps of the order of one or two cycles, not discovered in the triple-difference solution, remain in the observation set. The double-difference residuals will show a jump at the epoch where the slip occurred; that is, the residuals before epoch i and those at epoch i and later are shifted. The jumps can be located by analyzing the first differences of the residuals, that is, the differences of the residuals between adjacent epochs. The difference between the residuals of epochs i and $i - 1$ would indicate the cycle slip. All double differences must be corrected, starting with epoch i. If the solutions suffer from systematic errors, such as residual ionospheric effects, insufficiently accurate satellite ephemerides, or large errors in the coordinates of the fixed station, it is likely that the cycle slips may not be resolved at the one-cycle level. In such extreme cases, visual inspection of the residuals might be helpful in locating the problem.

If only two receivers observe, that is, if only one baseline is measured, it is not possible to determine the specific undifferenced phase sequence where the cycle slip occurred from analysis of the double-difference residuals. Consider the double differences

$$\varphi_{12}^{1p}(t) = [\varphi_1^1(t) - \varphi_2^1(t)] - [\varphi_1^p(t) - \varphi_2^p(t)] \tag{10.78}$$

for stations 1 and 2 and satellites 1 and p. The superscript p denotes the satellites 2 to S. Equation (10.78) shows that a cycle slip in φ_1^1 and φ_2^1 will affect all double differences for all satellites and cannot be separated. The slips Δ_1^1 and $-\Delta_2^1$ cause the same jump in the double-difference observation. The same is true for slips in the phase from station 1 to satellite p and station 2 to satellite p. However, a slip in the latter two phase sequences affects only the double differences containing satellite p. Other double-difference sequences are not affected.

For a session network, the double-difference observation is

$$\varphi_{1m}^{1p}(t) = [\varphi_1^1(t) - \varphi_m^1(t)] - [\varphi_1^p(t) - \varphi_m^p(t)] \tag{10.79}$$

The superscript p goes from 2 to S, and the subscript m runs from 2 to R. It is readily seen that a cycle slip in φ_1^1 affects all double-difference observations, an error in φ_m^1 affects all double differences pertaining to the baseline 1 to m, an error in φ_1^p affects all double differences containing satellite p, and an error in φ_m^p affects only one series of double differences, namely, the one that contains station m and satellite p. Thus, by analyzing the distribution of a blunder in the residuals of all double differences of the same epoch, we can identify the undifferenced phase observation sequence that contains the blunder. This identification gets more complicated if several slips occur at the same epoch. In session network processing it is always necessary to carry out cross checks. The same cycle slip must be verified in all relevant double differences before it can be declared an actual cycle slip. Whenever a cycle slip occurs in the undifferenced phase observations from the base station or to the base satellite, the cycle slip enters several double-difference sequences. It is not necessary that the undifferenced phase observations be corrected; it is sufficient to limit the correction to the double-difference phase observations.

Discrete Kalman filtering is one of several approaches suitable to automatically discover and eliminate cycle slips. Filtering can be applied to the undifferenced phases, the double differences and the triple differences of both L1 and L2 carriers, or any other observational sequence such as the ionospheric function (9.55), or the wide-lane function (10.24). Rather than filtering the rapidly changing carrier phases directly, it is preferable to process the misclosures $L = L_0 - L_b$, where, in the notation of adjustment, L_0 is the computed observation and L_b denotes the observation. The following standard formalism for Kalman filtering is taken from Gelb (1974):

$$X_{k+1} = \Phi_k X_k + W_k \tag{10.80}$$

$$Z_k = H_k X_k + V_k \tag{10.81}$$

$$\hat{X}_k(-) = \Phi_{k-1}\hat{X}_{k-1}(+) \tag{10.82}$$

$$P_k(-) = \Phi_{k-1}P_{k-1}(+)\Phi_{k-1}^T + Q_{k-1} \tag{10.83}$$

$$K_k = P_k(-)H_k^T[H_k P_k(-)H_k^T + R_k]^{-1} \tag{10.84}$$

$$\hat{X}_k(+) = \hat{X}_k(-) + K_k[Z_k - H_k\hat{X}_k(-)] \tag{10.85}$$

$$P_k(+) = (I - K_k H_k)P_k(-) \tag{10.86}$$

where $W_k \sim N_k(0, Q_k)$ and $V_k \sim N_k(0, R_k)$, and where k denotes the epoch of time, X is the system state vector, W is the random forcing function with a mean of zero and system driven noise matrix Q, Φ is the parameter transition matrix, Z is the vector of measurements, V is the measurement noise (residuals) with a mean of zero and measurement noise matrix R, P is the covariance matrix of the estimated vector, and K is the gain matrix. The $(-)$ indicates extrapolated quantities, and the $(+)$ indicates updated quantities. Equations (10.82) and (10.83) are the state and covariance extrapolation equations, and (10.84) through (10.86) are the measurement updating relations. The state vector

$$X = \begin{bmatrix} y \\ \dot{y} \\ \ddot{y} \end{bmatrix} \tag{10.87}$$

contains the respective function, for example, the L1 undifferenced carrier phase and its first and second derivative with respect to time. The applied dynamic system can be a simple third-order process corresponding to a quadratic polynomial with a parameter transition matrix (Goad, 1986),

$$\Phi_k = \begin{bmatrix} 1 & \Delta t & \frac{1}{2}\Delta t^2 \\ 0 & 1 & \Delta t \\ 0 & 0 & 1 \end{bmatrix} \tag{10.88}$$

and

$$H_k = [1 \quad 0 \quad 0] \tag{10.89}$$

Usually the system-driven noise matrix Q is assumed to be a diagonal 3×3 matrix, implying uncorrelated process noise. The measurement noise matrix is a scalar, because there is only one measurement per epoch. The idea of cycle slip detection with Kalman filtering is to compare the predicted value $\hat{X}_k(-)$ of (10.82) with the observed value Z_k. If the difference of both values exceeds the expected noise, then a cycle slip is suspected.

Analysis of dual-frequency carrier phase functions encounters additional difficulties because combinations of slips in L1 and L2 carrier phases can generate almost the same effects. For example, consider the ionospheric-free phase observable given in (9.51). Unfortunately, the ambiguities enter this function not as integers but in the combination of $\alpha N_1 + \beta N_2$, necessitating a search for a noninteger fraction in the residuals of the ionosphere-free phase. Note the simplified notation used here for the ambiguities in order to stress the fact that the following is valid for both the undifferenced and double-differences cases. Table 10.10 lists in column 1 small changes in the ambiguities and shows in columns 2 and 3 their effects on the ionospheric-free and the ionospheric phase observation respectively. The table shows that certain combinations of both integer ambiguities can produce almost identical changes in the ionospheric-free phase observation. It is interesting to note that the combination ($N_1 = 7$, $N_2 = 9$) causes a change of 0.033 cycles, whereas ($N_1 = 1$, $N_2 = 1$) causes a

TABLE 10.10 Effects of Small Jumps in Phase Ambiguity

ΔN_1	ΔN_2	$\alpha \Delta N_1 + \beta \Delta N_2$	$\Delta N_1 - \dfrac{f_1}{f_2} \Delta N_2$
±1	±1	± 0.562	∓ 0.283
± 2	± 2	± 1.124	∓ 0.567
± 1	± 2	∓ 1.422	∓ 1.567
± 2	± 3	∓ 0.860	∓ 1.850
± 3	± 4	∓ 0.298	∓ 2.133
± 4	± 5	± 0.264	∓ 2.417
± 5	± 6	± 0.827	∓ 2.700
± 6	± 7	± 1.389	∓ 2.983
± 5	± 7	∓ 1.157	∓ 3.983
± 6	± 8	∓ 0.595	∓ 4.267
± 7	± 9	∓ 0.033	∓ 4.550
± 8	± 10	± 0.529	∓ 4.833

TABLE 10.11 **Ambiguity Jumps with Similar Effects on the Ionospheric Phase Function**

ΔN_1	ΔN_2	$\Delta N_1 - \dfrac{f_1}{f_2} \Delta N_2$	ΔN_1	ΔN_2	$\Delta N_1 - \dfrac{f_1}{f_2} \Delta N_2$
-2	-7	6.9833	7	0	7.0000
-2	-6	5.7000	7	1	5.7167
-2	-5	4.4167	7	2	4.4333
-2	-4	3.1333	7	3	3.1500
-2	-3	1.8500	7	4	1.8667
-2	-2	0.5667	7	5	0.5833
-2	-1	-0.7167	7	6	-0.7000
-2	0	-2.0000	7	7	-1.9833
2	0	2.0000	-7	-7	1.9833
2	1	0.7167	-7	-6	0.7000
2	2	-0.5667	-7	-5	-0.5833
2	3	-1.8500	-7	-4	-1.8667
2	4	-3.1333	-7	-3	-3.1500
2	5	-4.4167	-7	-2	-4.4333
2	6	-5.7000	-7	-1	-5.7167
2	7	-6.9833	-7	0	-7.0000

change of 0.562 cycles, which, in turn, is almost identical with the change caused by ($N_1 = 8$, $N_2 = 10$). Assume for a moment that dual-frequency pseudoranges determine the ambiguities such that $|\Delta N_1| < 3$ and $|\Delta N_2| < 4$. In this case the slips could then be uniquely identified from the ionospheric-free phase combination (see entries in Table 10.10).

Table 10.11 shows another arrangement of integer ambiguities and their impact on the ionospheric phase combination. It is seen that, for example, the impact of the combinations $(-2, -7)$ and $(7, 0)$ differ by only 0.02 cycles. This amount is too small to be discovered reliably. Unfortunately, there is no unique combination of small $(\Delta N_1, \Delta N_2)$ that smoothes the ionospheric function if slips are present. As was mentioned above, there are other functions that are suitable for detecting cycle slips, for example, the functions W_k^p, R_k^p, $R_{k,1}^p$, which are listed in Table 10.2.

10.2.3 Ambiguity Fixing

When all cycle slips have been removed, one might attempt to fix the estimated real-valued ambiguities, denoted by $\hat{N}_{km}^{pq}$, to integer values. A simplified approach is to inspect the standard deviations of the adjusted $\hat{N}_{km}^{pq}$ parameters. If these are significantly smaller than one cycle, one can constrain the ambiguity parameters to the nearest integer and carry out a statistical test. There are, of course, $(R - 1)(S - 1)$ ambiguities to be inspected. The testing is conveniently done with sequential least squares as given in Table 4.2 (observation equation model with conditions between the parameters). For each ambiguity parameter a condition equation such as $\hat{N}_{km}^{pq} = N_{km}^{pq}$, where N_{km}^{pq} denotes the integer value, can be written. Each condition contributes to all adjusted parameters, including the other ambiguities, and to $\Delta V^T P V$. Because the ambiguity parameters are correlated, it is better to fix all ambiguity candidates in all combinations and to keep that combination which yields the smallest $\Delta V^T P V$. Imposing these conditions is a simple way of communicating to the adjustment our a priori knowledge that the phase ambiguities are, indeed, integers. For dual-frequency observations the phase ambiguities of each carrier are treated separately. One can first attempt to fix the wide-lane ambiguities, then adjust the ionospheric-free phase combination, allowing the L1 ambiguities to adjust, and finally fix the L1 ambiguities to integers.

Ambiguity fixing is routinely carried out for small baselines. The residual double-difference ionosphere and orbital errors make it difficult to fix ambiguities for long baselines. The ambiguity resolution for long baselines might be aided by simultaneously observing a network with significantly differing station separation, for example, a fiducial network spanning a continent, a regional network with 500-km station separations, and a couple of lines of 50–100 km (Schaffrin and Bock 1988). The idea is to fix the integer of small baselines first. Constraining some ambiguities to integers, even those pertaining to small baselines, introduces additional strength in the adjustment, which might make it possible to constrain some additional ambiguities during the next iteration. This procedure continues until no additional ambiguities can be fixed. The more ambiguities that are fixed (correctly), the better the expected result, because the ambiguity parameters are integers in the mathematical model and are not real numbers. Ambiguity fixing techniques are continuously being refined and are providing ways to fix integer ambiguities over longer and longer distances.

The computation effort for identifying and fixing the ambiguities is significantly reduced if so-called early exit strategies are implemented and a certain scheme is followed when selecting test candidates. The procedures discussed below are applicable to L1, L2, wide-lane, or narrow-lane ambiguities. It is assumed that the respective variance–covariance matrix of the estimated ambiguities is available. It is further assumed that for each ambiguity the set of trial values is known. The trial values can be derived from the float solution or an initial pseudorange solution.

10.2.3.1 An Ordering Scheme

Table 10.12 shows sets of trial integer ambiguities which are ordered as suggested by Frei and Beutler (1990). Without limiting the generality of the ordering scheme the table refers to a simplified situation. For example, we assume that there are only $s = 4$ ambiguities referring to five satellites and two stations. It is further assumed that the estimated real-valued ambiguities have the same value, say 4.8, and that the estimated integer range for each integer ambiguity is also the same, say $\{3, 4, 5, 6\}$. If the symbol d_i denotes the number of trial values for the ith ambiguity, then in this simplified case $d_i = 4$.

TABLE 10.12   Ordering the Trial Integer Ambiguity Sets

$N_{12}^{12}\ N_{12}^{13}\ N_{12}^{14}\ N_{12}^{15}$	$N_{12}^{12}\ N_{12}^{13}\ N_{12}^{14}\ N_{12}^{15}$	$N_{12}^{12}\ N_{12}^{13}\ N_{12}^{14}\ N_{12}^{15}$	$N_{12}^{12}\ N_{12}^{13}\ N_{12}^{14}\ N_{12}^{15}$
5 5 5 5	5 4 5 5	5 6 5 5	5 3 5 5
5 5 5 4	5 4 5 4	5 6 5 4	5 3 5 4
5 5 5 6	5 4 5 6	5 6 5 6	5 3 5 6
5 5 5 3	5 4 5 3	5 6 5 3	5 3 5 3
5 5 4 5	5 4 4 5	5 6 4 5	5 3 4 5
5 5 4 4	5 4 4 4	5 6 4 4	5 3 4 4
5 5 4 6	5 4 4 6	5 6 4 6	5 3 4 6
5 5 4 3	5 4 4 3	5 6 4 3	5 5 4 3
5 5 6 5	5 4 6 5	5 6 6 5	5 3 6 5
5 5 6 4	5 4 6 4	5 6 6 4	5 3 6 4
5 5 6 6	5 4 6 6	5 6 6 6	5 3 6 6
5 5 6 3	5 4 6 3	5 6 6 3	5 3 6 3
5 5 3 5	5 4 3 5	5 6 3 5	5 3 3 5
5 5 3 4	5 4 3 4	5 6 3 4	5 3 3 4
5 5 3 6	5 4 3 6	5 6 3 6	5 3 3 6
5 5 3 3	5 4 3 3	5 6 3 3	5 3 3 3

assumed real-valued ambiguities: $N_{12}^{1i} = 4.8$
range of trial integer ambiguities: 3-6

The first entry in column 1 contains the nearest integers to the real-valued ambiguities. For this example the entry is {5, 5, 5, 5} because these integers are closest to 4.8. For the next three sets in column 1 the last integer is varied by choosing the second nearest integer, the third nearest integer, etc., until the complete range for the last ambiguity has been used. Thus far we have created d_s trial sets by varying the sth ambiguity. Continuing down column 1 we see that the second to last ambiguity is now varied beginning with the second nearest integer, third nearest integer, etc. In each case the last ambiguity varies over its complete range. Thus, column 1 contains all trial sets for which the second last and the last ambiguities vary. There are $d_{s-1}d_s$ trial sets. In columns 2 to 4 the third last ambiguity is varied. For each trial value the second to last ambiguity varies over its range, and for each value of the second to last ambiguity, the last ambiguity varies over its range. Each trial value of the first ambiguity will generate a table like Table 10.12 as the third to last, the second to last, and the last ambiguities are varied. The variations of the first ambiguity are not shown in order to keep the table manageable. One can readily deduce from the above explanations that a generalization of the ordering scheme is the following: The last ambiguity varies fastest, the second to last ambiguity varies second-fastest, and so on, until the first ambiguity is reached, which varies the slowest. The total number of trial sets for s ambiguities is

$$N = \prod_{i=1}^{s} d_i \qquad (10.90)$$

The number N increases rapidly. For example, $N = 10^4$ if $s = 4$ and all $d_i = 10$. The number N increases rapidly as more satellites are simultaneously observed while decreases with a smaller range of trial values. To reduce computations when testing the integer ambiguity sets, it is important that N be as small as possible and that optimized processing strategies be used. The number N can be reduced by having good prior information on the station positions based on pseudorange solutions or by accumulating carrier phase observations over sufficiently long periods of time. The ordering of the trial ambiguity sets and the use of early exit strategies will lead to rapid identification of acceptable integer ambiguity sets.

10.2.3.2 Testing Complete Sets   Each trial ambiguity set is subjected to a general linear hypothesis test as described in Section 4.9.4. The zero hypothesis H_0 is

$$H_0: A_2 X^* + L_2 = 0 \qquad (10.91)$$

These are s conditions. The zero hypothesis states that a particular trial integer set is compatible with the estimated ambiguities. Assuming that the parameters are arranged as in (10.70), i.e., the ambiguity parameters are grouped at the beginning of the X vector, then the coefficient matrix A_2 is

$$A_2 = [{}_s I_s \quad {}_s 0_{u-s}] \qquad (10.92)$$

This special grouping is not a requirement but makes the simple form (10.92) possible. The vector X^* denotes the adjusted parameters of the float solution containing the real-valued estimates of ambiguities. The vector L_2 contains the trial integer ambiguity set to be tested. For example,

$$L_2^T = -[N_{12}^{12} \quad N_{12}^{13} \quad N_{12}^{14} \quad N_{12}^{15}]_k \qquad (10.93)$$

contains the kth trial set. The identifier k increases from 1 to N beginning with the first entry in column 1, proceeding downward, continuing with column 2, proceeding downward, etc., until the last set is reached. The zero hypothesis (10.91) for the kth set becomes

$$f \equiv \begin{bmatrix} \hat{N}_{12}^{12} - N_{12}^{12} \\ \hat{N}_{12}^{13} - N_{12}^{13} \\ \hat{N}_{12}^{14} - N_{12}^{14} \\ \hat{N}_{12}^{15} - N_{12}^{15} \end{bmatrix}_k = \begin{bmatrix} 0 \\ 0 \\ 0 \\ 0 \end{bmatrix} \qquad (10.94)$$

This condition gives $\Delta V^T PV$, which can be tested against the F distribution according to (4.285):

$$\frac{\Delta V^T PV}{V^T PV^*} \frac{n - u}{s} \sim F_{s, n-u, \alpha} \tag{10.95}$$

The quadratic form $V^T PV^*$ and the degrees of freedom $(n - u)$ refer to the float solution. If the statistic (10.95) is greater than the critical value $F_{s, n-u, \alpha}$, the kth trial set is rejected at the $(100 - \alpha)\%$ significance level. The test (10.95) must be applied to each trial set and a decision must be made to either accept or reject the set.

The computation of $\Delta V^T PV$ allows a high degree of optimization which reduces computational effort and saves time. According to (4.284) the respective expression is

$$\begin{aligned} \Delta V^T PV &= (A_2 X^* + L_2)^T T (A_2 X^* + L_2) \\ &= (A_2 X^* + L_2)^T LL^T (A_2 X^* + L_2) \\ &= f^T LL^T f \end{aligned} \tag{10.96}$$

with

$$T = (A_2 N^{-1} A_2^T)^{-1} \tag{10.97}$$

Since the T matrix is positive definite, it is factored as $T = LL^T$ according to the Cholesky factorization where L is a lower triangular matrix. Because the matrix T is the same for all ambiguity sets, the factorization is carried out only once, which saves computer time. Additional time savings are achieved by accumulating the product $f^T LL^T f$ in steps. If

$$g = L^T f \tag{10.98}$$

then $\Delta V^T PV$ can be written as

$$\Delta V^T PV = g^T g = \sum_{i=1}^{s} g_i^2 \tag{10.99}$$

As soon as the first element g_1 has been computed, it can be squared and taken as the first estimate of the quadratic form. Note that $\Delta V^T PV \geq g_1^2$. The value $\Delta V^T PV = g_1^2$ is substituted in (10.95) to compute the test statistic, which is then compared with the critical F value. If that test fails, the respective trial ambiguity set can immediately be rejected. There is no need in this case to compute the remaining g_i values. If the test passes, then the next value, g_2, is computed and the test statistic is computed based on $\Delta V^T PV = g_1^2 + g_2^2$. If this test fails, the ambiguity set is rejected; otherwise g_3 is computed, etc. This procedure continues until either the zero hypothesis is rejected or all g_i are computed and the complete sum of the g square terms is known. This early exit strategy rejects the zero hypothesis without necessarily computing all g_i coefficients. This is the approach used by Euler and Landau (1992) and Blomenhofer et al. (1993).

Using the early exit strategy in combination with the ordering scheme of Table 10.12 accelerates the testing even more. Closer inspection of the product

$$g^{\mathrm{T}} = [f_1 \ \ f_2 \ \ f_3 \ \ f_4] \begin{bmatrix} l_{11} & 0 & 0 & 0 \\ l_{21} & l_{22} & 0 & 0 \\ l_{31} & l_{32} & l_{33} & 0 \\ l_{41} & l_{42} & l_{43} & l_{44} \end{bmatrix} \tag{10.100}$$

reveals trial sets that need not be tested explicitly, depending on previous decisions. As soon as the F test (10.95) fails using all g terms, then all subsequent sets in that block can be skipped. Because of the special ordering, the absolute value of f_4 and, consequently, g_4^2 increases from one trial set to the next within the block. For example, if the set $\{5 \ \ 5 \ \ 4 \ \ 4\}$ fails the test, then the sets $\{5 \ \ 5 \ \ 4 \ \ 6\}$ and $\{5 \ \ 5 \ \ 4 \ \ 3\}$ also fail and need not be computed explicitly.

10.2.3.3 Testing Ambiguity Differences
Frei and Beutler (1990) test ambiguity differences instead of the complete ambiguity sets. They make ample use of the same ordering scheme of the trial ambiguity sets in order to reduce computation time. The procedure starts with testing the difference of the first and second ambiguity. Thus the zero hypothesis is

$$H_0: A_2 X^* + L_2 = 0 \tag{10.101}$$

with

$$A_2 = [1 \ \ -1 \ \ 0 \ \cdots \ 0] \tag{10.102}$$

$$L_2 = -(N_{12}^{12} - N_{12}^{13}) \tag{10.103}$$

$$f = (\hat{N}_{12}^{12} - \hat{N}_{12}^{13}) - (N_{12}^{12} - N_{12}^{13}) \tag{10.104}$$

Since matrix T of (10.97) is now of size (1×1) we obtain $\Delta V^{\mathrm{T}} P V = Tf^2$. The F test is performed according to (10.95) using $s = 1$. If this test fails, it can be deduced from Table 10.12 that the next

$$N_2 = \prod_{i=3}^{s} d_i \tag{10.105}$$

sets can be skipped (including the set being tested). If the test passes, then the differences between the first and third and the second and third are tested. If either of these tests fails, then the next

$$N_3 = \prod_{i=4}^{s} d_i \qquad (10.106)$$

sets can be skipped (including the set being tested). If both tests pass, the procedure continues with testing differences, with respect to the fourth, the fifth, and eventually the last ambiguity. Whenever sets are skipped, the procedure begins again with testing the differences between the first and the second ambiguity.

10.2.3.4 Sequential Reduction of Ambiguity Range

The integer ambiguity resolution accelerates if the number of candidates d_i for individual ambiguities is small. The smaller these search ranges are, the fewer ambiguity sets need to be tested. Chen (1993) proposes a method that reduces d_i by taking advantage of the effect of presumed fixed ambiguities in the search range. He chooses a Kalman filter implementation. Here the equivalent least-squares formulation is presented to indicate the basic idea of the method. Even though this method is not yet fully tested, it is outlined here because of its potential, and because it provides another interesting aspect to least-squares with conditions. Let us summarize, again, the principal expressions for imposing conditions:

$$H_0: A_2 X^* + L_2 = 0 \qquad (10.107)$$

While the changes in the parameters are

$$\Delta X = -N^{-1} A_2 T (A_2 X^* + L_2) \qquad (10.108)$$

the cofactor matrix of the parameters changes by

$$\Delta Q = -N^{-1} A_2^T T A_2 N^{-1} \qquad (10.109)$$

The matrix T is given in (10.97). See also the expressions in Table 4.2. As was discussed in Chapter 4, the diagonal elements of ΔQ are less than zero. Consequently, the diagonal elements of the cofactor matrix $Q + \Delta Q$, which includes the effects of the hypothesis (10.107), are smaller than those of the unconstrained case, $Q = N^{-1}$. This fact is used to reduce the search range for ambiguities. We also note that the matrix ΔQ needs to be computed only once if different values are constrained, i.e., L_2 is different, but A_2 is the same.

The system underlying the method is more easily understood if one assumes that the ambiguities are sequentially organized. Step 1 of the method begins with the search range and the number of trial integers d_1 for the first ambiguity N_{12}^{12}. These are usually derived from the float solution and the respective standard deviation. The true value of the N_{12}^{12} is one of these d_1 candidates. For each candidate the parameter correction $\Delta X_{1,i}(N_{12,i}^{12})$ is computed according to (10.108). Next the $\Delta Q(N_{12}^{12})$ matrix is computed. This matrix is the same for all d_1 candidate values. Using the thus computed ΔQ, we obtain from the matrix $Q + \Delta Q$ the standard deviation $\sigma(N_{12}^{13})$ for the second ambiguity. This standard deviation corresponds to that of the float solution but with the first ambiguity held fixed. Proper scaling with the a priori variance of unit weight is assumed.

In step 2 we compute d_2 search ranges for the second ambiguity. Each range is derived from $\sigma(N_{12}^{13})$ and the estimate for the second ambiguity $\hat{N}_{12,i}^{13}$ as given by $\Delta X_{1,i}(N_{12,i}^{12})$. For example,

$$\hat{N}_{12,i}^{13} - \sigma(N_{12}^{13}) \leq N_{12}^{13} < \hat{N}_{12,i}^{13} + \sigma(N_{12}^{13}) \qquad (10.110)$$

If no integer falls within this ith search region, then the first ambiguity candidate, $N_{12,i}^{12}$ is discarded. If $N_{12,i}^{12}$ is the true value, then the N_{12}^{13} integer will be included in the range (10.110). It is possible that the range contains several integers; all of them become candidates for N_{12}^{13}. At this point in the solution we have one or several candidates for the first ambiguity, and one or several candidates for the second ambiguity. For each pair of candidates we can compute the parameter corrections denoted by $\Delta X_{2,i,j}(N_{12,i}^{12}, N_{12,j}^{13})$ and the matrix $\Delta Q(N_{12}^{12}, N_{12}^{13})$, which is valid for all candidate pairs. Using the thus computed ΔQ, we obtain from the matrix $Q + \Delta Q$ the standard deviation $\sigma(N_{12}^{14})$ for the third ambiguity.

In step 3 we compute one search range of the third ambiguity for each pair $(N_{12,i}^{12}, N_{12,j}^{13})$. Each range is derived from $\sigma(N_{12}^{14})$ and the estimate for the third ambiguity $\hat{N}_{12,i,j}^{14}$ as given by $\Delta X_{2,i,j}(N_{12,i}^{12}, N_{12,j}^{13})$. If the range does not contain an integer, then that particular pair of first and second ambiguity candidates is ignored. If the ranges contain one or several integers then these become candidates for N_{12}^{14}. If $N_{12,i}^{12}$ and $N_{12,j}^{13}$ are the true values, then the range will contain N_{12}^{14}.

This process continues until the candidates for the last ambiguity are known. As this algorithm steps through the ambiguities sequentially from the first to the last, the number of candidates for each ambiguity is accumulated. An early exist strategy is used to abandon attempts to fix the ambiguities and to keep the float solutions as soon as the accumulating counter exceeds a specific value. The computation is efficient, because the search range decreases as more ambiguities have previously been fixed.

10.2.3.5 Eliminating All Position Parameters
Another algebraic variation with some degree of appeal is the elimination of all station parameters through appropriate transformation. Consider the $(S - 1)$ double-difference observation equations of the same epoch in the general form

$$_{S-1}V_1 = {}_{S-1}A_{33}X_1 + {}_{S-1}N_1 + {}_{S-1}L_1 \qquad (10.111)$$

The station coordinate parameters are contained in X, and N contains the ambiguities

$$N = [N_{12}^{12} \quad N_{12}^{13} \quad \cdots \quad N_{12}^{1S_1}]^T \tag{10.112}$$

There is a matrix E,

$$E = ({}_{S-1}F_3 \vdots {}_{S-1}G_{S-1-3}) \tag{10.113}$$

such that

$${}_{S-1-3}G_{S-1}^T {}_{S-1}A_3 = 0 \tag{10.114}$$

The columns of the matrix G span the null space of A or AA^T. See the discussion around equation (4.249) in Chapter 4 for another application. Using the property (10.114) the observation equation (10.111) can be transformed as follows:

$$G^T V = G^T N + G^T L \tag{10.115}$$

or

$$G^T(N - V + L) = 0 \tag{10.116}$$

Expression (10.116) represents $(S - 1 - 3)$ equations relating $(S - 1)$ ambiguities, residuals, and the misclosures. Double differences to five satellites generate one equation of the type (10.116). Adding another satellite adds another equation but also adds another ambiguity.

There are two ways to proceed with (10.116). If we assume that the residuals are small, that is, the unmodeled errors for the double differences are negligible, then we could set $V = 0$. In this case (10.116) represents $(S - 1 - 3)$ conditions for the $(S - 1)$ ambiguities. Through trial and error one could attempt to identify the correct set of ambiguities even with observations from just one epoch. Perhaps there might be several sets of ambiguities fulfilling these conditions. The identification process accelerates if the ranges of permissible ambiguities are known (from preprocessing). Each epoch adds another set of $(S - 1 - 3)$ equations to (10.116). The elements of G change with time as the coefficients in A change with the motion of the satellites. Eventually enough epochs will be available with different G matrices to allow a unique identification of the ambiguity. Only the correct set of ambiguities will always fulfill (10.116). In actual application where the residuals are not zero but are small, one would be looking for ambiguity values that are close to integers.

The search method described here does not require a least-squares solution;

this method is also discussed in Melbourne (1985). A least-squares formulation is readily obtained by recognizing that (10.115) is of the form of the mixed adjustment model. Refer to equation (4.41) of Chapter 4 for a discussion on the mixed adjustment model. Combining several epochs and adjusting leads to the same least-squares solution discussed above even though the station coordinate parameters are not present. Any of the search strategies discussed above could then be applied to the estimated set of ambiguities.

An equation like (10.116) can be written for L2 carrier phases. Thus, in case of dual-frequency data there are $(2S - 8)$ observations for $2(S - 1)$ ambiguities. For short baselines, where the residual ionosphere is negligible, the double-difference ionospheric phase function provides $(S - 1)$ conditions between the ambiguities.

10.2.3.6 Dual-Frequency Ambiguities for Short Baselines

The term short baseline generally implies the understanding that the ionospheric effects on the double differences are negligible. It further implies that the tropospheric effect is also negligible or that the observations have been corrected accordingly. The multipath is generally not corrected and is allowed to impose its signature on the residuals. Thus, the following observation equations apply:

$$\varphi_{km,1}^{pq}(t) = \frac{f_1}{c} \rho_{km}^{pq}(t) + N_{km,1}^{pq}(1) \qquad (10.117)$$

$$\varphi_{km,2}^{pq}(t) = \frac{f_2}{c} \rho_{km}^{pq}(t) + N_{km,2}^{pq}(1) \qquad (10.118)$$

$$\varphi_{km,1}^{pq}(t) - \frac{f_1}{f_2} \varphi_{km,2}^{pq}(t) = N_{km,1}^{pq}(1) - \frac{f_1}{f_2} N_{km,2}^{pq}(1) \qquad (10.119)$$

The first two observation equations follow from (8.41) as applied to the L1 and L2 carrier phase double differences. The shorthand notation has been used to express the double difference of the topocentric distances. The third equation is the ionospheric observable (9.55) appropriately double differenced. Because the double-differenced ionospheric term cancels, equation (10.119) takes on the character of a condition between the ambiguities; that is, the ambiguities must fulfill the function of the carrier phases given on the left-hand side. In terms of adjustment theory, we have the case of the observation equation model with conditions on the parameters. The general expressions of Table 4.5 of Chapter 4 apply. It is understood that the variance–covariance propagation must be applied to the undifferenced observation in order to obtain the stochastic model for the double-difference observations $\varphi_{km,1}^{pq}(t)$ and $\varphi_{km,2}^{pq}(t)$. Usually carrier phases of L1 and L2 are treated as uncorrelated. Rather than considering (10.119) as a condition, one can apply variance–covariance propagation to the ionospheric function and can compute a standard deviation for this function. Consequently, equation (10.119) changes its character from a condition equation to a regular observation equation or an "observed function" of some parameters. In any case, the usual least-squares formulations of Chapter 4 apply.

As was discussed in detail in Section 10.1.2, one can create many functions of the basic observables through appropriate transformations. Very popular are those functions that contain the wide lane. Consider the following example:

$$
\frac{f_1^2}{f_1^2 - f_2^2} \, \varphi_{km,1}^{pq}(t) - \frac{f_1 f_2}{f_1^2 - f_2^2} \, \varphi_{km,2}^{pq}(t)
$$

$$
= \frac{f_1}{c} \, \rho_{km}^{pq}(t) + \frac{f_1}{f_1 + f_2} \, N_{km,1}^{pq}(1) + \frac{f_1 f_2}{f_1^2 - f_2^2} \, N_{km,w}^{pq}(1)
$$

$$(10.120)$$

$$
\varphi_{km,1}^{pq}(t) - \frac{f_1}{f_2} \, \varphi_{km,2}^{pq}(t) = \left(1 - \frac{f_1}{f_2}\right) N_{km,1}^{pq}(1) - \frac{f_1}{f_2} \, N_{km,w}^{pq}(1) \quad (10.121)
$$

Equation (10.120) is the ionospheric-free observable (9.51) for double differences and (10.121) is, again, the ionospheric function. See also Table 10.2. The wide-lane ambiguity was introduced using the relation (10.20). Note that the L1 ambiguity is kept in the expression. Equation (10.121) can, again, be interpreted as a condition or as an observed function of the ambiguity parameters. Another interesting transformation is

$$
\frac{1}{2} \varphi_{km,1}^{pq}(t) + \frac{f_1}{2f_2} \, \varphi_{km,2}^{pq}(t)
$$

$$
= \frac{f_1}{c} \, \rho_{km}^{pq}(t) + \frac{f_1 + f_2}{2f_2} \, N_{km,1}^{pq}(1) - \frac{f_1}{2f_2} \, N_{km,w}^{pq}(1) \quad (10.122)
$$

$$
\varphi_{km,1}^{pq}(t) - \frac{f_1}{f_2} \, \varphi_{km,2}^{pq}(t) = \left(1 - \frac{f_1}{f_2}\right) N_{km,1}^{pq}(1) - \frac{f_1}{f_2} \, N_{km,w}^{pq}(1) \quad (10.123)
$$

or the pair (10.120) and (10.122). In fact, the latter functions are uncorrelated as is readily verified by applying the law of variance–covariance propagation.

All the transformations listed above are equivalent from the statistical point of view, provided that variance–covariance propagation is applied rigorously and that only random errors are present. In general, the complete list of ambiguity parameters in dual-frequency adjustments might look like

$$
N = [N_{12,w}^{12} \quad N_{12,w}^{13} \quad \cdots \quad N_{12,w}^{1S} \quad N_{12,1}^{12} \quad N_{12,1}^{13} \quad \cdots \quad N_{12,1}^{1S}] \quad (10.124)
$$

The testing for this extended set of ambiguities, i.e., identification of the integer values by statistical means, can proceed in the same manner as explained in previous sections for the single frequency case. For example, in the ordering scheme of Section 10.2.3.1 one should expect that the wide-lane ambiguities will not add many combinations, because their search range will generally be very small, e.g., one or two cycles. One could try to fix the wide-lane ambiguities first before attempting to fix the L1 ambiguities. Being able to fix the wide-lane ambiguities adds geometric strength to the solution by reducing the search range of the other ambiguities. The same line of thinking as explained in Section 10.2.3.4 applies. In fact, one would logically begin with the wide-lane ambiguity. The advantage of the wide-lane parameterization thus becomes obvious: the ability to add geometric strength by constraining the wide-lane ambiguities as early as possible in a sequential-type of solution, and thus, hopefully, also to be able to find the integers of the other ambiguities as fast as possible. This strategy is one of the pillars of the rapid static technique. Another advantage of the wide-lane parameterization is discussed in Section 10.4 for the case of long baselines.

10.2.3.7 Discernibility The ambiguity testing outlined above is essentially a repeated application of null hypotheses testing. Each ambiguity set constitutes a null hypothesis. The procedure tests the changes in the sum of the squares due to the constants. The decision to accept or to reject the null hypothesis is based on the probability of type I errors, which is usually taken to be $\alpha = 0.05$. In many cases several of the null hypotheses will pass the test identifying several qualifying ambiguity sets. This will happen whenever there is not enough information in the observations to determine the integers uniquely. Additional observations might help resolve this situation. However, in the presence of large model errors, it might not be possible to identify the integer ambiguities, regardless of the length of observation session. The ambiguity set that generates the smallest $\Delta V^{\mathrm{T}} P V$ fits the float solution best, and consequently, is considered to be the most valid fixed solution. The goal of additional statistical testing is to provide conditions that make it possible to discard all or some of the ambiguity sets that passed the null hypotheses tests. The type II error is used to make these additional decisions.

The alternative hypothesis H_a is always relative to the null hypothesis H_0. The formalism for the null hypothesis is given in Sections 4.9.3 and 4.9.4. The line of thought given in these two sections is applicable to completing the formalism for the alternative hypothesis. In general, the null and alternative hypotheses are

$$H_0: A_2 X^* + L_2 = 0 \tag{10.125}$$

$$H_a: A_2 X^* + L_2 + W_2 = 0 \tag{10.126}$$

Under the null hypothesis the expected value of the constraint is zero. See equation (4.275). Thus,

$$E(Z_{3,0}) \equiv E(A_2 X^* + L_2) = 0 \tag{10.127}$$

Because W_2 is a constant, it follows that

$$E(Z_{3,a}) \equiv E(A_2 X^* + L_2 + W_2) = W_2 \tag{10.128}$$

The random variable $Z_{3,a}$ is multivariate normal distributed with mean W_2, i.e.,

$$Z_{3,a} \sim N_{n-r}(W_2, \sigma_0^2 T^{-1}) \tag{10.129}$$

See equation (4.277) for the corresponding expression for the zero hypothesis. The matrix T has the same meaning as in Section 4.9.4, that is,

$$T = (A_2 N_1^{-1} A_2^T)^{-1} \tag{10.130}$$

The next step is to diagonalize the covariance matrix of $Z_{3,a}$ and to compute the sum of the squares of the transformed random variables. These newly formed random variables have a unit variate normal distribution with a nonzero mean. According to Section C.5.5 the sum of the squares has a noncentral chi-square distribution. Thus,

$$\frac{\Delta V^T PV}{\sigma_0^2} = \frac{Z_{3,a}^T T Z_{3,a}}{\sigma_0^2} \sim \chi_{n_2, \lambda}^2 \tag{10.131}$$

where the noncentrality parameter is

$$\lambda = \frac{W_2^T T W_2}{\sigma_0^2} \tag{10.132}$$

The reader is referred to the statistical literature, such as Koch (1988), for additional details on noncentral distributions and their respective derivations. Finally, the ratio

$$\frac{\Delta V^T PV}{V^T PV*} \frac{n_1 - r}{n_2} \sim F_{n_2, n_1 - r, \lambda} \tag{10.133}$$

has a noncentral F distribution with noncentrality λ. In summary, if the test statistic computed under the specifications of H_0 fulfills $F \leq F_{n_2, n_1 - r, \alpha}$, then H_0 is accepted with a type I error of α. The alternative hypothesis H_a can be separated from H_0 with the power $1 - \beta(\alpha, \lambda)$. The type II error is

$$\beta(\alpha, \lambda) = \int_0^{F_{n_2, n_1 - r, \alpha}} F_{n_2, n_1 - r, \lambda} \, dx \tag{10.134}$$

The integration is taken over the noncentral F-distribution function from zero to the value $F_{n_2, n_1 - r, \alpha}$, which is specified by the significance level.

Because the noncentrality is different for each alternative hypothesis according to (10.132), the type II error $\beta(\alpha, \lambda)$ also varies with H_a. Rather than using the individual type II errors to make decisions, Euler and Schaffrin (1990) propose a ratio of noncentrality parameters. The float solution is designated as the common alternative hypothesis H_a for all null hypotheses. In this case the value W_2 in (10.126) is

$$W_2 = -(A_2 X^* + L_2)$$ (10.135)

and the noncentrality parameter becomes

$$\lambda \equiv \frac{W_2^T T W_2}{\sigma_0^2} \approx \frac{\Delta V^T P V}{\sigma_0^2}$$ (10.136)

where $\Delta V^T P V$ is the change of the sum of squares due to the constraint of the null hypothesis.

Let the null hypothesis that causes the smallest change $\Delta V^T P V$ be denoted by H_{sm}. The change in the sum of the squares and the noncentrality are $(\Delta V^T P V)_{sm}$ and λ_{sm}, respectively. For any other null hypothesis the inequality $\lambda_j > \lambda_{sm}$ is true. If

$$\frac{(\Delta V^T P V)_j}{(\Delta V^T P V)_{sm}} \approx \frac{\lambda_j}{\lambda_{sm}} \geq \lambda_0(\alpha, \beta_{sm}, \beta_j)$$ (10.137)

then the two ambiguity sets comprising the null hypotheses H_{sm} and H_j are sufficiently discernible. Both hypotheses are sufficiently different to be distinguishable by means of their type II errors. Because of its better compatibility with the float solution, the ambiguity set of the H_{sm} hypothesis is kept, and the set comprising H_j is discarded. All null hypotheses for which $\lambda_j > \lambda_{sm} \lambda_0$ can be discarded. Figure 10.18 shows the ratio $\lambda_0(\alpha, \beta_{sm}, \beta_j)$ as a function of the

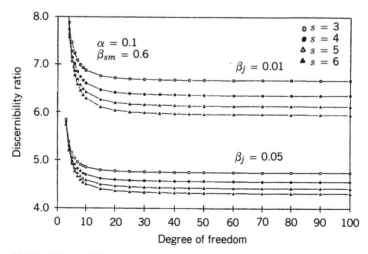

Figure 10.18 Discernibility ratio $\lambda_0(\alpha, \beta_{sm}, \beta_j)$. (After Euler and Schaffrin, 1990. Permission by Springer-Verlag, New York.) The symbol s corresponds to n_2 as used in (10.133).

degree of freedom and the number of conditions. Euler and Schaffrin (1990) recommend a ratio between 5 and 10. This leads to a relatively large β_{sm} and a smaller β_j. Recall that the type II error equals the probability of accepting the wrong null hypothesis. Since H_{sm} is the hypothesis with the least impact on the adjustment, i.e., the most compatible with the float solution, it is desirable to have $\beta_{sm} > \beta_j$. The larger that $\lambda_0(\alpha, \beta_{sm}, \beta_j)$ is, the more likely it is that the wrong zero hypothesis is accepted. From Figure 10.18 we see that it will take longer, i.e., more observations must be collected, to discard all but one of the null hypotheses. Observing more satellites reduces the ratio for the same type II errors. This is clearly a desirable feature and corresponds with the expectations that the ambiguities can be fixed faster, the more satellites that are observed.

In many software packages a fixed value for the ratio of the best and the second best solution, for example,

$$\frac{(\Delta V^T P V)_{2nd\,smallest}}{(\Delta V^T P V)_{sm}} > 3 \qquad (10.138)$$

is used to decide on discernibility. The explanations given above lend some theoretical justification to this commonly used practice, at least for a high degree of freedom.

10.2.4 Primary and Secondary Satellites

With classical static GPS techniques, a baseline can be determined from carrier phase observations using only four satellites, provided the time period of the observations is sufficiently long. The respective three double-difference equations contain the station coordinates and three ambiguity parameters. The change in the satellite geometry over time makes it possible to estimate all parameters. Any satellites in addition to the four satellites can be considered redundant, although these extra satellites improve the overall solution geometry, and thus help in accelerating the baseline solution. Hatch (1990) takes advantage of this redundancy of extra satellites in order to develop yet another procedure for rapidly identifying the integer ambiguities. Hatch discusses this method in connection with kinematic techniques.

The satellites are grouped into primary satellites, subsequently denoted by the subscript p (lowercase), and secondary satellites, denoted by the subscript s (lowercase). For example, the complete list of ambiguities can be written as

$$N = [N_{12,p}^{12} \quad N_{12,p}^{13} \quad N_{12,p}^{14} \vdots N_{12,s}^{15} \cdots N_{12,s}^{1S}] \qquad (10.139)$$

The observations are grouped similarly:

$$V_p = A_p X + L_p \qquad (10.140)$$

$$V_s = A_s X + L_s \qquad (10.141)$$

Each of the two groups may contain just one observation or several observations. It is important to note that the design matrices A_p and A_s do not contain columns for the ambiguities! The method assumes that the ambiguities are known and evaluates the effect of that assumption. The assumed ambiguities appropriately contribute to the misclosures L_p and L_s. For example, several trial sets for the three primary ambiguities might have been derived from the pseudorange solution taking the accuracy of that solution into account. Because only three primary ambiguities are varied, the term "search cube" is sometimes used. Details of determining trial ambiguity sets for a given physical search space are discussed in Section 10.3.2.1. Of course, one could get these trial sets from the float solution (when several epochs are available). For each of the primary trial ambiguity sets we can compute the change in position using the usual least-squares formulation

$$X_p = -(A_p^T P_p A_p)^{-1} A_p^T P_p L_p \tag{10.142}$$

This is a nonredundant solution because only three double-difference observations are involved. Each ambiguity trial set gives a different position X_p. The matrix A_p and the inverse matrix need to be computed only once. The coefficients of A_p are evaluated for the initial position obtained from the pseudorange or float solution. For each solution X_p, ambiguities for the secondary satellites can be derived from

$$N_{12,s}^{1q} = \varphi_{12,s}^{1q} - \rho_{12,p}^{1q}(X_p) \tag{10.143}$$

These estimates are rounded to the nearest integer. For each secondary trial ambiguity set, the correction to the position is computed using the sequential least-squares formulation (Table 4.2 of Chapter 4):

$$\Delta X_p = X_p - N_p^{-1} A_s^T P_s^{-1} (A_s X_p + L_s) \tag{10.144}$$

Note that the size of L_s is merely $(S - 4)$. If the set of primary ambiguities used to generate X_p is the correct one, then the respective secondary ambiguities $N_{12,s}^{1q}$ are correct and, consequently, ΔX_p should be zero. The residuals for the combined solution of primary and secondary observations are

$$\begin{bmatrix} V_p \\ V_s \end{bmatrix} = \begin{bmatrix} A_p \Delta X_p \\ A_s(X_p + \Delta X_p) + L_s \end{bmatrix} \tag{10.145}$$

The quadratic function

$$V^T P V = V_p^T P_p V_p + V_s^T P_s V_s \tag{10.146}$$

can be used to judge the potential solutions.

The degree of freedom is $(S - 1 - 3)$. The solution works better the more satellites observed. The secondary ambiguities (10.143) and the change (10.144) can be computed immediately after a primary position is computed by (10.142). In this way, it is possible to immediately reject primary solutions and to keep only potential solutions in computer memory.

10.2.5 Normal Points

Unlike in a geodetic network of angles and distances, the carrier phase observations are sequential in time. It is, therefore, useful to write the double-difference observation equation as a time series. For example,

$$v_{km}^{pq}(t) = \frac{\partial \varphi_{km}^{pq}(t)}{\partial u_k} du_k + \cdots + \frac{\partial \varphi_{km}^{pq}(t)}{\partial w_m} dw_m$$

$$+ N_{km}^{pq}(1) + \varphi_{km,0}^{pq}(t) - l_{km,b}^{pq}(t)$$

$$+ [I_{km,\varphi}^{pq}(t) + T_{km}^{pq}(t) + d_{km}^{pq}(t) + \epsilon_{km}^{pq}] \qquad (10.147)$$

The partial derivatives and the term $\varphi_{km,0}^{pq}(t)$ are a function of time. The variation in time results from the motion of the satellites. The coordinate corrections (du_k, dv_k, dw_k) and (du_m, dv_m, dw_m) are constant because the receivers do not move in static GPS applications. If the receiver location were accurately known then the partial derivative terms would not contribute to the residuals $v_{km}^{pq}(t)$. In that case, the variation over time of the residuals would equal the unmodeled ionospheric, tropospheric, and multipath delays, and, of course, the random observation noise. One would expect a smooth curve with a slope varying closely around zero. If the coordinate corrections are nonzero, one would expect a residual curve with almost constant slope, because the partial derivatives vary slowly with time. Hence, it should be possible to fit straight lines through short parts of this function. Thus, the residual function (10.147) is modeled by

$$v_{km}^{pq}(t) = a_{km}^{pq} + b_{km}^{pq}(t - t_0) \qquad (10.148)$$

For each double-difference series we fit a straight line and estimate the intercept a_{km}^{pq} and the slope b_{km}^{pq}. The time period $t - t_0$ should be short enough so that the straight-line model is a correct one.

The partial derivative of the residual line with respect to time follows from (10.147) as

$$\frac{\partial}{\partial t}[v_{km}^{pq}(t)] = \frac{\partial}{\partial t}\left[\frac{\partial \varphi_{km}^{pq}(t)}{\partial u_k}\right] du_k + \cdots + \frac{\partial}{\partial t}\left[\frac{\partial \varphi_{km}^{pq}(t)}{\partial w_m}\right] dw_m$$

$$+ [\dot{I}_{km,\varphi}^{pq}(t) + \dot{T}_{km}^{pq}(t) + \dot{d}_{km}^{pq}(t)] \qquad (10.149)$$

The partial derivatives with respect to the station coordinates are given in (10.63) and (10.64). Recognizing that only the satellite positions are functions of time, we obtain

$$\frac{\partial}{\partial t}\left[\frac{\partial \varphi_{km}^{pq}(t)}{\partial u_k}\right] = -\frac{f}{c}\left[\frac{\rho_k^p(t)\dot{u}^p - (u^p - u_k)\dot{\rho}_k^p(t)}{(\rho_k^p(t))^2}\right.$$

$$\left. - \frac{\rho_k^q(t)\dot{u}^q - (u^q - u_k)\dot{\rho}_k^q(t)}{(\rho_k^q(t))^2}\right] \qquad (10.150)$$

Similar expressions are valid for the other derivatives. All time derivatives in this expression can readily be computed from the given satellite ephemeris. Comparing equations (10.147) and (10.148) and using (10.149) we find the relation between the regression coefficients and the other parameters of interest:

$$\hat{a}_{km}^{pq} = N_{km}^{pq}(1) + \varphi_{km,0}^{pq}(t_0) - l_{km,b}^{pq}(t_0)$$

$$+ [I_{km,\varphi}^{pq}(t_0) + T_{km}^{pq}(t_0) + d_{km}^{pq}(t_0)] \qquad (10.151)$$

$$\frac{\partial}{\partial t} - [v_{km}^{pq}(t)] \equiv \hat{b}_{km}^{pq} = \frac{\partial}{\partial t}\left[\frac{\partial \varphi_{km}^{pq}(t)}{\partial u_k}\right]du_k + \cdots + \frac{\partial}{\partial t}\left[\frac{\partial \varphi_{km}^{pq}(t)}{\partial w_m}\right]dw_m$$

$$+ [\dot{I}_{km}^{pq}(t) + \dot{T}_{km}^{pq}(t) + \dot{d}_{km}^{pq}(t)] \qquad (10.152)$$

In the equations above, the caret has been used to indicate that the estimated slopes and intercepts from the straight-line fits should be used. Of course, this regression will also provide the full covariance matrix $\Sigma_{a,b}$.

Neglecting, as usual, the residual ionospheric, tropospheric, and multipath delay, equations (10.151) and (10.152) can be considered observation equations for determining coordinate corrections and integer ambiguities, if at least three double-difference series are available (i.e., four satellites have been observed). As is usual in relative positioning, the differentials du_k, dv_k, and dw_k are set to zero. When the full covariance matrix is used for the estimated intercept and slope observables, the solution will be identical to the double-difference float solution. According to (10.152) the coordinate corrections can be computed from the slope of three double-difference sequences without computing the ambiguities.

This method suffers from the effects of the ionosphere, troposphere, and multipath just like any other technique. If the double-difference series is too short, the estimated intercept and slope will have large standard deviations. If the series is too long the linear model might not be good enough. The quadruple ($\hat{a}_{km}^{pq}$, $\hat{b}_{km}^{pq}$, t_0, $\Sigma_{a,b}$) or the triplet ($\hat{b}_{km}^{pq}$, t_0, Σ_b) can be viewed as a compression of the double-difference carrier phase time series, or a data dimension reduction. If the series is split into several separate linear approximations, then the quadruples or triplets are often called normal points. Mader (1992) uses this method as a compression technique, and also in determining the approximate positions in rapid static application in order to initialize the ambiguity function procedure (which determines the accurate position coordinates). Weill (1993) discusses several of the statistical aspects.

10.2.6 Ambiguity Function

In all of the least-squares techniques discussed above, it is necessary to compute partial derivatives for the least-squares solution and to minimize $V^T P V$. The assumed approximate coordinates of the stations enter the derivatives and the misclosure term. The least-squares solution is iterated until the approximate receiver coordinates have converged to their true values. In the case of the ambiguity function technique the cosine of the residuals, after conversion to cycles, is maximized. Consider again the double-difference observation equation

$$
\begin{aligned}
v_{km}^{pq}(t) &= \varphi_{km,a}^{pq}(t) - \varphi_{km,b}^{pq}(t) \\
&= \frac{f}{c}\, \rho_{km,a}^{pq}(t) + N_{km,a}^{pq}(1) - \varphi_{km,b}^{pq}(t) \\
&\quad + [I_{km,\varphi}^{pq}(t) + T_{km}^{pq}(t) + d_{km}^{pq}(t)]
\end{aligned}
\tag{10.153}
$$

Like usual adjustment notation, the subscripts a and b denote the adjusted and the observed values, respectively. The terms in the brackets are the remaining unmodeled errors. The residuals in radians are

$$
\psi_{km}^{pq}(t) = 2\pi v_{km}^{pq}(t)
\tag{10.154}
$$

The key idea of the ambiguity function technique is to realize that a change in the integer $N_{km}^{pq}(1)$ changes the function $\psi_{km}^{pq}(t)$ by a multiple 2π. The cosine of this function is not affected by such a change because

$$
\cos\,[\psi_{km,L}^{pq}(t)] = \cos\,[2\pi v_{km,L}^{pq}(t)] = \cos\,\{2\pi[v_{km,L}^{pq}(t) + \Delta N_{km,L}^{pq}]\}
$$

$$
\tag{10.155}
$$

where $\Delta N_{km,L}^{pq}$ denotes the arbitrary integer. The subscript L, denoting the frequency identifier, has been added for the purpose of generality.

For dual-frequency observations there are $2(R-1)(S-1)$ double differences available. We assume that these observations are not correlated. If we further assume that all observations are equally weighted, then the sum of the squared residuals becomes

$$
\begin{aligned}
V^T P V(u_m, v_m, w_m, N_{km,L}^{pq}) &= \sum_{L=1}^{2} \sum_{m=1}^{R-1} \sum_{q=1}^{S-1} (v_{km,L}^{pq})^2 \\
&= \frac{1}{4\pi^2} \sum_{L=1}^{2} \sum_{m=1}^{R-1} \sum_{q=1}^{S-1} (\psi_{km,L}^{pq})^2
\end{aligned}
\tag{10.156}
$$

The variables that minimize this function are the coordinate shifts of station m and the ambiguities. The position of station k is held fixed as usual in order to achieve a geometrically strong solution. The ambiguity function is defined as

$$
\begin{aligned}
\mathrm{AF}\,(u_m, v_m, w_m) & \\
\equiv \sum_{L=1}^{2} \sum_{m=1}^{R-1} \sum_{q=1}^{S-1} & [\cos{(\psi_{km,L}^{pq})}] \\
= \sum_{L=1}^{2} \sum_{m=1}^{R-1} \sum_{q=1}^{S-1} & \cos\left\{ 2\pi \left[\frac{f_L}{c}\, \rho_{km,a}^{pq}(t) + N_{km,L,a}^{pq}(1) - \varphi_{km,L,b}^{pq}(t) \right] \right\} \\
= \sum_{L=1}^{2} \sum_{m=1}^{R-1} \sum_{q=1}^{S-1} & \cos\left\{ 2\pi \left[\frac{f_L}{c}\, \rho_{km,a}^{pq}(t) - \varphi_{km,L,b}^{pq}(t) \right] \right\}
\end{aligned}
\tag{10.157}
$$

The small double-difference ionospheric, tropospheric, and multipath terms are not listed explicitly in this equation, although they are present and will affect the ambiguity function technique just as they affect the other solution methods. Nevertheless, if we assume for a moment that these terms are negligible and that the receiver positions are perfectly known, then equation (10.157) shows that the maximum value of the ambiguity function is $2(R-1)(S-1)$ because the cosine of each term is 1. Observational noise will cause the value of the ambiguity function actually to be slightly below the maximum. Since the ambiguity function does not depend on the ambiguity, it is independent of cycle slips in the double differences. This invariant property is the most outstanding feature of the ambiguity function and is unique among all the other solution methods.

Because the function $\psi_{km,L}^{pq}$ is small, typically corresponding to several hundredth of a cycle, the cosine function is expanded in a series, and higher-order terms are neglected. Thus,

$$
\begin{aligned}
\mathrm{AF}\,(u_m, v_m, w_m) &= \sum_{L=1}^{2} \sum_{m=1}^{R-1} \sum_{q=1}^{S-1} \cos \psi_{km,L}^{pq} \\
&= \sum_{L=1}^{2} \sum_{m=1}^{R-1} \sum_{q=1}^{S-1} \left[1 - \frac{(\psi_{km,L}^{pq})^2}{2!} + \cdots \right] \\
&= 2(R-1)(S-1) - \frac{1}{2} \sum_{L=1}^{2} \sum_{m=1}^{R-1} \sum_{q=1}^{S-1} (\psi_{km,L}^{pq})^2 \\
&= 2(R-1)(S-1) - 2\pi^2 V^{\mathrm{T}} P V
\end{aligned}
\tag{10.158}
$$

The last part of this equation follows from (10.156). Clearly, the ambiguity function and the least-squares solution are equivalent, in the sense that at the point of convergence, the ambiguity function is maximized and $V^{\mathrm{T}} P V$ is minimized (Lachapelle et al. 1992).

There are several ways to obtain the ambiguity function solution. The simplest procedure is to assume a search volume, i.e., a cube or an ellipsoid, located around the initial estimate of the station position m. This estimate could be derived from a pseudorange solution, the normal point method, the triple-difference solution, or any other technique. This physical search volume is subdivided into a narrow grid of points with equal spacing. Each physical point is a candidate for the solution and is used to compute the ambiguity function (10.157). Note that the value of the double-difference range $\rho^{pq}_{km,0}(t)$, which is required in (10.157), is evaluated for the trial position. As the ambiguity function is computed by adding the individual cosine terms one double-difference observation at a time, early exit strategies can be implemented to accelerate the solution. For example, if the trial position differs sufficiently from the true position, the residual is likely to be much bigger than one would expect on the basis of the measurement noise and the unmodeled ionospheric, tropospheric, and multipath terms. An appropriate strategy could be to abandon the current trial position, i.e., to stop accumulating the ambiguity function for this particular trial position and to begin with the next trial position. This would occur as soon as one term is below the cutoff criteria, e.g.,

$$\cos \{2\pi [\varphi^{pq}_{km,0}(t) - \varphi^{pq}_{km,L,b}(t)]\}_i < \epsilon \tag{10.159}$$

The value of the cutoff criteria ϵ is critical, not only for accelerating solutions, but for assuring that correct solutions are not missed. This early exit strategy is unforgiving in the sense that once the correct (trial) position is rejected, the scanning of the remaining trial positions cannot yield the correct solution. The proper choice for ϵ is largely experimental, as are early exit strategies of the other methods. A matter of concern is that the trial positions are close enough to assure that the true solution is not missed. Very narrow spacing of the trial positions increases the computational load, despite the early exit strategy. The optimal spacing is typically related to the wavelength, although the relationship is not defined precisely, and to the number of satellites. Obviously, the smaller the original search volume, the faster the solution is obtained.

The ambiguity function technique can be modified in several ways to increase its speed. The technique can be applied to double-difference wide-lane observables. In this case, the trial positions can initially be widely spaced to reflect the wide-lane wavelength of 86 cm. Such solutions potentially serve to identify a smaller physical search space, which can then be scanned using narrowly spaced trial positions. Of course, one can apply the ambiguity function technique to the wide-lane and narrow-lane observables at the same time, or use it with any other function of the observables that maintains the integer nature of the ambiguities.

Instead of scanning equally spaced trial positions in a large physical search volume, one might first determine trial sets of ambiguities and might use them to identify a small number of trial positions. For example, one could use the primary and secondary satellite technique described in Section 10.2.4 and determine several trial triplets of primary ambiguities. The latter define the respective trial positions. The ambiguity function technique could then be applied to those trial positions that fall within the original search volume. This method will drastically reduce the number of trial positions evaluated. The trial points are separated in this case by the wavelength, instead of by arbitrary grid spacing.

The ambiguity function technique is by its very nature nonstatistical. It offers no opportunity to enter the correlation between the double-difference observables or correlations between other functions of the double differences. There is no direct accuracy measure for the final position that maximizes the ambiguity function, such as standard deviations of the coordinates. The quality of the solution is related to the spacing of the trial positions. If the trial positions, for example, have a one-centimeter spacing and a maximum of the ambiguity function is uniquely identified, then the point is said to be determined within centimeter accuracy. To arrive at such an accuracy measure, it might be necessary to take the position that maximizes the ambiguity function and to carry out another narrow-spaced search in a minicube centered at this location just to see up to what resolution the maximum can be uniquely identified. Alternatively, one can take the final position and solve the regular double-difference least-squares solution. Because the initial positions for this least-squares solution are very good, one iteration is sufficient and it should be possible to fix the integer using any of the methods described in the previous sections. The fixed solution would give the desired statistical information. The double-difference solution, of course, requires observations free of cycle slips.

The ambiguity function values for all trial positions are ordered by amplitude and are normalized (dividing by the number of observations). There could be several maxima that cannot be readily separated. Often the highest peak is surrounded by peaks of lesser value. Thus, there is a need to establish a rule to distinguish between the highest peak and the next highest. Such a rule should take into consideration the difference in amplitude of the ambiguity function and the separation of the respective trial positions. If the position separation is less than about one-half of the wavelength, then both peaks refer to the same solution, and the trial position with the highest amplitude should be considered the true solution. These situations typically happen when observational strength is lacking. As with other methods, more observations must be added to the solution. The strength depends, again, on the geometry of the satellite configuration, whether only L1 observations or L1 and L2 observations are available, etc. The more satellites are observed the shorter is the observation period for successful point positioning. Mader (1992) developed three-dimensional graphical displays to help in the identification of the maximum. Figures 10.19a–g illustrate how added observational strength helps with the identification of the maxima. The data consist of merely two epochs, separated by 2 hours, taken from a 1989 observation series of a 5-m baseline. Five satellites were observed in each epoch, and three of them were common to both epochs. In each case a 1-m cube was searched with a search resolution of 2 cm. On the computer screen candidate positions are marked with color-coded symbols indicating the relative amplitude of the ambiguity function. This color scheme

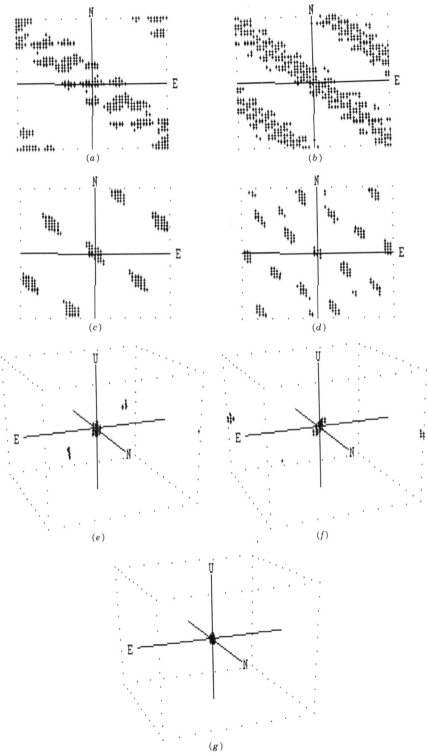

Figure 10.19 Ambiguity function maxima. (Courtesy of G. Mader, National Geodetic Survey).

is useful for tracking the maxima, but is lost in these figures. Figures (a)–(d) show the projection of all qualifying maxima within the search cube onto the northing–easting plane.

Figure (a) shows the maxima using L1 phase observations for just one epoch. The abundance of qualifying solutions confirms the lack of observational strength required for a unique solution in one epoch of single-frequency observations of five satellites. Figure (b) shows the same solution for the L2 carrier phases. Both figures indicate that the potential solutions are clustered along bands. The spacing of the bands in Figure (b) is smaller because the data sample happens to come from a receiver that uses the L2 frequency squaring technique, which halves the wavelength. Figure (c) shows the solutions for L1 phases at the second epoch, i.e., an instant that occurs about 2 hours after the first epoch. The second epoch solutions for the L2 carrier phases are shown in Figure (d). Comparing figures (a) and (b) with (c) and (d) one clearly sees that the solution patterns are distinctly different for both epochs. This is a result of the change in satellite geometry. In fact, in a graphical display of a sequential solution these patterns would change gradually from epoch to epoch, with some of the peaks disappearing and others appearing. Once sufficient geometric strength is reached, only one peak will remain. Figure (e) shows the solution of L1 observations for both epochs. Only three clusters of potential solutions remain. Figure (f) shows the L1 and L2 observations for the first epoch, i.e., just one epoch of dual-frequency observations only. The contribution of the second frequency is clearly visible, although multiple solutions are still present. Finally, Figure (g) shows the solution for both frequencies and both epochs. There is only one peak left.

The strength of the ambiguity function approach lies in the fact that the correct solution is obtained even if the data contain cycle slips. The ambiguity function formulation (10.157) is particularly simple when applied to double differences. Remondi (1984) discusses, among other interesting aspects, the application of the ambiguity function technique to single differences. As is the case with several other well-cherished "GPS ideas," the roots of the ambiguity function technique seem to point toward VLBI. Counselman and Gourevitch (1981) presented a very general ambiguity function technique, discussing in greater detail the patterns to be expected from trial solutions.

10.3 KINEMATIC POSITIONING

The basic idea of kinematic positioning is that the difference of observations between two epochs, collected at the same receiver, to the same satellite equals the change in topocentric range. It does not matter whether the receiver has moved between the epochs, nor does it matter which path the receiver followed to get from one point to the other. The integrated carrier phase observable cannot distinguish a moving satellite from a moving receiver's antenna. The result of a kinematic GPS survey is the trajectory of the moving antenna with respect to a stationary site. The antenna at the stationary site, also called the fixed site or reference station, remains stationary throughout the kinematic survey. The antenna of the other station, called the moving antenna or rover, moves to the points whose positions are to be determined (on the ground, in the air, or at sea). Both receivers must continuously track the satellites. As an extra measure of quality control, the moving antenna may return to its point of beginning or to some other station with known position upon completion of the survey. Kinematic GPS surveys are possible with pseudoranges, carrier phases, or a combination of both. As is always the case, the more accurate positions are derived from the carrier phase observables.

A kinematic survey can be initialized on the ground by occupying a pair of known stations, by performing an antenna swap, by a regular static survey or a rapid static survey, or while the moving receiver is in motion. In the latter case, we speak of on-the-fly (OTF) initialization. In fact, kinematic relative positioning is possible while both receivers are in motion. As was detailed in Section 8.3.4, the location of the reference receiver must only be known approximately in order to assure accurate relative positioning. The approximate initial positions of the moving reference receiver are continuously available from C/A-code pseudorange solutions. The same line of thought applies to monitoring orientation with GPS. In that case, at least three receivers observe continuously. Their relative locations are determined in three dimensions as a function of time and are typically parameterized by azimuth, pitch, and roll. In most orientation applications of GPS, the receivers are fixed to the platform; their spatial separation distances are constant and can, consequently, be constrained to strengthen the solution.

10.3.1 Initialization on the Ground

The simplest form of initialization is to occupy two known stations. For the ionospheric and tropospheric disturbances to be negligible, and for reasons of convenience, these stations should be relatively close. The relevant terms of the double-difference equation (8.41) are in this case

$$N_{km}^{pq}(1) = \varphi_{km}^{pq}(t) - \frac{f}{c}\left[\rho_k^p(t) - \rho_m^p(t)\right] + \frac{f}{c}\left[\rho_k^q(t) - \rho_m^q(t)\right] \quad (10.160)$$

Since the receiver locations are known, the ambiguity can be computed for each double difference. At least three ambiguities involving four satellites must be determined. The computed ambiguities will be close to an integer, provided that the baseline is short, the multipath effects are negligible, the receiver position coordinates are sufficiently accurate, and the satellite ephemeris is available. The computed values should be rounded to an integer.

Once the initial ambiguities are known, the kinematic survey can begin. Let the subscripts k and m denote the fixed and the moving receiver respectively;

then

$$\frac{f}{c}\,\rho_m^{pq}(t) = \frac{f}{c}\,[\,\rho_k^p(t) - \rho_k^q(t)] + N_{km}^{pq}(1) - \varphi_{km}^{pq}(t)$$

$$+ \left[I_{km,\varphi}^{pq}(t) + \frac{f}{c}\,T_{km}^{pq}(t) + d_{km,\varphi}^{pq}(t) \right] \qquad (10.161)$$

If four satellites are observed simultaneously, there are three equations like (10.161) available to compute the position of the moving receiver (u_m, v_m, w_m). If more than four satellites are available, the usual least-squares approach is applicable. The accuracy requirements for the fixed stations are the same as in any other GPS survey. See Section 8.3.4 for additional explanation of the accuracy requirements. Kinematic GPS, of course, is also affected by the ionosphere, troposphere, and multipath.

The kinematic technique requires that no cycle slips occur during the survey so long as only four satellites are observed. If five satellites are observed, then one cycle slip per epoch can easily be repaired. The three double differences that are free of cycle slips can be used to determine the receiver's position according to (10.161), while equation (10.160) is used to compute the new ambiguity for the double difference on which the slip occurred. This procedure works if the three remaining double differences have a good geometry. If six satellites are observed, two slips, at most, are permissible at the same time. Thus, an occasional cycle slip is not detrimental as long as there are three good double differences left to compute the position. With kinematic applications in mind, manufacturers equip their receivers with many channels for both the L1 and L2 bands. To obtain the most robust measurement system, it might be advantageous to combine the GPS receiver with an inertial measurement unit for the purpose of bridging those few seconds when the receiver has lost lock on all satellites, or when it can maintain lock only on one, two, or three satellites. Other special situations exist, such as constraining the movement of the receiver to one- or two-dimensional motion.

Remondi (1985b) introduced the antenna swap procedure in order to reduce the time required to carry out the ambiguity initialization. Assume that at epoch 1 receiver R_1 is located at station k and receiver R_2 is located at station m. If the antenna swap is executed, the antenna R_1 moves to station m and antenna R_2 moves to station k, and the epoch 2 observation is recorded. Using an expanded form of notation to identify the receiver and the respective observation, the double difference at epoch 1 can be written as

$$\varphi_{km}^{pq}(R_2 - R_1, 1) = \frac{f}{c}\,[\,\rho_k^p(R_1, 1) - \rho_k^q(R_1, 1)$$

$$- \rho_m^p(R_2, 1) + \rho_m^q(R_2, 1)] + N_{km}^{pq}(1) \qquad (10.162)$$

During data processing, it is assumed that antenna at receiver R_1 never moved from station k. Thus, for epoch t the double difference is

$$\varphi_{km}^{pq}(R_2 - R_1, t) = \frac{f}{c}[\rho_m^p(R_1, t) - \rho_m^q(R_1, t)$$
$$- \rho_k^p(R_2, t) + \rho_k^q(R_2, t)] + N_{km}^{pq}(1)$$

$$(10.163)$$

Differencing both observations gives

$$\varphi_{km}^{pq}(R_2 - R_1, t) - \varphi_{km}^{pq}(R_2 - R_1, 1)$$

$$= \frac{f}{c}[-\rho_k^p(t) + \rho_k^q(t) + \rho_m^p(t) - \rho_m^q(t)$$
$$- \rho_k^p(1) + \rho_k^q(1) + \rho_m^p(1) - \rho_m^q(1)]$$
$$\approx 2\frac{f}{c}[-\rho_k^p(t) + \rho_k^q(t) + \rho_m^p(t) - \rho_m^q(t)]$$

$$(10.164)$$

Note that the respective topocentric distances have the same sign in (10.164). These distances change during the antenna swapping operation because of the continuous motion of the satellites. Even if the satellites did not move at all, equation (10.164) could be solved for the vector separation from station k to m. The actual solution requires, as usual, observations of at least four satellites, giving three equations, which are solved for the position m holding station k fixed. Once the position of m is known, the ambiguities follow from (10.160). The usual initial accuracy requirements for station k apply.

It is instructive to compare (10.164) with the triple difference that could be computed if the antennas had not been swapped. According to (8.47) the triple difference is

$$\varphi_{km}^{pq}(t, 1) = \frac{f}{c}[\rho_k^p(t) - \rho_k^q(t) + \rho_m^p(t) - \rho_m^q(t)$$
$$- \rho_k^p(1) + \rho_k^q(1) + \rho_m^p(1) - \rho_m^q(1)] \qquad (10.165)$$

The respective topocentric distances in (10.165) have opposite signs. For a short time period these distance differences converge to zero, whereas in equation (10.164) these distances are added.

The moving antenna, at least in applications on the ground, is put over an unknown mark and will usually stay there for a short period of time. In theory, one phase observation to four satellites is sufficient to determine the relative positions. Following the reasoning leading to Table 4.8, it is recommended that more observation epochs be recorded, for example, 10, in order to improve the level of probability. The respective procedure is often called stop-and-go kinematic GPS surveying.

The quality of the kinematic GPS solution, as always, depends on the geometry of the satellite constellation and on the number of satellites. The more satellites observed, the stronger the solution. Since the satellite constellation is predictable well in advance and because the receiver locations are approximately known, one can simulate the accuracy of position determination in advance by computing the covariance matrix of the station position (or of the vector from the fixed to the moving receiver) on the basis of the satellite selection and the length of the station occupation. Analogous to the PDOP, which is popular in navigation, Goad (1989) introduced the RDOP (relative dilution of precision)

$$\text{RDOP} = [\text{Tr}(A^T P A)^{-1}]^{1/2} \tag{10.166}$$

where P denotes the unitless weight matrix of the double differences, and A is the design matrix containing the partial derivatives with respect to the baseline components as given in (10.63). The units of RDOP are cycles per meter.

10.3.2 Initialization on the Fly

Successful ambiguity resolution means in the most general case that the ambiguities are resolved in real time, hopefully within a few seconds. Until they are resolved, centimeter-level position accuracy will not be achieved. The conditions under which the ambiguities can be successfully resolved on the fly depends not only on software sophistication, but primarily on physical facts, such as distance between receivers, number of satellites, low multipath and other unmodeled errors, availability of full-wavelength dual-frequency carrier phases, and not too many cycle slips. All techniques have in common the need for an initial position estimate in order to define a physical search volume or to identify trial sets of ambiguities. All techniques eventually yield qualifying ambiguity sets and must provide ways for discernibility to separate the best and the second best solution, etc. OTF techniques recognize that the correct solution corresponds to integer ambiguities. If a cycle slip occurs, then the ambiguity for that double-difference sequence must be determined anew. As is the case for applications on the ground, if lock is maintained on 4 satellites, slips are not a problem. However, if lock is lost on all satellites, or all but 1 to 3, then the search must be reinitialized. If cycle slips repeatedly occur, then the ambiguities might not be resolved. Thus, algorithms must be able to resolve the ambiguities in a short period of time. Once the ambiguities are resolved, it is possible to determine the positions forward or backward in time using (10.161).

10.3.2.1 Search Volume and Trial Ambiguities
The initial position estimate and the search volume are most conveniently derived from the double-

differenced pseudoranges, using either the P-codes or the C/A-code. The more accurate this initial position is, the faster the search algorithm. If P-codes are not available, one should definitely use the C/A-code pseudoranges from modern narrow correlating receivers. The standard deviations of these code solutions serve to define the size of the search volume, which must contain the true position. Smoothing of the code observations can be helpful in defining a smaller search volume. Combining the code and carrier observations between epochs as

$$P_{km}^{pq}(1) = P_{km}^{pq}(t) - \frac{c}{f} [\varphi_{km}^{pq}(t) - \varphi_{km}^{pq}(1)] \qquad (10.167)$$

allows the computation of the epoch 1 pseudorange double difference from later epochs. This observable can be smoothed and used to determine the initial position.

Abidin (1993) proposes to use a search ellipsoid. The simplest form of the search volume, however, is the cube whose size equals a multiple of the average standard deviations of the initial position determination. See Figure 10.20. If the position of any corner of the cube is denoted by c, then the ambiguity can be computed from

$$N_{km,c}^{pq}(1) = \varphi_{km}^{pq}(t) - \frac{f}{c} [\rho_k^p(t) - \rho_{m,c}^p(t)] + \frac{f}{c} [\rho_k^q(t) - \rho_{m,c}^q(t)] \qquad (10.168)$$

An ambiguity value $N_{km,c}^{pq}(1)$ is computed for each corner of the cube; all eight of them determine the range of the ambiguities. This procedure gives the range of the L1 and/or the L2 ambiguities, or even the wide-lane and narrow-lane ambiguities as deemed necessary. The number of possible combinations forming ambiguity sets rises rapidly with increased range and number of satellites. In actual implementation of the OTF technique, it is necessary to resort to special strategies that will assure that only physically meaningful ambiguity combinations are used.

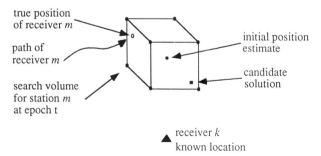

Figure 10.20 Search volume.

10.3.2.2 Propagation of Trial Positions The most general case in kinematic positioning is to compute the search volume and the ambiguity trial sets for each epoch. Assuming that the search algorithms are fast enough, there can be an independent search of the ambiguity space each epoch. Assuming also that there is sufficient geometric strength in the solution, the position of the moving receiver could be determined independently each epoch. Due to lack of geometric strength, several epochs must often be combined. Thus, the problem of propagating trial positions from one epoch to the next arises. Only those trial positions that have not yet been discarded should be propagated. The triple-difference carrier phase equation (8.47) can be rearranged as follows:

$$[\rho_m^p(t_2) - \rho_m^p(t_1)] - [\rho_m^q(t_2) - \rho_m^q(t_1)]$$

$$= [\rho_k^p(t_2) - \rho_k^p(t_1)] - [\rho_k^q(t_2) - \rho_k^q(t_1)] - \frac{c}{f}\,\varphi_{km}^{pq}(t_2, t_1) \quad (10.169)$$

In the notation of delta ranges (8.50) this equation is written in the compact form

$$\rho_m^{pq}(t_2, t_1) = \rho_k^{pq}(t_2, t_1) - \frac{c}{f}\,\varphi_{km}^{pq}(t_2, t_1) \quad (10.170)$$

or

$$\rho_m^{pq}(t_2) = \rho_m^{pq}(t_1) + \rho_k^{pq}(t_2, t_1) - \frac{c}{f}\,\varphi_{km}^{pq}(t_2, t_1) \quad (10.171)$$

If the same four satellites are observed for consecutive epochs, we can compute three triple differences and thus set up three equations like (10.171) and solve for the position of the moving receiver m at epoch t_2. Each trial position is propagated in this manner from one epoch to the next. During the time span over which the trial positions are propagated, there can be no cycle slips.

10.3.2.3 OTF Techniques To introduce the OTF idea let us first consider the following scenario. For simplicity only four satellites are involved (although in actuality OTF works much better with more satellites). The fixed receiver k is at a known location. Step 1 of the process is to determine the search volume as outlined in the previous section, to establish the range of the ambiguities following equation (10.168), and to set up trial sets (triplets) of the ambiguities. In step 2, for each element of a trial triplet, the topocentric range of the moving receiver is computed for epoch 1 by

$$\rho_{m,t}^{pq}(t_1) = \rho_k^{pq}(t_1) - \varphi_{km}^{pq}(t_1) + N_{km,t}^{pq}(1) \quad (10.172)$$

The subscript t indicates that the computations are based on the trial ambiguity. It is now possible to compute the trial position and to determine if it falls within the search volume. If it does not, then that ambiguity trial set is discarded. All trial sets are subjected to this test to identify the physically meaningful ambiguities for further consideration. In step 3, equation (10.171) is used to compute the topocentric range differences to the moving receiver at epoch t_2, $\rho_{m,t}^{pq}(t_2)$, using the triple differences and the trial ranges $\rho_{m,t}^{pq}(t_1)$. This step does not depend on the selection of ambiguities, because the triple differences are used. In step 4, these range differences are used to recompute the ambiguities as

$$N_{km,t}^{pq}(2) = \rho_{m,t}^{pq}(t_2) - \rho_k^{pq}(t_2) + \varphi_{km}^{pq}(t_2) \tag{10.173}$$

Step 4 is carried out for each trial set. Assuming that no additional cycle slips have occurred, then the respective set for which

$$N_{km,t}^{pq}(1) = N_{km,t}^{pq}(2) \tag{10.174}$$

is the correct ambiguity set. Steps 3 and 4 can be carried out for several epochs. In each case the correct ambiguity set fulfills equation (10.174). In actual application, equation (10.174) will only be approximately true due to unmodeled errors.

We can now check the various procedures for static and rapid static GPS surveying outlined in Section 10.2 regarding their suitability for OTF applications. All techniques are suitable, except the normal point approach discussed in Section 10.2.5. All ambiguity fixing methods of Section 10.2.3 require trial sets for the ambiguities and the covariance matrix for the initial estimate of the real-valued ambiguities. In principle, the search volume, the trial ambiguity sets, and the covariance matrix can be generated each epoch. For each epoch an independent attempt can be made to fix the ambiguities. Initially, several epochs can be combined in a sequential type of least-squares solution where the position of the moving receiver is estimated every epoch and the covariance portion of the ambiguities is accumulated from epoch to epoch until sufficient geometric strength is available to complete the ambiguity search. The Kalman filtering implementation, discussed below, is a generalization of this sequential least-squares solution. The ambiguity search techniques in Section 10.2.3 differ primarily as to how they implement early exit strategies.

The technique of primary and secondary satellites explained in Section 10.2.4 is readily adaptable to kinematic applications. The only measure to be taken is to propagate the trial position from one epoch to the next using (10.171). The same propagation of trial positions is required for the ambiguity function explained in Section 10.2.6.

In conclusion, the distinction between static and kinematic techniques become less meaningful as modern processing techniques are implemented. This is particularly true if the appropriate observables are available, and if enough satellites are visible to carry out the ambiguity resolution with just one epoch of data. Clearly, kinematic techniques are applicable to static situations and are the engine that makes "rapid static" out of "static." If not enough observational and geometric strength is available to resolve the ambiguities in one epoch, and several epochs must be used, the only difference between kinematic and static is the propagation of the trial positions.

10.3.3 Kalman Filtering Formulation

GPS issues would be simpler if there were just one algorithm applicable to all types of GPS positioning. The techniques discussed in previous sections certainly have many concepts in common, however, each might have its particular advantages given the right circumstances. Abidin (1993) takes advantage of this, and develops on OTF strategy based on the synergism of extrawide laning, ambiguity function, and least squares using primary and secondary satellites. The Kalman filtering implementation by Qin et al. (1992a) closely represents a unified formulation for static and kinematic GPS positioning. Some aspects of their algorithm are given here. Most of this section uses Kalman filter terminology directly, without making explicit references to the least-squares terminology used throughout this book.

The Kalman filtering algorithm as given in (10.80) to (10.86) also applies in this case. The system state vector X is grouped as follows:

$$X = (X_P, I^{pq}_{km,1,P}, N^{pq}_{km,1}, N^{pq}_{km,w})^{\mathrm{T}} \tag{10.175}$$

In the most general form, the parameter X_P contains the three-dimensional position coordinates, the velocity, and the acceleration of the moving receiver. $I^{pq}_{km,1,P}$ is the residual double-difference ionosphere. The remaining parameters refer to L1 and wide-lane ambiguities. Each group is treated independently, i.e., there are no correlations between the groups. Hence, the state transition matrix Φ and the process noise covariance matrix W are in block diagonal forms:

$$\Phi = \begin{bmatrix} \Phi_P & & & \\ & \Phi_I & & \\ & & I & \\ & & & I \end{bmatrix} \tag{10.176}$$

$$W = \begin{bmatrix} W_P & & & \\ & W_I & & \\ & & W_{N,1} & \\ & & & W_{N,w} \end{bmatrix} \tag{10.177}$$

Each group of state is described separately in the following sections.

Position, velocity, and acceleration vectors X_P are usually expressed in the local topocentric northing, easting, and height coordinate system. This coordinate system is preferred since the physical motion in this system is more intuitive. For example, for ship motion, the vertical motion obviously has different characteristics than the horizontal ones. One can model the average vertical component as a constant with horizontal components modeled as integrated velocity. We ignore any intercorrelation between each component, thus treating them as independent of each other. Also, for the majority of vehicle motion, the modeling of position and velocity in each dimension (second order system) is adequate. A second-order system equation in one dimension is

$$
\begin{bmatrix} x \\ v \end{bmatrix}_{k+1} = \begin{bmatrix} 1 & T \\ 0 & 1 \end{bmatrix} \begin{bmatrix} x \\ v \end{bmatrix}_k + w_P \tag{10.178}
$$

in which x is the position, v is the velocity, T is the observation period, and

$$
E(w_P w_P^T) = \begin{bmatrix} T^3/3 & T^2/2 \\ T^2/2 & T \end{bmatrix} q_P \tag{10.179}
$$

The symbol q_P denotes variance of the process noise. For static positioning, the state vector contains only the position, and q_P is set to a very small number. That is, the position is modeled as a constant. The small amount of process noise is for numerical stability. In summary, the state vector X_P, the transition matrix Φ_P and the process noise matrix W_P for a typical dynamic system, in the northing, easting, and height coordinate system, are

$$
X_P = \begin{bmatrix} n & v_n & e & v_e & h & v_h \end{bmatrix}^T \tag{10.180}
$$

$$
\Phi_P = \begin{bmatrix} 1 & T & & & & \\ 0 & 1 & & & & \\ & & 1 & T & & \\ & & 0 & 1 & & \\ & & & & 1 & T \\ & & & & 0 & 1 \end{bmatrix} \tag{10.181}
$$

$$
W_P = \begin{bmatrix} q_n T^3/3 & q_n T^2/2 & & & & \\ q_n T^2/2 & q_n T & & & & \\ & & q_e T^3/3 & q_e T^2/2 & & \\ & & q_e T^2/2 & q_e T & & \\ & & & & q_h T^3/3 & q_h T^2/2 \\ & & & & q_h T^2/2 & q_h T \end{bmatrix} \tag{10.182}
$$

The state vector X_P, the transition matrix Φ_P, and the process noise matrix W_P for a static system in the northing, easting, and height coordinate system are

$$X_P = [n \quad e \quad h]^T \tag{10.183}$$

$$\Phi_P = \begin{bmatrix} 1 & & \\ & 1 & \\ & & 1 \end{bmatrix} \tag{10.184}$$

$$W_P = \begin{bmatrix} q_n & & \\ & q_e & \\ & & q_h \end{bmatrix} \tag{10.185}$$

in which (n, e, h) are the vector components of northing, easting, and height in the geodetic horizon plane at the fixed station, (v_n, v_e, v_h) are the respective velocity components, and (q_n, q_e, q_h) are the variance of the process noise in the same components. All other empty entries are zeros.

When forming double-difference observables, the ionospheric delay effects are reduced. The residual ionosphere is spatially correlated. For a baseline length of less than 10 km, this residual ionosphere is normally less than 10% of an L1 wavelength (19 cm) and can be neglected. However, as baseline length increases, the residual ionosphere must be taken into account. Here the approach to handle the residual ionosphere is to model it, instead of removing it through ionosphere-free linear combination. A fundamental model is a first-order Gauss–Markov process. A discrete version of Gauss–Markov process is

$$\alpha_{k+1} = \varphi\alpha_k + w_k \tag{10.186}$$

in which

$$\varphi = e^{-T/\tau} \tag{10.187}$$

The symbol α denotes the residual ionospheric delay [meter] [denoted earlier by $I_{km,1,P}^{pq}(t)$], τ is the correlation time [sec], w_k is the perturbation [m], and T again denotes the data sampling period [sec]. Now

$$W_i = E(w_k w_k) = \frac{\tau}{2}(1 - e^{2T/\tau})q_i \tag{10.188}$$

with q_i being the variance of the residual ionosphere process noise for the correlation time τ. The case τ equals zero describes the pure white noise model with no correlation from epoch to epoch. Basically, it is an ionosphere constraint in each epoch of the data. It implies $\varphi = 0$. As τ approaches infinity, we obtain the pure random walk, which implies $\varphi = 1$. The variance q_i of the process noise and the initial estimate of residual ionosphere is determined through a linear function of the baseline length. See Section 10.4. Here each residual double-difference ionosphere is treated as independent from each other, even though it is obvious that this is not the case (since there is a common satellite in all double differences). In summary, the state vector X_I, the transition matrix Φ_I, and the process noise matrix W_I for n double differences are

$$X_I = [\alpha_1 \quad \alpha_2 \quad \cdots \quad \alpha_n]^T \tag{10.189}$$

$$\Phi_I = \begin{bmatrix} \varphi & & & \\ & \varphi & & \\ & & \ddots & \\ & & & \varphi \end{bmatrix} \tag{10.190}$$

$$W_I = \begin{bmatrix} W_i & & & \\ & W_i & & \\ & & \ddots & \\ & & & W_i \end{bmatrix} \tag{10.191}$$

The double difference integer ambiguities are simply modeled as constants. The states are transformed into L1 integer ambiguities and wide-lane integer ambiguities to take advantage of the longer wide-lane wavelength (86 cm). Again, these integer ambiguities are treated as independent from each other. Thus, $W_{N,1}$ and $W_{N,w}$ are diagonal matrices with small diagonal terms $(<10^{-10})$.

The observables are the codes and carrier phases listed in (10.1), but are double differenced. The matrix H, equation (10.81), contains the partial derivatives with respect to each state. For example, equation (10.63) contains the partials with respect to the position components. The variance–covariance matrix R of the observations, equation (10.84), is a block diagonal matrix:

$$R = \begin{bmatrix} R_{1,P} & & & \\ & R_{2,P} & & \\ & & R_{1,\varphi} & \\ & & & R_{2,\varphi} \end{bmatrix} \tag{10.192}$$

The submatrices are fully populated due to the correlated nature of the double differences at the same epoch.

To achieve ultimate positioning accuracy, the integer ambiguities in the carrier phase measurements must be resolved. The Kalman filter estimates the integer ambiguities as part the state vector. After an initial transition period, and when the estimation variances of integer ambiguities reach a predetermined criterion, an integer ambiguity search is launched using estimated values as initial guesses. Ambiguities will be fixed to integers when the ambiguity search criterion is met. For example, the procedure described in Section 10.2.3.2 can be readily applied.

10.4   LONG BASELINES

Long baselines require special attention. This section details a procedure for identifying independent baselines, introduces a weighting scheme for the double differences that depends on the baseline length as a way of extending the distance to which ambiguities can be fixed, and briefly introduces orbital relaxation. The accuracy of long vectors is primarily limited by uncertainties of ionospheric and tropospheric delays and by errors in assumed satellite positions. Although dual-frequency observations and water vapor radiometer measurements are effective in reducing the limitations caused by the propagation media, orbital errors must still be addressed.

10.4.1   Independent Baselines

In Section 10.2 an ordering scheme of the base station and the base satellite was introduced as a convenient means of identifying independent baselines in multistation solutions. It was further shown how the ambiguity parameters are transformed if the base station does not record observations temporarily, or if signals from the base satellite are not received. The goal in network solutions is always to identify $(R - 1)(S - 1)$ independent double-difference functions, where R is the number of receivers and S denotes the number of satellites. It is possible to form altogether $[R!(R - 2)! \times S!(S - 2)!]/2$ different double differences.

An important aspect in processing session networks, which contain a mixture of long and short baselines, is to use our knowledge of unmodeled errors. In the case of short baselines, for example, both the residual ionosphere in the double differences and the ephemeris errors cancel. This is the primary reason why it is not difficult to resolve ambiguities for short baselines. Fixing the ambiguities to integers adds strength to the solution. This additional strength (constraint) may just make it possible to also fix the ambiguities for the next longer baseline, even though the ambiguity search algorithms might not have been successful without this constraint. Such a sequential approach is conceptually similar to the one discussed in Section 10.2.3.4 on systematically reducing the ambiguity range. The difference here is that we consider several baselines simultaneously, and fix the ambiguities of the smaller baselines first.

Of course, as in fixing the ambiguities of the smaller baselines, one could proceed sequentially as well. The technique is sometimes referred to as "bootstrapping" from shorter to longer baselines. One takes advantage of the shorter baseline in order to resolve as many ambiguities as possible in the longer lines. Counselman and Abbot (1989) discuss this technique in connection with orbital determination. Incidentally, this is one of the rare cases in which network design actually matters. An appropriate procedure is to order the baselines (in all combination) by increasing length and to identify the set of independent baselines starting with the shortest.

There are several schemes available to accomplish the identification of independent baselines. One method is that of Bock et al. (1986). Here we follow the suggestion of Goad and Mueller (1988) primarily because it highlights yet another useful application of the Cholesky decomposition. Assume that matrix D of (10.53) reflects the ordering suggested here; i.e., the first rows of D refer to the double differences of the shortest baseline, the next set of rows to the second shortest baseline, and so on. The cofactor matrix (10.58) can be written as

$$Q_\Delta = \sigma_0^2 D D^T = \sigma_0^2 L\, L^T \qquad (10.193)$$

The symbol L denotes the lower triangular matrix obtained from the Cholesky algorithm (A.46). The symmetric cofactor matrix Q_Δ is an inner product matrix because

$$q_{ij} = \sum_k d_i(k) d_j(k) \qquad (10.194)$$

where $d_i(k)$ denotes the ith row of the matrix D. It is readily verified that the ith and jth columns of Q_Δ are linearly dependent if the ith and jth rows of D are linearly dependent. The matrix Q_Δ is singular in such a case. This situation exists exactly when two double differences are linearly dependent. The diagonal element j of the Cholesky factor L will be zero. Thus, one procedure for eliminating the dependent observations is to carry out the computation of L and to discard those double differences that cause a zero on the diagonal. The matrix Q_Δ can be computed row by row starting at the top; i.e., the double differences can be processed sequentially one at a time, from the top to the bottom. For each double difference the respective row of L can be computed. In this way, the dependent observations can immediately be discovered and removed. Only the independent observations remain. As soon as $(R - 1) \times (S - 1)$ double differences have been found, processing can end.

If all receivers observe all satellites for all epochs, this identification process needs to be carried out only once. The matrix L, since it is now available, can be used to decorrelate the double differences. The corresponding residuals might be difficult to interpret, but could be transformed to the original observational space using L again. As was mentioned above, there are various techniques available to identify the independent observations. Other techniques do not require the computation of L.

10.4.2 Constraining the Ionosphere

When measuring distances and performing leveling electronically, it is customary to express the stochastic model as a function of the distance. Recognizing that several errors in relative GPS positioning are more or less a function of receiver separation, Schaffrin and Bock (1988) introduced such a distance dependency in the stochastic model. The residual effect of the ionosphere on the double-difference observables is a well-known example of such errors. Their primary objective, therefore, is to develop a weighting scheme for the residual ionosphere in order to improve ambiguity resolution over longer baselines. The approach is general in that separate covariance matrices for the double difference and the residual ionosphere are established. The residual ionosphere is treated as a stochastic quantity; i.e., it is treated like a direct observation or like an observed parameter with weight. The "observed value" for the residual ionosphere on the double difference is zero, although it could be computed from an appropriate ionospheric model (Table 9.2).

The variance of the double-difference phase observable is modeled by the expression

$$\sigma^2(\Phi_{km}^{pq}) = 4(a^2 + b^2 s^2) \tag{10.195}$$

where a is a constant term (in units of millimeters), and b (without units) is proportional to the distance s between receivers. Note that the variance of the double difference is in units of distance squared. If the undifferenced phase observations are arranged in the order φ_k^p, φ_k^q, φ_m^p, φ_m^q then their cofactor matrix is assumed to have the general form

$$C_\Phi(\alpha, \beta, \delta) = \begin{bmatrix} \beta^2 & 0 & V_\delta \alpha \beta^2 & 0 \\ & \beta^2 & 0 & V_\delta \alpha \beta^2 \\ & & \beta^2 & 0 \\ \text{sym} & & & \beta^2 \end{bmatrix} \tag{10.196}$$

Taking the order of the undifferenced observations into consideration, one sees that the nonzero off-diagonal elements refer to the interstation correlation. The intersatellite correlation is set to zero. The symbols have the following meaning:

$$V_\delta \equiv \frac{2}{e^\delta + e^{-\delta}} \tag{10.197}$$

$$\beta^2 = a^2 + 0.3(b \times 10^4)^2 \tag{10.198}$$

$$\alpha = 1 - \frac{a^2}{\beta^2} \tag{10.199}$$

$$\delta = 0.56\lambda \tag{10.200}$$

The auxiliary quantities α, β, and δ are functions of the basic constants a, b, and the baseline length λ. The unit of λ is radians in this case. The expressions for the auxiliary quantities are valid for distances up to 10,000 km, and the numerical values were chosen such that the cofactor matrix is positive definite. If $b = 0$, then $\alpha = 0$, and the undifferenced observables become uncorrelated. Applying variance–covariance propagation and using (10.196) the variance of the double-difference observable becomes

$$\sigma^2(\Phi_{km}^{pq}) = 4\beta^2 \left(1 - \frac{2\alpha}{e^\delta + e^{-\delta}}\right) \tag{10.201}$$

For zero baselines, in which case $\lambda = 0$ and $\delta = 0$, the hyperbolic secant function $2/(e^\delta + e^{-\delta})$ reaches its maximum of 1. The variance of the double difference in this case is $4a^2$; i.e., the proportionality factor b is not involved as expected for short baselines.

Since we know that the ionosphere cancels in short baselines, the observed value should be zero, and the standard deviation of the residual ionospheric observable should be small. The cofactor matrix for the ionosphere should depend only on distance, i.e., $a = 0$ and $b \neq 0$. For this case equation (10.199) implies that $\alpha = 1$. Thus,

$$C_\kappa(\beta, \delta) = C_\Phi(\alpha = 1, \beta, \delta) \tag{10.202}$$

and

$$\sigma^2(I_{km,1,P}^{pq}) = 4\beta^2 \left(1 - \frac{2}{e^\delta + e^{-\delta}}\right) \tag{10.203}$$

This variance is zero for zero-length baselines. Equations (10.201) and (10.203) apply to any double differences. However, taking the construction of the cofactor matrix $C_\Phi(\alpha, \beta, \delta)$ into consideration, it follows that the covariance matrices of the double-differenced carrier phases and residual ionosphere are fully populated. The off-diagonal elements depend on the lengths of the baselines and, therefore, might be important for sessions of widely varying baseline lengths.

Once the covariance matrices of the double-differenced carrier phases and residual ionosphere are available, any of the standard adjustment models are applicable. For example, the adjustment model of observation equations and observed parameters (Table 4.3), or the mixed model can be used directly. An appropriate assumption is that the group of double-difference carrier phases and the ionosphere are uncorrelated. Usually the ionospheric observable is assumed to be zero simply because of the difficulty in obtaining a more accurate value. The adjustment can be performed with the original double-differenced observations or with the appropriate wide-lane and/or narrow-lane functions.

Schaffrin and Bock (1988) suggest the following transformation:

$$\varphi_{km,IF}^{pq}(t) \equiv \frac{f_1^2}{f_1^2 - f_2^2} \varphi_{km,1}^{pq}(t) - \frac{f_1 f_2}{f_1^2 - f_2^2} \varphi_{km,2}^{pq}(t)$$

$$= \frac{f_1}{c} \rho_{km}^{pq}(t) + \frac{f_1}{f_1 + f_2} N_{km,1}^{pq}(1) + \frac{f_1 f_2}{f_1^2 - f_2^2} N_{km,w}^{pq}(1)$$

(10.204)

$$\varphi_{km,IF2}^{pq}(t) \equiv \frac{1}{2} \varphi_{km,1}^{pq}(t) + \frac{f_1}{2f_2} \varphi_{km,2}^{pq}(t) - \frac{(f_1^2 + f_2^2) f_1}{2c f_2^2} I_{km,1,P}^{pq}(t)$$

$$= \frac{f_1}{c} \rho_{km}^{pq}(t) + \frac{f_1 + f_2}{2f_2} N_{km,1}^{pq}(1) - \frac{f_1}{2f_2} N_{km,w}^{pq}(1)$$

(10.205)

The first function (10.204) is identical to the ionospheric-free observable (9.51). The second function shows some resemblance to the ionospheric observable (9.55). However, there are two differences. First, the second carrier phase is added instead of subtracted. Second, the resulting ionospheric term is moved to the left-hand side and becomes part of the definition for this new function.

The variance–covariance matrices for the functions $\varphi_{km,IF}^{pq}(t)$ and $\varphi_{km,IF2}^{pq}(t)$ are obtained by propagating the stochastic models of the double-difference phases and the ionosphere. The functions $\varphi_{km,IF}^{pq}(t)$ and $\varphi_{km,IF2}^{pq}(t)$ are uncorrelated. If the ionospheric variances are set at zero, i.e., $\sigma^2(I_{km,1,P}^{pq})$ $= 0$, then the ionosphere is completely constrained. The stochastic model for $\varphi_{km,IF2}^{pq}(t)$ does not depend on the stochastic model of the ionosphere (because the latter is zero). In addition, if we take $I_{km,1,P}^{pq}(t) = 0$, then the observed values of $\varphi_{km,IF2}^{pq}(t)$ do not depend on the ionosphere. This is the case typically used for short baselines where the ionosphere cancels. The solution of (10.204) and (10.205) is identical to the combined solution of $\varphi_{km,1}^{pq}(t)$ and $\varphi_{km,2}^{pq}(t)$, and neglecting the ionosphere.

If the ionospheric variances are infinite, i.e., $\sigma^2(I_{km,1,P}^{pq}) = \infty$, then the function $\varphi_{km,IF2}^{pq}(t)$ does not contribute anything to the adjustment, because its variances are infinite as well, i.e., $\sigma^2(\Phi_{km,IF2}^{pq}) = \infty$. The solution is completely determined by the ionospheric-free observable (10.204).

The advantage of the ionospheric constraint approach is that the residual ionosphere can be weighted according to its expected size. For short baselines the weight will be high, for long lines it will be low. Usually the numerical value $I_{km,P}^{pq}(t) = 0$ is weighted. Weighting the residual ionosphere is a way to introduce knowledge about its size, which in turn strengthens the solution. For example, if we know that the ionosphere on the double difference is around 5 cm, then using a corresponding weight should help improve the wide-lane resolution.

10.4.3 Orbital Relaxation

A detailed discussion of the impact of a priori station and ephemeris errors as a function of the baseline length is given in Section 8.3.4. The objective of

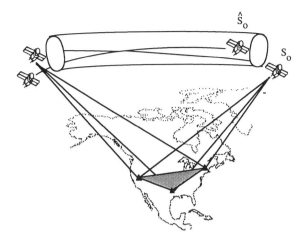

Figure 10.21 Orbital relaxation.

orbital relaxation is to improve the ephemeris accuracy in order to reduce orbital errors over long baselines. Figure 10.21 shows the concept of orbital relaxation. The uncertainty of the initial conditions S_0 on the orbit is indicated by the envelope around the predicted orbital path. Dedicated receivers should be as widely separated as possible to achieve the best geometric strength. Such GPS monitoring networks are already in operation. The receivers are usually operated by various national agencies that have an interest in serving the precise positioning community. The data are shipped electronically, for example via the Internet, to processing centers such as the National Geodetic Survey. However, the data can be retrieved from these databases by anyone.

Satellite motion is completely described by the force function detailed in Section 2.3.4. Because GPS satellites are at a high altitude above the earth, the earth's gravitational attraction can be accurately computed from a low-order spherical harmonic expansion of the gravity field. The farther the satellite is from the earth, the less critical are the inhomogeneities of the earth's gravitational field on its orbital motion. The accelerations caused by the sun and the moon can likewise be accurately computed. The solar radiation pressure effect is small, but accumulates with time. Thus, for longer orbits, solar radiation pressure parameters must be estimated for each satellite.

Orbital relaxation usually starts with initial conditions, for example, the Kepler elements or the Cartesian coordinates and their time derivatives, followed by the integration of the equations of motion, (2.105) and (2.106), over the desired time span. One relatively easy method for dealing with orbital errors is to assume that they result from errors in the initial conditions, and possibly from unknown solar radiation parameters. Usually no attempt is made to improve upon the other components of the force function.

Figure 10.22 shows two major options. The right branch is usually followed in surveying. An available ephemeris, such as the one broadcast by the navi-

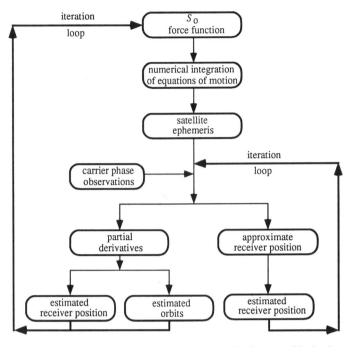

Figure 10.22 Carrier phase processing with and without orbital relaxation.

gation message or one obtained from any other ephemeris service, is used to compute the topocentric satellite distance as required for phase processing. Once the iteration, cycle slip detection, etc., have been completed, the results give accurate relative receiver positions. The left branch pertains to the orbital relaxation method. First, the partial derivatives are obtained with respect to the initial conditions and the solar radiation pressure parameters. Details are given in the next paragraph. Several options are now available. For example, one can estimate the initial conditions and the solar radiation pressure parameters for each satellite, or one can limit the estimation to the initial conditions only. For short arcs of 5–6 hours, (the time of one observation session), it may be useful to constrain some of the initial conditions, such as the semimajor axis or the eccentricity, the rationale being that short arcs do not contain much useful information about the shape of the orbital path (size and ellipticity). One can also attempt the "grand" solution in which satellite orbits are estimated together with at least some of the receiver positions.

The partial derivatives of the initial conditions and the solar radiation pressure parameters can be computed by numerical differentiation. For example, the initial value for the semimajor axis is changed by a small amount. The equations of motion are subsequently integrated numerically giving the so-called variant orbit. For a given epoch, the positions of the variant orbit and the original unperturbed orbit are differenced and divided by the increment. The general formulation for the numerical partial derivative is

$$\frac{\partial U}{\partial p_i} = \frac{U(p_1, p_2, \ldots, p_i + \Delta p_i, \ldots) - U(p_1, p_2, \ldots, p_i, \ldots)}{\Delta p_i} \quad (10.206)$$

This operation must be carried out for each orbital parameter and for each satellite. Thus, there might be as many as $8S$ numerical integrations required to provide all the variant orbits. The size of Δp_i should be about the same as the expected parameter correction. On completion of these computations, one can assemble a (3×8) matrix G of partial derivatives for each ephemeris epoch and satellite:

$$G^P = \frac{\partial(u^P, v^P, w^P)}{\partial S_0^P}$$

$$= \begin{bmatrix} \dfrac{\partial u^P}{\partial a_0} & \dfrac{\partial u^P}{\partial e_0} & \dfrac{\partial u^P}{\partial i_0} & \dfrac{\partial u^P}{\partial \Omega_0} & \dfrac{\partial u^P}{\partial \omega_0} & \dfrac{\partial u^P}{\partial M_0} & \dfrac{\partial u^P}{\partial s_0} & \dfrac{\partial u^P}{\partial Y_0} \\[2ex] \dfrac{\partial v^P}{\partial a_0} & \dfrac{\partial v^P}{\partial e_0} & \dfrac{\partial v^P}{\partial i_0} & \dfrac{\partial v^P}{\partial \Omega_0} & \dfrac{\partial v^P}{\partial \omega_0} & \dfrac{\partial v^P}{\partial M_0} & \dfrac{\partial v^P}{\partial s_0} & \dfrac{\partial v^P}{\partial Y_0} \\[2ex] \dfrac{\partial w^P}{\partial a_0} & \dfrac{\partial w^P}{\partial e_0} & \dfrac{\partial w^P}{\partial i_0} & \dfrac{\partial w^P}{\partial \Omega_0} & \dfrac{\partial w^P}{\partial \omega_0} & \dfrac{\partial w^P}{\partial M_0} & \dfrac{\partial w^P}{\partial s_0} & \dfrac{\partial w^P}{\partial Y_0} \end{bmatrix} \quad (10.207)$$

The parameters s and Y in (10.207) denote the solar radiation pressure and the Y bias. The subscript 0 denotes evaluation of the partials at the epoch of the initial conditions.

The numerical integration can be partially circumvented for short orbital arcs. In these cases, the solar radiation parameters may not be significant and can be deleted. Because in any adjustment the coefficients of the design matrix can tolerate some degree of approximation, it is permissible to compute the partial derivatives of the remaining six initial conditions from the analytical expressions of a Kepler orbit (normal orbit). The following equations, taken from Chapter 2, completely describe the normal orbit:

$$U = R_2(-x_p) R_1(-y_p) R_3(\text{GAST}) R_3(-\Omega) R_1(-i) R_3(-\omega) \mathbf{q} \quad (10.208)$$

$$\mathbf{q} = \begin{bmatrix} a(\cos E - e) \\ a\sqrt{1 - e^2} \sin E \\ 0 \end{bmatrix} \quad (10.209)$$

$$M = E - e \sin E \quad (10.210)$$

$$M = \sqrt{\frac{\mu}{a^3}} (t - t_0) \quad (10.211)$$

TABLE 10.13 Partial Derivatives for Normal Orbits

$$M_0 = R_2(-x_p)R_1(-y_p)R_3(GAST)R_3(-\Omega)R_1(-i)R_3(-\omega) \tag{a}$$

$$M_i = R_2(-x_p)R_1(-y_p)R_3(GAST)R_3(-\Omega)\frac{\partial R_1(-i)}{\partial i}R_3(-\omega) \tag{b}$$

$$M_\Omega = R_2(-x_p)R_1(-y_p)R_3(GAST)\frac{\partial R_3(-\Omega)}{\partial \Omega}R_1(-i)R_3(-\omega) \tag{c}$$

$$M_\omega = R_2(-x_p)R_1(-y_p)R_3(GAST)R_3(-\Omega)R_1(-i)\frac{\partial R_3(-\omega)}{\partial \omega} \tag{d}$$

$$\frac{\partial U}{\partial a} = M_0\begin{bmatrix} \cos E - e - a\sin E\dfrac{dE}{da} \\ \sqrt{1-e^2}\left(\sin E + a\cos E\dfrac{dE}{da}\right) \\ 0 \end{bmatrix} \tag{e}$$

$$\frac{\partial U}{\partial e} = M_0\begin{bmatrix} -a\left(1+\sin E\dfrac{dE}{de}\right) \\ a\left(-\dfrac{e\sin E}{\sqrt{1-e^2}} + \sqrt{1-e^2}\,a\cos E\dfrac{dE}{de}\right) \\ 0 \end{bmatrix} \tag{f}$$

$$\frac{\partial U}{\partial i} = M_i q \tag{g}$$

$$\frac{\partial U}{\partial \Omega} = M_\Omega q \tag{h}$$

$$\frac{\partial U}{\partial \omega} = M_\omega q \tag{i}$$

$$\frac{\partial U}{\partial t} = \begin{bmatrix} -a\sin E\dfrac{dE}{dt} \\ a\sqrt{1-e^2}\cos E\dfrac{dE}{dt} \\ 0 \end{bmatrix} \tag{j}$$

$$\frac{dE}{da} = -\frac{3M}{2a(1-e\cos E)} \tag{k}$$

$$\frac{dE}{de} = \frac{\sin E}{1-e\cos E} \tag{l}$$

$$\frac{dE}{dt} = \frac{1}{1-e\cos E}\sqrt{\frac{\mu}{a^3}} \tag{m}$$

The partial derivatives are listed in Table 10.13. They must be evaluated for the epoch of observations.

Finally, the design matrix is augmented to include the orbital parameters. The first step is to take the partial derivative of the double-difference phase equation with respect to the Cartesian satellite coordinates. Inspecting equation (8.41), we see that the topocentric distance is the only quantity dependent on the satellite position. Differentiating (8.41) gives

$$\frac{\partial \varphi_{km}^{pq}(t)}{\partial u^p} = \frac{f}{c}\frac{(u^p - u_k)}{\rho_k^p} - \frac{f}{c}\frac{(u^p - u_m)}{\rho_m^p} \tag{10.212}$$

$$\frac{\partial \varphi^{pq}_{km}(t)}{\partial u^q} = -\frac{f}{c} \frac{(u^q - u_k)}{\rho^q_k} + \frac{f}{c} \frac{(u^q - u_m)}{\rho^q_m} \tag{10.213}$$

Similar expressions follow for the partial derivative with respect to the v and w coordinates of the satellites. The satellite positions are estimated every epoch in the purely geometric approach, for which the satellite orbital dynamics, as implied by the equation of motion and the force function, is not utilized. In orbital relaxation, however, the satellite position parameters are related to the initial conditions by

$$\begin{bmatrix} du^p \\ dv^p \\ dw^p \end{bmatrix} = G^p \begin{bmatrix} da^p_0 \\ de^p_0 \\ di^p_0 \\ d\Omega^p_0 \\ d\omega^p_0 \\ dM^p_0 \\ ds^p_0 \\ dY^p_0 \end{bmatrix} \tag{10.214}$$

Recall that the partials in G are functions of time. The shorthand notation for equation (10.214) is

$$dU^p = G^p dS^p_0 \tag{10.215}$$

The subscript 0 indicates that the corrections for the orbital parameters refer to the epoch of the initial conditions. Thus, the number of orbital parameters has been reduced from $3TS$ to, at most, $8S$. For $S = 4$ satellites the complete orbital parameterization is

$$\begin{bmatrix} dU^1 \\ dU^2 \\ dU^3 \\ dU^4 \end{bmatrix} = \begin{bmatrix} G^1 \, dS^1_0 \\ G^2 \, dS^2_0 \\ G^3 \, dS^3_0 \\ G^4 \, dS^4_0 \end{bmatrix} \tag{10.216}$$

The design matrix is augmented by the parameters dS^1_0, dS^2_0, dS^3_0, and dS^4_0. Using the shorthand notation

$$\frac{\partial \varphi^{1q}_{12}(t)}{\partial (u^q, v^q, w^q)} = \begin{bmatrix} \dfrac{\partial \varphi^{1q}_{12}(t)}{\partial u^q} & \dfrac{\partial \varphi^{1q}_{12}(t)}{\partial v^q} & \dfrac{\partial \varphi^{1q}_{12}(t)}{\partial w^q} \end{bmatrix} \tag{10.217}$$

we obtain the portion of the design matrix pertaining to the orbital parameters and the observations of baseline 1 to 2:

$$
\frac{\partial\varphi\,{}^{1q}_{12}(t)}{\partial(S^{p}_{0})} =
\begin{bmatrix}
\dfrac{\partial\varphi\,{}^{12}_{12}(t)}{\partial(u^{1},\,v^{1},\,w^{1})}\,G^{1} & \dfrac{\partial\varphi\,{}^{12}_{12}(t)}{\partial(u^{2},\,v^{2},\,w^{2})}\,G^{2} & 0 & 0 \\[4mm]
\dfrac{\partial\varphi\,{}^{13}_{12}(t)}{\partial(u^{1},\,v^{1},\,w^{1})}\,G^{1} & 0 & \dfrac{\partial\varphi\,{}^{13}_{12}(t)}{\partial(u^{3},\,v^{3},\,w^{3})}\,G^{3} & 0 \\[4mm]
\dfrac{\partial\varphi\,{}^{14}_{12}(t)}{\partial(u^{1},\,v^{1},\,w^{1})}\,G^{1} & 0 & 0 & \dfrac{\partial\varphi\,{}^{14}_{12}(t)}{\partial(u^{4},\,v^{4},\,w^{4})}\,G^{4}
\end{bmatrix}
$$

(column labels: dS^{1}_{0}, dS^{2}_{0}, dS^{3}_{0}, dS^{4}_{0})

(10.218)

This is a matrix of size $(S - 1)S$; each additional satellite adds another row and column. Every independent baseline contributes a matrix similar to (10.218).

Table 10.14 summarizes all parameters involved with a simultaneous orbital and station adjustment. A good orbital software package should have options for weighting parameters and separately eliminating sets of parameters, such as receiver positions, clock parameters, orbital parameters, or solar radiation parameters.

10.5 ASPECTS OF REAL-TIME GPS

Transmitting the pseudorange and/or the carrier phase observations from the reference station to the moving receiver makes it possible to compute the location of the moving receiver in real time. This includes ambiguity resolution on the fly. On-site computations allow for real-time quality assurance of kinematic and rapid static applications, and allow precise navigation to a known location. Talbot (1993) describes an operational system for surveying applications that provides centimeter positioning accuracy in real time. Instead of transmitting the observation, one might prefer to transmit corrections which

TABLE 10.14 Partial Derivatives of Double Differences Including Orbital Parameters

$N^{12}_{12}N^{13}_{12}N^{14}_{12}$	$N^{12}_{13}N^{13}_{13}N^{14}_{13}$	$u_1 \quad v_1 \quad w_1$	$u_2 \quad v_2 \quad w_2$	$u_3 \quad v_3 \quad w_3$	orbit
I	0	$\dfrac{\partial\varphi^{1q}_{12}}{\partial(u_1,v_1,w_1)}$	$\dfrac{\partial\varphi^{1q}_{12}}{\partial(u_2,v_2,w_2)}$	0	$\dfrac{\partial\varphi^{1q}_{12}}{\partial(S_0)}$
0	I	$\dfrac{\partial\varphi^{1q}_{13}}{\partial(u_1,v_1,w_1)}$	0	$\dfrac{\partial\varphi^{1q}_{13}}{\partial(u_3,v_3,w_3)}$	$\dfrac{\partial\varphi^{1q}_{13}}{\partial(S_0)}$

are added to the observations of the moving receiver. The Radio Technical Commission for Maritime Services (RTCM), Special Committee 104, has prepared standards for transmission of carrier phases, pseudoranges, and corrections. If the position of the moving receiver is determined on the basis of the transmitted observation or observational corrections, and if the position is determined in real time, one speaks of differential GPS (DGPS). If a consistent strategy is followed, then DGPS and double differencing are the same. Because the accuracy of the corrections decrease with distance, the concept of wide-area differential GPS (WADGPS) is being given increasing attention.

10.5.1 Carrier Phase and Pseudorange Corrections

Transmitting correctors is less of a telemetry load than transmitting the raw observations, because the dynamic range of the corrections is small (Hatch 1992a). Following Taveira–Blomenhofer and Hein (1993) the first step in computing the carrier phase correction is to determine the integer K_k^p as

$$K_k^p \approx \frac{P_k^p(1) - \Phi_k^p(1)}{\lambda} \tag{10.219}$$

from the observations of the first epoch. In equation (10.219) the usual symbols are used for the pseudorange and the scaled carrier phase. Since the carrier phase measurement process cannot determine the ambiguity $N_k^p(1)$, its value is arbitrarily set by the counter register. We define the carrier phase range $\Omega_k^p(1)$ as

$$\Omega_k^p(1) = \Phi_k^p(1) + \lambda K_k^p \tag{10.220}$$

The numerical value of the carrier phase range is close to that of the pseudorange (because of the way the integer K_k^p is determined).

As usual, k denotes the fixed station and m denotes the moving receiver. For every satellite observed at the fixed station, we can compute the carrier phase range $\Omega_k^p(1)$ and the topocentric distance $\rho_k^p(1)$. The latter is derived from the adopted coordinates of the reference receiver and the satellite's location via the ephemeris. The misclosure for the carrier phase range is

$$L_k^p(1) = \rho_k^p(1) - \Omega_k^p(1)$$
$$= \Delta N_k^p + c\, dt_k - c\, dt^p - I_{k,\Phi}^p - T_k^p - d_{k,\Phi} - d_{k,\Phi}^p - d_\Phi^p \tag{10.221}$$

The term ΔN_k^p is present because (10.219) is only an approximate relation. There is no need to get the correct or accurate K_k^p (whatever correct or accurate might mean in this case). As long as ΔN_k^p is small, the misclosure $L_k^p(1)$ will be small, and the goal of getting a small quantity for transmission will be achieved. The mean misclosure for all satellites observed at the fixed site for epoch 1 is

$$\mu_k(1) = \frac{1}{n} \sum_{p=1}^{n} L_k^p(1) \tag{10.222}$$

where n denotes the number of satellites. The carrier phase correction at epoch 1 is

$$\Delta\Phi_k^p(1) = \rho_k^p(1) - \Omega_k^p(1) - \mu_k(1) \tag{10.223}$$

The change of the misclosure from one epoch to the next is

$$L_k^p(t, t-1) \equiv L_k^p(t) - L_k^p(t-1)$$
$$= [\rho_k^p(t) - \rho_k^p(t-1)] - [\Omega_k^p(t) - \Omega_k^p(t-1)] \tag{10.224}$$

The mean at epoch t is computed by

$$\mu_k(t) = \mu_k(t-1) + \frac{1}{n} \sum_{p=1}^{n} L_k^p(t, t-1) \tag{10.225}$$

where n denotes the number of satellites that are observed at epoch t. The phase correction of satellite p at epoch t is

$$\Delta\Phi_k^p(t) = \rho_k^p(t) - \Omega_k^p(t) - \mu_k(t)$$
$$= \rho_k^p(t) - [\Phi_k^p(t) + \lambda K_k^p] - \mu_k(t) \tag{10.226}$$

The second part of this equation follows after (10.220) has been substituted for the carrier phase range. This phase correction (10.226) is transmitted to the moving site where that site's carrier phase is corrected as

$$\overline{\Phi}_m^p(t) = \Phi_m^p(t) + \Delta\Phi_k^p(t) \tag{10.227}$$

To get the respective expressions that show the coordinates of receiver m, let us first write the between-receiver single-difference observable (8.34) in the form

$$\Phi_k^p(t) - \Phi_m^p(t) = [\rho_k^p(t) - \rho_m^p(t)] + \lambda N_{km}^p(1)$$
$$- c(dt_k - dt_m) + I_{km,\Phi}^p(t) + T_{km}^p(t)$$
$$+ d_{km,\Phi}(t) + d_{km,\Phi}^p(t) \tag{10.228}$$

Equation (10.226) can be solved for $\Phi_k^p(t)$ and substituted into (10.228). After rearrangement one obtains

$$-\Phi_m^p(t) - \Delta\Phi_k^p(t) = -\rho_m^p(t) + \lambda[N_{km}^p(1) + K_k^p]$$
$$- [c(dt_k - dt_m) - \mu_k(t)] + I_{km,\Phi}^p(t) + T_{km}^p(t) \quad (10.229)$$
$$+ d_{km,\Phi}(t) + d_{km,\Phi}^p(t)$$

The left side of this equation is equal to the negative of the corrected carrier phase $\overline{\Phi}_m^p(t)$. Differencing equation (10.229) between two satellites gives an expression corresponding to the double-difference observation as

$$\overline{\Phi}_m^{qp}(t) \equiv \overline{\Phi}_m^q - \overline{\Phi}_m^p$$
$$= \rho_m^q(t) - \rho_m^p(t) + \lambda[N_{km}^{pq}(1) + K_k^p - K_k^q] \quad (10.230)$$
$$+ I_{km,\Phi}^{pq}(t) + T_{km}^{pq}(t) + d_{km,\Phi}^{pq}(t)$$

The position m can now be computed at site m using the corrected observation $\overline{\Phi}_m^p(t)$ to at least four satellites and forming three equations like (10.230). This is the DGPS solution for carrier phases. It differs from the conventional double-difference solution due to the fact that the modified ambiguity

$$\overline{N}_{km}^{pq}(1) = N_{km}^{pq}(1) + K_k^p - K_k^q \quad (10.231)$$

is estimated instead of $N_{km}^{pq}(1)$. The position m computed from (10.230) is still relative to the reference position k because the coordinates of that station are used for computing the carrier phase correction $\Delta\Phi_k^p(t)$. However, the moving receiver's position can be determined without explicit knowledge of the location of the reference station k; it suffices to know the carrier phase corrections. The user m must use the same ephemeris and the same computational procedure used at station k to ensure consistency in the computation of the topocentric ranges $\rho_m^p(t)$ and $\rho_k^p(t)$.

One of the goals is to obtain small values for the misclosure $L_k^p(1)$ and, consequently, for the mean $\mu_k(t)$ and the carrier phase corrections $\Delta\Phi_k^p(t)$. This is accomplished through the specific choice of K_k^p in equation (10.219). This property will assure that the telemetry load is minimal. The telemetry load is even further reduced if the time between transmissions of the carrier phase corrections can be increased. For example, if the change in the misclosure from one epoch to the next is smaller than the measurement accuracy at the moving receiver, or if the variations in the misclosure are too small to adversely affect the required minimal accuracy for the moving receiver's position, it is possible to average carrier phase corrections over time and to transmit the averages. It might also be desirable to transmit the rate of correction $\partial\Delta\Phi/\partial t$. If t_0 denotes the reference epoch, the user can interpolate the correctors over time as

$$\Delta\Phi_k^p(t) = \Delta\Phi_k^p(t_0) + \frac{\partial\Delta\Phi_k^p}{\partial t}(t - t_0) \quad (10.232)$$

During the design of the DGPS system, precautions must be taken to assure that such a simple linear modeling of the phase correction is sufficiently accurate.

One way to reduce the size and the slope of the misclosure is to use the best available coordinates for the fixed receiver and the best satellite ephemeris. Clock errors affect the misclosure directly, as is seen in equation (10.221). The drift and variations of the receiver clock error dt_k can be effectively controlled by connecting a rubidium clock to the fixed receiver. The variability of the satellite clock error dt^P due to selective availability is likely the determining factor that limits the time over which averaging is permissible.

The derivation and explanations given above are also applicable to pseudorange corrections. In the case of pseudoranges, the respective equations are

$$L_k^p(1) = \rho_k^p(1) - P_k^p(1)$$

$$= c\,dt_k - c\,dt^P - I_{k,P}^p - T_k^p - d_{k,P} - d_{k,P}^p - d_P^p \quad (10.233)$$

$$\overline{P}_m^p(t) = P_m^p(t) + \Delta P_k^p(t) \quad (10.234)$$

$$\Delta P_k^p(t) = \Delta P_k^p(t_0) + \frac{\partial \Delta P_k^p}{\partial t}(t - t_0) \quad (10.235)$$

$$\overline{P}_m^{qp}(t) \equiv \overline{P}_m^q - \overline{P}_m^p$$

$$= \rho_m^q(t) - \rho_m^p(t) + I_{km,P}^{pq}(t) + T_{km}^{pq}(t) + d_{km,P}^{pq}(t) \quad (10.236)$$

The approach described here is applicable to the L1 and L2 carrier phases and to all three codes.

10.5.2 RTCM-104 Standard

For many years the Radio Technical Commission for Maritime Services (RTCM), a nonprofit corporation, has engaged scientists from the fields of GPS and other radionavigation systems to develop standards for various aspects of radionavigation. RTCM's Special Committee 104 has published several standards for differential GPS positioning. The most recent issue is Version 2.1 [RTCM-104 (1994)], which contains recommendations on the data message types (elements and formats) and the user interface (to be used with a variety of data links such as satellite or radiobeacon links).

The format of the differential message resembles that of the GPS broadcast navigation message discussed in Section 3.1.3.4. The same word length of 30 bits has been preserved. However, whereas the GPS broadcast messages use fixed length subframes, the differential messages utilize variable length formats. The current list of message types is found in Table 10.15. Reflecting the continued development of differential positioning, several of the message types

TABLE 10.15 Differential GPS Service Messages [RTCM-104 (1994)]

Message no.	Current status	Contents
1	fixed	pseudorange correction
2	fixed	delta pseudorange correction
3	fixed	reference station coordinates
4	retired	surveying
5	fixed	constellation health
6	fixed	null frame
7	fixed	beacon almanacs
8	tentative	pseudolite almanacs
9	fixed	partial satellite set differential correction
10	reserved	P-code differential correction (all)
11	reserved	C/A-code L1 and L2 delta correction
12	reserved	pseudolite station parameters
13	tentative	ground transmitter parameters
14	reserved	surveying auxiliary message
15	reserved	ionosphere (troposphere)
16	fixed	special message
17	tentative	ephemeris almanac
18	tentative	uncorrected carrier phases
19	tentative	uncorrected pseudoranges
20	tentative	carrier phase corrections
21	tentative	pseudorange corrections
22-58	--------	undefined
59	tentative	proprietary message
60-63	reserved	multipurpose usage

are only tentatively defined. Several empty slots exist for additional messages as needed.

Common Data   Each message type begins with a 2-word header which contains information that is relevant to all messages, such as the preamble, the message type ID, the reference station ID, the modified Z count, a sequence number, the length of frame, the station health, and parity checks. The 8-bit preamble, the modified Z count, and the sequence number help users with frame synchronization (similar to a GPS navigation message). The message type ID is 6 bits long, and thus can label 64 different message types. The length of the frame equals the number of words plus 2. The modified Z count also serves as reference time t_0 used in (10.235).

Pseudorange Correction   Messages 1, 2, and 9 pertain to the pseudorange corrections, i.e., the correction and its rate; see also equation (10.235). It might happen that the reference station and the user could decode different satellite messages immediately after an update of the broadcast ephemeris. To maintain consistency in using the same ephemeris, message 2 contains a special correction that relates the pseudorange correction as computed from the old and the new ephemeris. Message 9 is designed for use when the reference station is equipped with a highly stable clock. The latter message can provide additional updates for satellites whose pseudorange correction rate variation is too high (due to selective availability).

Radiobeacons   Message 7 applies to the use of radiobeacons to transmit differential GPS corrections. Radiobeacons operate at a low- to medium-frequency range and transmit ground wave signals to users for low-accuracy positioning. Radiobeacons are commonly used for aircraft and vessel navigation throughout the world because of their low cost. The U.S. Coast Guard has carried out extensive experiments with radiobeacons and is in the process of implementing such a differential service along the coastal United States and Alaska. Other countries are also pursuing such a radiobeacon-based GPS service.

Pseudolites   The suitability of pseudolites (pseudo-satellites) to transmit differential corrections and, at the same time, to improve solution geometry by acting as additional satellites has been discussed for many years. Pseudolites would be located at fixed positions on the ground, would transmit on the same L1 and L2 frequency as GPS satellites do, and would use the same code modulation technique. The transmissions of the pseudolite would include the differential GPS corrections. Thus, no additional frequencies would be required to carry the corrections, but the transmissions would be limited to line-of-sight. Messages 8 and 12 pertain to pseudolites.

Corrections for OTF   The major contribution of RTCM, Version 2.1, includes the new message types 18–21, which pertain to pseudorange and carrier phase transmission that satisfy the needs of modern positioning techniques such as OTF. Messages 18 and 19 pertain to the actual carrier phase and pseudorange observations, with option to include the ionospheric carrier phase observable and the ionospheric free pseudorange. Messages 20 and 21 contain the carrier phase and pseudorange corrections as outlined in the previous section.

Various   Message type 4, which was introduced to carry observational data relevant to surveying applications, is now superfluous in view of new message types 18–21. Messages 10 and 11 are reserved for P-code differential corrections and C/A-code L2 correction, in case C/A-codes are available on L2 in the future. The new types 19 and 21 can also be used for this purpose. The parameters of the ionospheric and tropospheric corrections of message 15 have not yet been defined.

10.5.3 WADGPS

The range corrections $\Delta\Phi_k^p(t)$ and $\Delta P_k^p(t)$ of (10.232) and (10.235) represent the combined effect of various error sources. The accuracy of these corrections degrades rapidly with distance. The basic idea of wide-area differential GPS (WADGPS) is to extend the range over which accurate corrections remain valid, and by doing so to minimize the number of reference stations to cover a particular region of the world. The approach is to analyze the individual error sources, attempt to model their variations in time and transmit respective corrections for each satellite to the user (Brown 1989; Kee et al. 1991; Ashkenazi et al. 1992). The user computes the effect of these corrections on the moving receiver's observables, taking the geographic location into account. WADGPS requires an extension of the current RTCM-104 standards.

Depending on the anticipated application, there are many ways and various degrees of sophistication in designing a WADGPS system. Most studies thus far have been carried out for the purpose of extending the "differential reach" of GPS for precision navigation. These studies typically use pseudoranges. Of course, the WADGPS concepts also apply to "extending the reach" of real-time relative positioning for surveying, in which case the carrier phases are part of the solution.

Actually, if the GPS control segment described in Chapter 3 could be redesigned to support civilian uses of GPS, and if SA could be turned off permanently, there probably would be no need for WADGPS. Since this is not possible, there is probably a need for a "Civilian GPS Control Segment" to support WADGPS. In the most general case, the control segment of WADGPS would consist of a network of globally distributed control stations which would transmit their observations immediately to a master station. Depending on the anticipated applications, the control network might cover merely a continent, the national geographic space, or even a smaller region. The master station would compute the corrections and would transmit these to the users via suitable communications links such as satellite, phone, radio, or dedicated navigation radiobeacons. All computations would be carried out sequentially in near real time in order to keep latency of the signal correctors at an acceptable level.

As discussed in Section 8.3.4, the effects of errors in the coordinates of the control stations and in the satellite ephemeris increase with the distance between receivers. The position of all control stations should, therefore, be well known. The ephemeris computed in real time at the master station should at least be better than the SA-degraded broadcast ephemeris. The orbital relaxation technique discussed in Section 10.4.3 is applicable. The new ephemeris would become part of the WADGPS message. Because of the spatial and temporal decorrelation of the ionosphere, it would be important that ionospheric delays be modeled, estimated, and transmitted to the user. The control station should preferably be equipped with dual-frequency receivers to map the ionosphere. See Chapter 9 for additional detail on the ionospheric effects and ways to observe the TEC. It is certainly desirable that the control stations drive their receivers with a rubidium frequency standards. This greatly reduces the jitters in the receiver clock and makes it possible to estimate better satellite clock errors which could be transmitted to the users as well. Recall that SA effectively degrades the stability of the satellite clocks.

WADGPS has the potential of negating the effect of SA over large regions, and of overcoming the spatial dependency associated with DGPS. Of particular interest are the possibilities of transmitting the corrections via satellites such as *Inmarsat* to cover large regions of the world. The WADGPS concepts and techniques are still being developed. Actually, not even the name is certain at this time. The Federal Aviation Administration appears to favor the term WAAS (Wide Area Augmentation Service) instead of WADGPS.

CHAPTER 11

NETWORK ADJUSTMENTS

This chapter begins with a discussion of minimal or inner constraint solutions for GPS vector networks. Despite the expected universal use of GPS in the near future, there will still be situations in which angular measurement by theodolite or distance measurement with an electronic distance meter will be economical or useful in supplementing GPS vectors. The second section, therefore, deals with the combination of terrestrial and GPS vector observations. Because GPS surveying yields three-dimensional vectors, the three-dimensional geodetic model is the most natural model to use. Also because GPS yields ellipsoidal height differences, one must use geoid undulation differences to obtain orthometric heights. The third section, therefore, deals with the computation of accurate geoid undulation differences. Brief remarks are made involving elements of physical geodesy in order to clarify existing limitations in determining geoid undulations from gravity. Since GPS gives the height dimension so accurately, a clear distinction between orthometric and ellipsoidal heights is important. As previously stated, the goal is to eventually replace much of today's expensive and error-prone leveling with GPS height determination. This, however, requires that accurate geoid undulations be available.

This chapter also contains several examples dealing with various aspects of vector adjustments. Terrestrial observations and vector observations are combined by means of scaling and rotating. The more general approach of integrated geodesy, which makes it possible to simultaneously adjust all types of observations (regardless of whether or not they are gravity dependent), is not discussed here. Examples of applications for the integrated geodetic model are found in Hein et al. (1988) and Milbert and Dewhurst (1992).

11.1 GPS VECTOR NETWORKS

The two immediate results of carrier phase processing are the vector between two stations, usually expressed in Cartesian coordinate differences in the WGS 84 or ITRF 92 geodetic system, and the 3×3 covariance matrix of the coordinate differences. The results of a GPS survey are obtained in this form if two receivers observe, and if carrier phase processing is carried out baseline by baseline. The complete covariance matrix is block diagonal, with 3×3 submatrices on the diagonal. In a session solution, in which the phases of R simultaneously observing receivers are processed together, the results are $(R - 1)$ independent vectors, and a $3(R - 1) \times 3(R - 1)$ covariance matrix. In this case, depending on software options available, the results can also be given by a set of $3R$ coordinates and the $3R \times 3R$ covariance matrix. The covariance matrix is still block diagonal, where the size of the nonzero diagonal matrices is a function of R. The purpose of the inner and/or minimal constraint vector solution is to combine the independently observed vectors, or session networks, into one network solution.

Like any other survey, a GPS survey that has determined the relative locations of a cluster of stations should be subjected to a minimal or inner constraint adjustment. To achieve meaningful control, the vector network should not contain unconnected vectors whose endpoints are not tied to other parts of the network. Every possible network should be connected to at least two stations (to avoid unconnected loops of vectors or loops with only one common point). At the network level, the quality of the derived vector observations can be assessed, the geometric strength of the overall network can be analyzed, internal and external reliability computations can be carried out, and blunders may be discovered and removed. For example, a blunder in an antenna height will not be discovered when processing baseline by baseline, but will be noticed in the network solution. Sequential adjustment techniques can also be used to adjust network components if necessary. Any covariance propagation for computing distances, angles, or any other function of the coordinates should be done with the minimal or inner constraint solution.

The mathematical model is the standard observation equation model

$$L_a = F(X_a) \tag{11.1}$$

in which L_a contains the adjusted vector of observations, and X_a denotes the adjusted station coordinates. This mathematical model is "naturally" linear if Cartesian coordinates are chosen for the parameterization of receiver positions. In this case the vector observation between stations k and m is written

$$\begin{pmatrix} \Delta u_{km} \\ \Delta v_{km} \\ \Delta w_{km} \end{pmatrix} = \begin{pmatrix} u_k - u_m \\ v_k - v_m \\ w_k - w_m \end{pmatrix} \tag{11.2}$$

A portion of the design matrix A for the model (11.2) is

$$
A_{km} = \begin{array}{cccccc} u_k & v_k & w_k & u_m & v_m & w_m \end{array} \\
\begin{bmatrix}
1 & 0 & 0 & -1 & 0 & 0 \\
0 & 1 & 0 & 0 & -1 & 0 \\
0 & 0 & 1 & 0 & 0 & -1
\end{bmatrix} \tag{11.3}
$$

The design matrix looks like one for a leveling network. The coefficients are either 1, -1, or 0. Each vector contributes three rows to the design matrix. Because vector observations contain information about the orientation and scale of the network, it is necessary only to fix the origin of the coordinate system. The minimal constraints for fixing the origin can be imposed by simply deleting the three coordinate parameters of one station from the set of parameters. Thus, that particular station is held fixed. In general, inner constraints must fulfill the condition

$$
EX = 0 \tag{11.4}
$$

according to (4.205), or, what amounts to the same thing,

$$
E^{\mathrm{T}}A = 0 \tag{11.5}
$$

It can readily be verified that

$$
E = [\ _3I_3 \quad _3I_3 \quad _3I_3 \quad \cdot \cdot \cdot\ ] \tag{11.6}
$$

fulfills these conditions. The matrix E consists of a row of 3×3 identity matrices. There are as many identity matrices as there are stations in the adjustment. The least-squares estimates for the inner constraint solution are based on the pseudoinverse

$$
N^+ = (A^{\mathrm{T}}PA + E^{\mathrm{T}}E)^{-1} - E^{\mathrm{T}}(EE^{\mathrm{T}}EE^{\mathrm{T}})^{-1}E \tag{11.7}
$$

of the normal matrix. See Chapter 4 for details on the derivation of the pseudoinverse.

If the approximate coordinates are taken to be zero, which can be done since the mathematical model is linear, then the origin of the coordinate system is put into the centroid of the cluster of stations. If nonzero approximate coordinates are used, the coordinate values for the centroid remain invariant; that is, their values, as computed from the approximate coordinates, are identical with those computed from the adjusted coordinates. The inner constraint solution is useful for analyzing the standard deviations of the adjusted coordinates or the standard ellipsoids. These quantities reflect the true geometry of the network and satellite constellation. Chapter 5 contains examples for the use of inner constraints in two-dimensional networks. The discussion in Section 4.8 on those quantities that remain invariant, with respect to different choices of minimal constraints, also applies to the GPS vector network.

The GPS determined coordinates refer to the coordinate system in which the satellite positions are given. This is now the WGS 84 geodetic datum. These positions can be transformed into another coordinate system, provided that the transformation parameters are available. In the next section additional details are given for the combination of GPS vector observations and terrestrial observations, such as angles and distances.

It must be stressed that the primary result of a typical GPS survey is a polyhedron of stations whose relative positions are accurately determined (to the centimeter or even the millimeter), but the translational position of the polyhedron is known only at the meter level (and worse). The orientation of the polyhedron is implied by the vector observations. Although it might be possible to determine geocentric positions with GPS more accurately by making special efforts, there will always be a significant and characteristic difference in accuracy for relative and geocentric positioning. This is indicated by Figure 11.1. The Cartesian coordinates (or coordinate differences) of a GPS survey can, of course, be converted to ellipsoidal latitude, longitude, and height. The implication is that the polyhedron of Figure 11.1 is referenced to the adopted ellipsoid and not to the geoid. When the geoid undulations are known, the polyhedron can be referenced to the geoid and orthometric heights can be computed. Because of the special importance of geoid undulations to supplement GPS vectors, Section 11.3 is completely devoted to geoid undulation computations.

Finally, the respective covariance components of the adjusted parameters can be transformed to the local geodetic coordinate system for ease of interpretation. This is accomplished by

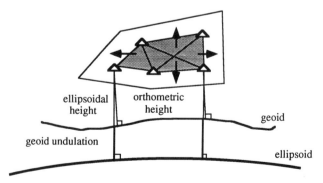

Figure 11.1 Band of uncertainty around polyhedron of receiver stations. GPS gives accurate relative locations only.

$$\Sigma_{n,e,h} = (HJ^{-1}) \, \Sigma_{u,v,w} \, (HJ^{-1})^{\mathrm{T}} \tag{11.8}$$

See Chapter 7 for the derivation of equation (11.8) and the definition of the H and J matrices.

11.1.1 A Simulated Example

A set of GPS vectors is given in Table F.2. This example is given to demonstrate quality control of GPS vector networks. All observations are simulated. With actual observations the weight matrix for the vectors would be of block-diagonal structure, with the correlations reflecting satellite geometry. Because vectors are oriented, the minimal constraints are imposed by fixing one station. Because the formation of the inner constraint solution as given by (11.3) and (11.7) is so straightforward, it is often the preferred solution. The inner constraint solution does have the advantage that the standard ellipses or standard ellipsoids are independent of the definition of the coordinate system.

The outcome of the inner constraint solution is summarized in Table 11.1. The vector observations are reduced to the station marker during the adjustment. With actual observations the a posteriori variance of unit weight is used with (4.267) to test whether or not the adjustment is distorted. It is critical that the best possible stochastic model be used when setting up the weight matrix. As one is dealing with a homogeneous set of observations, simple scaling the weight matrix can produce an a posteriori variance of unit weight close to one. Adjustment of such a vector network is very helpful in discovering blunders and in checking for consistency. For ease of interpretation, the residuals can be transformed to topocentric northing, easting, and height. In GPS networks, blunders can be caused not only by obvious things such as misinterpreting stations, or mistakenly determining antenna height, but also by neglecting to fix the integer ambiguities when it is possible to fix them.

11.2 COMBINATION THROUGH ROTATION AND SCALING

Assume a general situation in which a network of terrestrial observations, such as horizontal angles, slant distances, zenith angles, leveled height differences, and geoid undulations, are available. It is not required that each type of observation actually be present in every case. Assume further that the relative positions of a cluster of stations has been determined with GPS. The terrestrial observations can be those of an existing geodetic network in the neighborhood (if available) or part of any local survey, such as an engineering network or a simple traverse. The first step is to carry out separate minimal or inner constraint solutions for both the terrestrial observations and the GPS vectors. Once

TABLE 11.1 Inner Constraint GPS Vector Adjustment

St	St	Observation mark to mark	Residual	Adj. obs [m]
1	2	-4686.432	0.108	-4686.324
		-17722.135	0.046	-17722.088
		-14883.891	0.087	-14883.805
3	2	8499.881	-0.042	8499.839
		-5663.722	-0.001	-5663.724
		-8403.242	-0.032	-8403.273
4	2	14054.939	0.008	14054.947
		-1236.981	-0.019	-1237.000
		-6290.451	0.004	-6290.447
1	5	-20861.191	-0.135	-20861.326
		-3410.326	0.028	-3410.298
		4407.007	-0.055	4406.952
3	5	-7675.164	0.001	-7675.163
		8648.038	0.029	8648.067
		10887.410	0.073	10887.483
4	5	-2120.155	0.100	-2120.055
		13074.850	-0.060	13074.790
		13000.367	-0.057	13000.310
1	3	-13186.196	0.034	-13186.162
		-12058.312	-0.053	-12058.365
		-6480.517	-0.014	-6480.531
4	3	5555.129	-0.021	5555.108
		4426.708	0.016	4426.724
		2112.817	0.010	2112.826

No. of observations	24
No. of parameters	15
Inner constraints	3
Degree of freedom	12
A-posteriori variance of unit weight	0.60
$V^T P V$	7.21

each of these solutions can be declared correct, that is, once there is no evidence of blunders, and the a posteriori variance of unit weight passes the chi-square test, both data sets can be combined. One way of writing the mathematical model is

$$L_{1a} = F(X_a) \tag{11.9}$$

$$L_{2a} = F(\Delta, \xi, \eta, \alpha, X_a) \tag{11.10}$$

The model (11.9) pertains to the terrestrial observations and is thoroughly described in Chapter 7. In adjustment notation the parameters X_a denote station positions in the geodetic system (U) and nuisance parameters such as refraction coefficients. The observations for model (11.10) are the Cartesian coordinate differences between stations as obtained from GPS carrier phase processing. It is understood that the weight matrix for the vector observations is block diagonal; the size of the submatrices on the diagonal depends on the number of vectors observed simultaneously. The additional parameters in (11.10) are the scale factor Δ and three rotation angles. The rotation angles are small angles relating the geodetic coordinate system (U) implied in (11.9) to the satellite system (X), implied by (11.10). Because GPS yields vector components, that is, coordinate differences, there is no need to worry about the translation between the coordinate systems. If the (U) and the (X) coordinate systems are those of NAD 83 (supplemented with orthometric heights and geoid undulations) and WGS 84, then the least-squares estimates of the rotation angles should statistically be zero. This is because these coordinate systems coincide (see Chapter 14). If the estimates of the rotation angles and scale factor are not zero, then this could be an indication for local network distortions in the (U) system or a (remaining) systematic error in the geoid undulations. In the special case when L_{1a} in (11.9) contains no observations at all, some of the station coordinates in the (U) system can be treated as observed parameters. This is a simple way to get access to the coordinate system (U) and to implement the GPS vector observations into the existing network. It is important to note, however, that geodetic network coordinates in the (U) system are not required if the goal is to combine terrestrial and GPS vector observations over a small region (and not necessarily incorporate the GPS observations into an existing geodetic network such as the NAD 83). The coordinate system for adjusting only the terrestrial observations (11.9) can be defined locally by minimal constraints on the parameters. Any misalignment between the thus defined coordinate system and the WGS 84 will be absorbed by the rotation angles (ξ, η, α). One possible set of minimal constraints for the combined adjustment (11.9) and (11.10) could be the fixing in the (U) system of two horizontal positions, which would determine the horizontal translation, scale, and the azimuth rotation, and three heights, which would determine the remaining two rotations and the translation in height.

The mathematical model follows directly from the transformation expression (13.15) of Chapter 13:

$$T + U_0 + (1 + \Delta)M(\lambda_0, \varphi_0, \eta, \xi, \alpha)(U - U_0) - X = 0 \quad (11.11)$$

The vector notation has been dropped in (11.11). There is one such equation for every station occupied with a GPS receiver. The symbol T denotes the Cartesian translation coordinates between the origins of the (U) and (X) coordinate systems; U_0 is the position of the centroid (φ_0, λ_0) or any point in its vicinity; Δ denotes the difference in scale, and (ξ, η, α) are the rotation angles expressed in the local geodetic coordinate system at U_0. The rotation matrix

$$M(\lambda_0, \varphi_0, \eta, \xi, \alpha) = \alpha M_\alpha + \xi M_\xi + \eta M_\eta + I \quad (11.12)$$

is given by (13.18) through (13.20). Applying the transformation (11.11) to coordinate differences of two stations k and m yields

$$(1 + \Delta)M(\lambda_0, \varphi_0, \eta, \xi, \alpha)(U_k - U_m) - (X_k - X_m) = 0 \quad (11.13)$$

The translation T has been eliminated through this differencing operation, simply meaning that the relative positions of points in the two coordinate systems do not depend on the translatory components of the coordinate system's origin. The coordinate differences

$$\Delta X_{km} = X_k - X_m \tag{11.14}$$

denote the observed GPS vector between stations k and m as obtained from the carrier phase processing. Thus the mathematical model (11.10) can be written as

$$\Delta X_{km} = (1 + \Delta)M(\lambda_0, \alpha_0, \eta, \xi, \alpha)(U_k - U_m) \tag{11.15}$$

On substituting (11.12) into (11.15), we can readily give the partial derivatives of the design matrix. Table 11.2 lists the partial derivatives with respect to the station coordinates, assuming a Cartesian parameterization, a parameterization in terms of geodetic latitude, longitude, and height, and a parameterization in terms of the local geodetic coordinate systems. The transformation matrices J and H are given by (7.18) and (7.21). Table 11.3 contains the partial derivatives of the four transformation parameters. It is understood that the full co-

TABLE 11.2   Design Submatrix for Stations Occupied with GPS Receivers

Parameterization	Station m	Station k
(U)	$(1+\Delta)M$	$-(1+\Delta)M$
(φ, λ, h)	$(1+\Delta)MJ$	$-(1+\Delta)MJ$
(n, e, h)	$(1+\Delta)MJH^{-1}$	$-(1+\Delta)MJH^{-1}$

TABLE 11.3   Design Matrix for the Transformation Parameters

Δ	α	ξ	η
$U_k - U_m$	$M_\alpha(U_k - U_m)$	$M_\xi(U_k - U_m)$	$M_\eta(U_k - U_m)$

variance matrices for the observations are used whenever available. If the GPS polyhedron is tied to the existing geodetic network, say the NAD 83, then the old network stations should be treated as observables with a full weight matrix. If the GPS observations are processed in the session mode, then the full covariance matrix should be used for the GPS vectors as well.

11.2.1 A Simulated Example

The observations for this example are given in Tables F.1. and F.2. In Chapter 7 and Section 11.1, terrestrial observations and GPS vectors were adjusted separately. The purpose of this adjustment is to combine both data sets and to incorporate both sets into the existing geodetic network (U) such as NAD 83. This can be done by holding the given NAD 83 coordinates of stations 1, 3, and 5 fixed. The result is given in Tables 11.4 through 11.7. If the three NAD 83 station positions 1, 3, and 5 were distorted because of some regional distortion in the existing geodetic network, then the a posteriori variance of unit weight would be statistically different from 1, as verified by the chi-square test of (4.267). The residuals in Table 11.5 are almost identical with the residuals in Table 7.4 because the simulated terrestrial and GPS observations are consistent. The adjusted GPS vector components of Table 11.6 are given in the geodetic coordinate system (U). The standard deviations in Table 11.7 are in the local geodetic coordinate systems. Three rotation parameters and the scale parameter have been estimated.

Implementation of the GPS polyhedron into the geodetic network can be accomplished by several methods. If desired, the coordinates of all known horizontal stations can be introduced into the adjustment as weighted parameters to reflect the accuracy of the existing geodetic network. The geoid un-

TABLE 11.4 Summary of Observations for the Combined Solution

29	Slant distances (EDM)
0	Azimuths
28	Altitude (trigonometric)
15	Horizontal angles
0	Coordinate differences (relative positions)
0	Horizontal distances
11	Leveled height differences
5	GPS vector observations

No. of observations	98
No. of parameters	40
No. of observed parameters	6
Degree of freedom	64
A-posteriori variance of unit weight	0.61
$V^T P V$	39.0

TABLE 11.5   Combined Solution: Adjusted Terrestrial Observations

St	St	St	Code	Observation mark to mark	Residual	Adj. obs
6	1		1	4309.531	0.012	4309.543
9	1		1	10763.943	0.339	10764.283
15	1		1	6402.034	0.071	6402.105
14	6		1	11890.478	0.154	11890.632
9	6		1	6808.932	0.250	6809.181
6	1	15	4	72 44 7.179	2.886	72 44 10.065
1	15	14	4	176 20 53.542	-2.554	176 20 50.987
15	14	13	4	121 15 29.667	-1.638	121 15 28.029
14	13	12	4	229 40 25.360	-1.184	229 40 24.176
13	12	5	4	154 23 31.904	-1.140	154 23 30.765
6	1		3	7 37.374	10.889	7 48.263
15	1		3	23 47.950	0.573	23 48.523
1	15		3	-27 21.253	5.779	-27 15.474
14	15		3	-28 30.973	3.607	-28 27.366
9	15		3	-3 34.693	9.276	-3 25.417
7	6		7	1.439	-0.007	1.438
6	15		7	-36.309	-0.003	-36.312
15	14		7	48.185	-0.000	48.185
14	13		7	-39.119	0.001	-39.118
14	11		7	2.781	0.001	2.782

TABLE 11.6   Combined Solution: Adjusted GPS Vectors

St	St	Observation mark to mark - GPS -	Rotation	Scale	Res total	Res	Adj.obs [m] - Datum -
1	2	-4686.432	-0.067	0.048	0.003	0.022	-4686.429
		-17722.135	-0.034	0.181	0.139	-0.008	-17721.996
		-14883.891	0.062	0.152	0.199	-0.015	-14883.692
3	2	8499.881	-0.041	-0.087	-0.157	-0.029	8499.724
		-5663.722	-0.011	0.058	0.062	0.015	-5663.660
		-8403.242	-0.034	0.086	0.045	-0.007	-8403.196
4	2	14054.939	-0.033	-0.144	-0.147	0.030	14054.793
		-1236.981	-0.002	0.013	-0.006	-0.016	-1236.987
		-6290.451	-0.073	0.064	0.006	0.015	-6290.445
4	5	-2120.155	0.060	0.022	0.099	0.017	-2120.056
		13074.850	0.025	-0.134	-0.096	0.013	13074.754
		13000.367	-0.016	-0.133	-0.143	0.005	13000.223
4	3	5555.129	0.008	-0.057	-0.060	-0.011	5555.069
		4426.708	0.009	-0.045	-0.035	0.002	4426.673
		2112.817	-0.039	-0.022	-0.065	-0.005	2112.751

TABLE 11.7 Combined Solution: Adjusted Parameters

Stat	Latitude	St.dev [m]-north	Longitude	St.dev [m]-east	Height [m]	St.dev [m]
1	44 51 42.43980		291 10 03.11090		3.030	
2	45 03 02.86391	0.068	291 18 15.26273	0.068	31.219	0.084
3	44 56 38.69120		291 22 42.61260		7.250	
4	44 58 15.00643	0.043	291 27 52.31290	0.043	19.705	0.054
5	44 48 20.70280		291 25 44.38310		16.140	
6	44 53 02.44862	0.078	291 12 44.00068	0.106	14.269	0.026
7	44 56 00.00707	0.135	291 14 59.99736	0.136	15.700	0.020
8	44 55 54.03760	0.142	291 16 48.49974	0.131	21.060	0.024
9	44 53 22.17056	0.117	291 17 53.03665	0.106	47.873	0.024
10	44 54 00.00974	0.119	291 18 59.99769	0.118	17.187	0.020
11	44 51 19.69167	0.116	291 19 28.56723	0.102	-0.386	0.026
12	44 48 20.05087	0.033	291 23 34.69852	0.096	55.088	0.056
13	44 48 58.62901	0.064	291 21 39.42212	0.104	41.513	0.036
14	44 47 42.90773	0.083	291 17 46.32859	0.154	2.395	0.029
15	44 49 35.39887	0.069	291 13 53.52408	0.089	50.580	0.029

transformation   parameters

α [sec]	st.dev	ξ [sec]	st. dev	η [sec]	st.dev	Δ [ppm]	st.dev
-0.7D+00	0.8D+00	0.9D-02	0.1D+01	-0.1D+01	0.1D+01	0.1D+02	0.4D+01

model refraction correction applied

dulations should be computed for all stations with given orthometric heights to obtain ellipsoidal heights. The geoid undulations should be computed from the best available geoid model. The rotation parameters (ξ, η) will absorb a differential local geoid slope not computable from the global spherical harmonic solution. If the existing geodetic network is distorted, the new GPS derived positions will be distorted as well (as a result of weighting the coordinates of the horizontal stations). To avoid degrading the quality of the GPS vector observations, the preferred method is to simultaneously readjust GPS vectors and the terrestrial observations in the local region (see Figure 1.3).

11.3 ACCURATE GEOID UNDULATIONS

One of the functions of geodesy is to determine the gravity field of the earth and functions thereof, such as geoid undulations and deflections of the vertical. Geoid undulations are important for converting GPS-derived ellipsoidal height differences to orthometric height differences. There are various methods for inferring the magnitude of the geoid undulation based on physical observations. One approach is to estimate the geoid undulation by measuring the acceleration of gravity at the surface of the earth. More complicated approaches, normally limited to scientific research applications, include gravity gradiometry where the gradient of the gravity vector is observed, dynamic satellite geodetic techniques, which track the motion of a satellite as it orbits through the gravitational field of the earth, satellite-to-satellite tracking, and satellite altimetry. One of the preferred mathematical representations of the gravity field and its functionals is that of an infinite mathematical series with spherical latitudes and longitudes as arguments. The coefficients of this series are the spherical harmonic coefficients, sometimes simply referred to as the geopotential coefficients. Currently, solutions of degree and order 360 are available; the higher the degree and order, the more detail is represented by the mathematical function. Computing geoid undulations by any method requires deep understanding of geodetic theory. The interested reader can find details in the book by Moritz (1980). Geoid determination is still a subject of research; not only is the theory continuously refined, but new types of observations are becoming available. This section merely summarizes some of the methods of computing undulations, deemphasizing the fine elements of the theory. The global variations of the geoid are shown in Figure 6.12.

11.3.1 The Normal Gravity Field

The ellipsoid of revolution provides a simple yet accurate model for the geometric shape of the earth; it is useful as a mathematical device in formulating the three-dimensional and two-dimensional mathematical models of surveying and geodesy. Keeping in mind the goal of providing a comparatively simple mathematical model of the earth as a reference for the computations, we can extend the function of the ellipsoid by assigning to it a gravitational field that approximates the actual gravitational field of the earth. Ideally, the same body that is used in the geometric model should also be used in the physical model. This, in fact, is the case. Just as the earth can be described in terms of gravity and potential, so can the ellipsoid of revolution, with the help of a few appropriate specifications. In so modeling the gravity field of the ellipsoid, one must consider an appropriate mass for the ellipsoid and assume that the ellipsoid rotates with the earth. Furthermore, by means of mathematical conditions, the surface of the ellipsoid is defined to be an equipotential surface of its own gravity field, and thus the plumb lines of this gravity field intersect the ellipsoid perpendicularly. Because of this property, the gravity field of the ellipsoid is called the normal gravity field, and the ellipsoid itself is sometimes referred to as the level ellipsoid. It follows that

$$U = V + \Phi \tag{11.16}$$

$$W = T + V + \Phi \tag{11.17}$$

where the symbols (U, V, Φ, T) denote the normal gravity potential, the normal gravitational potential, the centrifugal potential, and the disturbing potential, respectively. The disturbing potential is the small difference between the gravity potential W of the earth and the normal gravity potential U. This definition of the disturbing potential differs from the one used in (2.108) for orbital computations. (The symbol U in (11.16) has a completely different meaning than the same symbol used in (11.11); confusion about this notation is not expected.)

It can be shown that the normal gravity potential is completely specified by four defining constants, which are symbolically expressed by

$$U = f(a, J_2, GM, \omega) \tag{11.18}$$

The symbol a denotes the semimajor axis of the ellipsoid, J_2 denotes the dynamic form factor, GM is the product of Newton's gravitational constant and the mass of the earth, and ω denotes the angular velocity of the earth. Note that GM is identical to k^2M of Section 2.3 and kM of Section 6.1; unfortunately, the symbolism is not unique in the literature. As a practical matter, the constants of the normal ellipsoid are defined by ongoing research in geodesy and related fields and are adopted by international convention. The most recent definitions constitute the Geodetic Reference System 1980 (GRS 1980). A full documentation of this reference system is given in Moritz (1984). The defining constants of the GRS 1980 are given in Table 11.8. The dynamic form factor is a function of the principal moments of inertia of the earth (polar and equatorial moment of inertia). It is essentially an alternative expression for the geometric flattening of the ellipsoid, but at the same time, it is the first and dominant geopotential coefficient in the expansion of the gravitational field.

On the basis of the defining elements, a number of secondary geometrical and physical constants that are useful in computations can be derived. Some of them are listed in Table 11.9. The symbols b and f respectively denote the semiminor axis and the flattening of the ellipsoid, m is an unnamed auxiliary quantity, and γ_e and γ_p denote the normal gravity at the equator and the pole, respectively.

The normal gravitational potential is given by a series of zonal spherical

TABLE 11.8 Defining Constants for the GRS 80

$$a = 6378\ 137\ \text{m}$$
$$GM = 3\ 986\ 005 \times 10^8\ \text{m}^3/\text{sec}^2$$
$$J_2 = 108\ 263 \times 10^{-8}$$
$$\omega = 7\ 292\ 115 \times 10^{-11}\ \text{rad/sec}$$

TABLE 11.9 Derived Constants for the GRS 80

$$b = 6\ 356\ 752.3141 \text{ m}$$
$$1/f = 298.257\ 222\ 101$$
$$m = 0.003\ 449\ 786\ 003\ 08$$
$$\gamma_e = 9.780\ 326\ 7715 \text{ m/sec}^2$$
$$\gamma_p = 9.832\ 186\ 3685 \text{ m/sec}^2$$

harmonics:

$$V = \frac{GM}{r}\left[1 - \sum_{n=1}^{\infty} J_{2n}\left(\frac{a}{r}\right)^{2n} P_{2n}(\cos\theta)\right] \tag{11.19}$$

Note that the subscript $2n$ is to be read "2 times n." The symbol θ denotes the spherical colatitude. The normal gravitational potential has no longitudinal dependency. The polynomial P_{2n} is given by (2.109); r denotes the geocentric distance of the surface point. The coefficients J_{2n} are expressed as a function of J_2 by closed mathematical expressions. Functionals of the normal gravity field can also be expressed by simple mathematical equations. For example, the normal gravity, defined to be the magnitude of the gradient of the normal gravity field, is given by Somigliana's closed formula

$$\gamma = \frac{a\gamma_e \cos^2\varphi + b\gamma_p \sin^2\varphi}{\sqrt{a^2 \cos^2\varphi + b \sin^2\varphi}} \tag{11.20}$$

as a function of the geodetic latitude. Another example is the normal gravity above the ellipsoid, which is given by (Heiskanen and Moritz 1967)

$$\gamma_h - \gamma = -\frac{2\gamma_a}{a}\left[1 + f + m + \left(-3f + \frac{5}{2}m\right)\sin^2\varphi\right]h + \frac{3\gamma_a}{a^2}h^2 \tag{11.21}$$

where h denotes the height above the ellipsoid.

11.3.2 Computing Geoid Undulations

Figure 11.2 shows the relationship of points on the reference ellipsoid and the geoid. We assume that both surfaces have the same potential per definition,

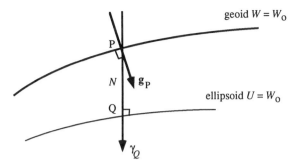

Figure 11.2 Potentials of the geoid and reference ellipsoid.

that is,

$$W_P = W(x, y, z)_P = W_0 \tag{11.22}$$

and

$$U_Q = U(x, y, z)_Q = U_0 \tag{11.23}$$

with

$$U_0 = W_0 \tag{11.24}$$

The distance between P and Q measured along the normal to the ellipsoid is the geoid undulation N. The difference in magnitude between the observed gravity on the geoid g_P and the normal gravity on the ellipsoid γ_Q is called the gravity anomaly Δg:

$$\Delta g = g_P - \gamma_Q \tag{11.25}$$

Figure 11.2 also shows that

$$U_P = U_Q + \frac{\partial U}{\partial n} N = U_Q - \gamma N \tag{11.26}$$

where symbol $(\partial U/\partial n)$ means partial differentiation in the direction of the ellipsoidal normal. This partial derivative equals the normal gravity. (The normal gravity vector is perpendicular to the ellipsoid at point Q.)

$$
\begin{aligned}
W_P &= U_P + T_P \\
&= U_Q - \gamma N + T_P
\end{aligned} \tag{11.27}
$$

Because $W_P = U_Q$ one obtains, dropping the subscript, $T = \gamma N$, or

$$N = \frac{T}{\gamma} \tag{11.28}$$

This is the Bruns equation, which relates the potential of disturbance, the geoid undulation, and the normal gravity. To obtain an expression for the geoid undulation N in terms of spherical harmonic coefficients, we first compute the disturbing potential. The complete gravitational potential is written

$$T + V = \frac{GM}{r}\left[1 + \sum_{n=2}^{\infty}\left(\frac{a}{r}\right)^n \sum_{m=0}^{n}(\overline{C}_{nm}\cos m\lambda + \overline{S}_{nm}\sin m\lambda)\overline{P}_{nm}(\cos\theta)\right]$$

$$\tag{11.29}$$

This expression differs from (2.108) in that the central gravity field term has been included. Also note the reversal in sign between the J coefficients of (11.19) and $\overline{C}$. Subtracting the normal gravity field (11.19) from (11.29) gives

$$T = \frac{GM}{r}\sum_{n=2}^{\infty}\left(\frac{a}{r}\right)^n \sum_{m=0}^{n}(\overline{C}_{nm}^* \cos m\lambda + \overline{S}_{nm}\sin m\lambda)\overline{P}_{nm}(\cos\theta) \tag{11.30}$$

where the asterisk denotes the difference between the respective coefficients of (11.29) and (11.19). Once the disturbing potential is available, the geoid undulations follow from Bruns equation (11.28) and the expression for the normal gravity (11.20). Using (11.28) and (11.30), we obtain the expression for the undulations in terms of spherical harmonic coefficients:

$$N = \frac{GM}{r\gamma}\sum_{n=2}^{n_{max}}\left(\frac{a}{r}\right)^n \sum_{m=0}^{n}(\overline{C}_{nm}^* \cos m\lambda + \overline{S}_{nm}\sin m\lambda)\overline{P}_{nm}(\cos\theta) \tag{11.31}$$

There is a simple mathematical relationship between the geoid undulation and the deflection of the vertical. Given the geopotential coefficients, it is possible to compute the deflections of the vertical as well. The deflections of the vertical are related to the undulations as follows (Heiskanen and Moritz 1967):

$$\xi = -\frac{1}{r}\frac{\partial N}{\partial\theta} = -\frac{1}{r\gamma}\frac{\partial T}{\partial\theta} \tag{11.32}$$

$$\eta = -\frac{1}{r\sin\theta}\frac{\partial N}{\partial\lambda} = -\frac{1}{\gamma r\sin\theta}\frac{\partial T}{\partial\lambda} \tag{11.33}$$

The Bruns equation (11.28) has been substituted for N. Differentiating (11.30) as required by (11.32) and (11.33) gives

$$\xi = -\frac{GM}{\gamma r^2} \sum_{n=2}^{n_{max}} \left(\frac{a}{r}\right)^n \sum_{m=0}^{n} (\overline{C}^*_{nm} \cos m\lambda + \overline{S}_{nm} \sin m\lambda) \frac{d\overline{P}_{nm}(\cos\theta)}{d\theta} \quad (11.34)$$

$$\eta = -\frac{GM}{\gamma r^2 \sin\theta} \sum_{n=2}^{n_{max}} \left(\frac{a}{r}\right)^n \sum_{m=0}^{n} m(-\overline{C}^*_{nm} \sin m\lambda + \overline{S}_{nm} \cos m\lambda) \overline{P}_{nm}(\cos\theta)$$

$$(11.35)$$

Modern solutions of the spherical harmonic coefficients are usually a result of combining surface gravity, satellite trajectory observations, and satellite altimeter data in one grand solution. The classical solution for the geoid undulations using only gravity data is given by the famous Stokes equation (Heiskanen and Moritz 1967):

$$N = \frac{R}{4\pi\gamma} \iint_{\sigma} \Delta g S(\psi) \, d\sigma \quad (11.36)$$

where R is the mean radius of the earth, Δg is the gravity anomaly, and σ is the surface area of the spherical earth. $S(\psi)$ is called the Stokes function:

$$S(\psi) = \frac{1}{\sin\left(\frac{\psi}{2}\right)} - 6\sin\left(\frac{\psi}{2}\right) + 1 - 5\cos\psi$$

$$- 3\cos\psi \ln\left(\sin\frac{\psi}{2} + \sin^2\frac{\psi}{2}\right) \quad (11.37)$$

The angle ψ is the spherical angle from the computation point (the point for which the undulation is computed) to the location of the differential surface element $d\sigma$. The derivation of the Stokes equation is based on a spherical approximation to the earth and the assumption that the geoid is the bounding surface of the earth; that is, there are no masses external to the geoid. The masses outside the geoid are mathematically shifted inside the geoid so that the assumptions hold. One of the principle difficulties with Stokes' equation is that it requires gravity anomalies over the whole globe; the integration is required over the whole earth. Various numerical techniques have been devised to carry out the integration.

The procedure currently used for high-precision geoid undulation computation is a combination of methods (11.31) and (11.36). This combination takes advantage of available spherical harmonic solutions to determine the global and regional trends in the geoid, but uses local gravity to determine the fine structure of the geoid. To assure that gravity anomalies are properly reduced, it is necessary that heights also be available for the gravity stations. The latter are usually derived from a digital terrain model (DTM). Modern procedures for computing the detailed geoid use fast Fourier transform (FFT) techniques. A general review of the mathematics of FFT and a broad range of applicability to problems in physical geodesy are given in Schwarz et al. (1990). van Hees (1990) discusses the use of the FFT with the spherical Stokes formula in order to circumvent planar approximations.

FFT techniques were used to compute the GEOID 90 geoid undulation model, which included 1.5 million ship and terrestrial gravity measurements (Milbert 1991). The improved follow-up model, GEOID 93, is shown in Figure 11.3. Although in surveying we are primarily interested in using the geoid undulation to relate orthometric and ellipsoidal heights, such an accurate geoid model contains interesting geophysical information. Milbert (1991) states,

> The figure of GEOID 90 provides an interesting amalgam of topographic relief and density variation. For example, the Rocky Mountains are prominent from Canada to Mexico. The Sacramento Valley in California is clearly visible, and one may discern the outline of the continental shelf in the Atlantic Ocean and the Gulf of Mexico. In the state of Washington one sees the Columbia River plateau, and just to the south east, the crescent shape of the Snake River plain is expressed in the geoid. It was noted by Christou et al. (1989), in their analysis of the UNB 86 (University of New Brunswick) geoid model for Canada, that a high-resolution geoid will display the structures that reflect density anomalies within the earth. Foremost among these features is the Mid-Continent gravity high, which runs south from Lake Superior. This pronounced feature is the expression of the igneous formations associated with the rift. The geoid has a local change of 5 m in this location. Further south toward Texas, a Y-shaped

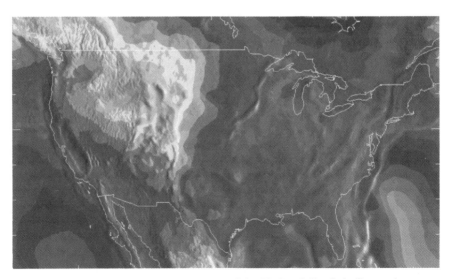

Figure 11.3 GEOID 93 (Courtesy of the National Geodetic Survey.)

feature portrays the Wichita Arbuckle System and the Quachita Orogenic Belt. At the north edge of the figure, a broad low is evident that reaches northward into Hudson Bay. This region was loaded by an ice sheet during the last great ice age, depressing the crust–mantle boundary. Not surprisingly, this area is now undergoing postglacial rebound.

11.4 DISCUSSION OF SELECTED GPS PROJECTS

This section contains four examples for GPS network applications. The Montgomery County (Pennsylvania) GPS survey was one of the first network densifications with GPS in the United States and now has primarily nostalgic value. It serves as an example in which an accurate geoid model is unavailable. It also demonstrates the alignment of the shapes of the ellipses of standard deviation caused by the lack of optimal satellite configuration. The Stanford Linear Collider (SLC) survey is relevant because it demonstrates the suitability of GPS for engineering surveys, and because an objective verification of the achieved accuracy was obtained by means of the realignment laser system. The Superconducting Super Collider (SSC) network allows the gathering of experience with an accurate geoid model and, finally, the 2000-plus Orange County network begs for using advanced concepts from least-squares for quality control.

For these examples the stochastic model does not include the mathematical correlation between simultaneously observed vectors. The variance–covariance matrix of the observed vectors is 3×3 block diagonal. If available, it is always best to include all correlations which theoretically should be included. Craymer and Beck (1992) discuss various aspects of session versus single-baseline processing. They also point out that inclusion of the trivial (dependent) baselines distorts the formal accuracy by artificially increasing the redundancy in the model, resulting in overly optimistic covariance matrices.

11.4.1 The Montgomery County Geodetic Network

This GPS network was contracted by the Pennsylvania Department of Transportation as part of an ongoing geodetic control densification program (Collins and Leick 1985). At the time of the survey the window of satellite visibility was about five hours, just long enough to allow two sessions with the then state-of-the-art static approach. The network shown in Figure 11.4 indicates that much liberty was taken in its design, thus taking advantage of GPS's insensitivity to the shape of small networks. Only the independent vectors between stations are shown; that is, if three receivers observe simultaneously, only two vectors are used. The longest line from station 2 to station 11 is about 42 km. Six horizontal stations and seven vertical stations were available for tying the GPS survey to the existing geodetic networks.

The vectors were computed individually with software provided by the manufacturer of the receivers. Because the survey took place during the "pioneer-

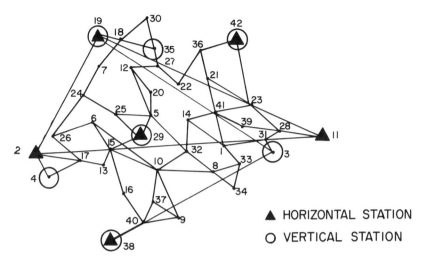

Figure 11.4 Existing control and independent baselines.

ing time'' of GPS surveying, the software available allowed only a vector-by-vector computation. Thus the correlations between vectors were not available even though more than two receivers observed simultaneously. However, the 3×3 covariance matrices for each vector were available. Figure 11.5 shows two intersections of the ellipsoid of standard deviation for the inner constraint least-squares solution: The top set of ellipses shows the horizontal intersection (i.e., the ellipses of standard deviation for horizontal positions), and the bottom set of ellipses shows the vertical intersection in the east–west direction. Figure 11.5 also shows the daily satellite visibility plot for the time of the project. The dots in that figure denote the directions of the semimajor axis of the ellipsoids of standard deviation for each station. These intersections tend to be located around the center of the satellite constellation. The standard ellipses clearly display a systematic orientation in both the horizontal and the vertical plane. This correlation between ellipses and satellite constellation enters into the adjustment through the 3×3 correlation matrices. With the completion of the satellite constellation, the satellites are distributed more evenly over the hemisphere, and alignments seen in Figure 11.5 for the horizontal ellipses would not now occur. Because satellites are observed above the horizon, the ellipses will still be stretched in the height direction.

With orthometric heights given for at least three common stations, an approximate geoid map can be computed with the help of the GPS polyhedron of stations. Recall that no accurate geoid model was available at that time. The polyhedron of stations is oriented with respect to the coordinate system in which the satellite positions were given; at the time of the Montgomery County survey this was WGS 72. Today this is the WGS 84 coordinate system. Actually, the positions of the polyhedron stations can be expressed in any ellipsoid as long as its location is specified by minimal constraints. For example,

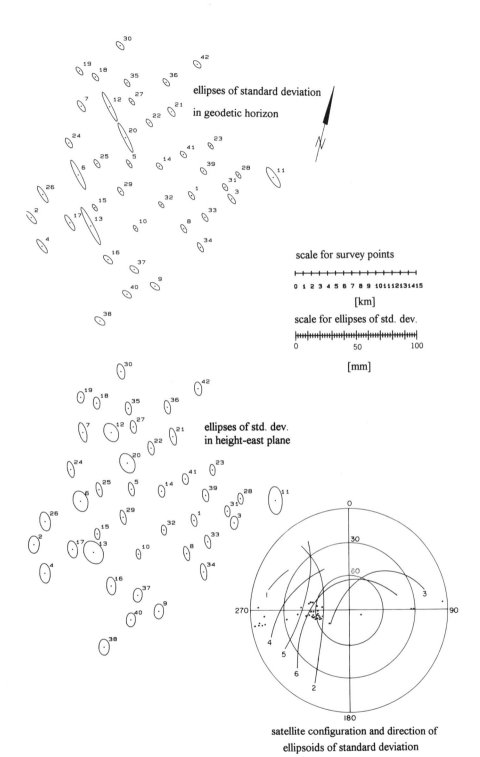

Figure 11.5 Inner constraint solution, ellipses of standard deviation, and satellite visibility plot.

the minimal constraints could be specified by equating the ellipsoidal latitude and longitude of one station, say station 29, to its astronomic latitude and longitude, and equating the ellipsoidal height with the orthometric height. The thus defined ellipsoid is tangent to the geoid at station 29. Note that merely fixing one station is sufficient to define minimal constraints. By comparing the resulting ellipsoidal heights with known orthometric heights, we can construct a geoid undulation map (with respect to the thus defined ellipsoid). The geoid undulations at other stations can be interpolated to give orthometric height from the basic relation $H = h - N$.

The method described above is even more general: The astronomic position for station 29 is not required. The ellipsoidal latitude and longitude, such as the NAD 83 positions, can just as well be used for specifying the minimal constraints. The adopted size for the ellipsoid also does not matter. The local ellipsoid is thus defined slightly differently, not being tangent to the geoid at station 29. The undulations with respect to such an ellipsoid are shown in Figure 11.6. Such an ellipsoid location does not present a problem for small areas so long as the goal is to compute undulations at stations with known orthometric heights, and to interpolate undulations in order to compute orthometric heights at other network stations.

Simple geometric interpolation of geoid undulations has its limits. For example, any error in the given orthometric heights will result inevitably in an erroneous geoidal feature, that is, a hill or a valley that is actually not part of the geoid. As a result, the orthometric heights computed from the interpolated geoid undulations will be in error. Depending on the size of the survey area

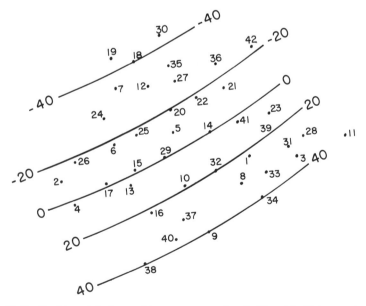

Figure 11.6 Geoid undulations with respect to a local ellipsoid. Units are in centimeters.

and the "smoothness" of the geoid in that region, such erroneous geoidal features might or might not be discovered. A way to avoid these ambiguities is to use modern geoid models.

A partial analysis (or discovery) of significant local features in the geoid might be possible provided that the orthometric heights are well distributed over the network. In this case the topocentric rotations (α, ξ, η) and the scale factor are estimated as implied by (11.7). There are 7 minimal constraints for such a solution: the ellipsoidal latitude and longitude for two stations and the heights for three stations distributed well over the network. If orthometric heights are used for these three stations, the angles (ξ, η) look like the deflection of the vertical angles. Using orthometric heights instead of ellipsoidal heights forces the ellipsoid to coincide locally with the geoid (as defined or implied by the orthometric heights). The rotation in azimuth α is determined by the rotational difference between the two stations held fixed and the GPS vector between the two stations. The scale factor is also determined by the two stations held fixed; it contains the possible scale error of the existing geodetic network and the effect of a constant but unknown undulation (i.e., geoid undulations with respect to the ellipsoid of the existing geodetic network).

Table 11.10 shows the result of several solutions. The first solution is a truly minimal constraint solution yielding the same $V^T PV$ as the inner constraint solution and serves as a basis for comparison. The values in parentheses denote the standard deviations assigned to the given orthometric heights. In subsequent solutions this standard deviation changes. The table shows that if the orthometric heights of stations 4 and 38 are weighted, then the quadratic form $V^T PV$ increases significantly. See specifically solutions 4 and 5. This simply means that the geoid at stations 4 and 38 seems to deviate significantly from the ellipsoidal surface fitted through it in the local area. The deviations

TABLE 11.10 Constraining Successively Existing Vertical Control

Sol	Observed height - adjusted height [m]							$V^T PV$
	3	4	19	29	35	38	42	
1	0					0	0	105
	(2)					(2)	(2)	
2	-2.9	-4.6	.6	1.6	-.4	3.9	.9	109
	(4)	(4)	(4)	(4)	(4)	(4)	(4)	
3	.1	-6.7	-.8	1.0	-.2	4.8	.9	106
	(2)		(2)	(2)	(2)		(2)	
4	-.4	-6.3	-.4	.5	-.1	5.2	.4	107
	(1)		(1)	(1)	(1)		(1)	
5	-.9	-1.4	.2	.4	0	1.2	.4	123
	(1)	(1)	(1)	(1)	(1)	(1)	(1)	

may indicate actual geoidal features or errors in orthometric heights. Errors in the GPS survey can be excluded because the inner constraint solution shows a consistent network accuracy at the centimeter level. If, indeed, a height accuracy of a couple of centimeters had been required for this project, it would have been necessary to run a level line to the stations in question and/or to compute the undulation differences from gravity data. In the case of station 38 the error could also be caused by a faulty antenna height because that station is connected to the network by only one vector.

11.4.2 The SLC Engineering Survey

A GPS survey was carried out in August 1984 to support construction of the Stanford Linear Collider (SLC) with the objective of achieving millimeter relative positional accuracy (Ruland and Leick 1985). This survey is of particular importance to engineering surveying because of the millimeter accuracy achieved with satellite surveying techniques and the combination of GPS vector and terrestrial observations. Because the network was only 4 km long, orbital errors of the satellites did not matter much. The ionospheric and tropospheric effects canceled, because the signals traveled through the same portions of the propagation media. The positional accuracy in such small networks is limited by the phase measurement accuracy and the ability of the antenna to precisely define the phase center and to minimize multipath.

The network is shown in Figure 11.7. Stations 1, 10, 19, and 42 are along the two-mile-long linear accelerator (linac); the remaining stations of the "loop" were to be determined with respect to the linac stations. The disadvantageous configuration of this network, in regard to terrestrial observations such as angles and distances, is obvious. To improve this configuration, one would have to add stations adjacent to the linac; this would have been costly because of the local topography and construction. For GPS positioning this network configuration is quite acceptable because the accuracy of positioning with GPS depends primarily on satellite configuration. Nevertheless, the network was also surveyed with "classical" instruments, such as theodolites, electronic distance measurement equipment, and levels, thus providing two independent determinations and an opportunity to compare both measurement techniques. The set of observations consisted of 18 GPS vectors, 106 slant

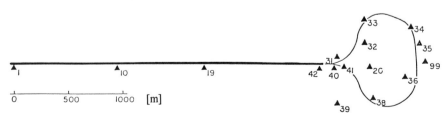

Figure 11.7 The SLC network configuration.

distances, and 93 horizontal angles. Figure 11.8 shows the horizontal ellipses of standard deviation and the satellite visibility plot for the inner constraint solution. The dark spot on the visibility plot represents the directions of the semimajor axes of the standard ellipsoids.

All stations were part of a high-precision level network. The leveling adjustment yielded a standard deviation of 0.3 mm for a 1-km double-run line. A comparison of GPS-derived ellipsoidal heights and leveled height differences was made. The ellipsoid used for referencing the GPS polyhedron was defined by the NAD 27 geodetic latitude and longitude of Station 41 and by equating ellipsoidal height and orthometric height for the same station. The line labeled "equipotential (observed)" in Figure 11.9 denotes the difference of GPS-derived ellipsoidal heights and leveled heights. In the context of an earlier survey for the construction of the linear accelerator, the National Geodetic Survey computed a geoid profile between stations 1 and 42 from the

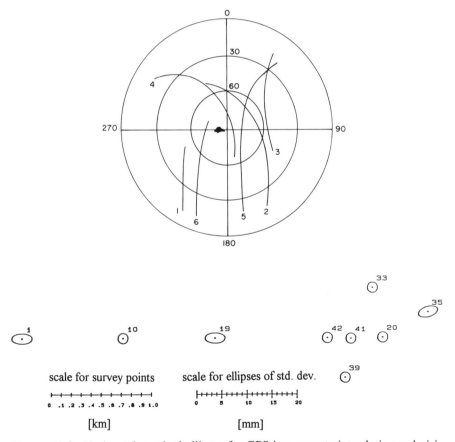

Figure 11.8 Horizontal standard ellipses for GPS inner constraint solution and visibility plot.

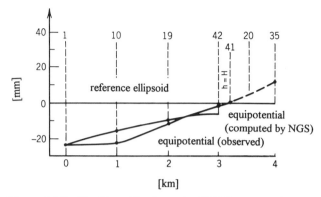

Figure 11.9 Geoid profile along the SLC linear accelerator.

visible topographic masses. Rice (1966) lists the components of the deflection of the vertical for stations 1 and 42 and for another point halfway between stations 10 and 19. From these values Rice computed a polynomial of the third order to represent the geoid along the accelerator. This function is shown in Figure 11.9 by the line labeled "equipotential (computed, NGS)," which has been translated to coincide with the observed equipotential surface at station 1. The deviation between both representations just barely exceeds, at station 10, the standard deviation for the GPS determined height difference from station 1 to station 10, and is within the standard deviation at stations 19 and 42. Incidentally, the overall slope of the observed equipotential profiles in Figure 11.9 is a result of adopting ellipsoidal rather than astronomic positions as minimal constraints for station 41. The east–west component of the deflection of the vertical at station 42 is given by Rice as 1. "85, which accounts for 27 mm of the 22-mm geoidal slope between stations 1 and 42. The remaining discrepancy of about 5 mm might be due to misalignment of the satellite coordinate system and the NAD 27 datum (or the lack of correcting for the misalignment).

This survey provided yet another interesting comparison. For the frequent realignment of the linear accelerator, the linac laser alignment system was installed. This system is capable of determining positions perpendicular to the axis of the linac to better than ± 0.1 mm over the total length of 3050 m. A comparison of the linac stations 1, 10, 19, and 42, as determined from the GPS vector solution with respect to the linac alignment system, was done by means of a transformation. The discrepancies did not exceed ± 1 mm for any of the four linac stations.

11.4.3 The SSC Ring

Even though the Superconducting Super Collider (SSC) might never be completed, the initial GPS survey to support land acquisition and construction is still of much interest since it demonstrates the utility of a modern geoid model.

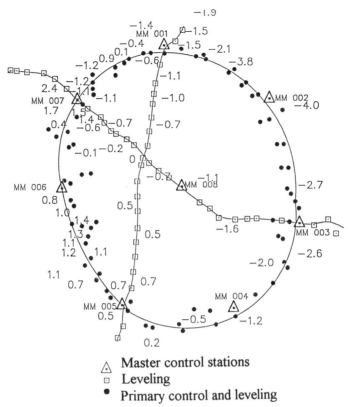

△　Master control stations
▢　Leveling
●　Primary control and leveling

Figure 11.10   The SSC geodetic network. The figure also shows some of the leveling corrections due to gravity.

The complete survey consisted of GPS observations, leveling, and gravity measurements. The specifications called for a GPS reference network to adhere to the Federal Geodetic Control Subcommittee (FGCS), formerly the Federal Geodetic Control Committee (FGCC), specifications for Order B and a primary control network to follow FGCS specifications for Order C-1 (FGCS 1988).

Figure 11.10 shows an overview of the SSC network. The reference network stations are spaced widely and cover the immediate area of the collider track and the four spokes radiating from the track as far as 20 km. The primary control network follows the collider track, with station spacing between 1 and 6 km. The collider track itself is of approximate elliptical shape with semiaxes of about 15 and 13 km. There were 91 GPS stations in the network. Details on the adjustment of the SSC network are found in Leick et al. (1992).

The vertical references were established through conventional leveling following the National Geodetic Survey (NGS) first-order leveling specifications. Leveling was carried out along the collider track and along the spokes. The collider portion of the network consisted of four loops. There were 175 leveling stations. Because gravity processing was involved in the adjustment of the first-order leveling network, it was decided that the leveling data be processed at NGS. NGS verified the precision of the level network as 0.72 mm/km standard deviation for a 1-km double-run section. Figure 11.10 shows the difference of Helmert (orthometric) heights, which NGS provided, and heights obtained from a leveling adjustment without using gravity data. The figure shows a smooth slope of about 4 mm in the east–west direction. The overall geoid slope in the east–west direction is about 60 cm.

For the initial minimal constraint solution of the GPS network, the NAD 83 position of the center station MM008 was held fixed. In the following minimal constraint solution, labeled solution 1 in Figure 11.11, the geodetic latitudes and longitudes of stations MM001 and MM003 of the previous solution, and the NGS-computed orthometric heights for MM001, MM003, and station 153 are held fixed; the orientation and scale parameters were estimated. With these specific minimal constraints, the estimates of the orientation in azimuth and the scale are statistically insignificant because the adjusted latitude and longitude from the initial adjustment are held fixed. The estimates of the other rotation parameters are $-6''.0 \pm 0''.06$ (around the local meridian) and $0''.8 \pm 0''.05$ (around local easting), confirming the primary east–west slope of the geoid. These rotation parameters indicate the average misalignment between the ellipsoidal surface and the geoid (as defined by the orthometric heights of stations MM001, MM003, and 153), and can be interpreted as average deflec-

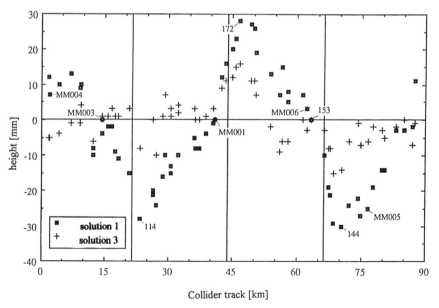

Figure 11.11 Unresolved features along the collider track.

tion of the vertical values. Since the rotation angles have been estimated, it follows that the values for solution 1 in Figure 11.11 indicate the deviation between the geoid and the ellipsoidal surface, as defined by the orthometric heights of the three stations held fixed. Of course, there is no natural law requiring that these two surfaces fit perfectly; however, the deviation is only about 60 mm.

Figure 11.12 shows another minimal constraint solution, solution 2, in which station MM001 is held fixed. The approximate (ellipsoidal) heights for the adjustment are obtained from NGS orthometric heights and GEOID 90 undulations according to $h = H + N$. Thus, the statistical expectation of the differences between adjusted and approximate heights is zero. However, a systematic variation is seen implying a residual east–west geoid slope. The magnitude of the rotations are $-0''.99 \pm 0''.04$ (east–west) and $0''.04 \pm 0''.04$ (north–south) as estimated from another solution, solution 3, for which the ellipsoidal heights (NGS orthometric plus GEOID 90 undulations) of stations MM001, MM003, and 153 are held fixed. The slope amounts to about 14 mm in east–west direction, which corresponds to 0.2 ppm. The remaining variation in height of solution 3 is shown in Figure 11.11.

The combination of GPS-derived ellipsoidal heights, orthometric heights obtained from leveling, and geoid undulations computed for the best available model should result in random discrepancies d, where $d = H + N - h$. According to solution 3 shown in Figure 11.11, the GEOID 90 model essentially

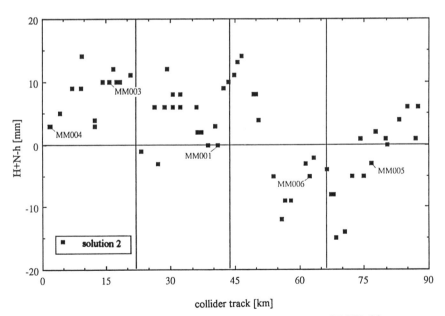

Figure 11.12 Height discrepancies with respect to GEOID 90.

follows the geoid correctly between stations MM004 and MM001. The negative feature around station 114 is completely modeled by GEOID 90. The remaining features around station 172 and 144 can be caused by actual geoid features not modeled by GEOID 90 or by systematic errors in the GPS survey or in leveling.

Finally, it is interesting to relate Figures 11.10 and 11.12. If the effect of gravity were ignored in the leveling adjustment, the value for the "height" of, for example, station MM003 would increase by 2.6 mm, according to Figure 11.10. There would be a corresponding increase for the respective value plotted in Figure 11.12. Considering all points along the collider track, it can be verified that the variation of the unmodeled east–west slope would increase from 14 mm to about 18 mm. Thus the use of gravity in the leveling adjustment reduces the unmodeled slope (as one expects).

11.4.3 Orange County Densification

The Orange County GPS survey comprises more than 7000 vectors linking more than 2000 stations. The purpose of the densification within Orange County was to provide a network densification of one-half mile to support cadastral activities. For such a large network it is beneficial to use graphical tools to analyze the observations, results of the adjustments, and other relevant dependencies. It is very difficult, if not impossible to mentally integrate and judge numerical data that enter such big adjustments, let alone data that are generated as part of the solutions. Simple two-variable plots of various functions are found to be useful for the analysis. Several of these plots indicate outliers (i.e., a deviation from an otherwise systematic variation). These outliers are prime candidates for in-depth studies. Redundancy numbers and internal reliability plots for individual vectors and for station averages appear useful in identifying weak portions of the network (which may result from automated deweighting for blunder detection). The variance–covariance matrix of vector observations is the "determining force" for shaping most of the functions.

The observations were made by private firms spanning several years. The Orange County Survey Division staff has been monitoring a reference network of 26 stations for crustal motion in cooperation with various institutions and agencies. These stations and other existing stations of the California high-precision geodetic network (CA HPGN) surrounding the County constitute a reference network to which the densification network was constrained. All parts of the network are seen in Figure 11.13. The reference network is marked with respective symbols. The 2000-plus stations of the densification network are marked with a dot (looking more like snowflakes).

The adjustment of the reference network is not discussed here, nor is there any discussion given on the possible time dependency of the coordinates in that region due to earthquakes. The following discussion refers to the densification solution only where the reference stations are held fixed. Additional detail is given in Leick and Emmons (1994).

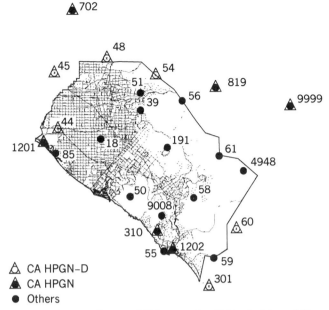

Figure 11.13 The Orange County GPS network. (After Leick and Emmons, 1994. Reprinted with permission of the American Society of Civil Engineers, 1994.)

A Priori Stochastic Information　When one is not familiar with the specific contents of a big data set, it is worthwhile to commence an analysis with a study of the variance–covariance matrices of the vectors as generated during the carrier phase processing step. For this purpose a simple function of the a priori statistics is computed as follows:

$$\sigma_b = \sqrt{\sigma_1^2 + \sigma_2^2 + \sigma_3^2} \tag{11.38}$$

where σ_1^2, σ_2^2, σ_3^2 are the diagonal elements of the 3×3 variance–covariance matrix of the vector observations. Note that σ_b does not equal the standard deviation of the observed vector length. Other simple functions such as the trace of the variance–covariance matrix can be used as well. Figure 11.14 displays σ_b as a function of the length of the vectors. For longer lines there appears to be a weak length dependency of about 1:200,000. Several of the shorter baselines show larger-than-expected values. It is not necessarily detrimental to include vectors with large variances in an adjustment, although they are unlikely to contribute to the strength of the network solution. Analyzing the averages of σ_b for all vectors of a particular station is useful in discovering stations which might be connected exclusively to low-precision observations.

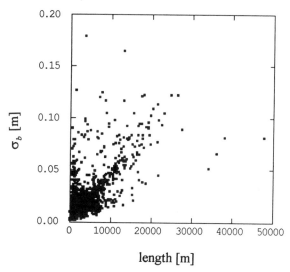

length [m]

Figure 11.14 A priori precision and length of baseline. (After Leick and Emmons, 1994. Reprinted with permission of the American Society of Civil Engineers, 1994.)

Variance Factor   Figures 11.15 and 11.16 show the estimated square root of the variance factor f_k for each vector k. This factor is computed from

$$f_k = \sqrt{\frac{(\bar{V}^T \bar{V})_k}{\bar{r}_{k1} + \bar{r}_{k2} + \bar{r}_{k3}}} \tag{11.39}$$

where $\bar{V}_k$ and $\bar{r}_k$ denote the decorrelated residuals and redundancy numbers of the vector. See equation (4.383) of Chapter 4 regarding the decorrelation of observations. Because the sum of the three redundancy numbers is small (since a factor is estimated for each vector), the estimates vary considerably around the expected values of 1. The estimates are shown in Figures 11.15 and 11.16 as a function of the baseline length and the a priori statistics σ_b. Both figures show the largest factors to be associated with the shortest baselines or the smallest σ_b. The value 0 is obtained if the residuals are zero (or less than $\frac{1}{2}$ mm because the last significant digits in the output of the residuals happens to be millimeter). In the solution discussed below the variance–covariance matrices of the observed vectors were scaled following the thoughts expressed in Section 4.11.3. The residuals are compared with the standard deviations of the observed vector. For large residuals, as compared to the respective standard deviations, a scaling factor F is computed from the residuals using an empirical rule. All the components of a vector are multiplied with the same factor. The result is shown in Figure 11.17.

Figures 11.14–11.17 show that the large scale factors tend to be associated

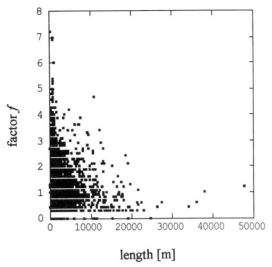

length [m]

Figure 11.15 Variance factors versus length of baseline. (After Leick and Emmons, 1994. Reprinted with permission of the American Society of Civil Engineers, 1994.)

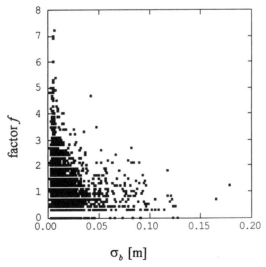

σ_b [m]

Figure 11.16 Variance factor versus precision of baseline. (After Leick and Emmons, 1994. Reprinted with permission of the American Society of Civil Engineers, 1994.)

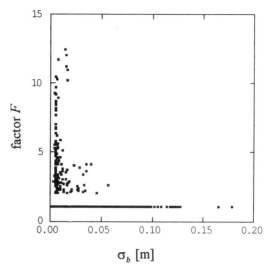

Figure 11.17 Applied variance scale factors. (After Leick and Emmons, 1994. Reprinted with permission of the American Society of Civil Engineers, 1994.)

with short baselines (which also tend to have the smallest σ_b). For short baselines the centering errors of the antenna and the separation of electronic and geometric center of the antenna are important, and neither is reflected by the stochastic model, i.e., the variance–covariance matrix. As a general rule, the variance–covariance matrices should not be scaled unconditionally with f-squared or F-squared. Rather, the proper procedure is to analyze the observations associated with the largest factors and to try to determine their cause. If proper justification can be found, the factors can be applied.

Redundancy Numbers   The vector redundancy number is the sum of the redundancy numbers for the three vector components,

$$R_k = \bar{r}_{k1} + \bar{r}_{k2} + \bar{r}_{k3} \tag{11.40}$$

The vector redundancy numbers R_k vary between zero and three. Values close to three indicate maximum contribution to the redundancy and minimum contribution to the solution, i.e., the observation is literally redundant. Such observations do not contribute to the adjustment because of the presence of other, usually much more accurate observations. A redundancy of zero implies an uncontrolled observation, e.g., a single vector to one station. A small redundancy number implies little contribution to the redundancy but a big contribution to the solution. Such observations "overpower" other observations and usually have small residuals. As a consequence of their "strength," blunders in those observations might not be discovered.

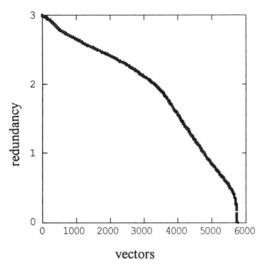

Figure 11.18 Ordered vector redundancy. (After Leick and Emmons, 1994. Reprinted with permission of the American Society of Civil Engineers, 1994.)

The ordered redundancy numbers in Figure 11.18 exhibit a distinctly sharp decrease as its smallest values are reached. There is one redundancy of zero. Inspection of the data indicates that these very small redundancies occur whenever there is only one good vector observation left to a particular station, and the other vectors to that station have been deweighted by scaling the variance–covariance matrices as part of the automatic blunder detection procedure. Typically, the scaled vectors have a high redundancy number, indicating their diminished contribution. The only remaining unscaled observation contributes the most; therefore, the respective residuals are very small, usually in the millimeter range. Thus, a danger of automated blunder detection and deweighting, is that parts of the network might become uncontrolled.

Figures 11.19 and 11.20 indicate that long vectors and vectors with large σ_b have large redundancy numbers. The shapes in these figures seem to suggest that it might be possible to identify all those vectors that can be deleted from the adjustment without affecting the strength of the solution. It would be interesting to analyze such figures when fixed and float vector solutions are included in one adjustment. Because the float solutions usually have larger standard deviations, one should expect the float solutions to receive a very high redundancy number, implying that nothing, or at least very little, is gained by including both types of vector solutions in the network adjustment. The steep slope seen in these figures suggests that the assembly of short baselines determines the shape of the network. Mixing short and long baselines is useful only if long baselines have been determined with an accuracy comparable to that of shorter lines. This can be accomplished through longer observation times, using dual-frequency receivers, and processing with a precise ephemeris.

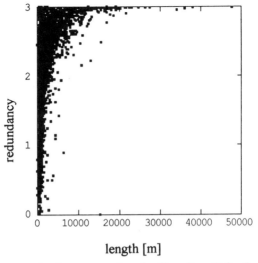

Figure 11.19   Vector redundancy vesus length of baseline. (After Leick and Emmons, 1994. Reprinted with permission of the American Society of Civil Engineers, 1994.)

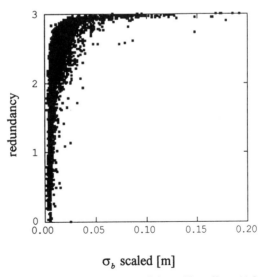

Figure 11.20   Vector redundancy versus precision of baseline. (After Leick and Emmons, 1994. Reprinted with permission of the American Society of Civil Engineers, 1994.)

Internal Reliability   Internal reliability values are shown in Figures 11.21 and 11.22. These values are a function of the internal reliability vector components as follows:

$$I_k = \sqrt{I_{k1}^2 + I_{k2}^2 + I_{k3}^2} \tag{11.41}$$

The internal reliability components are computed according to equation (4.372) for the decorrelated vector observations, and are then transformed back to the physical observation space. The plotted values are based on the factor $\delta_0 = 4.12$. There is a surprisingly linear relationship between internal reliability and the quality of the observations as expressed by σ_b. The slope essentially equals δ_0. The outliers in Figure 11.21 are all associated with small σ_b and pertain to a group of observations with a redundancy smaller than 0.5. The outliers include all those "single vectors" that result when the other vectors to the same station are deweighted. The linear relationship displayed makes it possible to easily identify the outliers for further inspection and analysis. Furthermore, this linear relationship nicely confirms that internal reliability is not a function of the shape of the GPS network.

Figure 11.22 shows the internal reliability as a function of the redundancy. There is a sharp decrease in internal reliability in the range of the largest redundancy numbers. There is an increase in internal reliability for smallest redundancy numbers. Constant internal reliability is a desirable property of a network. Whereas high internal reliability in the high-redundancy region in-

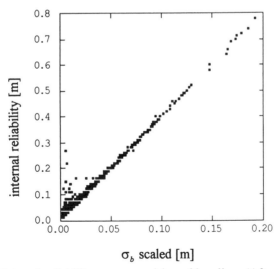

Figure 11.21   Internal reliability versus precision of baseline. (After Leick and Emmons, 1994. Reprinted with permission of the American Society of Civil Engineers, 1994.)

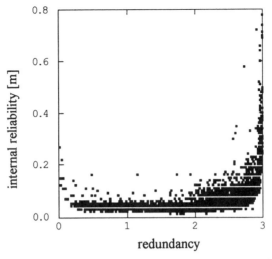

Figure 11.22 Internal reliability versus redundancy. (After Leick and Emmons, 1994. Reprinted with permission of the American Society of Civil Engineers, 1994.)

dicates potential savings by observing fewer vectors, it exposes weak portions of the network in the low redundancy region. In the latter case, additional observations are recommended.

Blunders and Absorption Figure 11.23 shows blunders as predicted by the respective residuals. As is detailed in (4.375), a relationship exists between

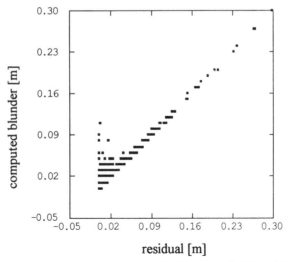

Figure 11.23 Computed blunders versus residuals. (After Leick and Emmons, 1994. Reprinted with permission of the American Society of Civil Engineers, 1994.)

computed blunders, residuals, and redundancies. The figure shows the blunder function

$$B_k = \sqrt{B_{k1}^2 + B_{k2}^2 + B_{k3}^2} \qquad (11.42)$$

versus the residual function

$$V_k = \sqrt{V_{k1}^2 + V_{k2}^2 + V_{k3}^2} \qquad (11.43)$$

The computed blunder and the residuals refer to the physical observation space. This relationship appears to be primarily linear with slope $1:1$ (at least for the larger residuals). The outliers seen for small residuals refer to the group of observations with the smallest redundancy numbers.

Figure 11.24 shows the absorption versus the redundancy. The absorption specifies that part of a blunder which is absorbed in the solution, i.e., absorption indicates falsification of the solution. The values

$$A_k = V_k + B_k \qquad (11.44)$$

are plotted. As expected, the observations with lowest redundancy tend to absorb most. At the extreme case, the absorption is infinite for zero redundancy, and zero for a redundancy of three (for vector observations). Clearly, very small redundancies reflect an insensitivity to blunders, which is not desirable.

Station Averages It is instructive to compute the averages per station for some of the functions discussed above. In these computations each vector is

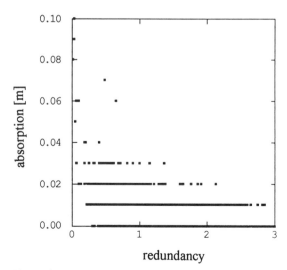

Figure 11.24 Absorption versus redundancy. (After Leick and Emmons, 1994. Reprinted with permission of the American Society of Civil Engineers, 1994.)

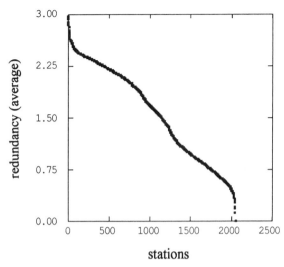

Figure 11.25 Ordered station redundancy. (After Leick and Emmons, 1994. Reprinted with permission of the American Society of Civil Engineers, 1994.)

used twice as it contributes to the average for each endpoint. Figure 11.25 shows the ordered station redundancy numbers. Except for one station with zero redundancy, the smallest redundancy number is 0.118. As is the case for individual vectors, the smallest values deserve extra attention. Roughly speaking, stations with redundancy numbers smaller than 0.5 are determined by only two vectors. Both of these vectors can be of high quality, i.e., they are not deweighted by automatic blunder detection. In fact, whenever the variance–covariance submatrix is scaled, the respective vector redundancy increases significantly. Stations containing such scaled vectors will show a misleadingly high station redundancy and cannot reliably be spotted in Figure 11.25. The stations with low redundancies in Figure 11.25, therefore, indicate portions of the network that would benefit from additional observations.

It is not sufficient to limit quality control to residuals and to plots or histograms of normalized residuals. It is equally important that the quality of the network be presented in terms of redundancy and reliability measures. These functions are, among other things, useful in judging the implications of deweighting. The consequences of deweighting are not always readily apparent in large networks.

CHAPTER 12

ELLIPSOIDAL AND CONFORMAL MAPPING MODELS

Computations on the ellipsoid and the conformal mapping plane have been available to surveyors since the early 1800s when K. F. Gauss developed the fundamentals of differential geometry. As is well known, Gauss used his many talents to develop geodetic computations on the ellipsoidal surface and on the conformal map. The problem presented itself to Gauss because he was in "responsible charge" of observing and computing a geodetic network in Northern Germany. Computations on a surface, whether on the ellipsoid or on the conformal map, are inherently two-dimensional. The third dimension, height, is being "used-up" for the reduction of the spatial observations onto these computation surfaces. Since the ellipsoidal surface has a changing curvature in latitude, the mathematics becomes complex, and the computations are labor intensive. Because the conformal map is flat in any sense, the formulas of plane trigonometry apply, thus reducing the computational work load considerably. Both approaches require a new theoretical element not discussed thus far, the geodesic. One of the definitions of the geodesic is that it is the shortest distance between two points on a surface. Unfortunately, the mathematical developments of the geodesic on the ellipsoid and its image on the map are fairly complex, requiring advanced mathematical skills. Because the solutions have been available for such a long time and belong to the most classical of all geodetic theories and are fully documented in the literature, this section contains only a list of the necessary formulas. The interested reader may consult Grossman (1976), Hristow (1955), Leick (1980a), Thomas (1952), and other texts of geodesy for the complete derivations.

The ellipsoidal and conformal mapping expressions are generally given in the form of a mathematical series. The formulas are the result of truncations at various steps along the mathematical development. The expressions given here are sufficiently accurate for typical applications in surveying where distances do not exceed a couple of hundred kilometers. Some of the expressions may even contain negligible terms when applied in small areas; however, with today's powerful desktop computers, one should not be overly concerned about a few unnecessary algebraic operations. Two types of observations are used for computations on a surface: azimuth (angle) and distance. This chapter lists only the expressions for azimuth and distance. The reductions, partial derivatives, and other quantities for angles can be obtained through differencing respective azimuth expressions.

12.1 REDUCTIONS TO THE ELLIPSOID

The ellipsoidal (geodetic) azimuth α of Chapter 7 is the angle between two normal planes containing the ellipsoidal normal at the observing station; an angle is defined similarly. These angles are obtained after the deflection of the vertical reduction is applied to the original observations. The spatial distance did not require any reduction for use in the three-dimensional model of Chapter 7. The goal of this section is to reduce these angles and distances further in order to obtain quantities on the ellipsoidal surface and on the mapping plane.

Figure 12.1 shows the next reduction to be applied to the azimuths and angles. The spatial azimuth α, which is the result of the reductions applied thus far, is shown in the figure as the azimuth of the normal plane containing the target point P_2. The intersection of this normal plane with the ellipsoid is denoted by the dashed line P_1 to P_2''. The representative of the space point P_2 on the ellipsoid is denoted by P_2', which is obtained intersecting of the ellipsoidal normal through the space point P_2 with the ellipsoid. The azimuth of the normal section that contains the ellipsoidal normal to P_1 and that goes

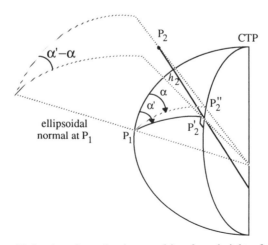

Figure 12.1   Angular reduction resulting from height of target.

**TABLE 12.1   Angular and Distance Reduction on
the Geodesic**

$$\left(\alpha_1' - \alpha_1\right) = 0.108^{\cdot} \cos^2 \varphi_1 \sin 2\alpha_1 h_{2[km]}$$

$$\left(\alpha_1^{\cdot} - \alpha_1'\right) = -0.028^{\cdot} \cos^2 \varphi_1 \sin 2\alpha_1 \left(\frac{S_{[km]}}{100}\right)^2$$

$$\frac{1}{R} = \frac{\cos^2 \alpha}{M} + \frac{\sin^2 \alpha}{N}$$

$$\overline{S} = \sqrt{\frac{s^2 - \Delta h^2}{\left(1 + \dfrac{h_1}{R}\right)\left(1 + \dfrac{h_2}{R}\right)}}$$

$$S = R\psi = 2R\sin^{-1}\left(\frac{\overline{S}}{2R}\right)$$

through the surface point P_2' is denoted by α'. The angular difference $(\alpha' - \alpha)$ is called the reduction due to target height and is given in Table 12.1. The height of the observing station at P_1 does not matter in this reduction, since α is the angle between planes.

The need for another angular reduction follows from Figure 12.2. Assume that two points P_1 and P_2 are located at different latitudes (P_2 could be P_2' of Figure 12.1). Line 1 is the normal section from P_1 to P_2 and line 2 indicates the normal section from P_2 to P_1. It can readily be seen that these two normal sections do not coincide, because of the changing curvature of the ellipsoidal meridian as a function of latitude. The question is, which of these two normal sections should be adopted for the computations? This dilemma is solved by introducing a curve connecting these two points in a unique way: the geodesic. There is only one geodesic from P_1 to P_2. As mentioned, the geodesic is the

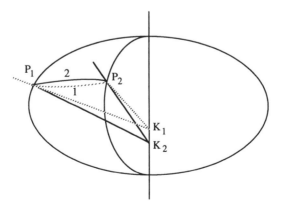

Figure 12.2   Normal sections on the ellipsoid.

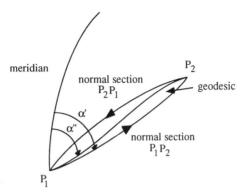

Figure 12.3 Angular reduction of the normal section to the geodesic.

shortest connection between two points on a surface. The derivation of the mathematical equation for the geodesic requires advanced application of differential geometry and elaborate solutions of differential equations. Figure 12.3 shows the approximate geometric relationship between the normal sections and geodesic. The angular reduction $(\alpha'' - \alpha')$ is required for computing the azimuth α'' of the geodesic. The expression is listed in Table 12.1; an approximate azimuth on the right-hand side is sufficient for the evaluation of the expressions.

The slant distance s must be reduced to the length of a geodesic. Figure 12.4 shows an ellipsoidal section along the line of sight. The expression for the length S of the geodesic given in Table 12.1 is based on a spherical approximation of the ellipsoidal arc. Within the accuracy of this approximation, there is no need to distinguish between the length of the geodesic and the normal section. The symbol R denotes the radius of curvature for the approx-

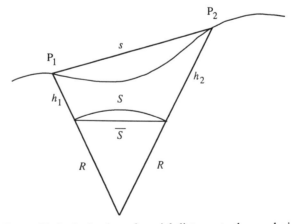

Figure 12.4 Reduction of spatial distance to the geodesic.

imating spherical arc; it is a function of the azimuth of the line of sight and the curvatures of the meridian and prime vertical at the center of the line. Note that computing the length of the geodesic requires ellipsoidal heights. Orthometric heights must first be corrected for the geoid undulations in order to avoid errors in the distance reduction.

The following four expressions are taken from Chapter 6; they serve as auxiliary expressions for the reductions listed in Table 12.1:

$$N = \frac{a}{\sqrt{1 - e^2 \sin^2 \varphi}} \tag{12.1}$$

$$M = \frac{a(1 - e^2)}{(1 - e^2 \sin^2 \varphi)^{3/2}} \tag{12.2}$$

$$e^2 = 2f - f^2 \tag{12.3}$$

$$f = \frac{a - b}{a} \tag{12.4}$$

12.2 THE ELLIPSOIDAL SURFACE MODEL

The result of the reductions is that angles and distances have been reduced to ellipsoidal surface quantities, such as azimuths of geodesic, angles between two geodesics, and the lengths of the geodesics. It is noted again that an angle observation can be expressed as a difference between two azimuths. The reductions of the azimuths can therefore readily be applied to angles as well.

At the heart of the ellipsoidal computations are the so-called direct and inverse problems, as indicated by Figure 12.5. For the direct problem, the given elements are the geodetic latitude and longitude of one station, say, $P_1(\varphi_1, \lambda_1)$, and the azimuth α_1'' and length S of the geodesic to another point P_2; the geodetic latitude φ_2, longitude λ_2, and the back azimuth α_2'', is required. For the inverse problem the geodetic latitudes and longitudes of $P_1(\varphi_1, \lambda_1)$ and $P_2(\varphi_2, \lambda_2)$ are given, and the forward and back azimuth and the length of the geodesic

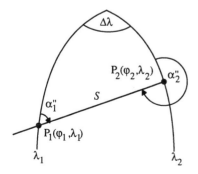

Figure 12.5 Direct and inverse solution on the ellipsoid.

TABLE 12.2   Inverse Solution on the Ellipsoid

$$\varphi = \frac{\varphi_1 + \varphi_2}{2}$$

$$t = \tan \varphi$$

$$V^2 = 1 + \eta^2$$

$$\eta^2 = \frac{e^2}{1 - e^2} \cos^2 \varphi$$

$$[1] = \frac{1}{M}$$

$$[2] = \frac{1}{N}$$

$$[3] = \frac{1}{24}$$

$$[4] = \frac{1 + \eta^2 - 9\eta^2 t^2}{24 V^4}$$

$$[5] = \frac{1 - 2\eta^2}{24}$$

$$[6] = \frac{\eta^2 (1 - t^2)}{8 V^4}$$

$$[7] = \frac{1 + \eta^2}{12}$$

$$[8] = \frac{3 + 8\eta^2}{24 V^4}$$

$$S \sin \alpha^{\cdot} = \frac{1}{[2]} \Delta\lambda \cos\varphi \left[1 - [3](\Delta\lambda \sin\varphi)^2 + [4](\Delta\varphi)^2 \right] \qquad \text{(a)}$$

$$S \cos \alpha^{\cdot} = \frac{1}{[1]} \Delta\varphi \cos\frac{\Delta\lambda}{2} \left[1 + [5](\Delta\lambda \cos\varphi)^2 + [6](\Delta\varphi)^2 \right] \qquad \text{(b)}$$

$$\Delta\alpha^{\cdot} = \Delta\lambda \sin\varphi \left[1 + [7](\Delta\lambda \cos\varphi)^2 + [8](\Delta\varphi)^2 \right] \qquad \text{(c)}$$

are required. Many solutions are available in the literature for the direct and inverse problems. Some of these solutions are valid for geodesics going all around the ellipsoid. Here the Gauss midlatitude solution is given.

Table 12.2 lists expressions for the inverse problem. Because the latitudes and longitudes are given, in this case all the auxiliary expressions at the upper part of the table can be computed. The elements (t, η, M, N, V) are evaluated for $\varphi = (\varphi_1 + \varphi_2)/2$. Furthermore, $\Delta\varphi = \varphi_2 - \varphi_1$ and $\Delta\lambda = \lambda_2 - \lambda_1$. The length and azimuth of the geodesic become

$$S = \sqrt{(S \sin \alpha'')^2 + (S \cos \alpha'')^2} \qquad (12.5)$$

$$\alpha'' = \tan^{-1} \left(\frac{S \sin \alpha''}{S \cos \alpha''} \right) \qquad (12.6)$$

$$\alpha_1'' = \alpha'' - \frac{\Delta\alpha''}{2} \qquad (12.7)$$

$$\alpha_2'' = \alpha'' + \frac{\Delta\alpha''}{2} \pm 180° \qquad (12.8)$$

The Gauss midlatitude direct solution requires iterations; the expressions are listed in Table 12.3. Given the position of the point $P_1(\varphi_1, \lambda_1)$ and the azimuth α_1'' and distance S to P_2, one can use approximate solutions of the first

TABLE 12.3   Direct Solution on the Ellipsoid

$$\lambda_2 - \lambda_1 = \Delta\lambda = [2]\frac{S\sin\alpha^{\cdot}}{\cos\varphi}\left[1 + [3](\Delta\lambda\sin\varphi)^2 - [4](\Delta\varphi)^2\right] \qquad \text{(a)}$$

$$\varphi_2 - \varphi_1 = \Delta\varphi = [1]\frac{S\cos\alpha^{\cdot}}{\cos\dfrac{\Delta\lambda}{2}}\left[1 - [5](\Delta\lambda\cos\varphi)^2 - [6](\Delta\varphi)^2\right] \qquad \text{(b)}$$

$$\Delta\alpha^{\cdot} = \Delta\lambda\sin\varphi\left[1 + [7](\Delta\lambda\cos\varphi)^2 + [8](\Delta\varphi)^2\right] \qquad \text{(c)}$$

two equations,

$$\lambda_2 = \lambda_1 + \frac{S\sin\alpha_1''}{N_1\cos\varphi_1}$$

$$\varphi_2 = \varphi_1 + \frac{S\cos\alpha_1''}{M_1}$$

to obtain a first estimate of the location of $P_2(\varphi_2, \lambda_2)$. Next, compute the latitude of the midpoint and the differences in latitude and longitude:

$$\varphi = \frac{\varphi_1 + \varphi_2}{2}$$

$$\Delta\varphi = \varphi_2 - \varphi_1$$

$$\Delta\lambda = \lambda_2 - \lambda_1$$

These values are entered into equation (c) of Table 12.3 to obtain an improved solution for the difference in azimuth $\Delta\alpha''$. This value, together with the given azimuth at P_1, yields

$$\alpha'' = \alpha_1'' + \frac{\Delta\alpha''}{2} \qquad (12.9)$$

Use α'' and the previously computed $\Delta\varphi$ and $\Delta\lambda$ and obtain improved values for $\Delta\varphi$ and $\Delta\lambda$ from the first two equations of Table 12.3. The improved positions for P_2 now become

$$\varphi_2 = \varphi_1 + \Delta\varphi$$

$$\lambda_2 = \lambda_1 + \Delta\lambda$$

Compute a new midlatitude

$$\varphi = \frac{\varphi_1 + \varphi_2}{2}$$

and loop back to equation (12.9) for another iteration. Repeat the iteration loop until $\Delta\alpha''$, $\Delta\varphi$, $\Delta\lambda$ do not change significantly. The last step is to compute the final positions of P_2:

$$\lambda_2 = \lambda_1 + \Delta\lambda$$

$$\varphi_2 = \varphi_1 + \Delta\varphi$$

$$\alpha_2'' = \alpha'' + \frac{\Delta\alpha''}{2} \pm 180°$$

The inverse solution constitutes the mathematical model for the adjustment on the ellipsoid. It is written in the general form

$$S = f(\varphi_1, \lambda_1, \varphi_2, \lambda_2)$$

$$\alpha_1'' = f(\varphi_1, \lambda_1, \varphi_2, \lambda_2)$$

The partial derivatives for the design matrix are listed in Table 12.4. Note that the adjustment does not require the use of the iterative direct solution. Only the inverse solution is required to compute the L_0 vector consisting of azimuths and lengths of geodesics. Extension of the mathematical model to include angle observations between geodesics is, once again, accomplished through differencing the azimuth expressions. Note also that the list of parameters does not contain heights. If there is any vertical information in the data set, it is of no use to the adjustment.

TABLE 12.4 Partial Derivatives for Adjustment on the Ellipsoid

	$d\varphi_1$	$d\lambda_1$	$d\varphi_2$	$d\lambda_2$
dS	$-M_1 \cos\alpha_1'$	$N_2 \cos\varphi_2 \sin\alpha_2'$	$-M_2 \cos\alpha_2'$	$-N_2 \cos\varphi_2 \sin\alpha_2'$
$d\alpha''$	$\dfrac{M_1 \sin\alpha_1'}{S}$	$\dfrac{N_2 \cos\varphi_2 \cos\alpha_2'}{S}$	$\dfrac{M_2 \sin\alpha_2'}{S}$	$\dfrac{-N_2 \cos\varphi_2 \cos\alpha_2'}{S}$

12.3 THE CONFORMAL MAPPING MODEL

In conformal mapping, the ellipsoidal surface is mapped conformally onto a plane. Conformality simply means preservation of angles! However, there is a minor pitfall to be avoided when interpreting a conformal map. An angle between two curves, say, two geodesics on the ellipsoid, is defined as the angle between the tangents on these curves. Thus, in conformal mapping the angle between the tangents of curves on the ellipsoid and the angle between tangents of the mapped curves are preserved. This preservation applies, of course, to the azimuth as well, since this is also an angle. The property of preserving angles makes conformal maps useful as a computation tool, because directional elements between the ellipsoid and the map have a known relationship. If the goal is to map the ellipsoid onto a plane for graphical demonstration on the computer screen or to assemble overlays of spatial data, any unique mapping from the ellipsoid to the plane can be used.

Users who prefer to work with plane mapping coordinates, such as state plane coordinates, rather than ellipsoid latitude and longitude can still use the three-dimensional adjustment procedures given in Chapter 7. The given mapping coordinates can be transformed to the ellipsoidal surface and then used, together with heights, in the formulation of the three-dimensional adjustment. The adjusted ellipsoidal positions can subsequently be mapped onto the conformal mapping plane. The user might not even be aware that ellipsoidal positions are used in the adjustment. Plane mapping coordinates might also have an advantage for the automatic generation of approximate positions in order to start the adjustment. When an initial pass is made through the file of original observations, the observations can be interpreted as observations on the mapping plane and plane coordinates can be generated automatically. Approximate height parameters can be computed separately with the help of zenith angles, distances, and leveled height differences. This procedure seems especially simple and appropriate for formulating the algorithm for the automatic generation of approximate positions in that the three-dimensional computations are replaced by two separate computations, one for the horizontal positions and one for the vertical positions.

Figure 12.6 shows the so-called mapping elements that link the ellipsoidal observations to corresponding quantities on the mapping plane. It is not useful and is indeed misleading and wrong to interpret the mapping plane of Figure 12.6 as the "horizon in which the local survey takes place." One should not even try to interpret the conformal map as a perspective projection of the ellipsoid, because it simply is not a perspective projection. The conformal map is merely a result of imposing the mathematical condition of conformality. In Figure 12.6, the Cartesian coordinate system in the mapping plane is denoted by (x, y). The points $P_1(x, y)$ and $P_2(x, y)$ are the images of corresponding points on the ellipsoid. Consider for a moment the geodesic connecting the points P_1 and P_2 on the ellipsoid. This geodesic can be mapped point by point;

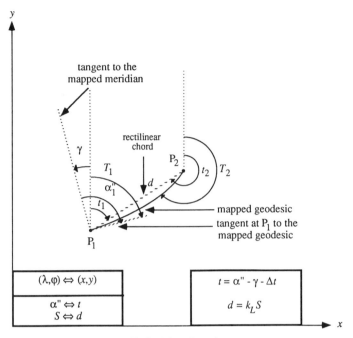

Figure 12.6 Mapping elements.

the result is the mapped geodesic as shown in the figure. This image is a smooth but mathematically complicated curve. The straight line between the images P_1 and P_2 is called the rectilinear chord. Next, the ellipsoidal meridian can be mapped; this image may or may not be a straight line on the map. To be general, Figure 12.6 shows the tangent on the mapped meridian. The angle between the y axis and the mapped meridian is called the meridian convergence; it is generally counted positive in the counterclockwise sense. Because of conformality, the ellipsoidal azimuth of the geodesic is preserved during the mapping and must be equal to the angle between the tangents on the mapped meridian and geodesic as shown. If the small angle $\Delta t = T - t$, called arc-to-chord correction, is known, then the grid azimuth t of the rectilinear chord can be readily computed from the meridian convergence and the azimuth of the geodesic. Next, is it realized that there is no specification in the conformal mapping condition as to the preservation of the length of the geodesic on the ellipsoid and the corresponding projected geodesic. In general, one is not interested in the length of the projected geodesic, but would like to obtain the length of the rectilinear chord. The line scale factor k_L, if multiplied by the length of the geodesic on the ellipsoid, gives the length of the rectilinear chord. Even though the term "map distortions" has many definitions, one associates a small Δt and a k_L close to 1 with small distortions of the map, meaning that the respective reductions in angle and distance are small and perhaps even negligible.

In summary, the conformal mapping elements are the meridian convergence γ, the arc-to-chord angular reduction Δt, and the line scale factor k_L. Once the grid azimuth t and the length of the rectilinear chord d have been computed from the ellipsoidal observations, with the help of the mapping elements, plane trigonometry applies for all computations. The same is, of course, true for angles between rectilinear chords. It is important to note that the mapping elements change in size and sign with the location of the line and with its orientation. To keep Δt small and k_L close to 1, we limit the area represented in a single mapping plane in size, thus the need for several mappings to cover large regions of the globe. In addition, the mapping elements are also functions of elements to be specified by the designer of the map. In this section the specifications leading to the transverse Mercator or the Lambert conformal mapping are given. Before the advent of computers, when it was difficult to compute reductions, much effort was spent minimizing the distortions patterns of the various conformal mappings. However, that need is not so critical anymore.

12.3.1   The Transverse Mercator Model

The specifications for the transverse Mercator mapping are as follows:

1. Use conformal mapping.
2. Adopt a central meridian λ_0 that passes more or less through the middle of the area to be mapped. For reasons of convenience, relabel the longitudes starting with $\lambda = 0$ at the central meridian.
3. Let the mapped central meridian coincide with the y axis of the map. Assign $x = 0$ to the image of the central meridian.
4. The length of the mapped central meridian should be k_0 times the length of the corresponding ellipsoidal arc, that is, at the central meridian $y = k_0 S(\varphi)$.

The general picture of the transverse Mercator map is shown in Figure 12.7. The image of the central meridian is a straight line; all other meridians are curved lines coming together at the pole and being perpendicular to the image of the equator, which coincides with the x axis. The mapped parallels are, of course, perpendicular to the mapped meridians but are not circles in the mathematical sense. The factor k_0 is an important element of design. If one were to analyze the magnitude of the mapping element Δt, one would find that the magnitude generally increases with separation from the central meridian, being essentially negligible at the central meridian. By choosing $k_0 < 1$, one allows some distortions at the central meridian for the benefit of having less distortions away from the central meridian. In this way, the longitudinal coverage of the area of one map can be extended for a given level of acceptable distortion.

The mapping equations for the direct mapping from $P(\varphi, \lambda)$ to $P(x, y)$ are given in Table 12.5. Note that λ is counted positive to the east, starting at the

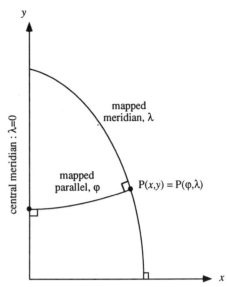

Figure 12.7   Transverse Mercator map.

TABLE 12.5   Transverse Mercator Mapping from Ellipsoid to the Plane

$$\frac{x}{k_0 N} = \lambda \cos\varphi + \frac{\lambda^3 \cos^3 \varphi}{6}\left(1 - t^2 + \eta^2\right)$$

$$+ \frac{\lambda^5 \cos^5 \varphi}{120}\left(5 - 18t^2 + t^4 + 14\eta^2 - 58t^2\eta^2\right)$$ (a)

$$\frac{y}{k_0 N} = \frac{S}{N} + \frac{\lambda^2}{2}\sin\varphi\cos\varphi + \frac{\lambda^4}{24}\sin\varphi\cos^3\varphi\left(5 - t^2 + 9\eta^2 + 4\eta^4\right)$$

$$+ \frac{\lambda^6}{720}\sin\varphi\cos^5\varphi\left(61 - 58t^2 + t^4 + 270\eta^2 - 330t^2\eta^2\right)$$ (b)

central meridian. The necessary auxiliary relations are

$$t = \tan\varphi \tag{12.10}$$

$$\eta^2 = \frac{e^2}{1 - e^2}\cos^2\varphi \tag{12.11}$$

where

$$S = \frac{a}{1 + n}\left(a_0\varphi - a_2\sin 2\varphi + a_4\sin 4\varphi - a_6\sin 6\varphi + a_8\sin 8\varphi\right)$$

$$\tag{12.12}$$

$$a_0 = 1 + \frac{n^2}{4} + \frac{n^4}{64} \qquad a_2 = \frac{3}{2}\left(n - \frac{n^3}{8}\right)$$

$$a_4 = \frac{15}{16}\left(n^2 - \frac{n^4}{4}\right) \qquad a_6 = \frac{35}{48}n^3$$

$$a_8 = \frac{315}{512}n^4 \qquad n = \frac{f}{2 - f}$$

Equation (12.12) gives the length of the ellipsoidal meridian from the equator to the latitude φ which is required in computing the y coordinate in (b) of Table 12.5. The symbols M and N denote the radius of curvature of the meridian and prime vertical, respectively. The elements (t, η, N) are evaluated for latitude φ; f denotes the flattening of the ellipsoid.

The inverse solution for transforming $P(x, y)$ to $P(\varphi, \lambda)$ is given in Table 12.6. All latitude-dependent terms in this table must be evaluated for the so-called footpoint latitude φ_f. The footpoint is a point on the central meridian obtained by drawing a parallel to the x axis through the point $P(x, y)$. Given the y coordinate, the footpoint latitude can be computed iteratively from (12.12). Because of condition 4 for the transverse Mercator mapping, the following relation holds:

$$\frac{y}{k_0} = S_f \tag{12.13}$$

where S_f denotes the length of the central meridian to the footpoint. Substitut-

TABLE 12.6 Transverse Mercator Inverse Mapping

$$\varphi = \varphi_f - \frac{t}{2}\left(1 + \eta^2\right)\left[\frac{x}{k_0 N}\right]^2$$

$$+ \frac{t}{24}\left(5 + 3t^2 + 6\eta^2 - 6\eta^2 t^2 - 3\eta^4 - 9t^2\eta^4\right)\left[\frac{x}{k_0 N}\right]^4 \tag{a}$$

$$- \frac{t}{720}\left(61 + 90t^2 + 45t^4 + 107\eta^2 - 162t^2\eta^2 - 45t^4\eta^2\right)\left[\frac{x}{k_0 N}\right]^6$$

$$\lambda = \sec\varphi_f\left\{\frac{x}{k_0 N} - \frac{1}{6}\left[\frac{x}{k_0 N}\right]^3\left(1 + 2t^2 + \eta^2\right)\right.$$

$$\left. + \frac{1}{120}\left[\frac{x}{k_0 N}\right]^5\left(5 + 28t^2 + 24t^4 + 6\eta^2 + 8t^2\eta^2\right)\right\} \tag{b}$$

All elements must be evaluated for the footpoint latitude φ_f

ing (12.12) in (12.13) gives

$$\frac{y}{k_0} = \frac{a}{1+n} \, a_0 \varphi_f$$

$$+ \frac{a}{1+n}(-a_2 \sin 2\varphi_f + a_4 \sin 4\varphi_f - a_6 \sin 6\varphi_f + a_8 \sin 8\varphi_f)$$

$$(12.14)$$

A first approximate solution is obtained by truncating (12.14) after the first term. The thus found value for φ_f is substituted into the other (smaller) terms, and an improved value is computed. This iterative procedure is symbolized by the following relations:

$$\varphi_f = y \, \frac{1+n}{aa_0 k_0} - f(\varphi_f) \qquad (12.15)$$

The mapping elements for the transverse Mercator mapping are given in Table 12.7. The symbol k in the second equation denotes the point scale factor. The point scale factor is the ratio of a differential distance on the map over the corresponding differential distance on the ellipsoid; thus it is the ratio of two corresponding differentially small distances. One of the properties of conformal mapping is that the point scale factor is independent of the orientation of

TABLE 12.7 Mapping Elements for the Transverse Mercator Mapping

$$\gamma = \lambda \sin\varphi \left[1 + \frac{\lambda^2 \cos^2\varphi}{3}(1 + 3\eta^2 + 2\eta^4) + \frac{\lambda^4 \cos^4\varphi}{15}(2 - t^2) + \cdots \right] \qquad (a)$$

$$k = k_0 \left[1 + \frac{\lambda^2}{2}\cos^2\varphi(1 + \eta^2) \right.$$

$$+ \lambda^4 \frac{\cos^4\varphi}{24}(5 - 4t^2 + 14\eta^2 + 13\eta^4 - 28t^2\eta^2 + 4\eta^6 - 48t^2\eta^4 - 24t^2\eta^6) \qquad (b)$$

$$+ \lambda^6 \frac{\cos^6\varphi}{720}(61 - 148t^2 + 16t^4) + \cdots \left. \right]$$

$$\Delta t_1 = \frac{x_1}{2k_0^2 R_1^2}\left(1 - \frac{x_1^2}{3k_0^2 R_1^2}\right)(y_2 - y_1) + \frac{1}{6k_0^2 R_1^2}\left(1 - \frac{3x_1^2}{2k_0^2 R_1^2}\right)(x_2 - x_1)(y_2 - y_1)$$

$$\approx \frac{1}{6k_0^2 R_1^2}(x_2 + 2x_1)(y_2 - y_1) \qquad (c)$$

$$\frac{1}{k_L} \equiv \frac{S}{d} = \frac{1}{6}\left(\frac{1}{k_1} + \frac{4}{k_m} + \frac{1}{k_2}\right) \qquad (d)$$

the differential line element; but k is a function of position. The line scale factor (last equation in Table 12.7) is expressed as a function of three point scale factors: k_1, at the beginning of the line; k_2, the end of the line; and k_m, at the middle of the line for which $\varphi = (\varphi_1 + \varphi_2)/2$ and $\lambda = (\lambda_1 + \lambda_2)/2$. This equation for the line scale factor is formally the same for all conformal mappings. One needs only to substitute the correct point scale factor. The other new symbol in Table 12.7 is the radius of curvature of the Gaussian sphere,

$$R = \sqrt{MN} \tag{12.16}$$

evaluated at P_1 or P_2 depending on the subscript.

12.3.2 The Lambert Conformal Model

The specifications for the Lambert conformal mapping are as follows:

1. Use conformal mapping.
2. Adopt a central meridian λ_0 that passes more or less through the middle of the area to be mapped. For convenience, relabel the longitudes, starting with $\lambda = 0$ at the central meridian.
3. Let the mapped central meridian coincide with the y axis of the map. Assign $x = 0$ for the image of the central meridian.
4. Map the meridians into straight lines passing through the image of the pole; map the parallels into concentric circles around the image of the pole. Select a standard parallel φ_0 that passes more or less through the middle of the area to be mapped. The length of the mapped standard parallel is k_0 times the length of the corresponding ellipsoidal parallel. Start counting $y = 0$ at the image of the standard parallel.

The general picture of the Lambert conformal map is shown in Figure 12.8. This mapping has a mathematical singularity at the pole, which is the reason why the angle of the meridian is λ' and not λ at the pole. The area of smallest distortion is along the image of the standard parallel in the east–west direction; as one departs from the standard parallel the distortions increase in the north–south direction. By choosing $k_0 < 1$ the distortions at the northern and southern extremities of the mapping area can be reduced by allowing some distortions in the vicinity of the standard parallel. Whenever $k_0 < 1$ there are two parallels, one south and one north of the standard parallel, along which the point scale factor k equals 1; that is, these two parallels are mapped without distortion in length. In the latter case, one speaks of a two-standard-parallel Lambert conformal mapping. The designer of the map has the choice of either specifying k_0 and φ_0 or the two parallels for which $k = 1$.

The Lambert conformal mapping equations are expressed in terms of the isometric latitude q. There is no need to detail the geometric and theoretical meaning of the isometric latitudes here; it suffices to say that the isometric

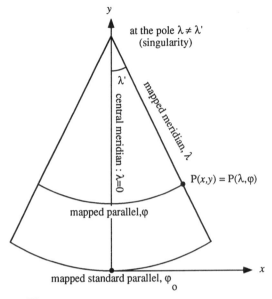

Figure 12.8 Lambert conformal mapping.

latitude and the geodetic latitude are related by

$$q = \ln \left[\tan \left(45° + \frac{\varphi}{2} \right) \left(\frac{1 - e \sin \varphi}{1 + e \sin \varphi} \right)^{e/2} \right] \qquad (12.17)$$

The symbol e denotes the first eccentricity of the ellipsoid given by (12.3).
Equation (12.17) gives q as a function of the ellipsoidal latitude φ. The inverse
solution, that is, computing φ from q,

$$\tan \left(45° + \frac{\varphi}{2} \right) = \epsilon^q \left(\frac{1 + e \sin \varphi}{1 - e \sin \varphi} \right)^{e/2} \qquad (12.18)$$

requires iterations. The symbol $\epsilon = 2.71828\ldots$ denotes the base of the natural
system of logarithm. As a first step, assume $\varphi = 0$ on the right hand side of
(12.18) and solve for a first approximation for the latitude φ. This value is
then substituted on the right-hand side, resulting in an improved solution for
φ. This iteration is continued until there is no change in φ.

The expressions for the direct Lambert conformal mapping are shown in
Table 12.8. Note that (k_0, φ_0) or, equivalently (k_0, q_0) specify the projections.
If two standard parallels φ_1 and φ_2 with $k = 1$ are specified, then k_0 and φ_0
follow from the expression of Table 12.9. The inverse solution is given in
Table 12.10. Once q has been found from the inverse solution, it can be con-
verted to geodetic latitude by (12.18). The mapping elements are found in
Table 12.11.

TABLE 12.8 Lambert Conformal Direct Mapping

$$x = k_0 N_0 \cot\varphi_0 \varepsilon^{-\Delta q(\sin\varphi_0)} \sin(\lambda \sin\varphi_0) \tag{a}$$

$$y = k_0 N_0 \cot\varphi_0 \left[1 - \varepsilon^{-\Delta q(\sin\varphi_0)} \cos(\lambda \sin\varphi_0) \right] \tag{b}$$

TABLE 12.9 Conversion from Two Standard Parallels to One Standard Parallel

$$k_0 = \frac{N_1 \cos\varphi_1}{N_0 \cos\varphi_0} \varepsilon^{(q_1 - q_0)\sin\varphi_0} = \frac{N_2 \cos\varphi_2}{N_0 \cos\varphi_0} \varepsilon^{(q_2 - q_0)\sin\varphi_0} \tag{a}$$

$$\varphi_0 = \sin^{-1}\left[\frac{\ln(N_1 \cos\varphi_1) - \ln(N_2 \cos\varphi_2)}{q_2 - q_1} \right] \tag{b}$$

TABLE 12.10 Lambert Conformal Inverse Mapping

$$\tan\lambda' = \frac{x}{k_0 N_0 \cot\varphi_0 - y} \tag{a}$$

$$r = \frac{k_0 N_0 \cot\varphi_0 - y}{\cos\lambda'} \tag{b}$$

$$\lambda = \frac{\lambda'}{\sin\varphi_0} \tag{c}$$

$$\Delta q = -\frac{1}{\sin\varphi_0} \ln\left(\frac{r}{k_0 N_0 \cot\varphi_0} \right) \tag{d}$$

$$q = q_0 + \Delta q \tag{e}$$

TABLE 12.11 Mapping Elements for Lambert Conformal Mapping

$$\gamma = \lambda' = \lambda \sin\varphi_0 \tag{a}$$

$$k = \frac{k_0 N_0 \cos\varphi_0}{N \cos\varphi} \varepsilon^{-(q - q_0)\sin\varphi_0} \tag{b}$$

$$\Delta t_1 = \frac{(2y_1 + y_2)(x_1 - x_2)}{6 k_0^2 R_0^2} \tag{c}$$

$$\frac{1}{k_L} \equiv \frac{S}{d} = \frac{1}{6}\left(\frac{1}{k_1} + \frac{4}{k_m} + \frac{1}{k_2} \right) \tag{d}$$

12.3.3   State Plane Coordinate Systems (SPC)

Each state and U.S. possession has a plane coordinate system defined for use in local surveying and mapping operations (Stem 1989). Many of the state plane coordinate systems are based on the transverse Mercator mapping. For states with large east–west extent, the Lambert conformal mapping is used. Some states utilize both the transverse Mercator and the Lambert conformal mapping for individual zones within the state system. The exception to this is the state plane system for the panhandle of Alaska. For this zone of the Alaska system, the oblique Mercator projection is used. As a result of the redefinition of the North American Datum, the National Geodetic Survey has published a set of defining constants for the U.S. state plane coordinate system, given in Tables 12.12 and 12.13. Table 12.13 also contains the adopted values of false north and false east and a four-digit code to identify the projection. Note that the "origin" as given in Table 12.13 is not identical with the origin of the coordinate system in Figures 12.7 and 12.8. Because of its importance, the specifications of the universal transverse Mercator (UTM) mapping are given in Table 12.14.

It is emphasized again that it is the *surveyor's* choice to use a state plane system or to generate his or her own mapping by merely specifying k_0 (usually 1) and the central meridian and/or the standard parallel of the surveying area.

TABLE 12.12   Legend for U.S. State Plane Coordinate System Defining Constants

Mapping	T — Transverse Mercator L — Lambert conformal O — Hotine oblique Mercator UTM — Universal Transverse Mercator 1:M — Scale Reduction at central meridian	

	Conversion factors	
Meters	U.S. survey feet	International feet
152400.3048	500,000.0	
213360.0		700,000.0
304800.6096	1,000,000.0	
609600.0		2,000,000.0
609601.2192	2,000,000.0	
914401.8289	3,000,000.0	
1.	3.28083333333	
1.		3.28083989501
0.3048		1.
1200/3937	1.	
0.30480060960	1.	

TABLE 12.13 U.S. State Plane Coordinate Systems

State Zone	SPCS Zone	Proj Type T/L/O	Latitude of Origin (DD-MM)	Longitude of Origin (DDD-MM)	False North (Meters)	False East (Meters)	Scale Factor (1:M)	Standard Parallels North (DD-MM)	Standard Parallels South (DD-MM)
Alabama									
East	0101	T	30°30'N	85°50'W	0.0	200000.0	25000		
West	0102	T	30 00	87 30	0.0	600000.0	15000		
Alaska									
Zone 1	5001	O	57 00	133 40	-5000000.0	5000000.0	10000	Axis Az	Tan⁻¹(-3/4)
Zone 2	5002	T	54 00	142 00	0.0	500000.0	10000		
Zone 3	5003	T	54 00	146 00	0.0	500000.0	10000		
Zone 4	5004	T	54 00	150 00	0.0	500000.0	10000		
Zone 5	5005	T	54 00	154 00	0.0	500000.0	10000		
Zone 6	5006	T	54 00	158 00	0.0	500000.0	10000		
Zone 7	5007	T	54 00	162 00	0.0	500000.0	10000		
Zone 8	5008	T	54 00	166 00	0.0	500000.0	10000		
Zone 9	5009	T	54 00	170 00	0.0	500000.0	10000		
Zone 10	5010	L	51 00	176 00	0.0	1000000.0	10000	51°50'N	53°50'N
Arizona									
East	0201	T	31 00	110 10	0.0	213360.0	10000		
Central	0202	T	31 00	111 55	0.0	213360.0	10000		
West	0203	T	31 00	113 45	0.0	213360.0	15000		
Arkansas									
North	0301	L	34 20	92 00	0.0	400000.0		34 56	36 14
South	0302	L	32 40	92 00	400000.0	400000.0		33 18	34 46
California									
Zone 1	0401	L	39 20	122 00	500000.0	2000000.0		40 00	41 40
Zone 2	0402	L	37 40	122 00	500000.0	2000000.0		38 20	39 50
Zone 3	0403	L	36 30	120 30	500000.0	2000000.0		37 04	38 26
Zone 4	0404	L	35 20	119 00	500000.0	2000000.0		36 00	37 15
Zone 5	0405	L	33 30	118 00	500000.0	2000000.0		34 02	35 28
Zone 6	0406	L	32 10	116 15	500000.0	2000000.0		32 47	33 53

TABLE 12.13 (continued)

State Zone	SPCS Zone	Proj Type T/L/O	Latitude of Origin (DD–MM)	Longitude of Origin (DDD–MM)	False North (Meters)	False East (Meters)	Scale Factor (1:M)	Standard Parallels North (DD–MM)	Standard Parallels South (DD–MM)
Colorado									
North	0501	L	39°20'N	105°30'W	304800.6069	914401.8289		39°43N	40°47'N
Central	0502	L	37 50	105 30	304800.6096	914401.8289		38 27	39 45
South	0503	L	36 40	105 30	304800.6096	914401.8289		37 14	38 26
Connecticut	0600	L	40 50	72 45	152400.3048	304800.6096		41 12	41 52
Delaware	0700	T	38 00	75 25	0.0	200000.0	200000		
Florida									
East	0901	T	24 20	81 00	0.0	200000.0	17000		
West	0902	T	24 20	82 00	0.0	200000.0	17000		
North	0903	L	29 00	84 30	0.0	600000.0		29 35	30 45
Georgia									
East	1001	T	30 00	82 10	0.0	200000.0	10000		
West	1002	T	30 00	84 10	0.0	700000.0	10000		
Hawaii									
Zone 1	5101	T	18 50	155 30	0.0	500000.0	30000		
Zone 2	5102	T	20 20	156 40	0.0	500000.0	30000		
Zone 3	5103	T	21 10	158 00	0.0	500000.0	100000		
Zone 4	5104	T	21 50	159 30	0.0	500000.0	100000		
Zone 5	5105	T	21 40	160 10	0.0	500000.0	0		
Idaho									
East	1101	T	41 40	112 10	0.0	200000.0	19000		
Central	1102	T	41 40	114 00	0.0	500000.0	19000		
West	1103	T	41 40	115 45	0.0	800000.0	15000		
Illinois									
East	1201	T	36 40	88 20	0.0	300000.0	40000		
West	1202	T	36 40	90 10	0.0	700000.0	17000		

	Code		Lat	Long					
Indiana			37°30'N	85°40'W	250000.0	100000.0	30000	42°04'N	43°16'N
East	1301	T	37 30	87 05	250000.0	900000.0	30000	40 37	41 47
West	1302	T							
Iowa			41 30	93 30	1000000.0	1500000.0		38 43	39 47
North	1401	L	40 00	93 30	0.0	500000.0		37 16	38 34
South	1402	L							
Kansas			38 20	98 00	0.0	400000.0		37 58	38 58
North	1501	L	36 40	98 30	400000.0	400000.0		36 44	37 56
South	1502	L							
Kentucky			37 30	84 15	0.0	500000.0		37 58	38 58
North	1601	L	36 20	85 45	500000.0	500000.0		36 44	37 56
South	1602	L							
Louisiana			30 30	92 30	0.0	1000000.0		31 10	32 40
North	1701	L	28 30	91 20	0.0	1000000.0		29 18	30 42
South	1702	L	25 30	91 20	0.0	1000000.0		26 10	27 50
Offshore	1703	L							
Maine			43 40	68 30	0.0	300000.0	10000		
East	1801	T	42 50	70 10	0.0	900000.0	30000		
West	1802	T							
Maryland	1900	L	37 40	77 00	0.0	400000.0		38 18	39 27
Massachusetts			41 00	71 30	750000.0	200000.0		41 43	42 41
Mainland	2001	L	41 00	70 30	0.0	500000.0		41 17	41 29
Island	2002	L							
Michigan			44°47'N	87°00'W	0.0	8000000.0		45°29'N	47°05'N
North	2111	L	43 19	84 22	0.0	6000000.0		44 11	45 42
Central	2112	L	41 30	84 22	0.0	4000000.0		42 06	43 40
South	2113	L							
Minnesota			46 30	93 06	100000.0	800000.0		47 02	48 38
North	2201	L	45 00	94 15	100000.0	800000.0		45 37	47 03
Central	2202	L	43 00	94 00	100000.0	800000.0		43 47	45 13
South	2203	L							

TABLE 12.13 (continued)

State Zone	SPCS Zone	Proj Type T/L/O	Latitude of Origin (DD-MM)	Longitude of Origin (DDD-MM)	False North (Meters)	False East (Meters)	Scale Factor (1:M)	Standard Parallels North (DD-MM)	South (DD-MM)
Mississippi									
East	2301	T	29 30	88 50	0.0	300000.0	20000		
West	2302	T	29 30	90 20	0.0	700000.0	20000		
Missouri									
East	2401	T	35 50	90 30	0.0	250000.0	15000		
Central	2402	T	35 50	92 30	0.0	500000.0	15000		
West	2403	T	36 10	94 30	0.0	850000.0	17000		
Montana	2500	L	44 15	109 30	0.0	600000.0		45 00	49 00
Nebraska	2600	L	39 50	100 00	0.0	500000.0		40 00	43 00
Nevada									
East	2701	T	34 45	115 35	8000000.0	200000.0	10000		
Central	2702	T	34 45	116 40	6000000.0	500000.0	10000		
West	2703	T	34 45	118 35	4000000.0	800000.0	10000		
New Hampshire	2800	T	42 30	71 40	0.0	300000.0	30000		
New Jersey	2900	T	38 50	74 30	0.0	150000.0	10000		
New Mexico									
East	3001	T	31°00'N	104°20'W	0.0	165000.0	11000		
Central	3002	T	31 00	106 15	0.0	500000.0	10000		
West	3003	T	31 00	107 50	0.0	830000.0	12000		
New York									
East	3101	T	38 50	74 30	0.0	150000.0	10000		
Central	3102	T	40 00	76 35	0.0	250000.0	16000		
West	3103	T	40 00	78 35	0.0	350000.0	16000		
Long Island	3104	L	40 10	74 00	0.0	300000.0	16000	40°40'N	41°02'N

Zone	Code	Type							
North Carolina	3200	L	33 45	79 00	0.0	609601.22		34 20	36 10
North Dakota									
North	3301	L	47 00	100 30	0.0	600000.0		47 26	48 44
South	3302	L	45 40	100 30	0.0	600000.0		46 11	47 29
Ohio									
North	3401	L	39 40	82 30	0.0	600000.0		40 26	41 42
South	3402	L	38 00	82 30	0.0	600000.0		38 44	40 02
Oklahoma									
North	3501	L	35 00	98 00	0.0	600000.0		35 34	36 46
South	3502	L	33 20	98 00	0.0	600000.0		33 56	35 14
Oregon									
North	3601	L	43 40	120 30	0.0	2500000.0		44 20	46 00
South	3602	L	41 40	120 30	0.0	1500000.0		42 20	44 00
Pennsylvania									
North	3701	L	40 10	77 45	0.0	600000.0		40 53	41 57
South	3702	L	39 20	77 45	0.0	600000.0		39 56	40 58
Rhode Island	3800	T	41°05'N	71°30'W	0.0	100000.0	160000		
South Carolina	3900	L	31 50	81 00	0.0	609600.0		32°30'N	34°50'N
South Dakota									
North	4001	L	43 50	100 00	0.0	600000.0		44 25	45 41
South	4002	L	42 20	100 20	0.0	600000.0		42 50	44 24
Tennessee	4100	L	34 20	86 00	0.0	600000.0		35 15	36 25
Texas									
North	4201	L	34 00	101 30	1000000.0	200000.0		34 39	36 11
North Cent.	4202	L	31 40	98 30	2000000.0	600000.0		32 08	33 58
Central	4203	L	29 40	100 00	3000000.0	700000.0		30 07	31 53
South Cent.	4204	L	27 50	99 00	4000000.0	600000.0		28 23	30 17
South	4205	L	25 40	98 30	5000000.0	300000.0		26 10	27 50

TABLE 12.13 (continued)

State Zone	SPCS Zone	Proj Type T/L/O	Latitude of Origin (DD-MM)	Longitude of Origin (DDD-MM)	False North (Meters)	False East (Meters)	Scale Factor (1:M)	Standard Parallels North (DD-MM)	Standard Parallels South (DD-MM)
Utah									
North	4301	L	40 20	111 30	10000000.0	500000.0		40 43	41 47
Central	4302	L	38 20	111 30	20000000.0	500000.0		39 01	40 39
South	4303	L	36 40	111 30	30000000.0	500000.0		37 13	38 21
Vermont	4400	T	42 30	72 30	0.0	500000.0	28000		
Virginia									
North	4501	L	37 40	78 30	20000000.0	3500000.0		38 02	39 12
South	4502	L	36 20	78 30	10000000.0	3500000.0		36 46	37 58
Washington									
North	4601	L	47 00	120 50	0.0	500000.0		47 30	48 44
South	4602	L	45 20	120 30	0.0	500000.0		45 50	47 20
West Virginia									
North	4701	L	38°30'N	79°30'W	0.0	600000.0		39°00'N	40°15'N
South	4702	L	37 00	81 00	0.0	600000.0		37 29	38 53
Wisconsin									
North	4801	L	45 10	90 00	0.0	600000.0		45 34	46 46
Central	4802	L	43 50	90 00	0.0	600000.0		44 15	45 30
South	4803	L	42 00	90 00	0.0	600000.0		42 44	44 04
Wyoming									
East	4901	T	40 30	105 10	0.0	200000.0	16000		
East Cent.	4902	T	40 30	107 20	100000.0	400000.0	16000		
West Cent.	4903	T	40 30	108 45	0.0	600000.0	16000		
West	4904	T	40 30	110 05	100000.0	800000.0	16000		
Puerto Rico	5200	L	17 50	66 26	200000.0	200000.0		18 02	18 26
Virgin Islands	5200	L	17 50	66 26	200000.0	200000.0		18 02	18 26

TABLE 12.14 UTM Mapping System Specifications

UTM zones	6° in longitude
Longitude of origin	central meridian of each zone
Latitude of origin	0° (equator)
Units	meter
False northing	0 meters for northern hemisphere 10,000,000 meters for southern hemisphere
False easting	500,000 meters
Central meridian scale	0.9996
Zone numbers	starting with zone 1 centered at 177°W and increasing eastward to zone 60 centered at 177°E
Limits of projection	S80° to N80°
Limits of zones and overlap	zones are bounded by meridians which are multiples of 6°W and 6°E of Greenwich
Reference ellipsoid	dependent on region for U.S.: Clarke 1866 (except Hawaii) for Hawaii: International Ellipsoid

In the latter case the mapping reductions Δt are small and k_L is close to 1. If, in addition, a local ellipsoid is specified, then most of the reductions can be neglected for small surveys. While these specifications might lead to a small reduction in the computational load, which in view of modern computer power, is no longer that critical, it increases the probability that reductions are inadvertently neglected when they should not be. Also, stressing this "plane view" of surveying increase the danger that the distinction between physical observations and model observation is eventually overlooked and with that the geodetic framework in which surveying is inevitably embedded.

CHAPTER 13

USEFUL TRANSFORMATIONS

Transformations between coordinate systems are routinely carried out in surveying. This chapter deals with transforming coordinates of nearly aligned Cartesian coordinate systems. One may think of the conventional terrestrial coordinate system (CTCS) and some other geocentric system. If the coordinates are given for a number of stations common to both coordinate systems, the transformation parameters can be estimated from a least-squares solution. The three-dimensional transformation requires that either ellipsoidal heights or orthometric heights and the geoid undulations are available for common stations. This chapter also contains a simplified transformation between latitudes and longitudes of two ellipsoidal surfaces. The transformation does not require height data explicitly, but it is valid only in small regions and must be considered to be only approximately valid. Transformation parameters between the WGS 84 and the most important local geodetic datums worldwide are given by the Defense Mapping Agency (1987).

13.1 SIMILARITY TRANSFORMATION FOR NEARLY ALIGNED COORDINATE SYSTEMS

The transformation of three-dimensional coordinate systems for the purpose of transforming geodetic datums has been given much attention, in particular since geodetic satellite techniques have made it possible to relate local geodetic datums to a geocentric datum. Some of the pertinent works are Veis (1960), Molodenskii et al. (1962), Badekas (1969), Vaniček and Wells (1974), Leick and van Gelder (1975), and Soler and van Gelder (1987). Figure 13.1 shows the coordinate system $(X) = (x, y, z)$, which is related to the coordinate system

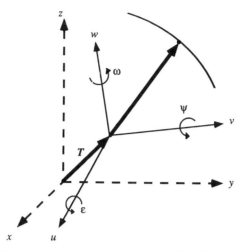

Figure 13.1 Differential transformation between Cartesian coordinate systems.

$(U) = (u, v, w)$ by the translation vector $T = (\Delta x, \Delta y, \Delta z)^{\mathrm{T}}$ between the origins of the two coordinate systems and the small rotations (ϵ, ψ, ω) around the (u, v, w) axes, respectively.

The transformation equation expressed in the (X) coordinate system can be seen from Figure (13.1):

$$T + (1 + \Delta)RU - X = 0 \tag{13.1}$$

where $1 + \Delta$ denotes the scale factor between the systems and R is the product of three consecutive orthogonal rotations around the axes of (U):

$$R = R_3(\omega) R_2(\psi) R_1(\epsilon) \tag{13.2}$$

The symbol R_i denotes the rotation matrix for a rotation around axis i as given in Appendix D. The angles (ϵ, ψ, ω) are positive for counterclockwise rotations about the respective (u, v, w) axes as viewed from the end of the positive axis.

For the purpose of distinguishing various approaches, the transformation (13.1) is labeled here as model 1. It defines the relation between the two systems in terms of the seven transformation parameters $(\Delta x, \Delta y, \Delta z, \Delta, \epsilon, \psi, \omega)$. Note again that the translation components are the shifts expressed in the (X) coordinate system and that the (U) coordinate system is rotated and scaled. The transformation parameters are solved by a least-squares solution. The Cartesian coordinates of both systems are the observations. Equation (13.1) has the form of the mixed adjustment model

$$F(L_a, X_a) = 0 \tag{13.3}$$

or

$$F(L_b + V, X_0 + X) = 0 \tag{13.4}$$

where L_a denotes the adjusted observation (coordinates), X_a denotes the adjusted transformation parameters, L_b denotes the observations [given coordinates in (U) and (X)], X_0 are the approximations to the transformation parameters (usually the first approximation is taken $X_0 = 0$), V denotes the residuals, and X is the parameter correction. Because the rotations between the systems (X) and (U) are small in geodetic network applications, it is permissible to simplify equation (13.2) as

$$R = I + Q = \begin{bmatrix} 1 & 0 & 0 \\ 0 & 1 & 0 \\ 0 & 0 & 1 \end{bmatrix} + \begin{bmatrix} 0 & \omega & -\psi \\ -\omega & 0 & \epsilon \\ \psi & -\epsilon & 0 \end{bmatrix} \tag{13.5}$$

By substituting (13.5) into (13.1) and neglecting second-order terms in scale Δ, rotation angles (ϵ, ψ, ω), and their products, we can write model (1) [equation (13.1)] as

$$T + U + \Delta U + QU - X = 0 \tag{13.6}$$

Each station P_i yields one such vector equation. The usual adjustment procedure $V^T P V = $ min, subject to the condition,

$$BV + AX + W = 0 \tag{13.7}$$

applies, where

$$B = \frac{\partial F}{\partial L}\bigg|_{L_b, X_0} \tag{13.8}$$

$$A = \frac{\partial F}{\partial X}\bigg|_{L_b, X_0} \tag{13.9}$$

$$W = F(L_b, X_0) \tag{13.10}$$

The symbol P is the weight matrix of the observations (the coordinates of the common stations). The observations and parameters are already linearly related in (13.6). Each station contributes three equations to (13.7). For example, for station P_i, taking the approximate parameters $X_o = 0$, one has

$$B_i V_i + A_i X + W_i = 0$$

or

$$
\begin{bmatrix} 1 & 0 & 0 & -1 & 0 & 0 \\ 0 & 1 & 0 & 0 & -1 & 0 \\ 0 & 0 & 1 & 0 & 0 & -1 \end{bmatrix}
\begin{bmatrix} v_u \\ v_v \\ v_w \\ v_x \\ v_y \\ v_z \end{bmatrix}
$$

$$
+ \begin{bmatrix} 1 & 0 & 0 & u & v & -w & 0 \\ 0 & 1 & 0 & v & -u & 0 & w \\ 0 & 0 & 1 & w & 0 & u & -v \end{bmatrix}
\begin{bmatrix} \Delta x \\ \Delta y \\ \Delta z \\ \Delta \\ \omega \\ \psi \\ \epsilon \end{bmatrix}
+ \begin{bmatrix} u - x \\ v - y \\ w - z \end{bmatrix} = 0
$$

$$(13.11)$$

A variation of (13.1), called model 2, is

$$T = U_0 + (1 + \Delta)R(U - U_0) - X = 0 \tag{13.12}$$

where U_0 is the positions vector in the system (U) of a point located some-where within the network to be transformed. A likely choice for U_0 is the centroid. All other notation is the same as in equation (13.1). If one follows the same procedure as described for the previous model, that is omitting sec-ond-order terms in scale and rotation and their products, model 2 [equation (13.12)] becomes

$$T + U + \Delta(U - U_0) + Q(U - U_0) - X = 0 \tag{13.13}$$

Equation (13.13) is the mathematical model for a least-squares solution. Each station contributes three equations of the form

$$
\begin{bmatrix} 1 & 0 & 0 & -1 & 0 & 0 \\ 0 & 1 & 0 & 0 & -1 & 0 \\ 0 & 0 & 1 & 0 & 0 & -1 \end{bmatrix}
\begin{bmatrix} v_u \\ v_v \\ v_w \\ v_x \\ v_y \\ v_z \end{bmatrix}
$$

$$
+ \begin{bmatrix} 1 & 0 & 0 & u - u_0 & v - v_0 & -(w - w_0) & 0 \\ 0 & 1 & 0 & v - v_0 & -(u - u_0) & 0 & w - w_0 \\ 0 & 0 & 1 & w - w_0 & 0 & u - u_0 & -(v - v_0) \end{bmatrix}
\begin{bmatrix} \Delta x \\ \Delta y \\ \Delta z \\ \Delta \\ \omega \\ \psi \\ \epsilon \end{bmatrix}
$$

$$
+ \begin{bmatrix} u - x \\ v - y \\ w - z \end{bmatrix} = 0
\tag{13.14}
$$

A third model was introduced by Veis (1960). It uses the same rotation point U_0 as model 2, but the rotations are about the axes (n, e, h) of the local geodetic coordinate system at U_0. The n axis is tangent to the geodetic meridian, but the positive direction is toward the south; the e axis is perpendicular to the meridian plane and it is positive eastward. The h axis is along the ellipsoidal normal with its positive direction upward, forming a right-handed system with n and e. Similar to equation (13.12) one obtains

$$
T + U_0 + (1 + \Delta)M(U - U_0) - X = 0
\tag{13.15}
$$

If (η, ξ, α) denote positive rotations about the (n, e, h) axes, and $(\varphi_0, \lambda_0, h_0)$ denote the geodetic coordinates for the point of rotation, then the M matrix is

$$
M = R_3^T(\lambda_0) R_2^T(90 - \varphi_0) R_3(\alpha) R_2(\xi) R_1(\eta) R_2(90 - \varphi_0) R_3(\lambda_0)
\tag{13.16}
$$

Since the rotation angles (η, ξ, α) are differentially small, the matrix M simplifies as

$$
M(\lambda_0, \varphi_0, \eta, \xi, \alpha) = \alpha M_\alpha + \xi M_\xi + \eta M_\eta + I
\tag{13.17}
$$

where

$$M_\alpha = \begin{bmatrix} 0 & \sin \varphi_0 & -\cos \varphi_0 \sin \lambda_0 \\ -\sin \varphi_0 & 0 & \cos \varphi_0 \cos \lambda_0 \\ \cos \varphi_0 \sin \lambda_0 & -\cos \varphi_0 \cos \lambda_0 & 0 \end{bmatrix} \quad (13.18)$$

$$M_\xi = \begin{bmatrix} 0 & 0 & -\cos \lambda_0 \\ 0 & 0 & -\sin \lambda_0 \\ \cos \lambda_0 & \sin \lambda_0 & 0 \end{bmatrix} \quad (13.19)$$

$$M_\eta = \begin{bmatrix} 0 & -\cos \varphi_0 & -\sin \varphi_0 \sin \lambda_0 \\ \cos \varphi_0 & 0 & \sin \varphi_0 \cos \lambda_0 \\ \sin \varphi_0 \sin \lambda_0 & -\sin \varphi_0 \cos \lambda_0 & 0 \end{bmatrix} \quad (13.20)$$

If, again, second-order terms in scale and rotations and their products are neglected, the model becomes

$$T + U + \Delta(U - U_0) + (1 + \Delta)(M - I)(U - U_0) - X = 0 \quad (13.21)$$

Thus, the only difference between model 2 and model 3 is that the rotations in model 3 can be readily visualized as rotations around the local geodetic coordinate axes of U_0. It can be shown that the three rotations (η, ξ, α) are related to the rotations (ϵ, ψ, ω) as follows:

$$\begin{bmatrix} \eta \\ \xi \\ \alpha \end{bmatrix} = R_2(90 - \varphi_0) R_3(\lambda_0) \begin{bmatrix} \epsilon \\ \psi \\ \omega \end{bmatrix} \quad (13.22)$$

Also, if $\Sigma_{\eta, \xi, \alpha}$ and $\Sigma_{\epsilon, \psi, \omega}$ denote the respective covariance matrices, then the principle of propagation of variances gives

$$\Sigma_{\eta, \xi, \alpha} = R_2(90 - \varphi_0) R_3(\lambda_0) \Sigma_{\epsilon, \psi, \omega} R_3^T(\lambda_0) R_2^T(90 - \varphi_0) \quad (13.23)$$

All three models are equivalent in the sense that once the transformation parameters are estimated, the direct transformation from one system to the other gives the same results for each model. This equivalence is also true for the variance–covariances of the transformed coordinates. The three models yield the same a posteriori variance of unit weight and the same residuals (corrections to the coordinates). The same scale factor is obtained in all cases;

model 1 and model 2 give the same rotation angles. The translation components for model 1 and model 2 follow from (13.1) and (13.12):

$$T_2 = T_1 - U_0 + (1 + \Delta)RU_0 \qquad (13.24)$$

Only T_1, that is, the translation vector as estimated from model 1, corresponds to the geometric vector between the origins of the coordinate systems (X) and (U). The translational component of model 2 is a function of U_0, as shown in (13.24). Because model 3 uses the same U_0 as model 2, both models yield identical translational components. More detail on the equivalence of these transformation models is given by Leick and van Gelder (1975).

If the common stations cover a small region of the earth's surface, as they usually do in network densifications, one should choose the centroid as the point of rotation. Taking $U_0 = (0, 0, 0)$ would result in large correlations between parameters and could cause a numerical singularity of the normal matrix. There is no need to limit the choices for the point of rotation to the centroid; any point within the cluster of stations is suitable. It is, of course, not necessary that all seven parameters be estimated for every transformation. In small areas, such as the size of a county, it might be sufficient to estimate only the translation components. Rather than solving so-called global translation parameters, that is, one set of seven parameters for a complete datum, one could solve sets of ''regional'' translation parameters. To obtain continuous expressions for the transformations, one could fit polynomials through the regional translation components. A similar procedure could be used for representing the scale factor. Such a detailed modeling of transformations could be used to model existing network distortions.

13.2 TRANSFORMING ELLIPSOIDAL POSITIONS IN A SMALL REGION

If the ellipsoidal heights (i.e., the orthometric heights and geoid undulations) are not available, and a three-dimensional transformation cannot be performed, it is still possible to use GPS observations for the densification of two-dimensional networks. The station in the satellite coordinate system, denoted here with subscript n for ''new system,'' as computed from the minimal or inner constraint solution, can be converted into the ellipsoidal latitude, longitude, and height following the procedures of Chapter 6. The ellipsoid used for this conversion should be of the same dimension as the ellipsoid of the geodetic system. Thus, one has two sets of ellipsoidal latitudes and longitudes for the common stations: (1) the given positions $P(\varphi_o, \lambda_o)$ in the geodetic system, and (2) the computed GPS positions $P(\varphi_n, \lambda_n)$ in the satellite system. The idea is to transform the GPS positions to the geodetic datum; that is, the transformation is from the system n to the system o. This transformation can also be

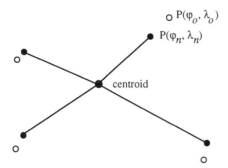

Figure 13.2 Transformation on the ellipsoidal surface.

applied for the transformation from the NAD 27 to the NAD 83 datum. In that case the NAD 83 positions would be denoted o, and the NAD 27 positions would be labeled n.

The two-dimensional transformation is done with the tools from classical geodesy, given in Chapter 12. First, one computes the centroid of the station in the n system. The centroid is located on the ellipsoidal surface, of course. Next, consider the geodesics that connect the centroid with the positions $P(\varphi_n, \lambda_n)$. The resulting central network is shown in Figure 13.2. The discrepancies $(\varphi_o - \varphi_n)$ and $(\lambda_o - \lambda_n)$ are used to compute the transformation parameters in a least-squares adjustment. Four transformation parameters are defined as follows: the translation of the centroid $d\varphi_c$ and $d\lambda_c$, the rotation $d\alpha$ of the geodesics at the centroid, and the scale factor $(1 - \Delta)$ of the geodesics. All geodesics from the centroid to the stations are rotated by the same amount in azimuth; all geodesics are scaled by one and the same scale factor.

Because the discrepancies $(\varphi_o - \varphi_n)$ and $(\lambda_o - \lambda_n)$ are small quantities, the coefficients that are traditionally used in the adjustment of two-dimensional networks on the ellipsoid are also appropriate for this transformation. The respective coefficients can be found in Table 12.4. The changes in length and azimuth of the geodesic from centroid to P_i are

$$dS_{ci} = -M_i \cos \alpha_{ic} d\varphi_i - M_c \cos \alpha_{ci} d\varphi_c - N_i \cos \varphi_i \sin \alpha_{ic} (d\lambda_i - d\lambda_c)$$

(13.25)

$$da_{ci} = \frac{M_i}{S_{ci}} \sin \alpha_{ic} d\varphi_i + \frac{M_c}{S_{ci}} \sin \alpha_{ci} d\varphi_c - \frac{N_i}{S_{ci}} \cos \varphi_i \cos \alpha_{ic} (d\lambda_i - d\lambda_c)$$

(13.26)

The symbol S_{ci} denotes the length of the geodesic from centroid to station i. Equations like (13.25) and (13.26) can be written for each station P_i. Since all geodesics emanating from the centroid are scaled and rotated by the same amount, the following conditions are valid:

$$dS_{ci} = \Delta S_{ci} \tag{13.27}$$

$$d\alpha_{ci} = d\alpha_c = \text{const} \tag{13.28}$$

where Δ denotes the scale change. The list of parameters to be estimated is

$$X = \{d\lambda_c, d\varphi_c, \Delta, d\alpha_c\} \tag{13.29}$$

The observations are the discrepancies in the ellipsoidal coordinates

$$d\lambda_i = \lambda_{io} - \lambda_{in} \tag{13.30}$$

$$d\varphi_i = \varphi_{io} - \varphi_{in} \tag{13.31}$$

Substituting (13.27) through (13.31) into (13.25) and (13.26) gives

$$\Delta S_{ci} = -M_i \cos \alpha_{ic}(\varphi_{io} - \varphi_{in}) - M_c \cos \alpha_{ci} d\varphi_c$$
$$- N_i \cos \varphi_i \sin \alpha_{ic}(\lambda_{io} - \lambda_{in}) + N_i \cos \varphi_i \sin \alpha_{ic} d\lambda_c \tag{13.32}$$

$$d\alpha_{ci} = \frac{M_c}{S_{ci}} \sin \alpha_{ci} d\varphi_c + \frac{M_i}{S_{ci}} \sin \alpha_{ic}(\varphi_{io} - \varphi_{in})$$
$$- \frac{N_i}{S_{ci}} \cos \varphi_i \cos \alpha_{ic}(\lambda_{io} - \lambda_{in}) + \frac{N_i}{S_{ci}} \cos \varphi_i \cos \alpha_{ic} d\lambda_c \tag{13.33}$$

In (13.32) and (13.33) the parameters and observations already appear in linear form. Because the parameters and observations are related by an implicit mathematical function, the mixed adjustment model is applied. The respective submatrices of B and A for station P_i are

$$B = \begin{matrix} & \varphi_{in} & \lambda_{in} & \varphi_{io} & \lambda_{io} \\ & \begin{bmatrix} M_i \cos \alpha_{ic} & N_i \cos \varphi_i \sin \alpha_{ic} & -M_i \cos \alpha_{ic} & -N_i \cos \varphi_i \sin \alpha_{ic} \\ -\dfrac{M_i}{S_{ci}} \sin \alpha_{ic} & \dfrac{N_i}{S_{ci}} \cos \varphi_i \cos \alpha_{ic} & \dfrac{M_i}{S_{ci}} \sin \alpha_{ic} & -\dfrac{N_i}{S_{ci}} \cos \varphi_i \cos \alpha_{ic} \end{bmatrix} \end{matrix}$$
$$\tag{13.34}$$

$$A = \begin{matrix} & d\varphi_c & d\lambda_c & \Delta & d\alpha_c \\ & \begin{bmatrix} -M_c \cos \alpha_{ci} & N_i \cos \varphi_i \sin \alpha_{ic} & -S_{ci} & 0 \\ \dfrac{M_c}{S_{ci}} \sin \alpha_{ci} & \dfrac{N_i}{S_{ci}} \cos \varphi_i \cos \alpha_{ic} & 0 & -1 \end{bmatrix} \end{matrix}$$
$$\tag{13.35}$$

The respective components of the W vector are

$$W = \begin{bmatrix} -M_i \cos \alpha_{ic}(\varphi_{io} - \varphi_{in}) - N_i \cos \varphi_i \sin \alpha_{ic}(\lambda_{io} - \lambda_{in}) \\ \dfrac{M_i}{S_{ci}} \sin \alpha_{ic}(\varphi_{io} - \varphi_{in}) - \dfrac{N_i}{S_{ci}} \cos \varphi_i \cos \alpha_{ic}(\lambda_{io} - \lambda_{in}) \end{bmatrix} \quad (13.36)$$

Once the adjusted parameters are available, the lengths and azimuths of the geodesics become

$$S_{ci,o} = S_{ci,n} + \Delta S_{ci} \quad (13.37)$$

$$\alpha_{ci,o} = \alpha_{ci,n} + d\alpha_c \quad (13.38)$$

The coordinates of the centroid are

$$\varphi_{co} = \varphi_{cn} + d\varphi_c \quad (13.39)$$

$$\lambda_{co} = \lambda_{cn} + d\lambda_c \quad (13.40)$$

With (13.37) through (13.40) the positions in the datum o can be computed by using the direct problem on the ellipsoid given in Table 12.3.

CHAPTER 14

DATUMS, STANDARDS, AND SPECIFICATIONS

While GPS requires additional refinements of the geodetic reference frame in order to fully utilize its measurement accuracy, it is also necessary to clarify and describe the outcome of positioning with GPS in terms readily understood by the ever increasing community of users. This last chapter, therefore, provides some remarks regarding the suitability of WGS 84 and NAD 83 to serve as the geodetic datum, examines the current accuracy standard for GPS positioning, discusses proposed changes, and provides some speculative thoughts on the development of GPS.

14.1 REFERENCE FRAME

Readers who have followed the enormous progress that GPS positioning has demonstrated over the last decade probably will appreciate the fact that questions concerning the reference frame are more relevant than ever. GPS not only brought positioning capability to new heights, but considerable progress in accuracy is also noted for related measurement systems such as very long base line interferometry (VLBI), lunar laser ranging (LLR), satellite laser ranging (SLR), and the measuring of relative and absolute gravity. The scientific view of modern reference frames and related information is available in the most current annual reports issued by the International Earth Rotation Service (IERS). In these reports the reader finds details about the IERS terrestrial reference frame (ITRF), the IERS celestial reference frame (ICRF), deficiencies of the current precession constants and of the IAU 1980 theory of nutation (see Table 2.1 in Chapter 2), extragalactic radio sources that define the ICRF, terrestrial sites that define the ITRF and their velocity fields, the latest in earth

486

rotation varitation, etc. Most of this information, it not all of it, is available via the Internet, of course.

If the broadcast ephemeris is used, GPS positions and vectors are referenced in the WGS 84 datum developed by the U.S. Defense Mapping Agency (Defense Mapping Agency 1987). This datum is practically equivalent for cartographic purposes to the NAD 83, which is the official datum for civil applications. Details on many practical and theoretical aspects of the NAD 83 datum are given in the final project report (Schwarz 1989). Table 14.1 lists the dimensions for the reference ellipsoids. The WGS 84 and the NAD 83 datums use the ellipsoid of the Geodetic Reference System 1980, whose definition is presented in Table 11.8. The slight difference between the effective flattening of both ellipsoids is a result of DMA constraining the normalized form of the second harmonic,

$$\overline{C}_{2,0} = -\frac{J_2}{\sqrt{5}} \tag{14.1}$$

after rounding the result to eight significant digits, instead of using the dynamical form factor J_2 directly. Thus, quantities that depend on the form factor, such as the flattening, generally differ after the eighth significant digit, while linear quantities such as the semiminor axis generally differ after the tenth significant digit (Schwarz 1989, p. 249). The positional variation resulting using these two different flattening terms is in the submillimeter range.

The NAD 83 and the WGS 84 are very nearly geocentric systems, with the z axis parallel to the direction of the conventional terrestrial pole (CTP), and with the x axis as the intersection of the Greenwich meridian as defined by the BIH (predecessor of IERS) and the plane of the terrestrial equator corresponding to the CTP. Both datums tried to realize an ITRF type of coordinate system. Even though the ellipsoids of both datums are identical for all practical purposes, a station may have slightly different coordinate values in both systems. This is a result of using independent sets of observations and applying different weighting schemes to the adjustment defining the WGS 84 and the NAD 83 coordinate system.

TABLE 14.1   Dimensions of Important Reference Ellipsoids

Datum	Reference ellipsoid	a [m]	$1/f$
NAD 27	Clarke 1866	6378206.4	294.9786982
WGS 72	WGS 72	6378135.0	298.26
NAD 83	GRS 80	6378137.0	298.257222101
WGS 84	WGS 84	6378137.0	298.257223563

Table 14.1 also lists the ellipsoids of the old datums, i.e., the WGS 72 and the NAD 27. The transformation between the WGS 72 and the currently adopted datums is (Soler and Hothem, 1988).

$$\mathbf{U}_{NAD83} \approx \mathbf{U}_{WGS84} = \begin{pmatrix} 0 \\ 0 \\ 4.5 \end{pmatrix} + (1 + 0.227 \times 10^{-6}) R_3(\omega) \mathbf{U}_{WGS72} \quad (14.2)$$

with $\omega = -0''.554$. The rotation matrix R_3 is given in Appendix D. The translation Δz is given in units of meters. Soler and van Gelder (1987) give an interesting discussion on this z shift.

Although the center of the WGS 84 and NAD 83 ellipsoids were placed to the geocenter as well as could possibly be done at the time, recent observational evidence suggests that the nongeocentricity of the NAD 83 is at the 2-m level when compared to ITRF stations (Soler et al. 1992a). The origin of the ITRF is located at the center of mass of the earth with an uncertainty of 10 cm. At the high level of accuracy, GPS vectors relate to the ITRF because the precise ephemeris is computed from observations taken at stations whose location is given in the ITRF, for example, the CIGNET, IGS, and VLBI stations. The nongeocentricity of the NAD 83 ellipsoid has a negligible effect on relative GPS positioning when expressed on the NAD 83 datum. However, the ellipsoidal heights will differ significantly (about half a meter) due to the nongeocentricity. (Note that this statement refers to ellipsoidal heights and not to ellipsoidal height differences.)

There is a need to assure that the ellipsoidal heights, the orthometric heights, and the geoid undulations are consistently related by $h = H + N$. The geoid undulations computed from expressions like (11.31) refer to an ellipsoid whose origin is at the earth's center of mass. To complicate matters even further, geoid undualtions computed from expression (11.31) refer to an "adopted ellipsoid" whose equatorial radius is now estimated to be 70 cm shorter than the GRS 80 ellipsoid (Rapp 1992). Thus, to assure consistency it is necessary that either the ellipsoidal heights or the geoid undulations are transformed to a compatible system if the NAD 83 is maintained as the reference datum. Because of the linear relation $h = H + N$ it does not matter mathematically whether h or N is altered to accommodate the nongeocentricity of the NAD 83 ellipsoid. This transformation could be incorporated in a future revision of GEOID 93. It would be confusing to speak of a shift of the geoid in this context, because the geoid is a physical equipotential surface and, as such, should remain uniquely defined. A shift of the geoid to accommodate a nongeocentric datum would effect the concept of "absolute" orthometric heights as conventionally defined and understood. No changes are expected to be applied to the orthometric heights of the recently completed North American Vertical Datum of 1988 (NAVD 88) (Zilkoski et al. 1992). Consequently, the introduction of a "best fitting geoid" to the NAD 83 datum should be appropriately explained to the scientific community in order to avoid unnecessary confusion when software developed at NGS to compute geoid heights is used by investigators working in a geocentric coordinate system.

14.2 STANDARDS AND SPECIFICATIONS

At the Fall 1991 ACSM convention, an ad hoc committee was established to assist and cooperate with the Federal Geodetic Control Subcommittee (FGCS) in their review and revision of the existing standards and specifications. A report on the recommendations of the ad hoc[1] committee is given in Leick (1993). This effort was undertaken in recognition of the fact that the objectives for the national geodetic reference frame must be expanded to include spatial referencing for Geographic Information Systems (GIS) applications, and in recognition of the fact that GPS makes new solutions possible. Because of the importance of this topic, some of the recommendations and background information are provided here.

It is generally anticipated that the geodetic network will serve as a spatial reference for an increasing number of applications. Many of these will typically range across the traditional spectrum of geodetic accuracy but also, perhaps even mostly, will encompass applications requiring merely low level accuracy. Because of the interdisciplinary nature of GIS applications and the variety of objects to be referenced spatially, we should expect many new users and uses of the geodetic reference frame. The standards should reflect this new and expanded role of the geodetic network.

The committee did not deal with specifications in great detail. To understand and separate the issues, the following definition of standards and specifications is given:

Standards: Standards specify the absolute and/or relative accuracy of a survey. Standards are independent of the measurement equipment and methodology. Ideally standards should not be rewritten as new technology becomes available. Rather, the standards should be derived from the objectives of the geodetic networks in terms of fulfilling the needs of professionals and society. Standards might require revision as the uses of geodetic networks change.

Specifications: Specifications contain the rules as to how the standards can be met. Typically, as new technology becomes available, the specifications might require modifications or additions.

[1]Members: A. Nicholas Bodner, Steven M. Briggs, Earl Burkholder, Robert Burtch, James Collins, Alan R. Dragoo, Roger W. Durham, Alfred Leick, Paul J. Hartzheim, Joseph V. R. Paiva, James P. Reilly, Gary W. Thompson, Ed McKay (NGS Liaison), Charles Challstrom (NGS Liaison).

14.2.1   Existing Classification Standards

The current classification standards are found in *Geometric Geodetic Accuracy Standards and Specifications for Using GPS Relative Positioning Techniques* (FGCS 1988) and *Standards and Specifications for Geodetic Control Networks* (FGCS 1984). These documents are available from the National Geodetic Survey Division. The GPS standards and specifications is only a preliminary document and is to be used only as a guide for planning and executing GPS surveys, according to a note printed on the cover. However, virtually all state highway departments and most counties use the GPS standards and specifications to produce GPS surveys. These standards are widely used, and have become the de facto standard for all GPS surveys. Because of the increased use of GPS for establishing control networks, the classical standards (FGCS 1984) are becoming less applicable.

A principal feature of the existing standards is the subdivision into orders and classes (Tables 14.2 and 14.3). The GPS standards define classes in terms of line accuracy, consisting of a base error and a line dependent error. The older classical standards and specifications contain only the distance dependent error. The subdivisions vary from Order AA (0.3 cm + 0.01 ppm) to Third-Order–Class II of 1:5000. These values refer to a 95% confidence level. Typically, the minimally constrained least-squares network adjustment is used to verify whether or not the standards of a survey have been fulfilled.

TABLE 14.2   Geometric Relative Positioning Accuracy Standards for Three-Dimensional Surveys (FGCS 1988)

Survey categories	Order	(95 percent confident level) Minimum geometric accuracy standard		
		Base error	Line-length Dependent error	
		e [cm]	p [ppm]	a [cm] [1:a]
Global-regional geodynamics	AA	0.3	0.01	1:100,000,000
National Geodetic Reference system, primary network	A	0.5	0.1	1: 10,000,000
National Geodetic Reference system, secondary networks	B	0.8	1	1:1,000,000
National Geodetic Reference system (terrestrial based)	C			
	1	1.0	10	1: 100,000
	2-I	2.0	20	1: 50,000
	2-II	3.0	50	1: 20,000
	3	5.0	100	1: 10,000

**TABLE 14.3 Distance Accuracy Standards
(FGCS 1984)**

Classification	Minimum distance accuracy
First-order	1:100,000
Second-order, class I	1: 50,000
Second-order, class II	1: 20,000
Third-order, class I	1: 10,000
Third-order, class II	1: 5,000

It has, to date, felt quite natural to express standards in terms of length-dependent elements because of the "physics" of the prevailing measurement tools. For example, the accuracy of an electronic distance meter very much depends on the length of line being measured. In least-squares adjustments, which is the favorite processing technique for geodetic networks, the stochastic model reflects this dependency on the line length.

The length-dependent principle has developed into a cornerstone in the philosophy and perception of geodetic networks. It is well suited for describing in simple terms the "principle of neighborhood" in geodetic networks. For example, Section 2.1 of the classical standards reads, "When a horizontal control point is classified with a particular order or class, NGS certifies that the geodetic latitude and longitude of that point bear a relation of specific accuracy to the coordinates of all other points in the horizontal control network" (FGCS 1984). The certification covers the relative location of all points, not just those connected by a direct measurement.

The length-dependent characterization of geodetic networks, surprisingly, tolerates translatory and even systematic errors. For example, a translation of the whole network does not affect the quality and shape of the internal geometry. If systematic errors build up slowly as the size of the network increases, large absolute position errors may eventually occur, even though in small regions the line-dependent accuracy might still be acceptable for many users. For many surveying projects the network accuracy in the immediate vicintiy of the project area is important in order to assure relative position accuracy.

Indeed, the length-dependent behavior of geodetic networks is primarily a local phenomenon. A geodetic network typically represents more than merely an assembly of distance observations. For example, the NAD 83 consists of different types of observations, each having its own stochastic model, such as the observations from the high-precision transcontinental traverse, astronomical positions and azimuths, Transit Doppler positions, and GPS vectors. The latter types of observations provide rigidity to the network. An immediate consequence of including these types of observations is the control of systematic errors and an increase of the line accuracy for longer distances. The ratio of error over distance now reflects the properties of these extra observations and constraints.

14.2.2   The Impact of GPS

In response to the challenge of GPS, the National Geodetic Survey is in the midst of completing the B-order network in the United States, more or less on a state-by-state basis. The 1993 status of this effort is shown in Figure 14.1 A major part of these solutions is a complete readjustment of the classical first- and second-order horizontal data. The result is a highly consistent solution for a particular state. The B-order network of each state is tied to several A-order stations to assure that translatory shifts of the network at state boundaries are kept small (Strange and Love 1991).

As theory predicts and experimentation with GPS confirms, the accuracy of GPS vectors only weakly depends on the distance between stations. Because station separations of thousands of kilometers can be determined very accurately, GPS is well suited to provide rigidity to geodetic networks of continental extent and to prevent the buildup of systematic errors. Soler et al. (1992b) compare and give a detailed analysis of GPS and VLBI positions in a network across the United States, spanning distances up to 2700 km. Both of these techniques are entirely independent, involving different hardware, software, and signal transmission sources. The GPS vector solutions were estimated in conjunction with orbital relaxation with five of the six Kepler elements estimated for each arc. The CIGNET stations in the United States served as fiducial sites. Note that these CIGNET stations are also co-located with VLBI stations. Table 14.4 shows a comparison between the GPS and the VLBI positions for 10

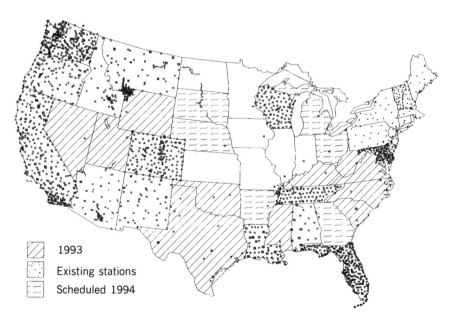

Figure 14.1   High-accuracy reference network (HARN).

**TABLE 14.4 Coordinate Differences in Local
Geodetic Coordinate Systems (East, North, Height)
Given in the Sense GPS-VLBI (Soler et al. 1992)**

Location	de [mm]	dn [mm]	dh [mm]
Austin (TX)	- 11	33	2
Bloomington (IN)	8	- 3	4
Carrolton (GA)	- 1	0	0
Fort Davis (TX)	- 4	12	- 3
Green Bank (WV)	13	6	- 15
Leonard (OK)	- 15	4	17
Maryland Point (MD)	8	- 8	- 34
Miles City (MT)	- 1	14	5
Platteville (CO)	4	14	- 3
Richmond (FL)	0	0	0

locations where VLBI antennas and GPS receivers were co-located. The minimal constraints for the GPS adjustment were defined by holding the VLBI position of Richmond (FL) fixed. The discrepancies are transformed to topocentric east, north, and height for easier interpretation. The discrepancies do not exceed 34 mm. The distances between the stations as computed separately for the GPS and the VLBI solutions were also compared. The respective distance differences are invariant with respect to potential discrepancies in the coordinate system definition. The largest relative accuracy was 10 ppb (part per billion) translating into a distance difference of just 19 mm.

With a remarkable tool such as GPS available to control the buildup of systematic errors in national geodetic networks and to assure homogeneity on a national scale, it becomes meaningful and practical to consider the use of position errors to describe the quality of station locations rather than using length-dependent formulations. A major change in the concept and perception of geodetic networks will occur as continuously tracking stations, also referred to as active control points (ACPs), come into use. The positions of new stations will be determined with respect to the ACP by either one single vector measurement, by repeated measurements of the same vector, or by single and/ or multiple measurements to several ACPs. Because GPS readily gives the whole spectrum from centimeter relative positioning accuracy to decameters, there will be pressure to use ACPs for accurate positioning as well as for lower-order accuracy positioning, as might be needed for GIS applications. In contrast to current practice, positioning with repsect to the ACP does not lend itself immediately to network-type solutions

Geodetic referencing is clearly undergoing several revolutionary developments propelled by two forces. GPS finally makes it possible to provide geodetic control at a suitable density and at accessible locations to make it attractive to use. Because there will be many new users and uses for geodetic data, it is appropriate to update and supplement the standards and specifications at this time. New features should be introduced into these updates as necessary, in order to achieve the best standards for this new phase in geodetic referencing. ''Old-time'' users such as trained surveyors and engineers will have no difficulty dealing with new (but related) elements contained in a revised version of the standards and specifications. Newcomers must start somewhere. They might just as well start with the revision.

14.2.3   Recommendations

Modern standards should reflect current and anticipated practices in geodesy and the uses of geodetic products. Certainly at present, and very likely over the next one to two decades, GPS will be the primary tool of choice for the establishment of geodetic networks. Soler et al. (1992a) verify that A-order GPS stations across the United States are determined to better than, say, 5 cm relative accuracy. See Table 14.4. Because these stations are distributed across the United States, this relative accuracy is practically equal to absolute positioning accuracy in the sense that the coordinates of these A-order stations will not change more than 5 cm in a possible future reoccupation (assuming no change in the datum definition). Strange and Love (1991) state that the relative accuracy of the B-order network stations is 1–3 cm.

14.2.3.1   *Revision of the GPS Standards*   The present fixed plus proportional error for GPS surveys is a significant improvement over the older conventional system that specified only the proportional part of the error (Table 14.2). The standards (and specifications) must comply with the physics of the measurement process as much as possible in order to be usable as a stochastic model in the least-squares adjustment. Including the constant part essentially eliminates the need to introduce minimum length of lines. The constant term models instrument setup errors caused by noncoincidence of the geometric and electronic center in the case of GPS antennas.

The existing relative accuracy standards, ranging from AA to C (Class 3) are considered adequate. No changes are recommended for these GPS standards (Table 14.2). Only station positions that are part of a network solution, i.e., the observations controlled by a minimal constraint network solution, should be considered for inclusion in the national geodetic network. Station positions obtained by other procedures, e.g., referencing to an ACP, should be flagged accordingly in the geodetic database. Thus, the geodetic network will be refined with only those observations that fulfill relative accuracy standards and have passed a minimal constraint network solution test.

14.2.3.2   *Position Accuracy*   Since the B-order densifcation of the national geodetic network will soon be completed, the proposed position classi-

fication reflects this situation. The proposed position accuracy is three-dimensional and essentially reflects the relative position accuracy of the new stations with respect to the B-order frame.

The procedure for computing position accuracy is straightforward. For each positioning event, compute the ellipsoid of standard deviation for the adjusted positions, holding the A and B order GPS stations fixed. Next, scale these ellipsoids to the 95% confidence level and compute the position accuracy from $SQRT(a^2 + b^2 + c^2)$, where a, b, and c denote the scaled semiaxes of the ellipsoid of standard deviation. This measure of position accuracy is not to be confused with distance accuracy as derived from rigorous application of variance–covariance propagation.

The defined position error reflects the quality of the position solution in a simple manner. Computing the elements of the ellipsoid of standard deviation does not require the full inverse of the normal matrix; thus, the computation load is manageable for statewide networks. As NGS completes the readjustment of the classical first- and second-order network stations as part of the establishment of the B-order frame, these first-, second-, and even third-order network stations can readily be assigned a position accuracy.

Since the A and B order stations are held fixed, the computed position errors are very meaningful, because they reflect the local accuracy of the network fairly well. The small translatory uncertainties of the cluster of A-order and B stations are not reflected by this position error. Possible small errors at the border between states or between clusters of A-order and B stations will be eliminated in a future nationwide readjustment. In the meantime, these minor "boundary conflicts" are the price to be paid in order to make meaningful measures of position accuracy available to the vast majority of users. There is no perfect solution. If it were not for the anticipated surge of users and uses of geodetic position data, such a nongeodetic solution would not be recommended.

Because correlations between stations are not available, the standard deviations of the adjusted coordinates (and thus the position accuracy value) must not be used for variance–covariance propagation. The fact that only one number per station is given to specify the quality of the position will, it is hoped, guard against misguided attempts to use these values in variance–covariance propagation.

Actually, no additional classification of the position accuracy is required. However, the quality of geodetic positions might be better understood and managed more successfully by new multidisciplinary users if a simple classification is introduced on the basis of a range in the position accuracy. The proposed classifications are listed in Table 14.5.

Note again that the proposed classification is three dimensional. During the readjustment of the classical data, as part of the regional solutions, the uncertainties in the height are expected frequently to be the determining factor for the classification. A separate horizontal and vertical classification is not recommended.

TABLE 14.5   Proposed Position Accuracy Standards

Class	Range of PA
millimeter	less than 0.005 m
centimeter	0.005 m to 0.05 m
decimeter	0.05 m to 0.20 m
submeter	0.20 m to 0.50 m
meter	0.5 m to 1.5 m
multimeter	1.5 m to 10 m
decameter	greater than 10 m

14.2.3.3   ACP Referencing   The classical procedure in geodetic network densification is that of the network mode. Typically more observations are made than are necessary to determine the network. The redundancy within the network is used to quality control the observations, i.e., to detect blunders and systematic errors. Because these additional checks are possible, surveying within a network is generally given higher ranking than stand-alone surveys such as open-ended traverses or positioning by just one GPS vector.

In the ACP mode, a new position is determined with respect to the active control station by means of just a single vector. Such a vector can, of course, be determined repeatedly. However, such repeat measurements do not reveal all the systematic errors, in particular those errors that are a function of the station location. Blunders such as station identification will remain undetected. The latter is also true in the expanded ACP mode when a new station is determined with respect to several ACPs. Station misidentifications and sloppy antenna centering still occur.

Considering the deficiency of quality control in the case of the ACP mode, it is recommended that points determined in the ACP mode be clearly flagged as such in the geodetic database, and that the reference station be identified specifically. These stations should not become members of the geodetic network. They can be classified according to Table 14.4 on the basis of the postadjustment statistics.

The procedure described above will maintain the highest quality control for network stations on the one side, but will provide a way to classify stations computed with respect to ACPs. It is expected that in the near future many positions will be determined with respect to ACPs simply for economic reasons and because many applications neither require highest accuracy nor the ultimate in controllability. Clearly, not every station whose position is determined with respect to the geodetic network should automatically become a member of the defining network stations.

14.2.4   Remarks Concerning Specifications

Even though the ACSM ad hoc committee did not comprehensively address the issue of specification, the topic surfaced naturally during the discussions.

It was understood that preparing specifications that fully incorporate recent and ongoing developments in measurement technology is a major and difficult task.

Several technical issues require clarification, such as making acceptable ties between conventional and GPS networks, establishing azimuth marks, and properly using geoid models. The specifications for GPS and conventional techniques should be combined in one report. The number of technical terms should be minimized, and established terminology from least-squares adjustment should be given preference. It is important to emphasize the central role of the minimal constraint least-squares solution for objective quality control for both GPS vector and conventional observations and to focus on the three-dimensional geodetic model as a unifying mathematical model applicable to both GPS vectors and conventional observations.

Because GPS provides additional vectors inexpensively, it was felt that the role of redundancy in networks could possibly be widened. Redundancy within the network has always been important in order to detect blunders and to strengthen the solution. Perhaps the specifications could be simplified by not listing how many angles must be turned and which distances or vectors must be observed, if sufficient redundancy (and good internal reliability) can be verified as part of the minimal constraint least-squares solution.

The role of software should also be widened. Traditionally, computers have been used for numerical computations. The emphasis should be shifted toward supporting quality control. Software should be able to verify reliably whether or not the relevant specifications are followed. Perhaps the new generation of software might have an artifical intelligence component and, at the very minimum, fully utilize computer graphics to achieve that goal. Modern software also fulfills an educational purpose. Appropriate software can be supplementary to many seminars and can replace much of the theory found in manuals. With such software in the hands of everyone trying to use the geodetic network, the manuals can become more simple and more readable, while the deep theory can be relegated to dedicated textbooks. Thus, in the new geodetic era (characterized by major advances in measurement technology, a vast increase in the number of users of geodetic products and individuals sharpening their skills in geodetic positioning), written specifications are not sufficient. Ideally, revised written specifications and advanced software would be issued at the same time.

The specification must address and possibly rank the various technologies available for positioning. A measurement technology should not be judged merely by the fact that a certain accuracy can be achieved. Other criteria, such as ease of use, robustness against blunders, and, most of all, the ability to formulate objective quality control, should be considered. The recommended technologies should allow most people to contribute the most to the geodetic network and the world of positioning in general.

Specifications should also address issues such as databases, data management, and data communication. Users might want to deposit their observations to a database across the country with a simple PUT command or retrieve data from a local competitor's database via a GET command. Others might want to query participating databases, which might be maintained by individual surveyors, government agencies, etc., to find out which data are available. To enable such communication, it might be necessary to maintain a central catalog (which can be updated by the users themselves). Such a catalog might allow queries on project name, generator of the data, observation types, and processing techniques, e.g., GPS network mode, referenced to ACP, photogeodesy, etc.

14.2.5 To Be Continued

The primary recommendation of ACSM's Ad Hoc Committee on Geodetic Accuracy Standards is to supplement current relative positioning standards, which are distance dependent, with point position accuracy standards. It is hoped that position accuracy standards will serve multidisciplinary users best. The proposed standards reflect the expanded role of the geodetic network for spatial referencing primarily for GIS applications, and build upon a highly accurate reference frame, as materialized by the Order A and B geodetic network stations being established by the National Geodetic Survey.

Meanwhile, the National Geodetic Survey requested the service of another ad hoc committee, called the National Spatial Reference System Committee. This group consisted of individuals from all levels of government, private industry, and academia, to review a draft NGS strategic plan, and to make specific comments and recommendations regarding the implementation of the National Spatial Reference System (NSRS). The worried reader should note that the geodetic base of the NSRS is exactly what is called the National Geodetic Reference System (NGRS). The new term is supposed to be more in harmony with the concept of a National Spatial Data Infrastructure (NSDI) and should insure that the "geo-referenced elements on the information highway" will be firmly anchored to the geodetic frame. Without going into detail on the recommendations, it might be worthwhile to point out additional, new terminology. For example, the Order A and B networks mentioned above might be referred to in the near future as the Federal Base Network (FBN). Orthometric heights are part of the definition of the FBN. The next level in network densification, having a spacing of about 25–30 km and to be established cooperatively, might be called the Cooperative Base Network (CBN). The proposed name for even higher density networks is User Densification Networks (UDNs). Finally, the automated control points (ACPs) might be called, in the future, continuously operating reference stations (CORS).

It is conceivable that there might be new GPS clients who will operate CORS nationwide. Yes, there could potentially be a "free ride" for surveying. The "capturing" of antispoofing combined with progress in OTF algorithmic developments surely will attract the attention of the Federal Aviation Administration (FAA) as useful tools for precision landings in the national airspace and worldwide. There might be a shift from the C/A-code pseudorange-type of DGPS solutions to OTF initialization. Because safety of passengers is involved, the best that GPS can offer is likely to be considered. This should tilt the scale in favor of dual-frequency P-code and carrier phase solutions. The prospect of such a large market will potentially reduce the price of dual-frequency P-code technology. Following the present trend in the pricing of electronic components, it is quite possible that P-code type receivers will soon be more affordable. Even today's prices would readily allow P-code receivers to become standard navigation aids at every airport, large or small. Thus, we might eventually have the pleasant situation of having dual-frequency receivers at every airport, i.e., a network of 1000 to 1600 or so CORS across the nation as part of a nationwide WADGPS service.

Airport receivers, of course, must be continuously quality controlled to insure safety of air traffic. Such a continuous monitoring might entail that the receivers be networked and that the observables be accessible via an electronic data network. Surveyors could tap into this information highway and get the observations needed for their solutions from the nearest airport receivers. Even if the airport receivers are not networked, the continuous observations at each site could readily be used to establish 1000 to 1600 stations of highest accuracy across the nation. If the data are not accessible in real time via the information highway, they will surely be stored temporarily to be preserved for potential evidence (if nothing else). These databases could be tapped by GPS users.

If the receivers at the airports are of the dual-frequency P-code type, they can serve the largest user commuity. Low-accuracy GIS data collectors would download the C/A-code observations to carry out the differential pseudorange solution. Others probably would download at least one carrier phase observable in order to smooth the codes. Some users might be happy with just the two P-codes. Surveryors, most likely, would prefer all observables in order to make static GPS as rapidly as possible or to use OTF initialization in kinematic applications. An example of the latter users are photogrammetrists who currently admire OTF for precision airplane navigation, but who still complain about the logistical difficulties and constraints of having their own stationary receivers at airports or near the project area. With the help of the FAA, and with continued research on OTF for long(er) baselines, it might, indeed, be conceivable that photogrammetric planes can fix the ambiguities with OTF initialization anywhere in the air at any time. At least it is imaginable.

APPENDIX A

ELEMENTS FROM LINEAR ALGEBRA

Surveying computations rest heavily upon concepts from linear algebra. In general, there is a nonlinear mathematical relationship between the observations and other quantities, such as coordinates, height, area, and volume. Seldom is there a natural linear relation between observations as there is in spirit leveling. Least-squares adjustment and statistical treatment require that nonlinear mathematical relations be linearized; that is, the nonlinear relationship is replaced with a linear relationship. Possible errors caused by neglecting the nonlinear portion are eliminated through appropriate iteration. The result of linearization is a set of linear equations that is subject to further analysis, thus the need to know the elements of linear algebra. The use of linear algebra in the derivation and analysis of surveying measurements fortunately does not require the memorization of all possible proofs and theorems. There is ample literature available on linear algebra. This appendix merely summarizes some elements from linear algebra for the sake of completeness.

A.1 DETERMINANTS

Each $(u \times u)$ square matrix A has a uniquely defined scalar called the determinant. The following notation is used for the determinant: det A or $|A|$. The determinant is said to be of order u. For a (1×1) matrix, the determinant is equal to the matrix element:

$$_1A_1 = a_{11} \tag{A.1}$$

$$|A| = a_{11} \tag{A.2}$$

In general, the determinant of a $(u \times u)$ matrix is expressed in terms of determinants of submatrices of size $(u - 1) \times (u - 1)$, $(u - 2) \times (u - 2)$, etc. Consider the matrix

$$A = \begin{bmatrix} a_{11} & a_{12} & a_{13} & \cdots & a_{1u} \\ a_{21} & a_{22} & a_{23} & \cdots & a_{2u} \\ \vdots & \vdots & \vdots & \cdots & \vdots \\ a_{u1} & a_{u2} & a_{u3} & \cdots & a_{uu} \end{bmatrix} \tag{A.3}$$

The minor can be computed for each element of the matrix. The minor is equal to the determinant of A after the respective row and column are deleted. The minor for $i = 1$ and $j = 2$ is

$$m_{12} = \begin{vmatrix} a_{21} & a_{23} & \cdots & a_{2u} \\ a_{31} & a_{33} & \cdots & a_{3u} \\ \vdots & \vdots & \cdots & \vdots \\ a_{u1} & a_{u3} & \cdots & a_{uu} \end{vmatrix} \tag{A.4}$$

The cofactor c_{ij} is equal to plus or minus the minor, depending on the subscripts i and j:

$$c_{ij} = (-1)^{i+j} m_{ij} \tag{A.5}$$

The determinant of A is

$$|A| = \sum_{j=1}^{u} a_{kj} c_{kj} \tag{A.6}$$

The subscript k is fixed in (A.6) but can be any value between 1 and u; that is, the determinant can be computed based on the minors for any one of the u rows or columns.

Determinants have many useful properties. One important property is that the determinant is zero if the columns are linearly dependent. In that case the matrix is singular. The rank of a matrix equals the order of the largest nonsingular square submatrix, that is, the largest order for a nonzero determinant that can be found.

A.2 EIGENVALUES AND EIGENVECTORS

Consider the equation

$$_uA_u {}_uX_1 = \lambda_u X_1 \tag{A.7}$$

A is a $(u \times u)$ matrix. The $(u \times 1)$ vector X is called the eigenvector, and the scalar λ is the respective eigenvalue. Equation (A.7) can be rewritten as

$$(A - \lambda I)X = 0 \qquad (A.8)$$

If X_0 denotes a solution of (A.8), then the product αX_0 is also a solution, where α is a scalar. Thus, the solution of (A.8) gives only the direction of the eigenvector. There exists a nontrivial solution for X if the determinant is zero; that is,

$$|A - \lambda I| = 0 \qquad (A.9)$$

This is the characteristic equation. It is a polynomial of the uth order in λ providing u solutions λ_i, with $i = 1, \ldots, u$. Some of the eigenvalues can be zero, equal (multiple solution), or even complex. Equation (A.8) provides an eigenvector X_i for each eigenvalue λ_i.

In least squares, the eigenvalues of symmetric and positive-definite matrices are of prime concern. For a symmetric matrix all eigenvalues are real. There may still be multiple solutions. The number of zero eigenvalues equals the rank defect of the matrix. Thus, computing the eigenvalues provides one method of finding out the rank defect of a matrix. The eigenvectors are mutually orthogonal; that is,

$$X_i X_j = 0 \qquad (A.10)$$

For positive-definite matrices all eigenvalues are positive. The normalized eigenvectors $X_i/\|X_i\|$ are denoted by the symbol E_i. If all u normalized eigenvectors are combined into one matrix

$$E = (E_1, E_2, \ldots, E_u) \qquad (A.11)$$

then E is an orthonormal matrix, or a rotation matrix, for which

$$E^\mathrm{T} = E^{-1} \qquad (A.12)$$

holds.

A.3 DIAGONALIZATION THEOREM

Eigenvalues and eigenvectors are useful elements for diagonalizing a matrix. Consider again a $(u \times u)$ matrix A and the respective matrix E that consists of the normalized eigenvectors. The product of these two matrices can be written as

$$AE = (AE_1 \quad AE_2 \cdots AE_u)$$

$$= (\lambda_1 E_1 \quad \lambda_2 E_2 \cdots \lambda_u E_u) \tag{A.13}$$

$$= E\Lambda$$

where Λ is a diagonal matrix with λ_i as elements at the diagonal. Multiplying this equation by E^T from the left and making use of equation (A.12), one gets

$$E^T A E = \Lambda \tag{A.14}$$

Taking the inverse of both sides by applying the rule (A.36) and using (A.12) gives

$$E^T A^{-1} E = \Lambda^{-1} \tag{A.15}$$

Equation (A.14) simply states that if a matrix A is premultiplied by E^T and postmultiplied by E, where the columns of E are the normalized eigenvectors, then the product is a diagonal matrix whose diagonal elements are the eigenvalues of A. Equation (A.14) is further modified by

$$\Lambda^{-1/2} E^T A E \Lambda^{-1/2} = I \tag{A.16}$$

Thus,

$$D^T A D = I \tag{A.17}$$

where

$$D = E\Lambda^{-1/2} \tag{A.18}$$

is a nonsingular matrix computed from the respective eigenvalues and eigenvectors.

If the $(u \times u)$ matrix A is positive-semidefinite with rank $R(A) = r < u$, an equation similar to (A.14) can be found. Consider the matrix

$$E = (_u F_r \quad _u G_{u-r}) \tag{A.19}$$

where the column of F consists of the normalized eigenvectors that pertain to the r nonzero eigenvalues. The submatrix G consists of $(u - r)$ eigenvectors which pertain to the $(u - r)$ zero eigenvalues. The columns of F and G span the column and null space, respectively, of the matrix A. Because of equation (A.7) it follows that

$$AG = 0 \tag{A.20}$$

Applying equations (A.19) and (A.20) gives

$$E^T AE = \begin{bmatrix} F^T \\ G^T \end{bmatrix} A(FG) = \begin{bmatrix} F^T AF & 0 \\ 0 & 0 \end{bmatrix} = \begin{bmatrix} \Lambda & 0 \\ 0 & 0 \end{bmatrix} \qquad (A.21)$$

The submatrix contains the r nonzero eigenvalues. If

$$D = (F\Lambda^{-1/2} | G) \qquad (A.22)$$

it follows that

$$D^T AD = \begin{bmatrix} I & 0 \\ 0 & 0 \end{bmatrix} \qquad (A.23)$$

where the symbol I denotes an $(r \times r)$ identity matrix.

A.4 QUADRATIC FORMS

Let X be a $(u \times 1)$ vector and A denote a $(u \times u)$ matrix. Then

$$_1V_1 = X^T AX \qquad (A.24)$$

is called a quadratic form. If, in addition,

$$X^T AX \geq 0 \qquad (A.25)$$

then the matrix A and the quadratic form are called positive-semidefinite. If

$$X^T AX > 0 \qquad (A.26)$$

then A and the quadratic form are called positive-definite. The following are some of the properties that hold for a $(u \times u)$ positive-definite matrix A:

1. $R(A) = u$ (full rank)
2. $a_{ii} > 0$ for all i
3. The inverse A^{-1} is positive-definite.
4. Let B be an $(n \times u)$ matrix with rank $u < n$. Then the matrix $B^T AB$ is positive-definite. If $R(B) = r < u$ then $B^T AB$ is positive-semidefinite.
5. Let D be a $(q \times q)$ matrix formed by deleting $(u - p)$ rows and the corresponding $(u - p)$ columns of A. Then D is positive-definite.

A necessary and sufficient condition for a symmetric matrix to be positive-definite is that the principal minor determinants be positive; that is,

$$a_{11} > 0, \quad \begin{vmatrix} a_{11} & a_{12} \\ a_{21} & a_{22} \end{vmatrix} > 0, \dots, |A| > 0 \qquad \text{(A.27)}$$

or that eigenvalues are all real and positive.

The eigenvectors and eigenvalues allow an important geometric interpretation for quadratic forms. If $_uA_u$ is positive-definite, then

$$X^T A X = V \qquad \text{(A.28)}$$

is the equation of a u-dimensional ellipsoid expressed in a Cartesian coordinate system (X). The center of the ellipsoid is at $X = 0$. Transforming (rotating) the coordinate system (X) by

$$X = EY \qquad \text{(A.29)}$$

gives the following expression for the quadratic form:

$$Y^T E^T A E Y = V \qquad \text{(A.30)}$$

The matrix E consists of u normalized eigenvectors. With (A.14) this expression becomes

$$Y^T \Lambda Y = V \qquad \text{(A.31)}$$

This expression can be written as

$$y_1^2 \lambda_1 + y_2^2 \lambda_2 + \cdots + y_u^2 \lambda_u = V \qquad \text{(A.32)}$$

or

$$\frac{y_1^2}{V/\lambda_1} + \frac{y_2^2}{V/\lambda_2} + \cdots + \frac{y_u^2}{V/\lambda_u} = 1 \qquad \text{(A.33)}$$

This is the equation for the u-dimensional ellipsoid in the principal axes form; that is, the coordinate system (Y) coincides with the principal axes of the hyperellipsoid, and the lengths of the principal axes are proportional to the reciprocal of the square root of the eigenvalues. All eigenvalues are positive because the matrix A is positive-definite. Equation (A.29) determines the orientation between the (X) and (Y) coordinate systems. If A has a rank defect, the dimension of the hyperellipsoid is $R(A) = r < u$.

A.5   INVERSE OF A NONSINGULAR MATRIX

Both singular and nonsingular matrices can be inverted. Inverting singular or even rectangular matrices requires the concept of generalized inverses. Extensive literature is available on the topic of generalized inverses. Examples are Rao and Mitra (1971), Bjerhammar (1973), and Grafarend and Schaffrin (1993). The least-squares technique of inner constraints can be formulated with the help of generalized inverses. In Section 4.8 this topic is dealt with in detail. Simple rules are given there for computing the generalized inverse for applications typical in surveying.

The inverse of a nonsingular square matrix A, denoted by A^{-1}, is defined by

$$AA^{-1} = I \tag{A.34}$$

$$A^{-1}A = I \tag{A.35}$$

If the matrices A, B, and C are also square and nonsingular, then

$$(ABC)^{-1} = C^{-1}B^{-1}A^{-1} \tag{A.36}$$

This is the reversal rule for inverse products. It is helpful for manipulating inverse expressions.

The inverse of partitioned matrices is helpful for algebraic computations during the development of least-squares algorithms. Partition the nonsingular square matrix N as follows:

$$N = \begin{bmatrix} N_{11} & N_{12} \\ N_{21} & N_{22} \end{bmatrix} \tag{A.37}$$

where N_{11} and N_{22} are square matrices, although not necessarily of the same size. Denote the inverse matrix by Q and partition it accordingly; that is,

$$Q = N^{-1} = \begin{bmatrix} Q_{11} & Q_{12} \\ Q_{21} & Q_{22} \end{bmatrix} \tag{A.38}$$

so that the size of N_{11} and Q_{11}, N_{12} and Q_{12}, etc., are respectively the same. Since Q is the inverse of N, the following equations must be fulfilled

$$NQ = I \tag{A.39}$$

$$QN = I \tag{A.40}$$

Each of matrix equations (A.39) and (A.40) is itself a set of four equations. Each set can be solved for the unknown submatrices $(Q_{11}, Q_{12}, Q_{21}, Q_{22})$ as a function of $(N_{11}, N_{12}, N_{21}, N_{22})$. Thus, there are two solutions for Q, each a different function of the submatrices N_{ij}. The actual solution for the submatrices Q_{ij} is carried out according to the standard rules for solving a system of linear equations, with the restriction that the inverse is defined only for square submatrices. The solution is

$$
Q = \begin{bmatrix} Q_{11} & Q_{12} \\ Q_{21} & Q_{22} \end{bmatrix}
$$

$$
= \left[\begin{array}{c|c} N_{11}^{-1} + N_{11}^{-1}N_{12}(N_{22} - N_{21}N_{11}^{-1}N_{12})^{-1}N_{21}N_{11}^{-1} & -N_{11}^{-1}N_{12}(N_{22} - N_{21}N_{11}^{-1}N_{12})^{-1} \\ -(N_{22} - N_{21}N_{11}^{-1}N_{12})^{-1}N_{21}N_{11}^{-1} & (N_{22} - N_{21}N_{11}^{-1}N_{12})^{-1} \end{array} \right]
$$

$$
= \left[\begin{array}{c|c} (N_{11} - N_{12}N_{22}^{-1}N_{21})^{-1} & -(N_{11} - N_{12}N_{22}^{-1}N_{21})^{-1}N_{12}N_{22}^{-1} \\ -N_{22}^{-1}N_{21}(N_{11} - N_{12}N_{22}^{-1}N_{21})^{-1} & N_{22}^{-1} + N_{22}^{-1}N_{21}(N_{11} - N_{12}N_{22}^{-1}N_{21})^{-1}N_{12}N_{22}^{-1} \end{array} \right]
$$

$$(A.41)$$

Usually the above partitioning technique is used to reduce the size of large matrices that must be inverted, or to derive alternative expressions during derivations. The respective submatrices in (A.41) must be equal because there is only one inverse Q.

Computation techniques for inverting nonsingular square matrices abound in linear algebra textbooks. In many cases the matrices to be inverted show a definite pattern and are often sparsely populated. For large adjustments it is necessary to take advantage of these patterns in order to reduce the computation load (George and Liu 1981). Other applications might produce very ill-conditioned (numerically near-singular) matrices. Even though the computation of the inverse is conceptually easy, there are several aspects to be considered when writing a quality computer program. In fact, software packages can be judged by the sophistication they employ for solving large systems of equations and inverting large matrices. Very useful subroutines are available in the public domain (Milbert 1984). For positive-definite matrices the so-called square root method, also known as the Cholesky method, is an efficient way to solve systems of equations and to compute the inverse. This algorithm is briefly summarized. Assume that the system of equations

$$
{}_u N_u \, {}_u X_1 = {}_u U_1 \tag{A.42}
$$

should be solved. Because the matrix N is the positive-definite it is written as the product of a lower triangular matrix L and an upper triangular matrix L^{T}:

$$
N = L L^{\mathrm{T}} \tag{A.43}
$$

Lower and upper triangular matrices have several useful properties. For example, the eigenvalues of a triangular matrix are equal to the diagonal elements, and the determinant of the triangular matrix is equal to the product of the diagonal elements. Because the determinant of a matrix product is equal to the product of the determinants of the factors, it follows that N is singular if one of the diagonal elements of L is zero. This fact can be used advantageously during the computation of L to eliminate parameters causing a singularity in the system (A.42) simply by deleting the respective rows and columns that generate zeros in the diagonal of L. The first step in the solution of (A.42) is to substitute (A.43) for N and premultiply with L^{-1} to obtain the reduced triangular equations:

$$L^T X = L^{-1} U$$
$$= C_U \tag{A.44}$$

In this special form the unknown elements of X follow from the backward solution, beginning with the last component of X. The Cholesky algorithm provides the instruction for computing the lower triangular matrix L. The elements of L are

$$l_{jk} = \begin{cases} \dfrac{1}{l_{kk}} \left(n_{jk} - \displaystyle\sum_{m=1}^{k-1} l_{jm} l_{km} \right) & \text{for } k < j \\[2em] \sqrt{n_{jj} - \displaystyle\sum_{m=1}^{j-1} l_{jm}^2} & \text{for } k = j \\[2em] 0 & \text{for } k > j \end{cases} \tag{A.45}$$

where $1 \le j \le u$ and $1 \le k \le u$. The matrix L does not necessarily have diagonal elements equal to unity. The Cholesky algorithm preserves the pattern of leading zeros in the rows and columns of N, as can be readily verified. For example, if the first x elements in row y of N are zero, then the first x elements in row y of L are also zero. Taking advantage of this fact speeds up the computation of L for large system that exhibits significant patterns of leading zeros. The algorithm (A.45) begins with the element l_{11}. Subsequently the columns (or rows) can be computed sequentially from 1 to u whereby previously computed columns (or rows) remain unchanged while the next one is computed. Equation (A.44) implies

$$L C_U = U \tag{A.46}$$

Thus, the vector C_U is obtained from the forward solution of (A.46) starting with the first element in C_U. The same result is obtained for C_U if the matrix

N is augmented by the row U^T and the algorithm (A.45) is executed over the range $j = u + 1$ and $1 \leq k \leq u$. Using L and C_U the backward solution (A.44) yields the parameters X.

When computing the sum of the squared residuals in least squares, the auxiliary quantity

$$l = -C_U^T C_U = -(L^{-1}U)^T (L^{-1}U)$$
$$= -U^T (L^{-1})^T L^{-1} U = -U^T N^{-1} U \qquad \text{(A.47)}$$

is used directly to compute $V^T P V$ to assess the quality of the adjustment. See Table 4.2 for additional detail. It is important to note that the Cholesky algorithm provides the solution for X and the quantity l without explicit use of the inverses N^{-1} and L^{-1}.

Computing the inverse requires a much bigger computational effort than solving a system of equations. The inverse is not required during the iterations of the least-squares solution but is typically computed after convergence in order to provide additional statistical information on the adjustment. The αth column q_α of the inverse N^{-1} satisfies the equation

$$N q_\alpha = U_\alpha \qquad \text{(A.48)}$$

where

$$U_\alpha^T = [0 \quad 0 \cdots 1 \cdots 0] \qquad \text{(A.49)}$$

The elements of U_α are equal to zero in all positions except location α, where the element is unity. The equation system (A.48) differs from (A.42) only in the right-hand side. Thus, the solution for C_α follows from (A.46) upon replacing the right-hand side U by the vector U_α and carrying out the forward solution. The vectors U_α and C_α have leading zeros in common. Column q_α follows from the backward solution of (A.44) using C_α on the right-hand side. Thus, there are u forward and u backward solutions required.

Premultiplying the matrix N in (A.43) with L^{-1} and postmultiplying it with the transpose gives the identity matrix. Thus, the Cholesky factor L can be used in ways similar to the matrix D in (A.17). A frequent application is the decorrelation of observations. In this case the inverse L^{-1} is not required explicitly. For example, if L denotes the Cholesky factor of the covariance matrix of the observations Σ_{L_b} and the a priori variance of unit weight is taken as unity, then the transformation (4.241) can be specified as

$$L^{-1} V = L^{-1} A X + L^{-1} L_{\text{mc}} \qquad \text{(A.50)}$$

To avoid a conflict in notation, the misclosure is labeled L_{mc}. Let us denote the decorrelated observations by

$$\overline{V} = \overline{A} X + \overline{L}_{\text{mc}} \qquad \text{(A.51)}$$

Then we see that

$$L\,\overline{L}_{mc} = L_{mc} \tag{A.52}$$

and

$$L\,\overline{A}_{\alpha} = A_{\alpha} \tag{A.53}$$

where the subscript α denotes the αth column of the matrices $\overline{A}$ and A, respectively. Thus, the transformed misclosure and the columns of the decorrelated design matrix can be computed from a series of forward solutions using L only. Upon completion of the adjustment the residuals follow from yet another forward solution

$$L\,\overline{V} = V \tag{A.54}$$

The same reasoning can be applied to the decorrelated redundancy numbers discussed in Section 4.10.6

A.6 DIFFERENTIATION OF QUADRATIC FORMS

Quadratic forms are multivariable functions that can be differentiated according to the standard rules of calculus. Matrix notation is very helpful in providing short and precise formulations. Consider the quadratic form

$$W = {}_1 X^T_{u\,u} A_{u\;u} Y_1 \tag{A.55}$$

The elements of the matrix A are constants. The variables are the u components of the X and Y vectors, respectively. Because (A.55) is a (1×1) matrix, it follows that

$$W = X^T A Y = Y^T A^T X \tag{A.56}$$

The total differential dW is obtained by taking the partial derivatives with respect to all variables. One can formally write

$$dW = \frac{\partial W}{\partial X}\,dX + \frac{\partial W}{\partial Y}\,dY \tag{A.57}$$

The vectors dX and dY contain the differentials of the X and Y components, respectively. From (A.56) and (A.57) it follows that

$$dW = Y^T A^T dX + X^T A\,dY \tag{A.58}$$

If the matrix A is symmetric, that is, if $A = A^T$, and if only one vector of variables is considered, it follows from equations (A.55) and (A.58) that the total differential of

$$V = {}_1X^T_u {}_uA_u {}_uX_1 \qquad (A.59)$$

is

$$dV = 2X^TA \, dX \qquad (A.60)$$

Equation (A.60) consists of one equation since dV is a (1×1) matrix.

In least-squares estimation it is required to compute the stationary point (minimum) for a quadratic function like (A.59). The usual procedure is to take the partial derivatives with respect to all variables and equate them to zero. It can readily be verified with the help of equations (A.59) and (A.60) that

$$\frac{dV}{dX} \equiv \frac{\partial V}{\partial X} \equiv \left[\frac{\partial V}{\partial x_1} \cdots \frac{\partial V}{\partial x_u} \right] = 2X^TA \qquad (A.61)$$

Expression (A.61) represents u equations; that is, each equation equals the partial derivative of V with respect to a specific component of X. It is useful to recognize the notation used in (A.61), since it seems to imply that the derivative of the scalar V has been taken with respect to the vector X. In this notation the factor dX in (A.60) is simply moved to the denominator of the left-hand side.

APPENDIX B

LINEARIZATION

Observations in surveying are usually related by nonlinear functions. To perform an adjustment, one must linearize these relationships. This is accomplished by expanding the functions in a Taylor series and retaining only the linear terms. Consider the nonlinear function

$$y = f(x) \tag{B.1}$$

which has only one variable x. The Taylor series expansion of this function is

$$y = f(x_0) + \frac{dy}{dx}\bigg|_{x_0} dx + \frac{1}{2!}\frac{d^2y}{dx^2}\bigg|_{x_0} dx^2 + \cdots \tag{B.2}$$

The linear portion is given by the first two terms as

$$\bar{y} = f(x_0) + \frac{dy}{dx}\bigg|_{x_0} dx \tag{B.3}$$

The derivatives are evaluated at the point of expansion x_0. At the point of expansion the linearized and the nonlinear functions coincide and are tangent. The separation

$$\epsilon = y - \bar{y} \tag{B.4}$$

increases as the parameter x departs from the expansion point x_0. The linearized form (B.3) is a sufficiently accurate approximation of the nonlinear relation (B.1) only in the vicinity of the point of expansion, see Figure B.1.

512

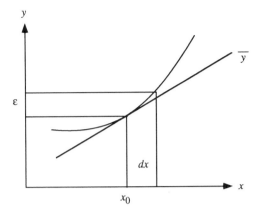

Figure B.1 Linearization.

The expansion of the two-variable function

$$z = f(x, y) \tag{B.5}$$

is

$$z = f(x_0, y_0) + \frac{\partial z}{\partial x}\bigg|_{x_0, y_0} dx + \frac{\partial z}{\partial y}\bigg|_{x_0, y_0} dy + \cdots \tag{B.6}$$

The point of expansion is $P(x = x_0, y = y_0)$. The linearized form

$$\bar{z} = f(x_0, y_0) + \frac{\partial z}{\partial x}\bigg|_{x_0, y_0} dx + \frac{\partial z}{\partial y}\bigg|_{x_0, y_0} dy \tag{B.7}$$

represents the tangent plane on the surface (B.5) at the expansion point. A generalization for the expansion of multivariable functions is readily seen. If n functions are related to u variables as in

$$Y = F(X) = \begin{bmatrix} f_1(X) \\ f_2(X) \\ \vdots \\ f_n(X) \end{bmatrix} = \begin{bmatrix} f_1(x_1, x_2, \ldots, x_u) \\ f_2(x_1, x_2, \ldots, x_u) \\ \vdots \\ f_n(x_1, x_2, \ldots, x_u) \end{bmatrix} \tag{B.8}$$

the linearized form is

$$\bar{Y} = F(X_0) + \frac{\partial F}{\partial X}\bigg|_{x_0} dX \tag{B.9}$$

where

$$\frac{\partial F}{\partial X} = {}_n G_u = \begin{bmatrix} \dfrac{\partial f_1}{\partial x_1} & \dfrac{\partial f_1}{\partial x_2} & \cdots & \dfrac{\partial f_1}{\partial x_u} \\[2ex] \dfrac{\partial f_2}{\partial x_1} & \dfrac{\partial f_2}{\partial x_2} & \cdots & \dfrac{\partial f_2}{\partial x_u} \\[2ex] \vdots & \vdots & \cdots & \vdots \\[2ex] \dfrac{\partial f_n}{\partial x_1} & \dfrac{\partial f_n}{\partial x_2} & \cdots & \dfrac{\partial f_n}{\partial x_u} \end{bmatrix} \tag{B.10}$$

The point of expansion is $P(X = X_0)$. Every component of Y is differentiated with respect to every variable. Thus, the matrix G has as many columns as there are parameters, and as many rows as there are components in Y. The components of $F(X_0)$ equal the respective functions evaluated at X_0. The vector dX contains the u differentials. Finally, if

$$Z = F(X, Y) \tag{B.11}$$

the linearized form becomes

$$Z = F(X_0, Y_0) + \left. \frac{\partial F}{\partial X} \right|_{X_0, Y_0} dX + \left. \frac{\partial F}{\partial Y} \right|_{X_0, Y_0} dY \tag{B.12}$$

The number of columns in the matrices $\partial F/\partial X$ and $\partial F/\partial Y$ corresponds to the number of components in X and Y, respectively. The number of rows equals the size of the vector Z.

APPENDIX C

ONE-DIMENSIONAL DISTRIBUTIONS

Brief explanations are given on the one-dimensional distributions and hypothesis testing. The material of this appendix can be found in the standard literature on statistics. The expressions for the noncentral distribution are given, e.g., in Koch (1988).

C.1 CHI-SQUARE DISTRIBUTION

The chi-square density function is given by

$$f(x) = \begin{cases} \dfrac{1}{\Gamma(r/2)2^{r/2}} x^{r/2 - 1} e^{-x/2} & 0 \leq x \leq \infty \\ 0 & \text{elsewhere} \end{cases} \tag{C.1}$$

The symbol r denotes a positive integer and is called the degree of freedom. The mean, that is, the expected value, equals r, and the variance equals $2r$. The symbol Γ denotes the standard gamma function, which is defined as

$$\Gamma(g) = (g - 1)! \tag{C.2}$$

and

$$\Gamma\left(g + \frac{1}{2}\right) = \frac{(2g)!\sqrt{\pi}}{g!2^{2g}} \tag{C.3}$$

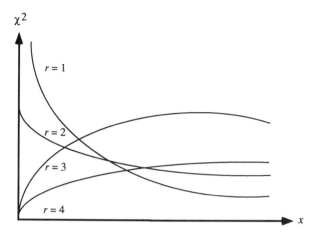

Figure C.1 Chi-square distribution for various degrees of freedom.

if g is a positive integer. The chi-square distribution is very unsymmetric for small degrees of freedom, as can be seen from Figure C.1.

The area under the curve expresses the probability, which is usually given in tabular form in statistics books:

$$P(\tilde{x} < w_\alpha) = \int_0^{w_\alpha} f(x)dx = 1 - \alpha \qquad (C.4)$$

Note the notation in (C.4), which implies that to the right of w_α there is the probability α. Integrating from w_α to infinity gives α. The degree of freedom is sufficient to completely describe the chi-square distribution. If $\tilde{x}$ has a chi-square distribution, then this is expressed by the notation $\tilde{x} \sim \chi_r^2$.

The distribution (C.1) is more precisely called the central distribution. The noncentral chi-square is a generalization of the central chi-square distribution. The density function does not have a simple closed form; it consists of an infinite sum of terms. If $\tilde{x}$ has a noncentral chi-square distribution, this is expressed by $\tilde{x} \sim \chi_{r,\lambda}^2$, where λ denotes the noncentrality parameter. The mean is

$$E(\tilde{x}) = r + \lambda \qquad (C.5)$$

as opposed to just r for the central chi-square distribution.

C.2 NORMAL DISTRIBUTION

The density function of the normal distribution is

$$f(x) = \frac{1}{\sigma\sqrt{2\pi}} e^{-(x-\mu)^2/2\sigma^2} \qquad -\infty < x < \infty \qquad (C.6)$$

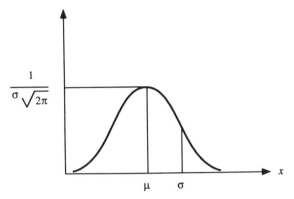

Figure C.2 Normal density function.

where μ and σ denote the mean and the variance, respectively (see Figure C.2). The notation $\tilde{x} \sim n(\mu, \sigma^2)$ is usually used. The two parameters μ and σ^2 completely describe the normal distribution. The normal distribution has the following characteristics:

1. The distribution is symmetric about the mean.
2. The maximum density is at the mean.
3. For small variances, the maximum density is larger and the slopes are steeper, than in the case of large variances.
4. The inflection points are at $x = \mu \pm \sigma$.

If $\tilde{x} \sim n(\mu, \sigma^2)$, then the transformed variable

$$\tilde{w} = \frac{\tilde{x} - \mu}{\sigma} \sim n(0, 1) \tag{C.7}$$

has a normal distribution with zero mean and unit variance. The random variable $\tilde{w}$ is said to have a standardized normal distribution. The density function for $\tilde{w}$ is

$$f(w) = \frac{1}{\sqrt{2\pi}} e^{-w^2/2} \qquad -\infty < w < \infty \tag{C.8}$$

Usually, the integral

$$P(\tilde{x} < w_\alpha) = \int_0^{w_\alpha} f(w)dw \tag{C.9}$$

is tabulated, because the normal density function is symmetric with respect to the mean. Table C.1 lists selected values that are frequently used. For a normal

TABLE C.1 Selected Values from the Normal Distribution

x	1σ	2σ	3σ	4σ	$0.674\,\sigma$	$1.645\,\sigma$	$1.960\,\sigma$	$2.326\,\sigma$
$N(x)-N(-x)$	0.6827	0.9544	0.9973	0.99994	0.5	0.90	0.95	0.98

distribution, in about 68% of all cases the observations fall within one standard deviation from the mean, and only every 370th observation deviates from the mean by more than 3σ. Therefore, 3σ is sometimes taken as the limit to what can be regarded as random error. Any larger deviation from the mean is usually considered a blunder. Statistically, large errors cannot be avoided, but their occurrence is unlikely. The 3σ criteria is not necessarily applicable in least-squares adjustments because the pertinent random variables have multivariate distributions and are correlated, thus reflecting the "geometry of the adjustment." Further details are given in Section 4.10.

C.3 t DISTRIBUTION

Assume that $\tilde{w} \sim n(0, 1)$ and $\tilde{v} \sim \chi_r^2$ are two stochastically independent random variables with unit normal and chi-square distribution, respectively; then the random variable

$$\tilde{t} = \frac{\tilde{w}}{\sqrt{\tilde{v}/r}} \tag{C.10}$$

has a t distribution with r degrees of freedom:

$$f(t_r) = \frac{\Gamma[(r+1)/2]}{\sqrt{\pi r}\,\Gamma(r/2)} \frac{1}{[1+(t^2/r)]^{(r+1)/2}} \qquad -\infty < t < \infty \tag{C.11}$$

The density function (C.11) is symmetric with respect to zero (see Figure C.3). Furthermore, if r equals infinity, the t distribution is identical to the standardized normal distribution; that is,

$$t_\infty = n(0, 1) \tag{C.12}$$

The density in the vicinity of the mean (zero) is smaller than for the unit normal distribution, whereas the reverse is true at the extremities of the distribution. The t distribution converges rapidly toward the normal distribution. If the random variable $\tilde{w} \sim n(\delta, 1)$ is normal distributed with unit variance but with a nonzero mean, then the function (C.10) has a noncentral t distribution with r

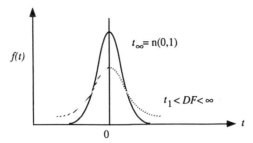

Figure C.3 The probability density function of the t distribution.

degrees of freedom and a noncentrality parameter δ. The distribution function is mathematically complicated.

C.4   F DISTRIBUTION

Consider two stochastically independent random variables, $\tilde{u} \sim \chi^2_{r_1}$ and $\tilde{v} \sim \chi^2_{r_2}$, distributed with r_1 and r_2 degrees of freedom, respectively; then the random variable

$$\tilde{F} = \frac{\tilde{u}/r_1}{\tilde{v}/r_2} \tag{C.13}$$

has the density function

$$f(F_{r_1, r_2}) = \frac{\Gamma\left(\dfrac{r_1 + r_2}{2}\right)\left(\dfrac{r_1}{r_2}\right)^{r_1/2}}{\Gamma(r_1/2)\Gamma(r_2/2)} \frac{F^{(r_1/2) - 1}}{\left(1 + \dfrac{r_1 F}{r_2}\right)^{(r_1 + r_2)/2}} \quad 0 < F < \infty \tag{C.14}$$

This is the F distribution with r_1 and r_2 degrees of freedom (see Figure C.4). The mean, or the expected value, is

$$E(F_{r_1, r_2}) = \frac{r_2}{r_2 - 2} \tag{C.15}$$

Care should always be taken to identify the degrees of freedom properly since the density function is not symmetric in these variables. It can also be shown that the following relationship holds

$$F_{r_1, r_2, \alpha} = \frac{1}{F_{r_2, r_1, 1 - \alpha}} \tag{C.16}$$

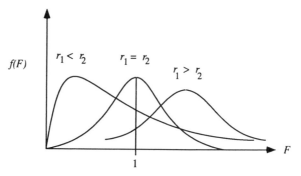

Figure C.4   The F distribution.

Figure (C.4) shows the density function for various degrees of freedom. The F distribution is related to the chi-square and the t distributions as follows:

$$\frac{\chi_r^2}{r} \sim F_{r,\infty} \tag{C.17}$$

$$t_r^2 \sim F_{1,r} \tag{C.18}$$

If $\bar{u} \sim \chi_{r_1,\lambda}^2$ has a noncentral chi-square distribution with r_1 degrees of freedom and a noncentrality parameter λ, then the function F in (C.13) has a noncentral F distribution with r_1 and r_2 degrees of freedom and noncentrality parameter λ. The mean for the noncentral distribution is

$$E(F_{r_1,r_2,\lambda}) = \frac{r_2}{r_2 - 2}\left(1 + \frac{\lambda}{r_1}\right) \tag{C.19}$$

C.5   DISTRIBUTION OF SELECTED FUNCTIONS

The following functions of random variables are required in the derivation of distributions of key random variables in least-squares estimation.

C.5.1   The Sum of Normal Distributed Variables

Assume that $(\tilde{x}_1, \tilde{x}_2, \ldots, \tilde{x}_n)$ are n stochastically independent variables, each having a normal distribution, with different means μ_i and variances σ_i^2. Then the linear function

$$\tilde{y} = k_1\tilde{x}_1 + k_2\tilde{x}_2 + \cdots + k_n\tilde{x}_n \tag{C.20}$$

is distributed as

$$\tilde{y} \sim n\left[\sum_i^n (k_i\mu_i), \sum_i^n (k_i^2\sigma_i^2)\right] \tag{C.21}$$

C.5.2 The Square of Standardized Normal Distribution

If the random variable $\tilde{w}$ has a standardized normal distribution, that is, $\tilde{w} \sim n(0, 1)$, then

$$\tilde{\nu} = \tilde{w}^2 \sim \chi_1^2 \tag{C.22}$$

has a chi-square distribution with one degree of freedom.

C.5.3 The Sum of Chi-Square Distributed Variables

Assume that $(\tilde{x}_1, \tilde{x}_2, \ldots, \tilde{x}_n)$ are n stochastically independent random variables, each having a chi-square distribution. The degrees of freedom r_i can differ. Then the random variable

$$\tilde{y} = \tilde{x}_1 + \tilde{x}_2 + \cdots + \tilde{x}_n \tag{C.23}$$

is distributed

$$\tilde{y} \sim \chi_{\Sigma r_i}^2 \tag{C.24}$$

The degree of freedom equals the sum of the individual degrees of freedom.

C.5.4 The Sum of Squares of Standardized Normal Distributed Variables

Assume that $(\tilde{x}_1, \tilde{x}_2, \ldots, \tilde{x}_n)$ are n stochastically independent random variables, each having a normal distribution. The means are nonzero. Then

$$\tilde{y} \sim \sum \tilde{w}^2 = \sum_{i}^{n} \left(\frac{\tilde{x}_i - \mu_i}{\sigma_i} \right)^2 \sim \chi_n^2 \tag{C.25}$$

C.5.5 The Sum of Squares of Normal Distributed Variables

Assume that $(\tilde{x}_1, \tilde{x}_2, \ldots, \tilde{x}_n)$ are n stochastically independent normal random variables with different means μ_i and variables σ_i^2. Then the sum of squares

$$\tilde{y} = \sum \tilde{x}_i^2 \sim \chi_{n,\lambda}^2 \tag{C.26}$$

has a noncentral chi-square distribution. The degree of freedom is n and the noncentrality parameter is

$$\lambda = \sum \frac{\mu_i^2}{\sigma_i^2} \tag{C.27}$$

C.6 HYPOTHESIS TESTS

A hypothesis is a statement about the parameters of a distribution (population). If a statistical hypothesis completely specifies all parameters of the distribution, it is called a simple hypothesis; if it specifies less than is necessary to describe the distribution, it is called a composite statistical hypothesis. The so-called parameter space is a convenient vehicle for visualizing the specification of hypothesis. Consider one unknown parameter μ, which can have any value. The parameter space is

$$\Omega = (\mu: -\infty < \mu < \infty) \tag{C.28}$$

The zero hypothesis H_0: $\mu = \mu_0$ would be a simple hypothesis, whereas the alternative hypothesis H_1: $\mu \neq \mu_0$ is composite because it refers to the whole line from $-\infty$ to ∞, except for point μ_0. With two unknown parameters μ and σ^2 the parameter space is

$$\Omega = (\mu, \sigma^2): -\infty < \mu < \infty, 0 < \sigma^2 < \infty \tag{C.29}$$

The null hypothesis H_0 from above specifies only μ, but says nothing about σ^2. The null hypothesis defines the following subspace ω of Ω.

$$\omega = (\mu, \sigma^2): \mu = \mu_0, 0 < \sigma^2 < \infty \tag{C.30}$$

The subspace ω is a straight line in the Ω plane. If the null hypothesis would also specify $\sigma^2 = \sigma_0^2$, then ω would define a single point in the Ω plane, and the null hypothesis would be of simple type. The alternative hypothesis always represents the subspace $\Omega - \omega$ which is in this case half the plane of Ω minus the line specified by ω. Usually alternative hypotheses are of the composite type.

   A test of a statistical hypothesis is a rule that based on the sample values leads to a decision to accept or to reject the null hypothesis. A test statistic is computed from the sample values (observations) and from the specifications of the null hypothesis. The test statistic is compared with either a normal, chi-square, t, or F distribution. If the test statistic falls within a *critical region*, the null hypothesis is rejected. Thus, the value of the test statistic is used to decide on the hypothesis that the sample (observations) comes from a certain population with specified parameters. The distribution of the test statistic and the distribution of the sample (observations) do not have to be the same. Figure C.5 displays the probability density functions for the specifications of the null hypothesis H_0 and the alternative hypothesis H_1. The figure also shows the critical region for which the null hypothesis is rejected, and the alternative hypothesis is accepted if the computed sample statistic $t|H_0$ falls in that region. Thus, reject H_0 if

$$t > t_\alpha \tag{C.31}$$

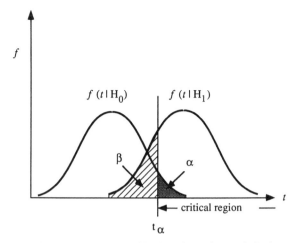

Figure C.5 Example of critical region of a statistical test.

Because the sample statistic t is computed from sample values (observations), the computed value may fall within the critical region even though the null hypothesis H_0 is true. There is a probability α that this can happen. One speaks of a *type I error* if the hypothesis H_0 is rejected although it is true; the probability of a type I error is α, which, incidently, is also called the significance level of the test. However, there is a probability that the sample statistics fall in the critical region when H_0 is false (and hence H_1 is true). The probability is denoted by $1 - \beta$ in Fig. C.5. If the sample statistics does not fall in the critical region, but the alternative hypothesis H_1 is true, one would mistakenly accept H_0 and commit a *type II error*. The probability of committing a type II error is β.

The alternative hypothesis is generally of a composite type. The implications of this fact can readily be visualized from Figure C.5. The shape and location of the alternative density function varies. Thus, the probability of an error of the second kind, β, depends on the specifications of the specific H_1 hypotheses. A desirable approach in statistical testing would be to minimize the probability of both kinds of errors. However, this is not practical, because all distributions of the alternative hypotheses, which, in general, are of the noncentral type, would have to be computed. Figure C.5 shows that the probability β increases as α decreases. A conventional procedure is to fix the probability of a type I error, say, $\alpha = 0.05$, and to compute β. However, much effort has gone into research as to how the magnitude of β can be controlled (Baarda 1968). After all, committing a type II error implies accepting the null hypothesis even though the alternative hypothesis is true. For deformations this could mean that it has been concluded that no deformation took place, even though actual deformations occurred. Such an error could be costly in many respects.

The rule (C.31) is referred to as a one-tail test in the upper end of the

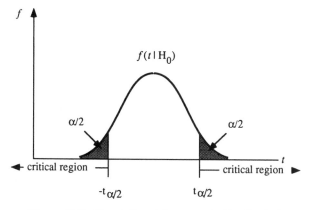

Figure C.6 Two-tail test (symmetric distribution).

distribution. Frequently, a two-tail test is employed. In that case the null hypothesis is rejected if

$$|t| > t_\alpha \tag{C.32}$$

That is,

$$P(|t| > t_\alpha) = \alpha \tag{C.33}$$

The rules (C.32) and (C.33) apply when the probability distribution under the null hypothesis is symmetric, as shown in Figure C.6. The critical regions are at both tails of the distribution, each tail covering a probability area of $\alpha/2$. If the distribution under the null hypothesis is not symmetric, the rule (C.33) is replaced by

$$t > t_{\alpha/2} \tag{C.34}$$

$$t < t_{1-\alpha/2} \tag{C.35}$$

This rule ensures that there is the probability $\alpha/2$ at each of the tails (the symbol t_α denotes the value of t to the right of which is the probability α).

C.7 GOODNESS-OF-FIT TEST

Assume that we have a series of observations and we wish to test whether the observations come from a certain population with a specified distribution. We subdivide the observation series into n classes. Then, n_i is the number of observations in class i. The subdivision should be such that $n_i \geq 5$. Compute for each class the expected number d_i of observations based on the hypothetical distribution. It can be shown that

$$\chi^2 = \sum_{i=1}^{n} \frac{(n_i - d_i)^2}{d_i} \tag{C.36}$$

is distributed approximately as χ^2_{n-1}. If the population parameters are estimated from the sample, then the degree of freedom $(n - 1)$ should be reduced for each parameter estimated. The zero hypothesis states that the sample is from the specified distribution. Reject H_0 at a $100\alpha\%$ significance level if

$$\chi^2 > \chi^2_{n-1,\alpha} \tag{C.37}$$

APPENDIX D

ROTATION MATRICES

Rotations between coordinate systems are conveniently expressed in terms of rotation matrices. The rotation matrices

$$R_1(\theta) = \begin{bmatrix} 1 & 0 & 0 \\ 0 & \cos\theta & \sin\theta \\ 0 & -\sin\theta & \cos\theta \end{bmatrix} \tag{D.1}$$

$$R_2(\theta) = \begin{bmatrix} \cos\theta & 0 & -\sin\theta \\ 0 & 1 & 0 \\ \sin\theta & 0 & \cos\theta \end{bmatrix} \tag{D.2}$$

$$R_3(\theta) = \begin{bmatrix} \cos\theta & \sin\theta & 0 \\ -\sin\theta & \cos\theta & 0 \\ 0 & 0 & 1 \end{bmatrix} \tag{D.3}$$

describe rotations by the angle θ of a right-handed coordinate system around the first, second, and third axis, respectively. The rotation angle is positive for a counterclockwise rotation, as viewed from the positive end of the axis about which the rotation takes place. The result of successive rotations depends on the specific sequence of the individual rotations. An exception to this rule is differentially small rotations for which the sequence of rotations does not matter.

APPENDIX E

SELECTED EXPRESSIONS FROM SPHERICAL TRIGONOMETRY

A spherical triangle is defined by great circles on the sphere. A great circle is the intersection of the sphere with a plane that contains the center of the sphere. Figure E.1 shows a spherical triangle with the usual labels for the sides and angles. Note that the sides of the spherical triangle are given in angular units.

Law of sine

$$\frac{\sin a}{\sin \alpha} = \frac{\sin b}{\sin \beta} = \frac{\sin c}{\sin \gamma} \tag{E.1}$$

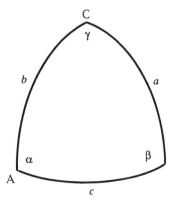

Figure E.1 Defining elements of a spherical triangle.

Law of cosine for sides

$$\cos a = \cos b \cos c + \sin b \sin c \cos \alpha \qquad (E.2)$$

$$\cos b = \cos c \cos a + \sin c \sin a \cos \beta \qquad (E.3)$$

$$\cos c = \cos a \cos b + \sin a \sin b \cos \gamma \qquad (E.4)$$

Law of cosine for angles

$$\cos \alpha = -\cos \beta \cos \gamma + \sin \beta \sin \gamma \cos a \qquad (E.5)$$

$$\cos \beta = -\cos \gamma \cos \alpha + \sin \gamma \sin \alpha \cos b \qquad (E.6)$$

$$\cos \gamma = -\cos \alpha \cos \beta + \sin \alpha \sin \beta \cos c \qquad (E.7)$$

APPENDIX F

SAMPLE NETWORK

To demonstrate some of the least-squares adjustment principles, we require a data set for a representative survey. Figure F.1 shows a network upon which a set of observations have been simulated. The traditional observations (angles, distances, and height differences) are shown in Table F.1. Simulated GPS baseline observations are given in Table F.2. Finally, the approximate station coordinates to be used in the various example adjustments are listed in Table F.3. Because these data are simulated, they will not be used as a basis of analysis, for example, adding a blunder and trying to locate it.

The stochastic model for the sample data assumes that all observations are uncorrelated. The standard deviations of the EDM distances and the GPS baseline vectors are computed from

$$\sigma = 0.002 + \frac{s}{20,000}$$

and

$$\sigma = 0.005 + \frac{s}{200,000}$$

respectively (all units are in meters; s denotes the observed distance). Even the GPS vector components are assumed uncorrelated, although in practice these components are correlated and that correlation can also exist between GPS vectors depending on the observation procedure. It is assumed that the geoid undulation differences have been computed using techniques explained in

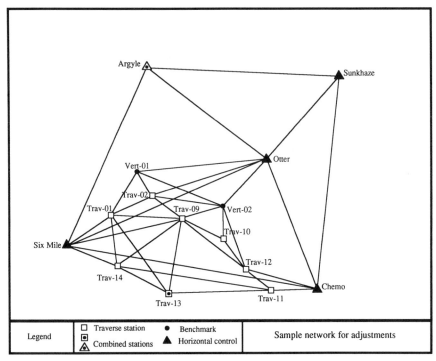

Figure F.1 Sample network.

TABLE F.1 Network Observations

Bs	Inst.	Fs	Bs ht.	Hi	Fs ht.	Observation	Obs. code	Std dev	Obs. number
6	1		2.050	1.980		4309.532	1	0.217	1
9	1		1.860	1.980		10763.946	1	0.540	2
15	1		1.780	1.980		6402.034	1	0.322	3
14	6		4.230	1.890		11890.482	1	0.597	4
9	6		2.320	1.890		6808.936	1	0.342	5
8	6		1.660	1.890		7537.958	1	0.379	6
7	6		3.220	1.890		6240.467	1	0.314	7
8	7		1.990	1.780		2386.367	1	0.121	8
10	7		2.140	1.780		6437.029	1	0.324	9
3	7		3.550	1.780		10212.794	1	0.513	10
9	8		4.320	2.320		4897.435	1	0.247	11
10	8		3.040	2.320		4550.697	1	0.230	12
3	8		1.890	2.320		7885.082	1	0.396	13
15	9		1.890	3.020		8755.456	1	0.440	14
14	9		1.780	3.020		10473.668	1	0.526	15
13	9		1.430	3.020		9534.070	1	0.479	16
11	9		2.050	3.020		4323.494	1	0.218	17
10	9		2.430	3.020		1877.258	1	0.096	18
13	10		4.220	2.840		9940.456	1	0.499	19
11	10		3.250	2.840		4988.542	1	0.251	20
13	11		3.040	2.580		5217.340	1	0.263	21
12	11		2.670	2.580		7745.107	1	0.389	22

TABLE F.1 (continued)

Bs	Inst.	Fs	Bs ht.	Hi	Fs ht	Observation	Obs. code	Std dev	Obs. number
5	11		6.430	2.580		9933.981	1	0.499	23
13	5		1.680	1.760		5508.348	1	0.277	24
12	5		3.110	1.760		2850.338	1	0.145	25
13	12		2.380	1.920		2799.123	1	0.142	26
14	13		2.860	3.270		5630.528	1	0.284	27
15	13		1.990	3.270		10298.657	1	0.517	28
15	14		2.040	1.750		6183.209	1	0.311	29
6	1	15	3.120	1.890	2.660	72 44 07.18	4	3.0	30
1	15	14	2.060	3.430	2.550	176 20 53.54	4	3.0	31
15	14	13	1.890	2.380	2.050	121 15 29.67	4	3.0	32
14	13	12	2.210	3.420	4.660	229 40 25.36	4	3.0	33
13	12	5	1.670	2.340	3.110	154 23 31.90	4	3.0	34
12	5	11	1.660	2.430	1.330	34 13 22.98	4	3.0	35
5	11	10	1.680	2.110	2.430	229 01 28.39	4	3.0	36
11	10	9	1.820	1.670	1.770	58 44 45.47	4	3.0	37
10	9	8	1.540	2.990	1.560	291 41 42.65	4	3.0	38
9	8	7	1.330	2.990	1.280	111 14 57.95	4	3.0	39
8	7	6	1.770	1.390	2.630	114 09 16.53	4	3.0	40
7	6	1	2.040	2.550	2.330	206 30 12.21	4	3.0	41
1	6	9	1.930	2.040	3.020	209 47 32.99	4	3.0	42
15	9	6	2.110	1.940	2.850	47 57 39.94	4	3.0	43
1	15	9	3.020	2.000	4.220	89 05 27.31	4	3.0	44
6	1		5.030	2.590		09 45.36	3	10.0	45
15	1		4.210	2.590		24 56.79	3	10.0	46
1	15		3.020	2.440		- 26 45.92	3	10.0	47
14	15		2.850	2.440		- 28 01.22	3	10.0	48
9	15		2.950	2.440		- 02 59.91	3	10.0	49
15	14		2.800	2.330		25 33.64	3	10.0	50
13	14		1.760	2.330		22 09.02	3	10.0	51
14	13		1.050	3.220		- 26 32.01	3	10.0	52
12	13		1.780	3.220		14 13.33	3	10.0	53
13	12		2.340	5.320		- 20 59.83	3	10.0	54
5	12		1.760	5.320		- 51 48.82	3	10.0	55
12	5		1.660	2.050		45 58.74	3	10.0	56
11	5		2.330	2.050		- 07 53.07	3	10.0	57
5	11		1.050	2.330		02 58.59	3	10.0	58
10	11		3.220	2.330		11 26.41	3	10.0	59
11	10		2.330	1.870		- 12 53.25	3	10.0	60
9	10		3.220	1.870		58 11.81	3	10.0	61
10	9		4.320	3.750		- 55 33.23	3	10.0	62
8	9		3.440	3.750		- 20 14.37	3	10.0	63
15	9		4.020	3.750		- 50.90	3	10.0	64
6	9		3.710	3.750		- 18 41.64	3	10.0	65
9	8		2.710	4.010		16 41.29	3	10.0	66
7	8		3.040	4.010		- 09 42.72	3	10.0	67
8	7		1.760	3.220		05 12.58	3	10.0	68
6	7		2.310	3.220		- 02 36.54	3	10.0	69
9	6		1.940	2.080		15 21.13	3	10.0	70
7	6		1.440	2.080		- 01 04.00	3	10.0	71
1	6		2.980	2.080		- 09 32.93	3	10.0	72
7	6					1.438	7	0.03	73
6	15					- 36.309	7	0.03	74
15	14					48.185	7	0.03	75
14	13					- 39.118	7	0.03	76
14	11					2.781	7	0.03	77
11	10					- 17.579	7	0.03	78
10	9					- 30.707	7	0.03	79
9	8					26.796	7	0.03	80
8	7					5.333	7	0.03	81
7	3					8.429	7	0.03	82
3	10					- 9.952	7	0.03	83

TABLE F.2 Network GPS Baseline Observations

Fs	Inst	Fs ht.	Hi	ΔU	ΔV	ΔW	Covariance matrix		
1	2	1.320	2.030	-4686.615	-17721.671	-14884.397	0.015145	0.000000	0.000000
								0.015145	0.000000
									0.015145
3	2	1.760	2.030	8499.814	-5663.546	-8403.435	0.005060	0.000000	0.000000
								0.005060	0.000000
									0.005060
4	2	1.830	2.030	14054.892	-1236.850	-6290.594	0.006763	0.000000	0.000000
								0.006763	0.000000
									0.006763
1	5	1.320	1.990	-20861.369	-3409.884	4406.536	0.012761	0.000000	0.000000
								0.012761	0.000000
									0.012761
3	5	2.220	1.990	-7675.107	8647.889	10887.576	0.007125	0.000000	0.000000
								0.007125	0.000000
									0.007125
4	5	3.220	1.990	-2119.837	13074.044	13001.240	0.009564	0.000000	0.000000
								0.009564	0.000000
									0.009564
1	3	4.120	3.000	-13185.916	-12059.058	-6479.730	0.010007	0.000000	0.000000
								0.010007	0.000000
									0.010007
4	3	2.430	3.000	5554.984	4427.085	2112.415	0.001768	0.000000	0.000000
								0.001768	0.000000
									0.001768

Chapter 11. A standard deviation of 3 cm is therefore assigned to the geoid undulations. The latter value is used in the three-dimensional adjustment when leveled orthometric heights and computed undulation differences are combined to give "observed" ellipsoidal height differences.

The horizontal and vertical angles of Table F.1 are considered reduced to the ellipsoidal normal. In practice, expressions (6.39) and (6.42) would be

TABLE F.3 Network Station Positions

Station name	Station number	Latitude	Longitude	Ellipsoidal height	Orthometric height
Six Mile	01	44 51 42.44	291 10 03.11	3.030	25.87
Argyle	02	45 03 02.86	291 18 15.25	31.250	53.95
Otter	03	44 56 38.69	291 22 42.61	7.250	30.02
Sunkhaze	04	44 58 15.00	291 27 52.31	19.740	42.19
Chemo	05	44 48 20.70	291 25 44.38	16.140	39.23
Trav-01	06	44 53 02.45	291 12 44.01	14.230	36.88
Vert-01	07	44 56 00.01	291 15 00.01	15.670	38.06
Trav-02	08	44 55 54.03	291 16 48.50	21.070	44.17
Trav-09	09	44 53 22.16	291 17 53.04	47.860	70.68
Vert-02	10	44 54 00.00	291 19 00.00	17.150	39.14
Trav-10	11	44 51 19.69	291 19 28.57	-0.430	22.23
Trav-11	12	44 48 20.05	291 23 34.69	55.110	78.05
Trav-12	13	44 48 58.63	291 21 39.42	41.470	64.51
Trav-13	14	44 47 42.91	291 17 46.33	2.350	25.46
Trav-14	15	44 49 35.50	291 13 53.52	50.540	73.24

	Legend
Observation Code	Observation
1	Distance
3	Vertical angle
4	Horizontal angle
7	Ellipsoidal height Difference

	Notes
Bs	Backsight station
Bs ht.	Height of backsight target
Fs	Foresight station
Fs ht.	Height of foresight target
Hi	Height of instrument
Std. dev	Given in units compatible with observations:meters for heights, and distances arc seconds for angles, directions

used to compute the reduction due to the deflection of the vertical. The reader is reminded that the deflection of the vertical effect cancels for angles according to (6.39), if the vertical angles of the two sightings are close to zero. One must also be aware that vertical angle observations often provide height information for the slant distances (corresponding to the reduction of the slant distances to the horizon in the classical two-dimensional model), and are not generally used in precision height determination. The respective deflection of the vertical reductions (6.42) is usually negligible. However, the surveyor should always ensure that shortcuts in reductions, if made at all, are not detrimental to the solution in terms of creating systematic errors. Also, recall the discussion of the local ellipsoid of Chapter 7, whereby the deflections of the vertical are further reduced locally.

APPENDIX G

GLONASS

The Russian counterpart to GPS, called Global Navigation Satellite System (GLONASS) is expected to become operational during 1995–1996. Kazantsev et al. (1992) discuss the status of the GLONASS. Presently, there are several efforts underway to build receivers that can receive signals from both the GPS and the GLONASS satellites. Initial results are already available in the literature. There is, of course, a strong interest to incorporate GLONASS satellites into ambiguity fixing on-the-fly solutions because of the potential increase in the number of simultaneously visible satellites. As has been emphasized previously, the more satellites available at the same time, the better, faster, and more reliable are the OTF techniques discussed in Chapter 10. This appendix is limited to a brief overview of GLONASS and some aspects that are peculiar to that system. A more complete presentation of GLONASS in combination with GPS and other satellites systems to form the integrated Global Navigation Satellite System (GNSS) of tomorrow is a subject of vast interest.

Table G.1 contains a brief comparison of GPS and GLONASS data. Both satellite systems are expected to consist of 24 satellites. The GLONASS satellites are located in three orbital planes separated by 120°. The nominal inclination is 64.8° and the nominal eccentricity is zero. The GLONASS satellites are about 1000 km higher than their GPS counterparts, resulting in an orbital period of about 11^h15^m.

A major difference with respect to GPS is that GLONASS uses frequency division multiple access (FDMA) technology to discriminate signals received at the antenna. Consequently, all GLONASS satellites transmit the same P-codes. Just as with GPS, there is a C/A-code on L1 and P-codes on both carriers L1 and L2. The carrier frequency, in MHz, is derived from the fol-

TABLE G.1   Comparison of GPS and GLONASS

	GPS	GLONASS
satellites		
number of satellites	24	24
number of orbital planes	6	3
satellites per orbital plane	4 (uneven)	8 (even)
orbital inclination	55 degrees	64.8 degrees
nominal eccentricity	0	0
orbital radius	26,560 km	25,510 km
orbital period	$\approx 11^h 58^m$	$\approx 11^h 15^m$
signals		
signal separation technique	CDMA	FDMA
carrier	L1:1575.42 MHz	L1:1602.5625-1615.5 MHz
	L2:1227.60 MHz	L2:1246.4375-1256.5 MHz
C/A-code (L1)	1.023 MHz	0.511 MHz
P-code (L1, L2)	10.23 MHz	5.11 MHz
navigation message		
duration (min)	12.5	2.5
capacity (bits)	37,500	7,500
word duration (sec)	0.6	2.0
word capacity (bits)	30	100
number of words within a frame	50	15
ephemeris	modified Kepler elements	geocentric cartesian coord., velocities and accelerations
general		
time reference	UTC(USNO)	UTC(SU)
geodetic datum	WGS 84	SGS 85
selective availability	yes	not planned
antispoofing of P-codes	yes	possible

lowing expression

$$f_L^p = L(178.0 + p \times 0.0625) \tag{G.1}$$

where p is the GLONASS channel number, which takes the value 1 to 24, i.e., each satellite is assigned a particular GLONASS channel number. The value $p = 0$ is reserved for testing. The factor L,

$$L = \begin{cases} 9 & \text{for L1 band} \\ 7 & \text{for L2 band} \end{cases} \tag{G.2}$$

completes the specifications for the L1 and L2 carriers. It is readily verified that the frequency spacing at L1 and L2 is 0.5625 and 0.4375 MHz, respectively. The ratio of the P-code to the C/A-code chipping rate is 10 for both systems.

Whereas the structure of the navigation message is similar in both systems, its rate of transmission of 50 bits per second is identical. The literature seems to indicate that there are currently no plans to implement selective availability degradation, and that antispoofing capability exists. Gouzhva et al. (1994) give additional detail on the GLONASS signal structure and make reference to the GLONASS Interface Control Document (published by the Scientific/Production Group of Applied Mechanics, GLAVKOSMOS, Institute of Space Device Engineering, 1991).

Parts of the frequency regions on which GLONASS operates are reserved for, or are highly contested by other users. Recently, the range 1610–1626.5 MHz was allocated to low earth orbiting (LEO) satellites whose primary mission might be global voice communication. The region 1610.6–1613.8 MHz is traditionally used by radio astronomers. Daly et al. (1993) address some of these issues and report on experiments and proposed solutions that entail the shifting of GLONASS channels that lie directly in the radio astronomy band, and/or reusing the same frequency for antipodal satellites. In the latter case, the required bandwidth can be reduced and yet satellites on opposite sides of the earth do not interfere with each other.

Success in geodetic applications of using GLONASS as a standalone satellite system or in combination with GPS depends primarily on the control of clock errors. For example, consider again the double-difference equation (8.41) as it applies to GLONASS. The relevant terms are

$$\varphi_{km}^{pq}(t) = \frac{f^p}{c}\left[\rho_k^p(t) - \rho_m^p(t)\right] - \frac{f^q}{c}\left[\rho_k^q(t) - \rho_m^q(t)\right]$$

$$- (f^p - f^q)(dt_k - dt_m) + N_{km}^{pq}(1) + I_{km,\varphi}^{pq}(t) + \frac{f}{c}T_{km}^{pq}(t)$$

$$+ d_{km,\varphi}^{pq}(t) + \epsilon_{km,\varphi}^{pq} \tag{G.3}$$

The superscripts identify the satellites and the GLONASS channel numbers. Since the carrier frequencies differ for satellites, the large station clock error term

$$\Delta\varphi_{km,c}^{pq} = -(f^p - f^q)(dt_k - dt_m) \tag{G.4}$$

with

$$f^p - f^q \leq 24 \times 0.5625 = 13.5 \text{ MHz} \tag{G.5}$$

does not cancel. A receiver clock synchronization of 1 μsec will cause a double-difference observable error of 13.5 cycles. The receiver clock errors must be known at the nanosecond level for the observable not to be affected more than the carrier phase resolution of 0.01 cycles. It might be necessary to estimate the receiver clock errors per epoch or model them as a function of time. In any case, the estimated receiver clock parameters will absorb residual ionospheric and tropospheric effects and the double-differenced multipath. The receiver clock errors also make it difficult to estimate ambiguity parameters. It is still better to use double differences than single differences because the latter depend on receiver clock errors by

$$\Delta\varphi^p_{km,c} = -f^p(dt_k - dt_m) \tag{G.6}$$

This effect is at least a hundred times bigger than the change in the double differences (G.4).

The satellite clock errors dt^p and dt^q cancel in the single difference just as they do for GPS. Thus, regarding the satellite clock errors, the GLONASS and GPS satellites can readily be used together to form a double-difference observable. In the navigation solution, discussed in Section 8.1.1, it is important in both systems that the correct signal emission time is computed using the respective clock correction terms of the broadcast message. One must, however, keep in mind that the corrected GLONASS time tags refer to UTC(SU) whereas the GPS time system refers to UTC(USNO). However, both national time systems are coordinated and monitored by the IERS to determine the difference between both time scales to 1 μsec and better. See Section 2.2 for additional information on time systems.

Because the GLONASS frequencies differ from satellite to satellite there might be some additional concern with receiver hardware delays if these delays are frequency dependent. In the latter case, the double differenced receiver hardware delays will not be distinguishable from the multipath term $d^{pq}_{km,\varphi}(t)$ unless both have different time signatures.

REFERENCES

Abidin, H. Z. (1993). "On-the-fly ambiguity resolution: formulation and results," *Manuscripta Geodaetica*, **18**(6):380–407.

Ascheid, G., and Meyr, H. (1986). "Cycle slips in phase-locked loops: A tutorial survey," in: Lindsey and Chie (eds.), *Phase-locked loops*. IEEE Press, New York, 29–42.

Ashby, N. (1993). "Relativity and GPS," *GPS World* **4**(11):42–47.

Ashjaee, J., and Lorenz, R. (1992). "Precision GPS surveying after Y-code," *Proc. ION-GPS-92*, 657–659.

Ashkenazi, V., Hill, C. J., and Nagel, J. (1992). "Wide area differential GPS: A performance study," *Proc. ION-GPS-92*, 589–598.

Baarda, W. (1967). *Statistical concepts in geodesy*, Publication on Geodesy, New Series, **2**(4), Netherlands Geodetic Commission.

Baarda, W. (1968). *A testing procedure for use in geodetic networks*, Publication on Geodesy, New Series, **2**(5), Netherlands Geodetic Commission.

Badekas, J. (1969). *Investigations related to the establishment of a World Geodetic System*, OSUR 124.

Barndt, L. (1993). "U.S. Coast Guard GPS Information Center (GPSIC) and its function within the civil GPS service (CGS)," *Proc. ION-GPS-93*, **1**:27–36.

BIH (1986). *Annual Report*, Bureau International de L'Heure, Paris.

Bishop, G. L., Klobuchar, J. A., and Doherty, P. H. (1985). "Multipath effects on the determination of absolute ionospheric time delay from GPS signals," *Radio Science*, **20**(3):388–396.

Bissell, C. C., and Chapman, D. A. (1992). *Digital signal transmissions*, Cambridge University Press, Cambridge, UK.

Bjerhammar, A. (1973). *Theory of errors and generalized matrix inverses*, Elsevier Scientific, New York.

Black, H. D. (1978). "An easily implemented algorithm for the tropospheric range correction," *Journal of Geophysical Research* 83(B4):1825–1828.

Blaha, G. (1971). *Investigations of critical configurations for fundamental range networks*, OSUR 150.

Blair, B. E., ed. (1974). *Time and frequency: Theory and fundamentals*, NBS Monograph 140, U.S. Department of Commerce, Washington, DC.

Blanchard, A. (1976). *Phase-locked loops: Application to coherent receiver design*, Wiley, New York.

Blewitt, G. (1993). "Advances in Global Positioning System technology for geodynamics investigations: 1978–1992," in: Smith, D. E., and Turcotte, D. L. (eds.), *Contribution of space geodesy to geodynamics: Technology*. Geodynamics Series, Vol. 25, AGU, Washington, DC.

Blomenhofer, H., Hein, G., and Walsh, D. (1993). "On-the-fly phase ambiguity resolution for precise aircraft landing," *Proc. ION-GPS-93*, 2:821–830.

Bock, Y., Abbot, R. I., Counselman, C. C., Gourevitch, S. A., and King, R. W. (1985). "Establishment of three-dimensional geodetic control by interferometry with the Global Positioning System," *Journal of Geophysical Research*, 90(B9):7689–7703.

Bock, Y., Gourevitch, S. A., Counselman, C. C., King, R. W., and Abbot, R. I. (1986). "Interferometric analysis of GPS phase observations," *Manuscripta Geodaetica*, 11(4):282–288.

Bossler, J. D., Goad, C. C., and Bender, P. (1980). "Using the Global Positioning System (GPS) for geodetic positioning," *Bulletin Géodésique*, 54(4):553–563.

Braasch, M. S. (1990). "A signal model for GPS," *Navigation*, 37(4):363–377.

Braasch, M. S., Fink, A., and Duffus, K. (1993). "Improved modeling of GPS selective availability," *Proceedings: 1993 National Technical Meeting*, 121–130, The Institute of Navigation, Alexandria, VA.

Brown, A. (1989). "Extended differential GPS," *Navigation*, 36(3):265–285.

Bruns, H. (1878). *Die Figure der Erde*. Publication of the Preuss. Geod. Institute, Berlin.

Chen, D. (1993). Fast ambiguity search filter (FASF): a novel concept for GPS ambiguity resolution," *Proc. ION-GPS-93*, 2:781–787.

Christou, N. P., Vaniček, P., and Ware, C. (1989). "Geoid and density anomalies," *EOS Transactions*, 70(22):625.

Coco, D. S., Coker, C., Dahlke, S. R., and Clynch, J. R. (1991). "Variability of GPS satellite differential group delay biases," *IEEE Transactions on Aerospace and Electronic Systems*, 27(6):931–938.

Cohen, C. E., Pervan, B., and Parkinson, B. W. (1992). "Estimation of absolute ionospheric delay exclusively through single-frequency GPS measurements," *Proc. ION-GPS-92*, 325–330.

Collins, J., and Leick, A. (1985). "Analysis of Macrometer networks with emphasis on the Montgomery (PA) County survey," *Proc. Pos. with GPS*, 2:667–693.

Counselman, C. C. (1973). "Very long baseline interferometry techniques applied to problems of geodesy, geophysics, planetary sciences, astronomy, and general relativity," *Proceedings of the IEEE*, 61(9):1225–1230.

Counselman, C. C. (1987). "Method and system for determining position using signals from satellites," *U.S. Patent 4,667,203.*

Counselman, C. C., and Abbot, R. I. (1989). "Method of resolving radio phase ambiguity in satellite orbit determination," *Journal of Geophysical Research,* **94**(B6):7058–7064.

Counselman, C. C., and Gourevitch, S. A. (1981). "Miniature interferometer terminals for earth surveying: Ambiguity and multipath with Global Positioning System," *IEEE Transactions on Geoscience and Remote Sensing,* Vol. **GE-19**(4):244–252.

Counselman, C. C., and Shapiro, I. I. (1979). "Miniature interferometric terminals for earth surveying," *Bulletin Géodésique,* **53**(2):139–163.

Craymer, M. R., and Beck, N. (1992). "Session versus single-baseline GPS processing," *Proc. 6th DMA,* **2**:995–1004.

Daly, P., Riley, S., and Raby, P. (1993). "Recent advances in the implementation of GNSS," *Proc. ION-GPS-93,* **1**:433–440.

Davies, K. (1990). *Ionospheric radio,* IEE Electromagnetic Wave Series 31.

Davis, J. L. (1986). *Atmospheric propagation effects on radio interferometry,* AFGL-TR-86-0243, Air Force Geophysics Laboratory.

Defense Mapping Agency (1987). *Department of Defense World Geodetic System 1984: Its definition and relationships with local geodetic systems,* DMA TR 8350.2, Defense Mapping Agency, Washington, D.C.

van Dierendonck, A. J., Fenton, P., and Ford, T. (1992). "Theory and performance of narrow correlator spacing in a GPS receiver," *Navigation,* **39**(3):265–283.

Eeg, J. (1986). *On the adjustment of observations in the presence of blunders,* Geodetic Institut, Report No. 1, Copenhagen, Denmark.

Engelis, T., Rapp, R., and Bock, Y. (1985). "Measuring orthometric height differences with GPS and gravity data," *Manuscripta Geodaetica,* **10**(3):187–194.

Eren, K. (1987). *Geodetic network adjustment using GPS triple difference observations and a priori stochastic information,* Tech. Report No. 1, Institute of Geodesy, University of Stuttgart, Stuttgart, Germany.

Escobal, P. R. (1965). *Methods of orbit determination,* Wiley, New York.

Essen, L., and Froome, K. D. (1951). "The refractive indices and dielectric constants of air and its principal constituents at 24000 Mc/s," *Proceedings of Physical Society of London,* **64**(B):862–875.

Eubanks, T. M. (1993). "Variation in the orientation of the earth," in: Smith, D. E., and Turcotte, D. L. (eds.), *Contribution of space geodesy to geodynamics: Technology.* Geodynamics Series, Vol. 24, AGU, Washington, DC.

Euler, H. J., and Landau H. (1992). "Fast ambiguity resolution on-the-fly for real-time applications," *Proc. 6th DMA,* **2**:650–658.

Euler, H. J., and Schaffrin, B. (1990). "On a measure for the discernibility between different ambiguity solutions in the static-kinematic GPS mode," *Proceedings: Kinematic systems in geodesy, surveying, and remote sensing,* 285–295, Springer, Banff, Alberta, Canada.

Fenton, P. C., Falkenberg, W. H., Ford, T. J., Ng, K. K. and Van Dierendonck, A. J. (1991). "NovTel's GPS receiver: The high performance OEM sensor of the future," *Proc. ION-GPS-91,* 49–58.

FGCS, 1984. *Standards and specifications for geodetic control networks*, National Geodetic Information Center, Silver Spring, MD.

FGCS, 1988. *Geometric Geodetic Accuracy Standards and Specifications for using GPS Relative Positioning Techniques*, National Geodetic Information Center, NGS, Silver Spring, MD.

van Flandern, T. C., and Pulkkinen, K. F. (1979). "Low precision formulae for planetary positions," *The Astronomical Journal Supplement Series*, **41**:391–411.

Fliegel, H. F., and Gallini, T. E. (1989). "Radiation pressure models for Block II GPS satellites," *Proc. 5th DMA*, **2**:789–798.

Fliegel, H. F., Fees, W. A., Layton, W. C., and Rhodus, N. W. (1985). "The GPS radiation force model," *Proc. Pos. with GPS*, **1**:113–119.

Fliegel, H. F., Gallini, T. E., and Swift, E. R. (1992). "Global Positioning System radiation force model for geodetic applications," *Journal of Geophysical Research*, **97**(B1):559–568.

Frei, E., and Beutler, G. (1990). "Rapid static positioning based on the fast ambiguity resolution approach 'FARA': Theory and first results," *Manuscripta Geodaetica*, **15**(6):325–356.

Gelb, A. (1974). *Applied optimal estimation*, The MIT Press, Cambridge, MA.

George, A., and Liu, J. W-H. (1981). *Computer solutions for large sparse positive definite matrices*, Prentice-Hall, Englewood Cliffs, NJ.

Georgiadou, Y., and Kleusberg, A. (1988). "On carrier signal multipath effects in relative GPS positioning," *Manuscripta Geodaetica*, **13**(3):172–179.

Goad, C. C., ed. (1985). *Proc. Pos. with GPS*.

Goad, C. C. (1986). "Precise positioning with the Global Positioning System," *Proceedings: Third International Symposium on Inertial Technology for Surveying and Geodesy*, Surveying Engineering, University of Calgary, Publication 60005, Alberta, Canada.

Goad, C. C. (1989). "Kinematic survey of Clinton Lake Dam," *Journal of Surveying Engineering*, **115**(1):67–77.

Goad, C. C. (1992). "Robust techniques for determining GPS phase ambiguities," *Proc. 6th DMA*, **1**:245–254.

Goad, C. C., and Goodman, L. (1974). "A modified Hopfield tropospheric refraction correction model," Presented at the Fall Annual Meeting of the American Geophysical Union, San Francisco, December 1974.

Goad, C. C., and Mueller, A. (1988). "An automated procedure for generating an optimum set of independent double difference observables using Global Positioning System carrier phase measurements," *Manuscripta Geodaetica*, **13**(6):365–369.

Gouzhva, Y. G., Koudryavtsev, I., Korniyenko, V., and Pushkina, I. (1994). "GLONASS receivers: An outline," *GPS World*, **5**(1):30–36.

Grafarend, E. (1992). "The modeling of free satellite networks in spacetime," *Proceedings: Global Positioning System in geosciences*, 1–20, Department of Mineral Resources, Technical University of Crete, Greece.

Grafarend, E., and Schaffrin, B. (1993). *Ausgleichungsrechnung in linearen Modellen*, BI Wissenschaftsverlag, Mannheim.

Greenspan, R. L., Tetwsky, A. K., Donna, J. I., and Klobuchar, J. A. (1991). "The effects of ionospheric errors on single-frequency GPS users,'' *Proc. ION-GPS-91*, 291–297.

Greenwalt, C. R., and Shultz, M. E. (1962). *Principles of error theory and cartographic applications*, ACIC Technical Report No. 96, Aeronautical Chart and Information Center, St. Louis, MO.

Grossman, W. (1976). *Geodaetische Rechnungen und Abbildungen in der Landesvermessung*, Witter Verlag, Stuttgart.

Hargreaves, J. K. (1992). *The solar-terrestrial environment*, Cambridge University Press, Cambridge, UK.

Hatch, R. R. (1990). ''Instantaneous ambiguity resolution,'' *Proceedings: Kinematic systems in geodesy, surveying and remote sensing*, 299–308, IAG symposium No. 107, Springer, New York.

Hatch, R. R. (1992a). ''Current issues in kinematic navigation,'' *Proc. ION-GPS-92*, 1063–1070.

Hatch, R. R. (1992b). *Escape from Einstein*, Kneat Kompany, Wilmington, CA.

Hatch, R. R., Keegan, R., and Stansell, T. A. (1992). ''Kinematic receiver technology from Magnavox,'' *Proc. 6th DMA*, **1**:174–181.

van Hees, S. G. (1990). ''Stokes formula using fast Fourier techniques,'' *Manuscripta Geodaetica*, **15**(4):235–239.

van Hees, S. G. (1992). ''Practical formulas for the computation of the orthometric, dynamic, and normal heights,'' *Zeitschrift für Vermessungswesen*, **117**(11):727–734.

Hein, G. W. (1986). ''Integrated geodesy: State-of-the-art 1986 reference text,'' in: Sünkel, H. (ed.), *Lecture Series in Earth Sciences*, Vol. 7, *Mathematical and Numerical Techniques in Physical Geodesy*, New York.

Hein, G. W., Leick, A., and Lambert, S. (1988). ''Integrated processing of GPS and gravity data,'' *Journal of Surveying Engineering*, **115**(1):15–33.

Heiskanen, W. A., and Moritz, H. (1967). *Physical geodesy*, Freeman, San Francisco, CA.

Hermann, B. R. (1992). ''Five years of absolute positions at the Naval Surface Warfare Center,'' *Proc. 6th DMA*, **2**:552–529.

Hofmann-Wellenhof, B., Lichtenegger, H., and Collins, J. (1992). *GPS Theory and Practice*, Springer, New York.

Hopfield, H. S. (1969). ''Two-quadratic tropospheric refractivity profile for correcting satellite data,'' *Journal of Geophysical Research*, **74**:4487–4499.

Hotine, M. (1969). *Mathematical geodesy*. ESSA Monograph 2, U.S. Department of Commerce, Washington, D.C.

Hradilek, L. (1984). *Three-dimensional terrestrial triangulation*, Konrad Wittwer, Stuttgart.

Hristow, W. K. (1955). *Die Gausschen und Geographischen Koordinaten auf dem Ellipsoid von Krasssowsky*, VEB Verlag Technik, Berlin.

ICD-GPS-200 (1991). *Interface control document ICD-GPS-200* (updated April 1993). GPSIC, U.S. Coast Guard, 7323 Telegraph Road, Alexandria, VA 22310-3998.

Janes, H. W., Langley, R. B., and Newby, S. P. (1991). ''Analysis of tropospheric delay prediction models: Comparisons with ray-tracing and implications for GPS relative positioning,'' *Bulletin Géodésique*, **65**(3):151–161.

Jorgensen, P. S. (1986). "Relativity correction in GPS user equipment," *Proc. position location and navigation system 1986 (PLANS)*, 177–183, Las Vegas, Nevada, 4–7 November 1986, IEEE, New York.

Kaplan, G. H., ed. (1981). "The IAU resolutions on astronomical constants, time scales, and the fundamental reference frame," United States Naval Observatory Circular, No. 163, Washington, D.C.

Kaula, W. K. (1962). "Development of the lunar and solar disturbing functions for a close satellite," *The Astronomical Journal*, **67**(5):300–303.

Kaula, W. M. (1966). *Theory of satellite geodesy*, Blaisdell, Waltham, MA.

Kazantsev, V. N., Reshetnev, M. F., Kozlov, A. G., and Cheremisin, V. F. (1992). "Current status, development program and performance of the GLONASS system. *Proc. ION-GPS-92*, 139–144.

Kee, C., Parkinson, B. W., and Axelrad, P. (1991). "Wide area differential GPS," *Navigation*, **38**(2):123–143.

Keegan, R. (1990). "P-code aided Global Positioning System receiver," *U.S. Patent 4,972,431*.

King, R. W., Masters, E. G., Rizos, C., Stolz, A., and Collins, J. (1985). *Surveying with GPS*, Monograph 9, School of Surveying, University of New South Wales, Australia.

Klobuchar, J. A. (1987). "Ionospheric time-delay algorithm for single-frequency GPS users," *IEEE Transactions on Aerospace and Electronic Systems*, **AES-23**(3):325–331.

Klobuchar, J. A. (1991). "Ionospheric effects on GPS," *GPS World*, **2**(4):48–50.

Koch, K. R. (1988). *Parameter estimation and hypothesis testing in linear models*, Springer, New York.

Kok, J. (1984). *On data snooping and multiple outlier testing*, NOAA Technical Report NOS NGS 30, National Geodetic Information Center, NOAA, Silver Spring, MD.

Krabill, W. B., and Martin, C. F. (1987). "Aircraft positioning using Global Positioning System carrier phase data," *Navigation*, **34**(1):1–21.

Kunches, J. M., and Hirman, J. W. (1990). "Predicted solar flare activity for the 1990s: Possible effects on navigation systems," *Navigation*, **37**(2):169–179.

Kursinski, R. (1994). "Monitoring the earth's atmosphere with GPS," *GPS World* **5**(3):50–54.

Lachapelle, G., Cannon, M. E., and Erickson, C. (1992). "High precision C/A-code technology for rapid static DGPS surveys," *Proc. 6th DMA*, **1**:165–173.

Lachapelle, G., Sun, H., Cannon, M. E., and Lu, G. (1994). "Precise aircraft-to-aircraft positioning using multiple receiver configuration," *Presented at the National Technical Meeting, Institute of Navigation*, San Diego, 24–26 January.

Ladd, L. W., Counselman, C., and Gourevitch, S. A. (1985). "The Macrometer II dual-band interferometer surveyor," *Proc. Pos. with GPS*, **1**:175–180.

Lear, W. M., Montez, M. N., Rater, L. M., and Zyla, L. V. (1992). "The effect of selective availability on orbit space vehicles equipped with SPS GPS receivers," *Proc. ION-GPS-92*, 825–840.

Leick, A. (1980a). *Geometric geodesy, 3D-geodesy and conformal mapping*. Dept. of Surveying Engineering, University of Maine, Orono, ME.

Leick, A. (1980b). *Adjustment computations*, Dept. of Surveying Engineering, University of Maine, Orono, ME.

Leick, A. (1984). GPS surveying and data management, *Proceedings: 22nd Annual Conference of the Urban and Regional Information Systems Association (URISA)*, Seattle, August.

Leick, A. (1993). "Accuracy standards for modern three-dimensional geodetic networks," *Surveying and Land Information Systems*, **53**(2):111–116.

Leick, A., and Emmons, M. (1994). "Quality control with reliability for large GPS networks," *Journal of Surveying Engineering*, **120**(1):26–41.

Leick, A., and Mueller, I. I., (1979). "Defining the celestial pole," *Manuscripta Geodaetica*, **4**(2):149–183.

Leick, A., and van Gelder, B. H. W. (1975). *On similarity transformation and geodetic network distortions based on Doppler satellite observations*, OSUR 235.

Leick, A., Aiken, C., and Kor, J. (1992). "Definition of the vertical for the Superconducting Super Collider using GPS, leveling and gravity," *Surveying and Land Information Systems*, **52**(2):69–74.

Lichten, S. M., and Border, J. S. (1987). "Strategies for high precision GPS orbit determination," *Journal of Geophysical Research*, **92**(B12):12751–12762.

Lindsey, W. C., and Chie, C. M. (1986). "A survey of digital phase-locked loops," in: Lindsey and Chie (eds.), *Phase-locked loops*, 296–317, IEEE Press, New York.

Lorenz, R. G., Helkey, R. J., and Abadi, K. K. (1992). "Global Positioning System receiver digital processing technique," *U.S. Patent 5,134,407*.

Mader, G. L. (1986). "Dynamic positioning using GPS carrier phase measurements," *Manuscripta Geodaetica*, **11**(4):272–277.

Mader, G. L. (1992). "Rapid static and kinematic Global Positioning System solutions using the ambiguity function technique," *Journal of Geophysical Research* **97**(B3):3271–3283.

Malys, S., Bredthauer, D. Hermann, B. R., and Clynch, J., (1992). "Geodetic point positioning with GPS: A comparatively evaluation of methods and results," *Proc. 6th DMA*, **2**:550–562.

Mannucci, A. J., Wilson, B. D., and Edwards, C. D., (1993). "A new method for monitoring the earth's ionospheric total electronic content using the GPS global network," *Proc. ION-GPS-93*, **2**:1323–1351.

McKinnon, J. A. (1987). *Sunspot numbers: 1610–1985; based on "The Sunspot Activity in the Years 1620–1960,"* National Academy of Sciences, Report UAG-95, Washington, D.C.

Meehan, T. K., and Young, L. E. (1992). "On-receiver signal processing for GPS multipath reduction," *Proc. 6th DMA*, **1**:200–208.

Meehan, T. K., Srinivasan, J. M., Spitzmesser, D. J., Dunn, C. E., Ten, J. Y., Thomas, J. B., Munson, T. N., and Duncan, C. B. (1992), "The TurboRogue GPS receiver," *Proc. 6th DMA*, **1**:209–218.

Melbourne, W. (1985). "The case for ranging in GPS-based geodetic systems," *Proc. Pos. with GPS*, **1**:373–386.

Milbert, D. G. (1984). *Heard of Gold: Computer routines for large, sparse, least squares computations*, NOAA Technical Memorandum NOS NGS-39, National Technical Information Center, NOAA, Silver Spring, MD.

Milbert, D. G. (1991). "GEOID 90: A high-resolution geoid for the United States," *EOS Transactions*, **72**(49):545.

Milbert, D. G., and Dewhurst, W. T. (1992). "The Yellowstone–Hebgen Lake geoid obtained through the integrated geodesy approach," *Journal of Geophysical Research*, **97**(B1):545–557.

Miranian, M., and Klepczynski, W. J. (1991). "Time transfer via GPS at USNO," *Proc. ION-GPS-91*, 215–221.

Molodenskii, M. S., Eremeev, V. F., and Yurkina, M. I. (1962). *Methods for study of the external gravitational field and figure of the earth.* Translation from Russian, National Technical Information Services, Springfield, VA.

Moritz, H. (1980). *Advanced physical geodesy*, Wichman, Karlsruhe.

Moritz, H. (1984). "Geodetic Reference System 1980," *Bulletin Géodésique*, **58**(3):388–398.

Moritz, H., and Mueller, I. I. (1987). *Earth rotation; Theory and observation*, Ungar, New York.

Mueller, I. (1969). *Spherical and Practical Astronomy as Applied to Geodesy*, Ungar, New York.

Nolan, J., Gourevitch, S., and Ladd, J. (1992). "Geodetic processing using full dual band observables," *Proc. ION-GPS-92*, 1033–1941.

Pope, A. J. (1971). "Transformation of covariance matrices due to changes in minimal control." Paper presented at the American Geophysical Union Fall Meeting, San Francisco, CA.

Pope, A. J. (1972). "Some pitfalls to be avoided in the iterative adjustment of non-linear problems," *Proceedings: 38th Annual Meeting, American Society of Photogrammetry.*

Pope, A. J. (1976). *The statistics of residuals and the detection of outliers.* NOAA Technical Report NOS 65 NGS 1, National Technical Information Center, NOAA, Silver Spring, MD.

Qin, X., Gourevitch, S., and Kuhl, M. (1992). "Very precise GPS: Development status and results," *Proc. ION-GPS-92*, 615–624.

Qin, X., Gourevitch, S., Ferguson, K., Kuhl, M., and Ladd, J. (1992b). "Dynamic short baseline calibration and attitude determination using Ashtech 3DF system," *Proc. 6th DMA*, **1**:190–199.

Rao, C. R., and Mitra, S. K. (1971). *Generalized inverses of matrices and its applications*, Wiley, New York.

Rapp, R. H. (1992). Computation and accuracy of global geoid undulation models," *Proc. 6th DMA*, **2**:865–872.

Rapp, R. H., Wang, Y. M., and Pavlis, N. K. (1991). "*The Ohio State 1991 geopotential and sea surface topography harmonic coefficient models*," OSUR 410.

Remondi, B. W. (1984). *Using the Global Positioning System (GPS) phase observable for relative geodesy: Modeling, processing, and results.* NOAA, Rockville, MD 20852, 1984, Reprint of doctoral dissertation, Center for Space Research, University of Texas at Austin.

Remondi, B. W. (1985a). "Global Positioning System carrier phase: description and use," *Bulletin Géodésique*, **59**(4):361–377.

Remondi, B. W. (1985b). "Performing centimeter-level surveys in seconds with GPS carrier phase: initial results," *Navigation* **32**(4):386–400.

Rice, D. (1966). "Vertical alignment: Stanford linear accelerator," in: *Earth Movement Investigations and Geodetic Control for Stanford Linear Accelerator Center*, Report ABA 106.

Robinson, S. E. (1986). "A new algorithm for microwave delay estimation from water vapor radiometer data," *TDA Progress Report 42-87*, 149–157, Jet Propulsion Lab., Pasadena, CA.

RTCM-104 (1994). *RTCM recommended standards for differential NAVSTAR GPS services*, Version 2.1. Radio Technical Commission for Maritime Services, 655 Fifteenth Street, NW, Suite 300, Washington, D.C. 20005.

Ruland, R., and Leick, A. (1985). "Application of GPS to a high precision engineering survey network," *Proc. Pos. with GPS*, **1**:483–493.

Saastamoinen, J. (1972). "Atmospheric correction for the troposphere and stratosphere in radio ranging of satellites," in: *Use of artificial satellites for geodesy*, Geophysical Monograph 15, American Geophysical Union, Washington, D.C.

Saastamoinen, J. (1973). "Contribution to the theory of atmospheric refraction," *Bulletin Géodésique*, **107**(1):13–14.

Salzberg, I. M. (1973). "Tracking the Apollo lunar rover with interferometry techniques," *Proceedings of the IEEE*, **61**(9):1233–1235.

Santerre, R. (1991). "Impact of GPS satellite sky distribution," *Manuscripta Geodaetica*, **16**(1):28–53.

Santerre, R., and Beutler, G. (1993). "A proposed GPS method with multi-antenna and single receiver," *Bulletin Géodésique*, **67**(4):210–223.

Schaffrin, B., and Bock, Y. (1988). A unified scheme for processing GPS dual-band phase observations," *Bulletin Géodésique*, **62**(2):142–160.

Schaffrin, B., and Grafarend, E. (1986). "Generating classes of equivalent linear models by nuisance parameter elimination," *Manuscripta Geodaetica*, **11**(4):262–271.

Schreiner, W. S. (1990). "A covariance study for orbit accuracy improvement of the GPS satellites using fiber optics tracking," *Proc. ION-GPS-90*, 563–568.

Schwarz, C. R. (ed.), (1989). *North American Datum 1983*, NOAA Professional Paper NOS 2. NOAA, National Technical Information Service, Silver Springs, MD.

Schwarz, K. P., Sideris, M. G., and Forsberg, R. (1990). The use of FFT techniques in physical geodesy," *Geophysical Journal International*, **100**:485–514.

Schwarze, V. S., Hartmann, T., Leins, M., and Soffel, M. H. (1993). "Relativistic effects in satellite positioning," *Manuscripta Geodaetica*, **18**(5):306–316.

Seeber, G. (1993). *Satellite geodesy: Foundation, methods, and applications*, Walter de Gruyter, Berlin.

Seeber, G., and Wuebbena, G. (1989). "Kinematic positioning with carrier phases and 'on-the-way' ambiguity resolution," *Proc. 5th DMA*, **II**:600–609.

Soler, T., and Hothem, L. D. (1988). "Coordinate systems used in geodesy: Basic definitions and concepts," *Journal of Surveying Engineering*, **114**(2):84–97.

Soler, T., and van Gelder, B. H. W. (1987). "On differential scale changes and the satellite Doppler system z-shift," *Geophys. J. R. Astr. Soc.*, **91**:639–656.

Soler, T., Carlson, A. E., and Evans, A. G. (1989). "Determination of vertical deflections using the Global Positioning System and geodetic leveling,'' *Geophysical Research Letters*, **16**(7):695–698.

Soler, T., Love, J. D., Hall, L. W., and Foote, R. H. (1992a). ''GPS results from statewide high precision networks in the United States,'' *Proc. 6th DMA*, **2**:573–582.

Soler, T., Strange, W. E., and Hothem, L. (1992b). ''Accurate determination of Cartesian coordinates at geodetic stations using the Global Positioning System,'' *Geophysical Research Letters*, **19**(6):533–536.

Spilker, J. J. (1980). ''Signal structure and performance characteristics,'' *Global Positioning System*, **I**:29–54, The Institute of Navigation, Alexandria, VA.

Stem, J. E. (1989). *State Plane Coordinate systems of 1983*, NOAA Manual NOS NGS 5.

Strange, W. E., and Love, J. D. (1991). ''High accuracy reference networks: A national perspective,'' Presented at the ASCE Specialty Conference on Transportation Application of GPS Positioning Strategies,'' Sacramento, California, 18–21 September.

Talbot, N. C. (1993). ''Centimeters in the field: A users perspective of real-time kinematic positioning in a production environment,'' *Proc. ION-GPS-93*, **2**:589–598.

Taveira–Blomenhofer, E., and Hein, G. W. (1993). ''Investigations on carrier phase corrections for high precision DGPS navigation,'' *Proc. ION-GPS-93*, **2**:1461–1467.

Thomas, P. D. (1952). *Conformal projections in geodesy and cartography*. Special Publication No. 251, U.S. Department of Commerce, Washington, D.C.

Tralli, D. M., and Lichten, S. M. (1990). ''Stochastic estimation of tropospheric path delays in Global Positioning System geodetic measurements,'' *Bulletin Géodésique*, **64**(2):127–159.

Tranquilla, J. M., and Al-Rizzo, H. M. (1993). ''Theoretical and experimental evaluation of precise relative positioning during periods of snowfall precipitation using the Global Positioning System,'' *Manuscripta Geodaetica*, **18**(5):362–379.

Tranquilla, J. M., and Colpitts, B. G. (1989). ''GPS antenna design characteristics for high-performance applications,'' *Journal of Surveying Engineering*, **115**(1):2–14.

Vaniček, P., and Krakiwsky, E. (1982). *Geodesy: The concepts*, North-Holland, New York, NY.

Vaniček, P., and Wells, D. E. (1974). ''Positioning of horizontal geodetic datums,'' *The Canadian Surveyor*, (5).

Vaniček, P., Langley, R. B., Wells, D. E., and Delikaraglou, D. (1984). ''Geometrical aspects of differential GPS positioning,'' *Bulletin Géodésique*, **53**(1):37–52.

Veis, G. (1960). ''Geodetic use of artificial satellites,'' *Smithsonian Contributions to Astrophysics*, **3**(9):95–161.

Vincenty, T. (1979). *The HAVAGO three-dimensional adjustment program*, NOAA Technical Memorandum NOS NGS 17. National Technical Information Service, NOAA, Silver Spring, MD.

Wahr, J. M. (1981). ''Body tides on an elliptical rotating, elastic and oceanless earth,'' *Geophys. J. R. Astr. Soc.*, **64**:705–727.

Wanninger, L. (1993). ''Effects on the equatorial ionosphere on GPS,'' *GPS World*, **4**(7):48–54.

Weill, L. R. (1993). "Optimal rapid static positioning without ambiguity resolutions," *Proc. ION-GPS-93*, **2**:1069–1080.

Winch, R. G. (1993). *Telecommunication transmission systems*, McGraw-Hill, New York.

Winkler, G. (1986). "The U.S. Naval Observatory (USNO) PTTI Data Service," *Proceedings: 40th Annual Frequency Control Symposium*, Philadelphia, IEEE.

Withington, F. N. (1985). "Role of telecommunication in timekeeping," *Journal of Surveying Engineering*, **111**(1):36–42.

Wolf, H. (1963). "Die Grundgleichungen der dreidimensionalen Geodäsie in elementarer Darstellung," *Zeitschrift für Vermessungswesen*, **88**(6):225–233.

Xia, R. (1992). "Determination of absolute ionospheric error using a single frequency GPS receiver," *Proc. ION-GPS-92*, 483–490.

Yan, H. J., Sauermann, K., and Groten, E. (1992). "The ray bending corrections in tropospheric refraction," *Proc. 6th DMA*, **1**:291–301.

Zilkoski, D. B., Richards, J. H., and Young, G. M. (1992). "Results of the general adjustment of the North American Vertical Datum of 1988," *Surveying and Land Information Systems*, **52**(3):133–149.

ABBREVIATIONS FOR FREQUENTLY USED REFERENCES

Proc. Pos. with GPS   Proceedings of the first international symposium on precise positioning with the Global Positioning System, Positioning with GPS—1985, Washington, D.C., April 15–19, National Geodetic Information Center, NOAA, Silver Spring, MA.

OSUR NNN   Report of the Department of Geodetic Science No. *NNN*, Ohio State University, Columbus, OH.

Proc. 5th DMA   Proceedings of the 5th International geodetic symposium on satellite positioning, Santa Cruces, New Mexico, March 13–17, 1989, Defense Mapping Agency, Washington, D.C.

Proc. 6th DMA   Proceedings of the 6th international geodetic symposium on satellite positioning, Columbus, OH, March 17–20, 1992, Defense Mapping Agency, Washington, D.C.

Proc. ION-GPS-90   Proceedings of the ION-GPS-90, Colorado Springs, Colorado, September 19–21, 1990, The Institute of Navigation, Alexandria, VA.

Proc. ION-GPS-91   Proceedings of the ION-GPS-91, Albuquerque, New Mexico, September 11–13, 1991, The Institute of Navigation, Alexandria, VA.

Proc. ION-GPS-92   Proceedings of the ION GPS-92, Albuquerque, New Mexico, September 16–18, 1992, The Institute of Navigation, Alexandria, VA.

Proc. ION-GPS-93   Proceedings of the ION GPS-93, Salt Lake City, Utah, September 22–24, 1993, The Institute of Navigation, Alexandria, VA.

AUTHOR INDEX

SUBJECT INDEX